Photosynthetic
Light-Harvesting Systems

Photosynthetic Light-Harvesting Systems

Organization and Function

Proceedings of an International Workshop
October 12–16, 1987
Freising, Fed. Rep. of Germany

Editors
Hugo Scheer · Siegfried Schneider

W DE G

Walter de Gruyter · Berlin · New York 1988

Editors

Prof. Dr. Hugo Scheer
Botanisches Institut der Universität München
Menzinger Straße 67
D-8000 München 19
Federal Republic of Germany

Prof. Dr. Siegfried Schneider
Institut für Physikalische und Theoretische Chemie
der Technischen Universität München
Lichtenbergstraße 4
D-8046 Garching
Federal Republic of Germany

Library of Congress Cataloging in Publication Data

Photosythetic light-harvesting systems.
Includes bibliography and indexes
1. Photosynthesis--Congresses. 2. Bacteria, Photosynthetic--Congresses.
I. Scheer, Hugo, 1942- . II. Schneider, Siegfried, 1940- .
QR88.5.P483 1988 589.9'013342 88-7083
ISBN 0-89925-427-6 (U.S.)

CIP-Kurztitelaufnahme der Deutschen Bibliothek

Photosynthetic light harvesting systems : organization and function ; proceedings of an internat. workshop, october 12-16, 1987, Freising, Fed. Rep. of Germany / ed. Hugo Scheer ; Siegfried Schneider. - Berlin ; New York : de Gruyter, 1988
ISBN 3-11011531-X
NE: Scheer, Hugo [Hrsg.]

 Printing: Gerike GmbH, Berlin. Binding: Lüderitz & Bauer GmbH, Berlin. - Printed in Germany.

PREFACE

The present volume contains the proceedings of an international workshop on "Organization and Function of Photosynthetic Antennas" held at the Bildungszentrum der Erzdiözese München und Freising, in Freising, West Germany, from October 12 to 16, 1987. The enthusiasm of the participants, the high quality of the research presented, and the attractiveness of the conference center made it a very successful and enjoyable meeting.

Photosynthesis is the basic and most important biological process of solar energy conversion and storage. Because of this fact and because of the present focus on alternative energies, research in this field has attracted many excellent scientists from diverse disciplines and has also received considerable funding. Progress has therefore been rapid.

The light harvesting antenna complexes comprise the largest part of the photosynthetic apparatus and are mainly responsible for the adaption of photosynthetic organisms to different light conditions. Despite their importance antenna systems have, however, rarely been treated as the prime subject during conferences on photosynthesis.

Two major reasons for this are (i) the enormous structural diversity of antenna systems, resulting in considerable specialization on selected systems; and (ii) the sophisticated experimental techniques required for an investigation of the extremely rapid primary photophysical processes that lead to an efficient funneling of the absorbed photon energy to the reaction center.

Reports on the enormous progress achieved during the last years have been scattered among a variety of conferences, where usually only one aspect of the interdisciplinary research described in this book was among the main topics to be discussed in detail. A meeting devoted exclusively to the various aspects of research on light harvesting complexes seemed to be, therefore, urgently needed.

The aim of this workshop was to bring together researchers working on different antenna systems who are applying different methods to study their organization and function. The interdisciplinary character is apparent from the various contributions ranging from botany, microbiology and molecular genetics to advanced spectroscopic techniques and quantum mechanics.

The positive response of all participants also pointed to the need for presenting in written form the current knowledge on this subject. Because there is no other monography available, we hope that these proceedings can also help to fill a regretable gap by providing an overview of recent progress and present knowledge in this area.

We would like this book to be of use both to the active researcher in the field, and to other readers, who might be interested in the subject. Therefore, we have given ample space to each participant to present his work in all necessary detail and asked several colleagues to write additional introductory chapters for each of the major sections. Finally we have compiled a comprehensive subject index, by which even the non-expert should be able easily to find a relevant reference.

No refereeing of the contributed papers was carried out, in order to comply with the general desire for rapid publication. This goal could be achieved thanks to the prompt delivery of the original manuscripts. We are particularily indebted to the authors of the introductory chapters and the concluding remarks, who kept to very stringent deadlines.

Many people worked hard to make the conference a success. Special thanks are due Mrs. E. Feigel, P. Kahlfuß and S. Schoy, for their help in organizing the meeting and preparing the proceedings. We gratefully acknowledge financial support by the Deutsche Forschungsgemeinschaft, Bonn, given to us via the Sonderforschungsbereich 143: "Primärprozesse der bakteriellen Photosynthese".

Thanks are also due the staff of W. de Gruyter Verlag for their help in preparing these proceedings.

München, December 1987

Hugo Scheer

Siegfried Schneider

CONTENTS

SECTION II: ORGANIZATION: MOLECULAR GENETICS AND CRYSTALLOGRAPHY

LIST OF PARTICIPANTS

J. Amesz, Huygens Laboratory, State University of Leiden, Leiden, The Netherlands

F. Baumann, Institut für Physikalische und Theoretische Chemie, Technische Universität München, Garching, FRG

A. Blumen, Physikalisches Institut, Universität Bayreuth, Bayreuth, FRG

J. Breton, Department de Biologie, CEN-Saclay, Gif-Sur-Yvette-Cedex, France

D.C. Brune, Department of Chemistry, Arizona State University, Tempe, Arizona, USA

R. Brunisholz, Institut für Molekularbiologie und Biophysik, ETH-Hönggerberg, Zürich, Switzerland

D.A. Bryant, Department of Molecular and Cell Biology, Penn State University, University Park, Pennsylvania, USA

R.J. Cogdell, Department of Botany, University of Glasgow, Glasgow, U.K.

G. Drews, Institut für Biologie 2, Universität Freiburg, Freiburg, FRG

R. Feick, Max-Planck-Institut für Biochemie, Martinsried, FRG

R. Fischer, Botanisches Institut, Universität München, München, FRG

J. Friedrich, Physikalisches Institut, Universität Bayreuth, Bayreuth, FRG

E. Gantt, Department of Botany, University of Maryland, College Park, Maryland, USA

N.E. Geacintov, Chemistry Department, New York University, New York, New York, USA

P. Geiselhart, Institut für Physikalische und Theoretische Chemie, Technische Universität München, Garching, FRG

R. Ghosh, Institut für Planzenbiologie, University, Zürich, Switzerland

T. Gillbro, Department of Physical Chemistry, Umea University, Umea, Sweden

Z. Gombos, Biological Research Center, Hungarian Academy of Sciences, Szeged, Hungary

J. Grabowski, Institute of Environmental Engineering, Poznan Technical University, Poznan, Poland

R. van Grondelle, Department of Biophysics, Free University De Boelelaan, Amsterdam, The Netherlands

P. Hoffmann, Department of General Botany, Humboldt University of Berlin, Berlin, GDR

A.R. Holzwarth, Max-Planck-Institut für Strahlenchemie, Mülheim a.d. Ruhr, FRG

R. Huber, Max-Planck-Institut für Biochemie, Martinsried, FRG

C.N. Hunter, Department of Pure and Applied Biology, Imperial College, London, U.K.

R. Knox, Department of Physics and Astronomy, University of Rochester, Rochester, New York, USA

W. Kühlbrandt, The Blackett Laboratory, Biophysics Section, Imperial College, London, U.K.

W. Kufer, Botanisches Institut, Universität München, München, FRG

A.W.D. Larkum, School of Biological Sciences, The University of Sydney, Sydney, Australia

M. Mimuro, National Institute for Basic Biology, Okazaki, Aichi, Japan

E. Mörschel, Fachbereich Biologie, Philipps-Universität, Marburg, FRG

J.M. Olson, Department of Chemistry, Biochemistry, University of Odense, Odense, Denmark

R.M. Pearlstein, Physics Department, Indiana-Purdue University, Indianapolis, Indiana, USA

B. Robert, Department de Biologie, CEN-Saclay, Gif-Sur-Yvette, Cedex, France

R. Rümbeli, Institut für Molekularbiologie und Biophysik, Eidgenössische Technische Hochschule, Zürich, Switzerland

K. Sauer, Department of Chemistry and Microbiology, University of California, Berkeley, California, USA

C. Scharnagl, Institut für Physikalische und Theoretische Chemie, Technische Universität München, Garching, FRG

H. Scheer, Botanisches Institut, Universität München, München, FRG

A. Scherz, Department of Biochemistry, Weizmann Institut, Rehovot, Israel

T. Schirmer, Laboratory of Molecular Biology, Cambridge, U.K.

G. Schmidt, Botanisches Institut, Universität München, München, FRG

S. Schneider, Institut für Physikalische und Theoretische Chemie, Technische Universität München, Garching, FRG

W. Sidler, Institut für Molekularbiologie und Biophysik, Eidgenössische Technische Hochschule, Zürich, Switzerland

S. Siebzehnrübl, Botanisches Institut, Universität München, München, FRG

V. Sundström, Department of Physical Chemistry, University of Umea, Umea, Sweden

B. Szalontai, Biological Research Center, Hungarian Academy of Sciences, Szeged, Hungary

J.P. Thornber, Department of Biology, University of California, Los Angeles, California, USA

J. Ullrich, 3. Physikalisches Institut, Universität Stuttgart, Stuttgart, FRG

W. Wehrmeyer, Fachbereich Biologie, Philipps-Universität, Marburg, FRG

W. Welte, Institut für Biophysik und Strahlenbiologie, Universität Freiburg, Freiburg, FRG

P. Westhoff, Botanisches Institut, Universität München, München, FRG

C. Wilhelm, Institut für Allgemeine Botanik, Johannes-Gutenberg-Universität, Mainz, FRG

B. Wittmershaus, Department of Chemistry, Arizona State University, Tempe, Arizona, USA

H. Zuber, Institut für Molekularbiologie und Biophysik, Eidgenössische Technische Hochschule, Zürich, Switzerland

Section I

Organization: Biochemical Methods

THE BIOCHEMISTRY OF LIGHT-HARVESTING COMPLEXES

R.J. Cogdell

Department of Botany, University of Glasgow
Glasgow G12 8QQ, UK

The function of the light-harvesting system of photosynthetic organisms is to capture radiant solar energy and to transfer that energy, as efficiently as possible, to the reaction centres. If the main pigments involved in these antenna functions (chlorophylls, bacteriochlorophylls, carotenoids and phycobilins) were just mixed in solution the photochemical reactions that we associate with light-harvesting would only occur, if at all, to a very limited extent. Historically the important point about the photosynthetic light-harvesting system is that the antenna pigments are bound to specific apoproteins. The aim of the biochemist who studies light-harvesting pigment-protein complexes is to understand the detailed structure, ultimately to atomic resolution, of their particular complex of interest. Hopefully, this will then reveal the important details of the pigment-protein and the pigment-pigment interactions within the complex, since it is these interactions which are thought to control the photochemical behaviour of the pigments. Once this has been achieved it is then necessary to continue, and to begin to study how the different antenna complexes interact together, and with the reaction centres, to form the complete photosynthetic unit, and to investigate how these complexes are synthesised and assembled.

Determination of the structure of an antenna complex usually proceeds through several discrete stages. Initially the composition of the complex must be determined (that is the number and types of the pigments and proteins present). Comparatively speaking most of the research into the structure of the higher plant and algal light-harvesting complexes is only at this first stage (for example, see the chapter by Thornber et al.). The different types of complex present are still being catalogued and attempts are continuing to try and define their

Photosynthetic Light-Harvesting Systems

composition. So far only the LHC 11_β (i.e. the well-known chlorophyll a/b protein) has been extensively studied and there is still some controversy on its pigment content. In this regard the reader should compare the chapter by Thornber et al. with that by Kühlbrandt.

Once this stage is completed then the primary structure of the apoprotein(s) is(are) required. If you are fortunate enough to be able to sequence a range of apoproteins (by either direct amino acid sequencing or indirectly by sequencing the structural gene) from a series of homologous complexes isolated from several different species, it is possible at this stage to look for highly conserved regions. These, of course, may have functional significance. Studies on the primary structures of a range of phycobiliproteins and purple bacterial antenna apoproteins provide good examples of how this type of comparative approach, based upon primary structural data, can be exploited (the reader who is interested in this type of approach should consult the chapters by Sidler and Zuber and by Brunisholz and Zuber). This is especially elegant in the case of the purple bacteria where the primary structure of more than twenty antenna apoproteins has been determined (1). Conserved histidine residues have been identified as possible ligands to the magnesium atom at the centre of the bacteriochlorophyll molecule and structural models have been proposed. A detailed analysis of the distribution of aromatic residues near the conserved histidines has allowed the origin of the spectral shifts of the bound bacteriochlorophylls and their circular dichroism characteristics to be predicted. These types of subtle predictions, based upon primary structure, provide excellent targets for molecular biologists who wish to use site-directed mutagenesis to investigate the relationship between structure and function in these complexes. The final stage in this process is to try and crystallize the complex and then to use X-ray diffraction in order to determine its three-dimensional structure to high resolution. If it is not possible to obtain highly ordered three-dimensional crystals, a low resolution structure can sometimes be determined from two-dimensional crystalline arrays of the complex.

When the reader goes through and examines the chapters of this book that describe studies aimed at understanding the structure of the antenna complexes, he will see work at all of the stages indicated above.

It is now becoming clear that there are two main types of antenna complexes found in photosynthetic organisms. These are the 'core' antenna complexes and the peripheral or 'variable' antenna complexes. The core complexes come closely packaged with the reaction centres, they are usually found in a constant amount with respect to the concentration of reaction centres present and they can be best thought of as 'dedicated' to that reaction centre. These 'core' complexes are most easily visualised in the membranes of Bchl b-containing species of purple bacteria, such as Rhodopseudomonas viridis (2,3), where they are organised into a regular, two-dimensional, paracrystalline array. The peripheral antenna types are present in the photosynthetic unit in variable amounts with respect to reaction centre concentration, and are able to transfer their absorbed radiant energy to several different reaction centres. The higher plant chlorophyll a/b protein is an example of this type of antenna complex. Without exception the core light-harvesting complexes are all integral membrane proteins. On the other hand, some of the photosynthetic peripheral antenna complexes are not only extramembrane complexes, such as in the chlorosomes of the green photosynthetic bacteria, but also water soluble, as in the case of the phycobiliproteins of the cyanobacteria and the red algae. The chlorosome light-harvesting system deserves a further mention at this stage in order to emphasise a point which is considered in several of the subsequent chapters in this book. An organic solvent soluble, abundant, low molecular weight polypeptide has been isolated from chlorosomes of both Chloroflexus and Chlorobium, and indeed sequenced (4 and Gerola et al., this book). It has been suggested quite frequently that this is the apoprotein of a Bchl c-containing pigment-protein complex (4) and putative structural models have even been presented. However, as yet, nobody has been able to isolate such a complex in a native pigmented form and so its existence must be questioned. The state of the Bchl c in the chlorosome and the role of protein in its functioning as an accessory light-harvesting pigment continues to be a very controversial topic of ongoing research.

The aim of this introductory chapter is to put the more specialised chapters in this book, which deal with the biochemical characterisation of the light-harvesting complexes, into a broader context. In such a short review it is impossible to be comprehensive, so I am just going to concentrate upon the phycobilisome antenna structure and use it to point out some general features of how antenna systems appear to work. The phycobiliproteins are probably the best characterised of all the light-harvesting complexes. This is because, in contrast to most antenna complexes, they are water soluble and because their chromophores are covalently attached to their apoproteins.

The primary structure of a large number of phycobiliproteins has been described (1) and the sites of bilin attachment have been determined (1,5). The bilin chromophores are linked to their apoproteins by way of one or two rather stable thio-ether bonds to specific cysteine residues. The tight-binding of the chromophore in the correct conformation is very important for its functioning in light-harvesting (5,6,7). When the biliproteins are denatured and the bilin chromophore is essentially 'free', it is extremely flexible. In this state bilins only have moderate extinction coefficients (8) and rather low ($\sim 10^{-3}$) fluorescence quantum yields (6,9). A low fluorescence quantum yield means a very fast decay from the first excited singlet state and this would make singlet-singlet energy transfer from one bilin to another (i.e. the antenna function) very unfavourable. When the chromophores are correctly bound in a rather rigid conformation, in the native phycobiliprotein, the rapid deactivation of the first excited singlet state is severely hindered (7,9). Now the absorption extinction coefficients are increased by about a factor of five and the fluorescence quantum yield rises to about 50% (10). Both these changes greatly enhance the capacity of the chromophores to act as efficient light-harvesters. This illustrates how important the antenna apoproteins in general are in being able to select for those photochemical properties which optimise the ability of the bound-chromophores to act as efficient light-harvesters.

Although numerous studies have been presented in which the excited state dynamics within phycobilisomes have been investigated in the psec time domain (and several more will be discussed in this volume), still perhaps the easiest to understand, and the most informative as regards demonstrating how a heterogeneous antenna system can act to funnel the absorbed energy towards the reaction centres, is the study described by Searle et al. (11). These workers, using phycobilisomes isolated from Porphyridium cruentum, made use of the fact that the different phycobiliprotein types absorb and fluoresce at different wavelengths.

Using psec excitation they were able to put light energy specifically into phycoerythrin and look for energy transfer by monitoring the fluorescence emitted by phycoerythrin itself, then from phycocyanin, allophycocyanin and finally from chlorophyll a. Searle et al. (11) were thus able to see the sequential flow of excitation energy from PE ——> PC ——> APC ——> Chla. If PC is excited energy is transferred to APC not back to PE. The phycobilisome acts, therefore, like a funnel, where any absorbed light energy is transferred inwards and downwards to the chlorophyll-containing thylakoid membrane. The more antenna systems are studied in detail the clearer it is becoming that this wavelength programming of the light-harvesting array, whereby peripheral antenna complexes absorb light at shorter wavelengths than those closer to the reaction centre, is an important general feature of light-harvesting. This downhill energy transfer gives light-harvesting a directionality and increases the efficiency of the system making it more probable that any absorbed photon will reach a reaction centre.

The organisation of the phycobiliproteins within the phycobilisome and how phycobilisomes are assembled is now quite well understood (5,12). Glazer (5) has presented a diagrammatic model of the phycobilisome from the cyanobacterium Synechocystis 6701 and it serves nicely to illustrate the general principles of phycobilisome construction. At the centre is a core of three units of allophycocyanin, where each unit is a hexamer of an $\alpha\beta$ monomer. The different hexamers differ mainly in their content of colourless linker polypeptides. These linker polypeptides are important for the correct assembly of the whole phycobilisome structure and for subtle modulations of the spectral properties of the bilin chromophores (5,17). Rod-like structures are joined onto this core. These rods consist of phycoerythrin and phycocyanin hexamers stacked one on top of each other. The rods also contain their own specific linker polypeptides. Table 1 summarizes the composition and some of the spectral and molecular characteristics of a range of phycobilisome components.

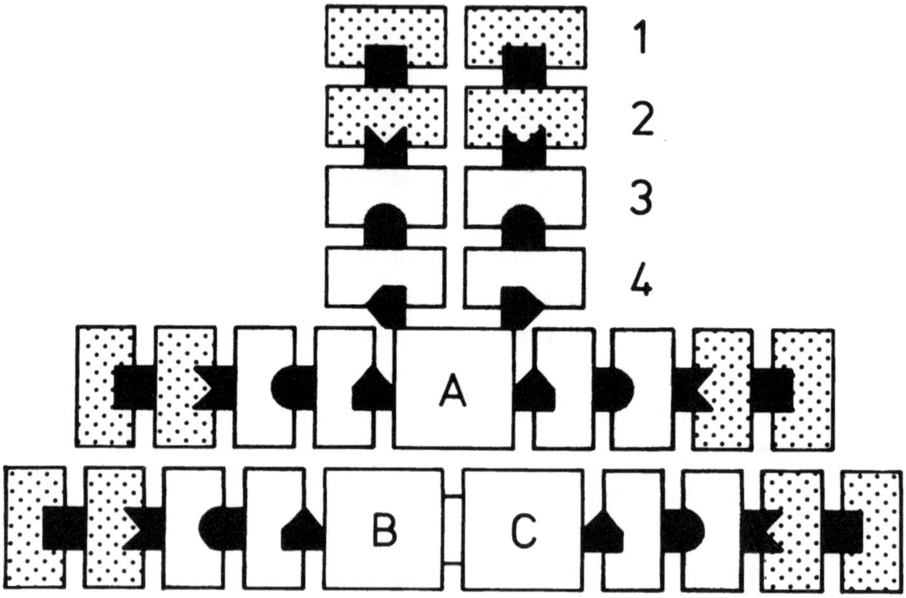

Fig.14 A schematic representation of the Synechocystis 6701 phycobilisome.

This Figure has been adapted and re-drawn from data presented in Glazer (5). The phycobilisome is constructed of 'rods' connected to three 'core' cylinders. A rod is made up of four hexameric biliprotein complexes, each of which is attached through its specific linker polypeptide to the component adjacent to it in the structure. The following abbreviations are used AP - allophycocyanin, PC - phycocyanin,

PE - phycoerythrin; α^{AP}, β^{AP} etc., stand for the β polypeptides of allophycocyanin; L is the designation for the linker polypeptides, with the subscript denoting its location in the phycobilisome i.e. R - in the rod, RC - rod/core junction, C - core and CM - core/membrane junction, while the superscript denotes the apparent size in KD. The rods are the large rectangles, the linker polypepticles are the dark symbols and the three, large squares represent the allophycocyanin core.

$\underline{1}$: $(\alpha^{PE}\ \beta^{PE})_6\ L_R^{30.5}$, $\underline{2}$: $(\alpha^{PE}\ \beta^{PE})_6\ L_R^{31.5}$,

$\underline{3}$: $(\alpha^{PC}\ \beta^{PC})_6\ L_R^{33.5}$, $\underline{4}$: $(\alpha^{PC}\ \beta^{PC})_6\ L_{RC}^{27}$,

$\underline{A}$: $((\alpha^{AP}\ \beta^{AP})_3)_2 + ((\alpha^{AP}\ \beta^{AP})_3\ L_c^{10})_2$,

$\underline{B}$ and $\underline{C}$: $(\alpha^{AP}\ \beta^{AP})_3 + (\alpha^{AP}\ \beta^{AP})_3\ L_c^{10} + (\alpha^{AP}\ \beta^{AP})_2\ \beta^{18.5}\ L_{cm}^{99}$ +

$(\alpha_1^{APB}\ \alpha_2^{AP}\ \beta_3^{AP})L_c^{10}$

Each core component is a tetramer of trimers. It is not yet known how the core cylinders are held together. The linker polypeptides hold the whole phycobilisome structure together and modulate the absorption and fluorescent properties of the pigments so as to promote excitation energy transfer down through the structure towards the chlorophyll-containing membrane.

The crystal structure of two phycocyanin molecules has now been determined to a resolution of 3Å, (C-phycocyanin from Mastigocladus laminosus (a trimer) and C-phycocyanin from Agmenellum quadruplicatum (a hexamer)) (18,19). As a result of this a lot of detailed work is now going on to try to pinpoint the molecular details of the interactions of the bilin chromophores with the individual amino acid side chains within these complexes.
Hopefully now as work with other antenna systems progresses to this level of sophistication more of the intimate details of the molecular mechanisms involved in efficient light-harvesting will become clear.

Table 1. Typical Phycobilisome Components, their Spectral and Molecular Characteristics[1]

A.

Coloured Apoproteins[2]	Room Temp. Absorption Peaks (nm)	Room Temp. Fl.Emission Peak (nm)	M_r (KD)	Subunit Structure	Chromophore Content[3]
R-PE	498,542,565	578	260	$(\alpha\beta)_6\gamma$	-
B-PE	498,545,563	575	260	$(\alpha\beta)_6\gamma$	2PEB$_1$ 4PEB 2PEB^1+ 2PUB
C-PE	565	578	230	$(\alpha\beta)_6\gamma$	2PEB, 4PEB
R-PC	553,615	640	130	$(\alpha\beta)_3\gamma$	1PCB, 1PCB + 1PEB
C-PC	620	650	120	$(\alpha\beta)_3\gamma$	1PCB, 2PCB
APC-B	618,673	680	98	$(\alpha\beta)_3$	1PCB, 1PCB
APC-I	654	680	145	$(\alpha\beta)_3\gamma$	-
APC-II	650	660	105	$(\alpha\beta)_3$	1PCB, 1PCB

B.

uncoloured (linker) polypeptides-M_r (KD)	Cyanobacterial Phycobilisomes e.g. *Synechococcus* 6301	Red Algal Phycobilisomes e.g. *Porphyridium cruentum*
95	-	present
75	present	-
45-75	-	present
35-45	present	present
30-35	present	present
25-30	present	present
10-25	present	present

[1] This table has been compiled from data contained in the following references: 13, 14, 15, 16.
[2] R = Rhodophytan, B = Bangiophycean and C = Cyanophycean
[3] Chromophore abbreviations: PEB - phycoerythrobilin, PCB - phycocyanobilin, PUB - phycourobilin.

Before the reader leaves this chapter it is worthwhile to point out a few cautionary remarks. These are aimed at the more biophysical researchers in this area and should be viewed rather like 'government health warnings.' Research aimed at understanding the molecular details of the functioning of the light-harvesting systems has progressed so well in recent years that a range of sophisticated physical techniques are now being used that probe very precisely the fine details of these systems. These techniques include, among others, 'hole-burning,' psec flash photolysis and absorbance- and fluorescence-detected magnetic resonance. However, the biological system is quite delicate. It is not only very difficult as a biochemist, to provide a completely homogeneous sample but it is also quite clear that the isolated samples have a limited stability and must be correctly handled at all times. The results obtained with these detailed physical measurements are really only as good as the samples used in them. It is clear that some of the arguments currently doing the rounds in the more physical studies undoubtedly arise from differences in the quality of the samples used as well as interspecies differences, where similar antenna types show subtly different behaviour. These possible sources of error must be much more carefully controlled in the future if the potential of these elegant physical techniques is to be fully realised.

References

1. Zuber, H. 1986. Encl.Pl.Physiol. (New series) 19, 238.
2. Stark, W., W. Kühlbrandt, I. Wildhaber, K. Mühlethaler. 1984. EMBO J. 3, 777.
3. Engelhardt, H., R. Gukenherger, R. Hegerl, W. Baumeister. 1985. Ultramicroscopy 16, 395.
4. Wechsler, T., F. Suter, R.C. Fuller, H. Zuber. 1985. FEBS Lett. 181, 173.
5. Glazer, A.N. 1985. Ann.Rev.Biophys.Biophys.Chem. 14, 47.
6. Scheer, H. 1982. In: Primary Processes in Photosynthesis (F.K. Fong, ed.). Springer-Verlag, New York, Berlin, Heidelberg, p.7.

7. Scheer, H. 1986. Encl.Pl.Physiol. (New series) 19, 327.

8. Brandlmeir, T., H. Scheer, W. Rudiger. 1981. Naturforsch. 36c, 431.

9. Braslavsky, S.E., A.R. Hozwarth, K. Schaffner. 1983. Agnew.Chem. 22, 670.

10. Grabowski, J. and E. Gantt. 1978. Photochem.Photobiol. 28, 39.

11. Searle, G.F.W., J. Barber, G. Porter, C.J. Treadwell. 1978. Biochim.Biophys.Acta 501, 246.

12. Gantt, E. 1986. Encl.Pl.Physiol. (New series) 19, 260.

13. Bogarad, L. 1975. Ann.Rev.Pl.Physiol. 26, 369.

14. Cohen-Bazire, G. and D.A. Bryant. 1981. In: The Biology of the Cyanobacteria (N.Carr and B. Witton, eds.). Blackwell, New York, p.143.

15. Glazer, A.N. 1977. Mol.Cell.Biochem. 18, 135.

16. Gantt, E. 1981. Ann.Rev.Pl.Physiol. 32, 327.

17. Tandeau de Marsac, N. 1983. Bull.Inst.Pasteur 81, 201.

18. Schirmer, T., W. Bode, R. Huber, W. Sidler, H. Zuber. 1985. J.Mol.Biol. 184, 257.

19. Schirmer, T., R. Huber, M. Schneider, W. Bode, M. Miller, M.L. Hacket. 1986. J.Mol.Biol. 188, 651.

PHYCOBILISOME-THYLAKOID INTERACTION: THE NATURE OF HIGH MOLECULAR WEIGHT POLYPEPTIDES

E. Gantt, C.A. Lipschultz, F.X. Cunningham, Jr.

Smithsonian Institution/Botany Department, University of Maryland
College Park, Maryland 20742

Introduction

Specific high molecular weight polypeptides appear to serve as conduits in energy transfer from phycobilisomes to thylakoids. Only those isolated from phycobilisomes have been partially characterized until now. They range in molecular weight from ca. 75 kD to 120 kD as measured by their migration upon polyacrylamide gel electrophoresis (1-3). Their location in the core of the phycobilisome has been clearly established in several species (3,4), and at least two molecules are present per phycobilisome. These polypeptides are blue in color, and have long been regarded as being the terminal pigment of the phycobilisome because their fluorescence maxima are longer (>660 nm) than that of allophycocyanin (1-6). An absorbance maximum at 665 nm and an emission maximum at 683 nm are characteristic of the 94 kD polypeptide isolated from *Nostoc* phycobilisomes (7). Energy transfer from the phycobilisome to photosystem II through this polypeptide is considered to be more likely than through allophycocyanin-B (8).

Immuno-cross-reactivity studies have suggested that the high molecular weight polypeptide structure is highly conserved. Phycobilisomes from many red algae and cyanobacteria have been shown to cross-react with antisera to the 94 kD polypeptide from *Nostoc* (1,3). High molecular weight polypeptides in thylakoid membranes, free of phycobilisomes, also have been shown to be cross-reactive with antisera to the phycobilisome high molecular weight polypeptides (8,9). We are in the process of

Photosynthetic Light-Harvesting Systems

investigating the nature of, and possible functional relationship of these thylakoid polypeptides in the phycobilisome-thylakoid linkage. In Nostoc the amino acid analysis suggests that those proteins derived from thylakoids, though similar in molecular weight, are different than the cross-reactive proteins in phycobilisomes (10). Recent results, with limited protease digestions and further immuno-cross-reactivity studies carried out on the high molecular weight polypeptides purified from phycobilisomes, and from thylakoids, will be discussed.

Materials and Methods

High molecular weight polypeptides were prepared from isolated phycobilisomes of Nostoc sp. and Porphyridium cruentum and separated by SDS (or LiDS) gel electrophoresis (5,8). The polypeptide bands were excised and the electrophoresis step was repeated. Antisera were prepared in rabbits against the gel-purified polypeptides.

Thylakoid membranes were prepared as in (10) at 4°C by passing them twice through a French pressure cell (10,000 psi) in 10mM HEPES/0.3 M sucrose/10 mM EDTA (pH 7.5). Thylakoid membranes remained in the supernate following a 10,000 g centrifugation (10 min.) to remove cell debris. Magnesium chloride was added to a final concentration of 15 mM for concentrating the thylakoids, which were pelleted (27,000 g 20 min) and twice more suspended in the buffer plus $MgCl_2$, and finally concentrated between layers of 0.75 M and 2.0 M sucrose (95,000 g 1 h). Phycobiliproteins were no longer detectable by absorbance spectroscopy.

Samples were analyzed by SDS gel electrophoresis, and stained with Coomassie Blue. Gels were photographed with Polaroid P/N-55 film. Negatives were scanned, in the transmittance mode, with a Gilford spectrophotometer; the scans were converted to absorbance when areas or peak heights were desired. Trypsin digestion on the high molecular weight polypeptides was carried out at 0.05% and 0.50% trypsin (w/w) at 25°C for 30 min.

Relative immuno-cross-reactivity was ascertained on nitrocellulose membranes with a Bio-Dot microfiltration apparatus, according to BioRad bulletin 1141 and adaptation of the method according to Hawkes (11). The antibody reaction was identified with horse radish peroxidase coupled to goat-antirabbit serum and color was developed with 4-chloro-1-naphthol.

Results and Discussion

The existence of a 94 kD polypeptide in phycobilisomes (N-PBS-94) of Nostoc has been previously shown. Further it was shown by crossed-immunoelectrophoresis that the antiserum to this polypeptide was monospecific and did not cross-react with other polypeptides in phycobilisomes (1,3). Subsequently, we found that thylakoids, stripped of phycobilisomes, contained 94 kD and 92 kD (N-TH-94, N-TH-92 respectively) polypeptides which were recovered from immuno-precipates of Triton X-100 solubilized thylakoids (8). Since the N-PBS-94 is blue we attempted to ascertain the blue color in immuno-precipitates from stripped thylakoids. We have been unable to accomplish this, and in fact, have only obtained spectra of chlorophyll a. Immunoprecipitates from phycobilisomes under the same conditions were always blue and had characteristic long wavelength fluorescence emissions.

Phycobiliproteins are no longer spectrally detectable after repeated rinsing (Materials and Methods) (Fig. 1), as shown by the trough in the phycoerythrin-absorbing region (540-560 nm). Since the absorbance spectrum of N-PBS-94 has a broad shoulder at 610-640 nm and a peak at 665 nm, the absorbance peak at 630 nm in thylakoids (Fig. 1 A) is of interest. That the absorbance region around 630 nm is mainly due to chlorophyll of photosystem I is shown by the spectral comparison with a purified PS I preparation (Fig. 1 B) which does not contain polypeptides in the 92-94 kD region (12).

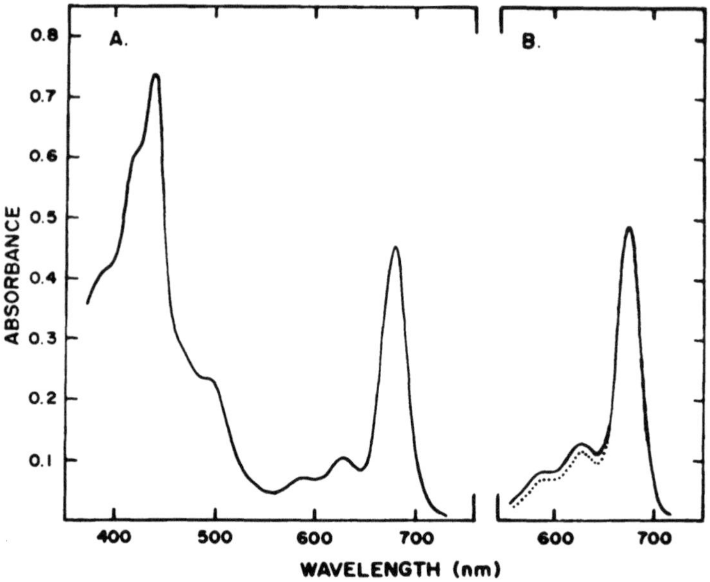

Figure 1. Absorption spectra of Nostoc sp.. A. Thylakoids rinsed and purified, suspended in 0.01 M HEPES/0.3 M sucrose/0.01 M EDTA (pH 7.5). B. Thylakoids (—), and (...) photosystem I preparation suspended in 0.1 % Triton X-100 and 0.05 M phosphate.

The polypeptides from thylakoids and phycobilisomes could not be distinguished by immunocrossreactivity studies. In Fig. 2 there is a continuous precipitate line of identity between thylakoids (wells # 1,3), phycobilisomes (wells # 2,4), and antigen (N-PBS-94) when reacted against anti-N-PBS-94. The reaction pattern suggests that the polypeptides from the two sources have common epitopes, otherwise spurring would be expected.

The 94 kD polypeptides from phycobilisomes and thylakoids from Nostoc do not exhibit any gross differences when exposed to limited digestion with trypsin (Fig. 3), or with chymotrypsin (not shown). However, when the Nostoc polypeptides are compared with the 92 kD phycobilisome (PC-PBS-92) polypeptide from Porphyridium the digest patterns are clearly different. It is interesting to note that in an earlier comparison of polypeptides derived from thylakoids and phycobilisomes of Porphyridium there was also no apparent difference in the limited protease digest patterns, and both cross-reacted with the same antiserum (9).

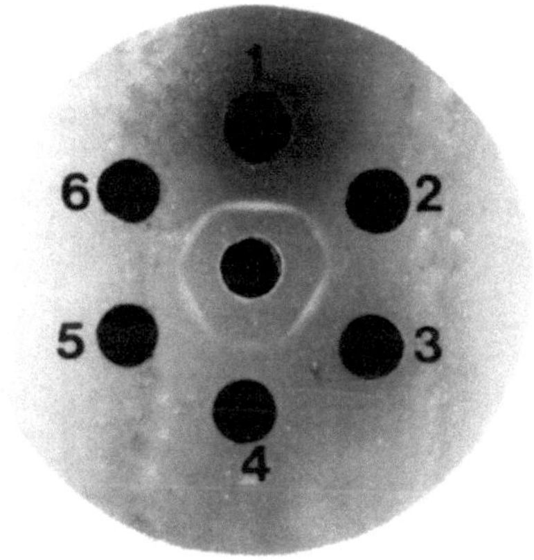

Figure 2. Ouchterlony immunodiffusion plate with antiserum to N-PBS-94 kD in center well. Thylakoids: Triton-solubilized # 1, and # 3 (1:1 dil.). Phycobilisomes: dissociated # 2, and # 4 (1:1 dil.). Antigen: # 5 (1:1 dil.), and # 6.

The above results clearly indicate that the high molecular weight polypeptides from Nostoc phycobilisomes and from thylakoids are very similar. However, their amino acid composition has been shown to be different. Of 1,000 residues 4 were methionine residues in N-T-94 while none were found in N-PBS-94, and these polypeptides contained 11 and 5 histidine residues respectively. Also, the thylakoid polypeptides had a greater proportion of negatively charged residues and a greater glycine content. Furthermore, the thylakoid polypeptides were blocked at the N-terminus, unlike those from phycobilisomes (10).

The relative amounts of high molecular weight polypeptides were estimated in phycobilisomes and in thylakoids (N-TH-94 and N-TH-92) from scans of Coomassie Blue stained gels (Table 1). The thylakoid polypeptides appear to represent a substantial portion of the photosystem II membrane component. In detergent fractionated Porphyridium thylakoids the high molecular weight polypeptide isolated together with what appears to be the photosystem II antenna polypeptide of ca. 48 kD (data not shown), which is not surprising since the direct connection between photosystem II and phycobilisomes had been previously established (13).

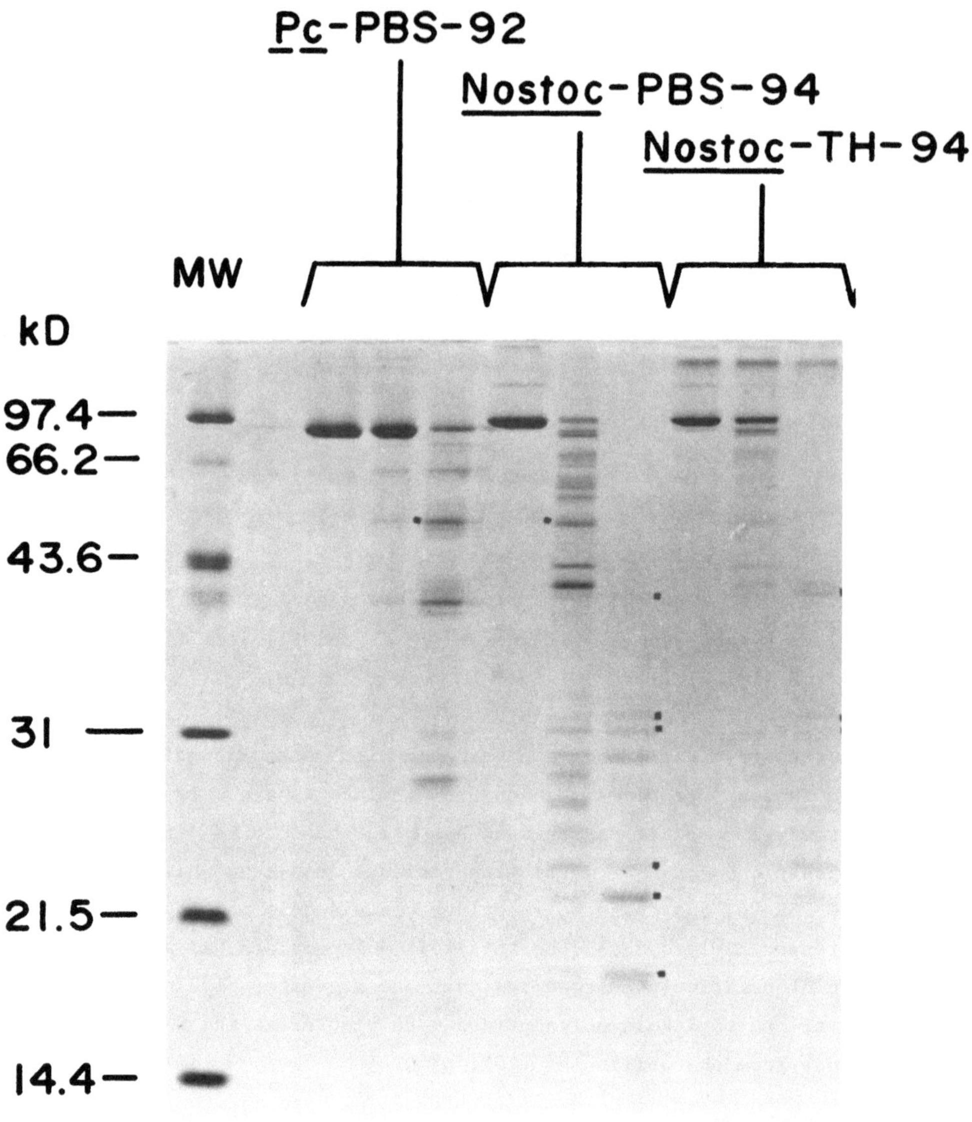

Figure 3. Trypsin digestion of high molecular weight polypeptides from Nostoc and Porphyridium. Left lane of each set, undigested control. Second and third lanes of each set 0.05% and 0.50 % trypsin, respectively, 30 min., 25°C. Dots suggest digestion complexes of similar size.

Table 1. Percentage of Coomassie Blue staining proteins.

	% of Total Phycobilisome	% of Total Thylakoid
Nostoc:		
N-PBS-94	1.5	--
N-TH-94 & 92	--	6.1
Porphyridium:		
PC-PBS-92	1.3	--
PC-TH-92	--	3.1

For a more direct comparison of the approximate amounts of high molecular weightcomponents we used a nitrocellulose dot blot procedure with antisera to N-PBS-94 (Fig. 4). A positive reaction was obtained with dissociated phycobilisomes, phycobilisome-thylakoid vesicles, thylakoids alone, and purified N-PBS-94. Visual inspection shows a gradation of purple color from the higher to the lower concentrations in all but spinach. An additive color reaction is obvious only in the phycobilisome-thylakoid vesicles and is best seen in this black-and-white photograph in lanes 5-9. Calculations indicate that in lane 9 the sample contained 0.5 nanograms of high molecular components, equivalent to 0.25 ng in phycobilisomes and 0.25 ng in thylakoids. Although the results depend on surface reacting sites only, at the low sample concentrations the reaction is probably quantitatively reliable. The color reaction with the purified antigen is less than expected, and is probably due to interference by detergent (SDS) which is required to keep the purified N-PBS-94 in solution. Spinach thylakoids, used as a control, did not give a positive color reaction, which is consistent with the absence of high molecular weight polypeptides (greater than 75 kD) in fully denatured samples.

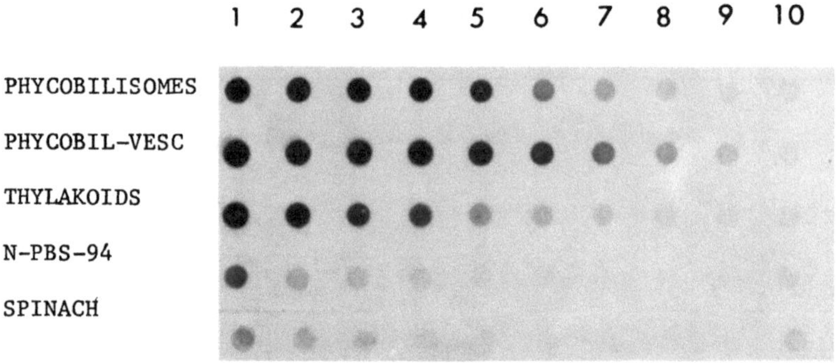

Figure 4. Immuno dot blots representing a dilution range in lanes 1-9 (1 to 1/256) of phycobilisomes, phycobilisome-vesicles, thylakoids, and N-PBS-94 of Nostoc, and spinach thylakoids. Bovine serum albumin in lane 10. The phycobiliprotein content was equal in phycobilisomes and phycobilisome-vesicles, and the chlorophyll content was equalized in phycobilisome-vesicles, thylakoids of Nostoc and of spinach.

In Nostoc the combined results suggest existence of at least two separate high molecular weight types; one in thylakoids, the other in phycobilisomes, and that they may occur in approximately a 1:1 ratio. The phycobilisome polypeptide contains phycocyanobilin in a special environment, while the thylakoid polypeptides probably contains chlorophyll a. It will be of interest to ascertain if the polypeptides are associated with one another. Such an association, together with specific orientation of the chromophores may be significant in regulating distribution of energy from the phycobilisomes to the photosystems, or possibly the regulation within the photosystem II core.

A consistent ratio between 94 kD polypeptides of phycobilisome and thylakoids, specifically of photosystem II, would predict that phycobilisome binding sites are determined by the thylakoid structure. This idea is supported by the phycobilisome to thylakoid reassociation studies carried out by Kirilovsky et al. (14). They found that energetically functional reassociations occurred only up to the point where the native phycobilisome number to thylakoid area was not exceeded. It will be of interest to explore if these polypeptides are encoded in different genes, or if they are differently processed proteins of the same gene product.

Acknowledgement

This research was supported in part by Contract AS05-76ER-04310, and Grant DE-FG05-87ER13652 from the Department of Energy.

References

1. Gantt, E., C.A. Lipschultz, T. Redlinger. 1985. In: Molecular Biology of the Photosynthetic Apparatus (K.E. Steinback et al., eds). Cold Spring Harbor, New York, p. 223.

2. Zilinskas, B.A. 1982. Plant Physiol. 70, 1060.

3. Zilinskas, B.A. D.A. Howell. 1986. Plant Physiol. 80, 829.

4. Glazer, A.N. 1985. Ann. Rev. Biophys. Biophys. Chem. 14, 47.

5. Redlinger, T., E. Gantt. 1981. Plant Physiol. 68, 1375.

6. Lundell, D.J., G. Yamanaka, A.N. Glazer. 1981. J. Cell. Biol. 91, 315.

7. Mimuro, M., E. Gantt. 1986 Photosyn. Res. 10, 201.

8. Mimuro, M., C.A. Lipschultz, E. Gantt. 1986. Biochim. Biophys. Acta 852, 126.

9. Redlinger, T., E. Gantt. 1982. Proc. Natl. Acad. Sci. USA. 79, 5542.

10. Gantt, E., F.X. Cunningham, C.A. Lipschultz, M. Mimuro. 1986. Plant Physiol. (submitted).

11. Hawkes, R. 1986. In: Methods in Enzymology vol. 121. Academic Press. p. 484.

12. Cunningham, F.X., E. Gantt. 1987. Plant Physiol. 83, 59.

13. Chereskin, B.M., J.D. Clement-Metral, E. Gantt. 1985. Plant Physiol. 77, 626.

14. Kirilovsky, D., M. Kessel, I. Ohad. 1983. Biochim. Biophys. Acta. 724, 416.

ON THE STRUCTURE OF PHOTOSYSTEM II-PHYCOBILISOME COMPLEXES OF CYANOBACTERIA

E. Mörschel

Fachbereich Biologie-Botanik der Philipps-Universität, Karl-von Frisch-Str., D - 3550 Marburg

G.-H. Schatz

Max-Planck-Institut für Strahlenchemie, Stiftstr. 34-36, D - 4330 Mülheim 1

Introduction

The phycobilisomes are the dominant light-harvesting pigment-protein complexes of cyanobacteria and red algae containing the biliproteins and linker polypeptides, which are necessary to assemble these aggregates (1, 2). They are bound to the external surface of the thylakoids and transfer the absorbed light energy mainly to photosystem (PS) II. As the efficiency of excitation energy transfer approaches 95% of the initially absorbed light quanta (3), it was postulated that phycobilisomes and the reaction centres of PSII should be closely spatially arranged. This relationship was corroborated by showing the alignment of EF freeze-fracture particles, purported PSII complexes, with phycobilisomes (4), by the isolation of allophycocyanin containing PSII (5) and of PSII-phycobilisome complexes (6). It is the aim of this study to examine the hypothesis that the 10 nm exoplasmic freeze-fracture particles correspond to PSII complexes. For this purpose we compared "in situ" EF-particles of cyanobacteria with oxygen evolving PSII complexes which were isolated and reconstituted into liposomes and visualized by electron microscopy. The information gained on the arrangement of PSII and the phycobilisomes indicates the structural foundations of the very effective energy transport.

Photosynthetic Light-Harvesting Systems

Supramolecular structure of thylakoids.

The supramolecular organization of thylakoids from the cyanobacteria Mastigocladus laminosus and Synechococcus spec. was investigated by freeze-fracture analysis of whole cells. The protoplasmic fracture face (PF) contained densely packed particles with sizes of 7.0-9.0 nm. The exoplasmic (EF) -particles were scattered on the thylakoids (Fig. 1 e) or aligned in short or long rows often running parallel to each other as shown in Fig. 1a,b. In this case, the spacing of neighbouring EF-particle rows was 45 to 75 nm, thus matching the centre-to-centre distance of parallel phycobilisome rows at the outer surface of the thylakoids. The EF-particles within one row were packed with little or no space between them; their centre to centre periodicity was 10-12 nm and thus corresponded to the centre-to-centre distance of phycobilisomes (4). The size of the EF-particles was determined by measuring the axes running parallel and perpendicular to the direction of the rows; the latter axis could only be determined accurately when the particle rows were shadowed parallel to their direction (Fig. 1a), as is shown clearly by comparing Figs. 1a and 1f. The EF-particles were 20 nm long and 10 nm wide and were aggregated linearly with their longitudinal faces; the scattered EF-particles measured 10 x 10 nm. Some particles within the rows measured also 10 nm; they probably originated from the dissociation of the 20 x 10 particles. The rowed 10 x 20 nm particles were divided by a central furrow perpendicular to their long axis into two side-by-side domains of 10 x 10 nm each (Fig. 1). These domains showed an additional faint central furrow parallel to the long axis of the whole particle (Fig. 1a,d,c).This furrow was also observed in the EF-particles which were scattered on the thylakoids (Fig. 1 e). We therefore conclude that the 10 x 20 nm EF-particles are dimers of the 10 nm EF-particles.

The relationship of the intramembrane protein-complexes with phycobilisomes was analysed by combined freeze-fracturing and freeze-etching analysis, which simultaneously exposes the exo-

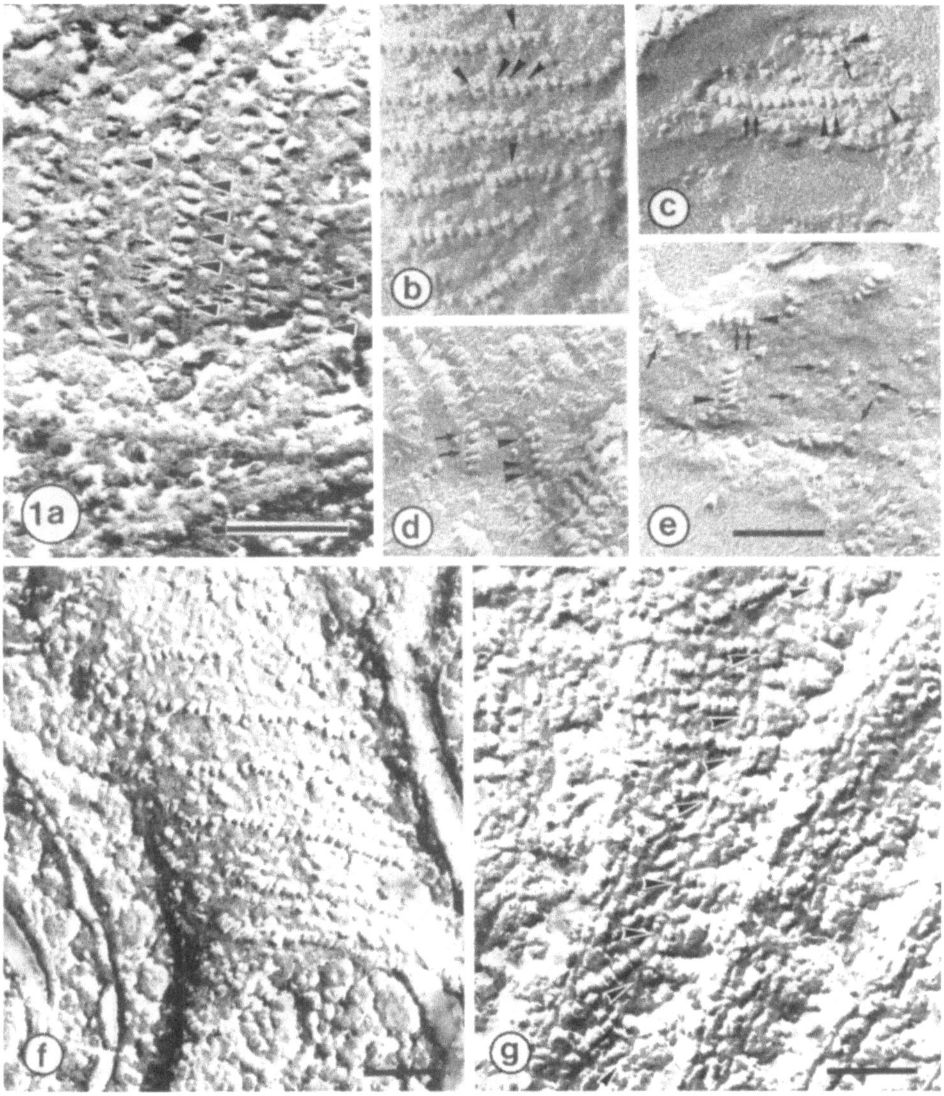

Fig. 1. Freeze-fracture through thylakoids of Synechococcus exposing EF-faces. a,b,c,e: show an extensive alignment of EF-particles; e: rowed and scattered EF-particles are visible. The rowed EF-particles are 10 x 20 nm and divided into two parts by a central lace perpendicular to their long axis (arrowheads). A further furrow (arrows) divides the particles parallel to their long axis (a-e) and also the scattered 10 nm EF-particles in (c). f: EF-particle rows shadowed perpendicular to their direction. g: Fracture through a thylakoid of Mastigocladus exposing EF-face and phycobilisomes. The EF-particle rows are directed towards the centre of the phycobilisome bases. Bars: 100 nm.

plasmic face and the protoplasmic surface of the thylakoids. These preparations revealed the linear arrangements of the integral EF-particles as well as the peripherally associated phycobilisomes as shown in Fig. 1g. A strict alignment of EF-particles with phcobilisomes is always observed. The micrographs show that the rows of EF-freeze-fracture particles are directed towards the center of the phycobilisome bases and they are directly continued by the lines of the phycobilisomes on the outer surface of the thylakoids. The phycobilisome pattern on the surface of the thylakoids reflects the pattern of the EF-particles within the membrane in alignment and periodicity. From these findings we conclude, that the phycobilisomes are bound on top of the rowed intramembrane EF-particles in a 1:1 ratio.

Analysis of photosystem II particles.

In order to test the correlation of the EF-particles with photosystem II, we isolated water splitting PSII particles from Synechococcus spec. and analysed these particles biochemically and structurally after incorporation into artificial lipid bilayer membranes. The PSII particles were prepared after Schatz and Witt (5) and further purified by density gradient centrifugation in SPC-buffer (6) without any loss in PSII activity. The high degree of enrichment of PSII was indicated by chlorophyll/PSII ratios between 70 to 90. The absorbance spectra of cells and of PSII complexes are shown in Fig. 2. The main pigments of the photosystem II complexes were allophycocyanin and chlorophyll (Chl) as revealed by their absorption maxima at 652 and 680 nm, respectively. The low temperature fluorescence emission spectra of whole cells compared with those of photosystem II preparations are given in Fig. 4a,b. Whole cells showed a strong emission peak at 730 nm and minor peaks at 660, 685 and 695 nm when excited at 445 nm (Fig. 4a). The 730 nm emission peak was attributed to photosystem I. Purified PSII preparations lacked this 730 nm emission maximum and

2

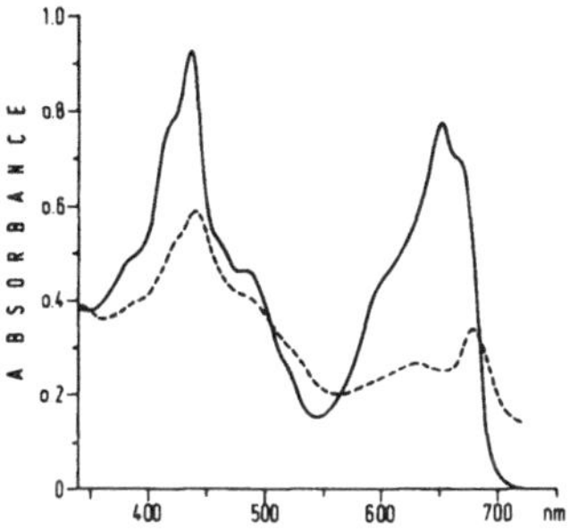

3

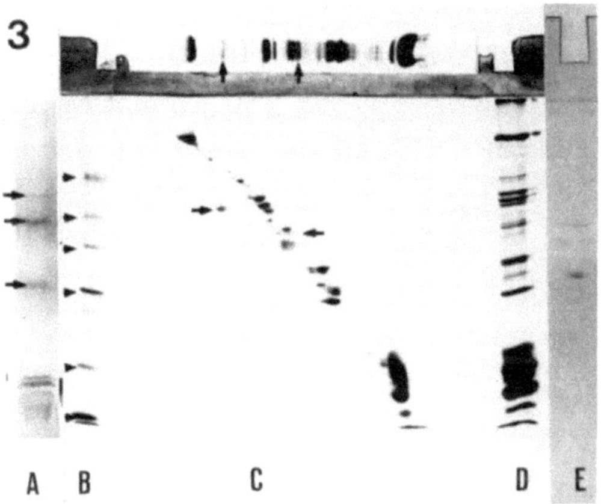

4

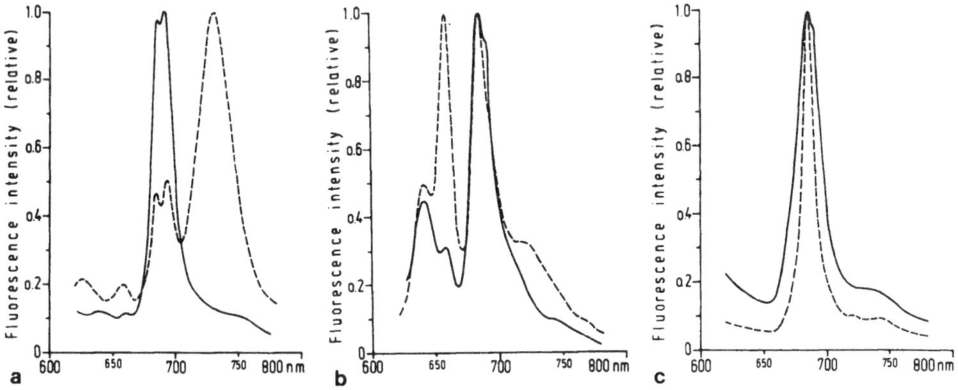

Fig. 2. Absorption spectrum of PSII complexes (--) compared to that of whole cells (--). Fig. 3. Photosystem II particles after polyacrylamide gel electrophoresis (PAGE). A:PSII complexes solubilized in LDS. The gel is not stained; from the top: chlorophyll-proteins CP IIa,b,c (arrows) and biliproteins. C: 2-dim. PAGE of PSII complexes. 1st dim.: Separation of chlorophyll-proteins after LDS-solubilization as in A. 2nd dim.: Separation in SDS after complete denaturation; the apoproteins of CP IIb,c are marked by arrows. E: Separation of CP IIa after SDS PAGE. D: Denatured photosystem II complexes separated by SDS PAGE. B: Marker proteins from the top (arrowheads): bovine serum albumin (67 000), ovalbumin (45 000), glyceraldehyde-3-phosphate dehydrogenase (36 000),carbonic anhydrase (29 000), trypsinogen (24 000), α-lactalbumin (14 200). Fig. 4 a,b: 77°K fluorescence emission spectra of PSII complexes (--) compared to those of whole cells (--). Excitation wavelengths: 445 nm (a,c), 590 nm (b). c: 77°K fluorescence emission spectra of CP IIa and CP IIb which are very similar (--) and CP IIc (--). CP IIa,b have a shoulder at 692 nm which is absent in CP IIc.

consequently photosystem I. This corresponds well with the high Chl/PSI ratio of 1500 of the PSII preparations compared to ratios of 300 to 350 in unfractionated thylakoids (5). When excited at 445 nm the PSII preparations showed a minor maximum at 660 nm characteristic for allophycocyanin and two major maxima at 685 and 693 nm (Fig. 4a). The fluorescence at 685 nm was attributed to an interplay of allophycocyanin B, the large membrane-phycobilisome linker (L_{CM}) and a chlorophyll antenna, whilst the 693 nm fluorescence belonged to a chlorophyll antenna alone. The phycobilisome constituents that were not dissociated from the complex were tightly coupled to photosystem II, as was evident from excitation within the absorption range of phycocyanin at 590 nm (Fig. 4b). Most of the energy was transferred to both the final phycobilisome emitters and the photosystem II antennae as shown from the emission peak at 685 nm and the shoulder at 692 nm. Only a minor part of the fluorescence was emitted by uncoupled phycocyanin and allophycocyanin at 640 nm and 660 nm, respectively. Purification of the PSII complexes in 0.1 M phosphate or 20 mM MES/NaOH, 10mM $MgCl_2$, 2mM K_2HPO_4 (MMP) buffer resulted in a loss of phycocyanin and part of allophycocyanin; this is further evidence that phycobilisomes were bound to PS II complexes.

The chlorophyll-protein (CP) and polypeptide pattern of photosystem II complexes were determined by two dimensional gel electrophoresis (Fig.3). In the first dimension, run in the presence of LDS for optimal yield of intact chlorophyll-complexes, three chlorophyll-proteins termed CP II a,b,c as well as the biliproteins were resolved (Fig.3A; 3C, 1. dim.). Apparent molecular weights of 110, 79 and 45 kDa were determined for the three chlorophyll-proteins. In the second dimension, the polypeptide pattern after complete denaturation by SDS was determined (Fig. 3C). Prominent polypeptides could be attributed to the α-and ß-biliprotein subunits of allophycocyanin (14, 16 kDa) and phycocyanin (15, 18 kDa) and at least four phycobilisome-linker polypeptides corresponding to 29, 31, 34 and 120 kDa. After SDS electrophoresis, the chlorophyll-complexes CP IIa and b were resolved as apoproteins of 47 kDa and CP IIc as one of 41 kDa. The chlorophyll-complexes CP II b,c contained no other proteins. Complex CP IIa (Fig. 3E) revealed an additional faint band of 80 kDa. After separation in the first dimension, the chlorophyll-proteins were cut out of the gels and analysed by fluorescence emission spectroscopy at liquid nitrogen temperature (Fig. 4c). CP IIa and b were very similar in their spectroscopic properties and showed maxima at 685 nm and shoulders at 693 nm, whilst CP IIc exhibited only one peak at 686 nm, when excited at 445 nm. Thus the isolated chlorophyll complexes had similar emission properties as the "in situ" photosystem II antennae.

The purified photosystem II complexes were incorporated into liposomes made from soybean phosphatidylcholine. Fig. 5a shows a typical PSII-proteoliposome fraction after negative staining. The liposome fraction was nearly uniform and consisted mainly of unilamellar vesicles with diameters in the range of 30 - 300 nm. The protein particles within the plane of the liposome membranes were clearly visible as electron translucent dots. The diameter of the incorporated particles was about 10 nm. Dissociated biliprotein aggregates were observed as soluble discs or double discs in the suspension between the proteoliposomes (Fig.5a).

Fig. 5b shows a typical negatively stained fraction of PS II-particles solubilized by Triton X-100 from proteoliposomes which were separated from free biliproteins. Two particle classes are distinguished: spherical-ellipsoidal and binary particles. The dominant structures are the binary particles, which are preferably deposited on the support with their broad faces. They measure 20-25 x 10-14 nm and are divided perpendicular to their long axis by a lace into two spherical-ellipsoidal parts of about 10-13 nm. Aggregations of these binary particles to rows also occurred as shown in Fig. 5b (inset). Within these rows, particles are visible in "top view" and measure about 24 x 12 nm; the central division, separating the particles into two parts of about 12 nm size, is clearly visible. These binary

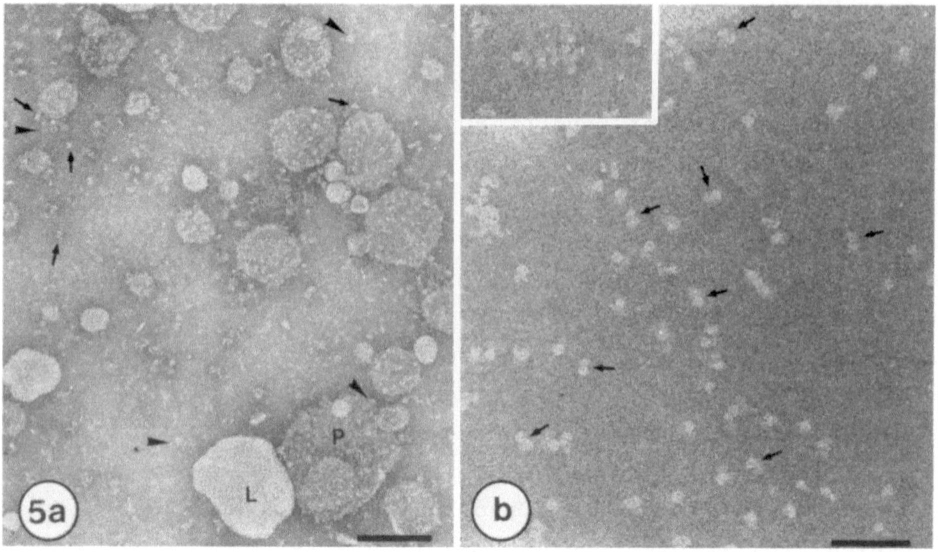

Fig. 5 a: PSII proteoliposome fraction. Soluble biliproteins are visible in face view (arrowheads) or in profile (arrows). Liposome (L), proteoliposome (P). b:Isolated PSII complexes negatively stained with 1% phosphotungstic acid (pH 7); binary complexes marked by arrows. Inset: Binary complexes assembled to a short row. Bars: 100 nm.

particles are very similar in their appearance compared to the "in situ" EF-particles of Fig. 1. The second particle class is represented by spherical-ellipsoidal particles of about 10-12 x 14-16 nm; they are supposed to be the building blocks of the binary particles.

The photosystem II proteoliposomes were analyzed by freeze-fracture analysis. In general the size of the proteoliposomes was more homogeneus with this technique than with negative staining; their diameters were between 80-170 nm. This is due to the fact that disruption and drying effects did not occur. Fig. 6a gives an overview over the proteoliposome fraction. The freeze-fracture particles were randomly distributed and their diameter was about 10.3 nm on the concave and convex fracture faces as analyzed by the size distribution measurements (not shown). Some particles showed a central furrow, that was also present in the EF-particles "in situ"; side by side aggregations of two particles were observed too.

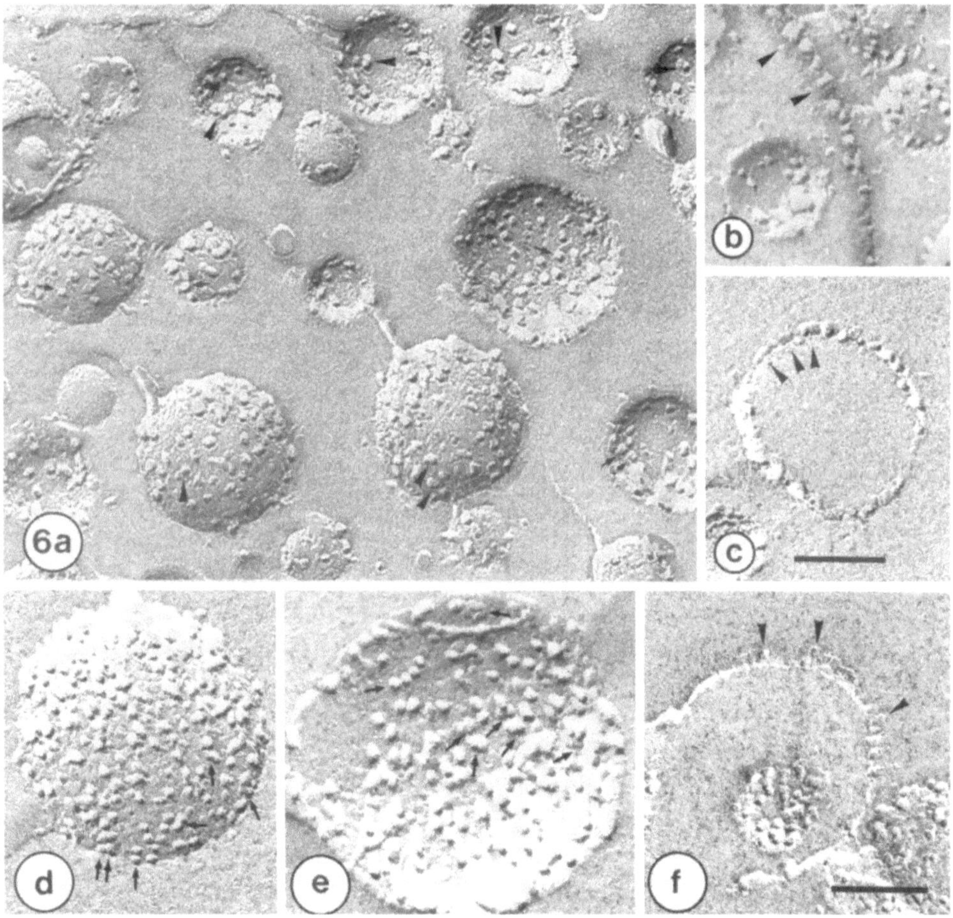

Fig 6a: Freeze-fractured PSII proteoliposomes without addition of cholesterol. Particles with a central furrow are marked by arrows, binary particles by arrowheads. b,c: Cross fractured proteoliposome showing the transmembrane PSII complexes; putative dimers are visible (arrows). Preparation without (b) and addition of cholesterol (c). d,e: PSII-proteoliposomes prepared with the addition of cholesterol showing binary PSII particles (arrows). f: Cross fractured PS II-proteoliposome with putative triangular phycobilisome centres (arrows). Bars: 100 nm.

Proteoliposomes prepared with the addition of cholesterol looked very similar, the major difference being however, that the capability for oxygen evolution (460 μmol O_2 x mg Chl^{-1} x h^{-1}) was completely preserved when cholesterol was present. Cholesterol is known to suppress lipid phase transitions and acts as a stabilizing factor in native membranes (7). Parallel to the high oxygen evolution, significantly more (30-50 %) of the PS II-particles occurred in dimeric configurations (Fig. 6 d-e). Sometimes parts of phycobilisomes still bound to the incorporated photosystem II particles were observed on the outer surface of the proteoliposomes (Fig. 6f). Cross fractures running perpendicular to the plane of the liposome membrane showed that the particles were spanning the membrane (Fig. 6 b,c). With a height of 13-16 nm they were considerably larger than the lipid bilayer and thus were exposed to the hydrophilic environment on both sides of the membrane. The width of the particles was about 10 nm; also putative dimeric particles were visible in these micrographs, dependent on the exposure of their longitudinal face parallel to the fractured surface.

Conclusions

This study demonstrates that isolated photosystem II complexes from Synechococcus spec. belong to the 10 nm size family of freeze-fracture particles. Particles of this group are the dominant species of exoplasmic fracture faces of cyanobacterial thylakoids (4,8,9,10,11,12). Therefore we conclude that the 10 nm EF-particles correspond to PSII complexes. This conclusion is supported by several lines of evidence: 1. A correlation between PSII activity and the number of 10 nm EF-particles was observed in the wild type and a phycobilisome deficient mutant of the red alga Cyanidium caldarium (13). 2. Thylakoids of heterocysts exhibit no PSII activity and concomitantly lack the 10 nm EF-particle class (14). 3. Thylakoids from which PSII was extracted resemble the particle pattern of PF-faces (15). The 10 nm particles contain at least the reaction centre complex, the water splitting system, the chlorophyll antennae with apoproteins of 41 and 47 kDa and a phycobilisome binding polypeptide (L_{CM}). A functional PS II-unit has a molecular weight of about 400 000 as estimated from the minimum requirement for oxygen evolution (16) including the phycobilisome linker L_{CM}. This value corresponds well to the weight of 418 000 for a spheric particle of 10.3 nm diameter calculated with a partial specific volume of 0.73 ml/g for protein.

In thylakoids of Synechococcus spec., a large population of EF-particles is aggregated side by side yielding binary structures of 10 x 20 nm, which are attached at their longitudinal faces to form rows of variable length. We propose that these binary particles are transmembrane dimeric units made up of two individual 10 nm PSII complexes. This is confirmed by the following observations: 1. the central lace dividing the particles perpendicular to their long axes, 2. occasional dissociations releasing individual 10 nm particles, and the occurrence of 10-12 nm particles in negative staining and freeze-fracture preparations, 3. a large population of dimeric photosystem II particles observed in proteoliposomes made of phospholipids and cholesterol in which oxygen evolution was almost completely preserved. Deep-etch studies by Giddings and coworkers (8) showed that the EF-particles protrude from the thylakoids with two domains. The formation of PS II dimers "in vivo" and their organization into rows appears to be a unique feature of cyanobacteria and red algae. Possible explanations are that either the photosystem II complexes might have developed specific binding sites to form these aggregations, or that the phycobilisomes may induce this organization. The photosystem II complexes serve as binding sites for the light harvesting system, the hemi-discoidal phycobilisomes. This structural relationship is shown by the tightly bound biliproteins and the phycobilisome linker polypeptides that could not be removed by using conditions conserving the phycobilisome structure. Isolation experiments, as also energy transfer and structural studies, support the concept of a supramolecular relationship between phycobilisomes and photosystem II. About 95 % of the light energy initially absorbed by phycobilisomes is transferred to photosystem II (3,17). Gantt and coworkers (6,18)

7

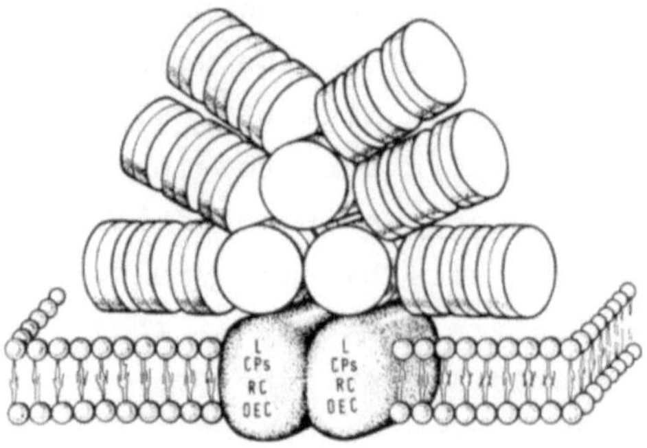

Fig. 7 Schematic model of the supramolecular structure of PSII-phycobilisome-complexes in cyanobacteria. L: Linker polypeptide, CPs: Chlorophyll binding proteins (apoproteins 41 and 47 kDa); RC: Reaction centre, OEC: Oxygen evolving complex.

isolated water-splitting phycobilisome-photosystem II particles from the red alga Porphyridium cruentum and water splitting photosystem II particles with bound biliproteins were isolated from Synechococcus spec. (5) and Anacystis nidulans (19). Structural studies showed that the rowed EF-particles are associated with phycobilisomes in a 1 to 1 ratio (4). The idea that the 10 x 20 nm EF-particles are built up of two photosystem II complexes is supported by phycobilisome morphology, because the two basal phycobilisome core cylinders fit directly in size to the two PS II monomers. It is supposed that each of the basal core cylinders contains two terminal energy emitters, namely α-allophycocyanin B and the core-membrane linker L_{CM} (20,21). The bigger hemi-ellipsoidal phycobilisomes of Porphyridium cruentum and Griffithsia pacifica are probably linked to a series of 2-3 neighboring EF-particles and hence 4-6 photosystem II units (22,23,24,25).

Thus we propose that cyanobacteria contain phycobilisome-PSII supercomplexes composed of two photosystem II complexes and one hemi-discoidal phycobilisome, as shown in the model illustrated in Fig. 7. The rowed phycobilisome-PS II complexes may act as energy conducting tracks. If this is true, energy is not only funneled within the complex to the photosystem II reaction centres, but also transferred to neighbouring photosystem II complexes and phycobilisomes. Energy transferred to photosystem II complexes may be distributed within the dimer and transferred to neighbouring complexes within the same row via the chlorophyll a antennae or via the phycobilisome core after reverse transfer. The resulting energy conducting "fibre system" would allow for an efficient energy distribution along the plane of the thylakoid by connecting many PS II reaction centres.

Acknowledgements

The study was generously supported by the Deutsche Forschungsgemeinschaft.

References

1. Gantt, E. 1980. Int. Rev. Cytol. 66, 45

2. Mörschel, E., E. Rhiel. 1987. In: Electron microscopy of proteins. Macromolecular structure and function (Harris, J.R., Horne, R.W., eds.), Academic Press, London, New York, p. 209.

3. Ley, A.C., W.L. Butler. 1977. Biochim. Biophys. Acta 592, 349.

4. Mörschel, E., K. Mühlethaler. 1983. Planta 158, 451.

5. Schatz, G.H., H.T. Witt. 1984. Photobiochem. Photobiophys. 7, 1.

6. Clement-Metral, J.C., E. Gantt. 1983. FEBS Lett. 156, 185.

7. Harrison, R., G. G. Lund. 1980. Biological membranes. Their structure and function. Blackie, Glasgow, London

8. Giddings, T.H., C. Wasman, L.A. Staehelin. 1983. Plant. Physiol. 71, 409.

9. Lefort-Tran, M., G. Cohen-Bazire, M. Pouphile. 1973. J. Ultrastr. Res. 44, 199.

10. Golecki, J.R., G. Drews. 1982. In: The Biology of Cyanobacteria, (Carr, N.G., Whitton, B.A., eds.) Blackwell Scientific Publications, Oxford, London, Edinburgh, Boston, Melbourne, p. 125.

11. Lichtle, C., J.C. Thomas. 1976. Phycologia 15, 393.

12. Neushul, M. 1970. Am. J. Bot. 57, 1231.

13. Wollman, F.A. 1979. Plant Physiol. 63, 375.

14. Giddings, T.H., L.A. Staehelin. 1979. Biochim. Biophys. Acta 546, 373.

15. Mörschel, E., Schatz, G. H. 1987. Planta (in press)

16. Ohno, T., K. Satoh, S. Katoh. 1986. Biochem. Biophs. Acta 852, 1.

17. Khanna, R., J.-R., Graham, J. Myers, E. Gantt. 1983. Arch. Biochem. Biophys. 224, 534.

18. Chereskin, B.M., J.D. Clement-Metral, E. Gantt. 1985. Plant Physiol. 77, 626.

19. Pakrasi, H.B., L.A. Sherman. 1984. Plant Physiol. 74, 742.

20. Glazer, A.N. 1984. Biochim. Biophys. Acta 768, 29.

21. Redlinger, T., E. Gantt. 1983. Plant Physiol. 73, 36.

22. Kursar, T.A., R.S. Alberte. 1983. Plant Physiol. 72, 409.

23. Ley, A.C. 1984. Plant. Physiol. 74, 451.

24. Ohki, K., Y. Fujita. 1987. In: Progress in Photosynthesis Research 2, (Biggins, J. ed.), Martinus Nijhoff, Dordrecht, Boston, Lancaster, p. 157.

25. Stevens, C.L.R., J. Myers. 1976. J. Phycol. 12, 99.

STRUCTURE OF CRYPTOPHYTE PHOTOSYNTHETIC MEMBRANES

W. Wehrmeyer
Fachbereich Biologie der Universität, Karl-von-Frisch-Straße
D-3550 Marburg, Federal Republic of Germany

Introduction

General characteristics of the photosynthetically active Cryptophytes were summarized by E. Gantt (1) and phycobiliprotein antennae of Cryptophytes have also been reviewed (2). Even in comprehensive up-to-date reviews on photosynthetic membranes and light harvesting systems, however, the unique organization of cryptophytan chloroplasts is only briefly referred to (3-6) or has been excluded altogether (7). Cryptophytes are unique in that they combine peripheral biliprotein antennae and integral chlorophyll a/c_2 light harvesting pigment protein complexes (chl a/c LHC) in their photosynthetic apparatus. The biliproteins Cryptophytes have in common with cyanobacteria and red algae, the chl a/c LHC they share with eight other algal classes. Most species of the Cryptophytes are obligate phototrophs. The colorless members, which exist as obligate heterotrophs, are not considered here; they may still contain plastids as leukoplasts but not thylakoids (8). Excluded are also such Cryptophytes which are endosymbionts in other organisms (9, 10).

Intrathylakoidal Material

The thylakoids of Cryptophytes are unique as the intrathylakoidal lumen is filled with electron dense material which was suggested to be the site of phycobiliproteins (11, 12). Indirect evidence for this assumption was provided by trypsin

Photosynthetic Light-Harvesting Systems

treatment, deoxycholate/pigment extraction and electron microscopy (13, 14). The amount of this material could be reduced by exposure of high energy fluence rates (15-17) and by nitrogen deprivation (16, 18). These treatments however do not exclusively act on phycobiliproteins. A direct and specific visualization of cryptophytan biliproteins is possible by applying the immunogold labeling technique. By this method it could be demonstrated that the phycoerythrin 545 of Cryptomonas maculata is localized in the thylakoid lumen. The additional positive gold label of thylakoid membranes can be interpreted as an indication of phycoerythrin species associated with the thylakoid membrane or, less probably, of those in transit through the thylakoid membrane (19).

Freeze Fracture Studies on Thylakoid Membranes in situ

The structural analysis of the supramolecular organisation of thylakoids of Cryptophytes was performed predominantly by freeze-fracturing techniques (18, 20-24). Freeze-fractured thylakoid membranes of Cryptophytes revealed, at first glance, faces almost similar to the well-known pattern of thylakoids from chloroplasts in green plants. They exhibit protoplasmic (PF) and exoplasmic (EF) fracture faces with a low frequency of larger particles in EF and a high frequency of smaller particles in PF faces. Table 1 summarizes the results reported by different authors. It is not easy to decide whether the size differences for the PF and EF face particles of the various organisms reflect real differences in composition or simply variability of shadow-casting and measuring techniques. In general, however, the up-to-date values of PF and EF particles are almost 3 nm smaller than the values reported earlier (20, 22). It can hence be suggested that the particles in EF faces of chl c-containing algae in fact are smaller than the EF particles of chl b-containing green algae and higher plants (27, 28). In cyanobacteria and red algae EF face par-

Table 1. Particle Sizes and Mean Particle Frequencies of Cryptophytes as Determined by Different Authors. The last two lines refer to values obtained from a Dinophyte and a brown alga.

Species	Modal particle size [nm] PF_u/PF_s	Mean particle size $PF_u (PF_1)$ [nm]	$PF_s (PF_2)$	Mean particle frequency $PF_u(PF_1)$ [number μm^{-2}]	$PF_s (PF_2)$	Reference
Chroomonas sp.	13 / 11	-	-	2112 ± 288	2800 ± 409	Dwarte et Vesk (1983)
Cryptomonas sp.	14 / 11			2390 ± 410	3500 ± 420	"
Cryptomonas mac.	7.5 et 10	8.7 ± 2	8.3 ± 1.7	2814 ± 297	4138 ± 646	Rhiel et al. (1985)
" (-N)	7.0	7.3 ± 1.5		3371 ± 524		"
Rhodomonas lens	10 / 9	10.88 ± 2.7	9.98 ± 2.2	4240 ± 117	2566 ± 235	Spear-Bernstein, Miller (1985)
Cryptomonas rufesc.	7.5 - 9.5 / 5 - 7			2600 ± 100	4950 ± 150	Lichtlé et al. (1986)
(HL)	7.5 9.5 / 5 - 6			2500 ± 100	6100 ± 150	"
	EF_u/EF_s	$EF_u (EF_1)$	$EF_s (EF_2)$	$EF_u (EF_1)$	$FF_s (EF_2)$	
Chroomonas sp.	15 / 15			291 ± 124	825 ± 179	Dwarte et Vesk (1983)
Cryptomonas sp.	15 / 15			112 ± 63	1013 ± 205	"
Cryptomonas mac.	10 et 12.5	9.7 ± 2.1	10.9 ± 1.8	156 ± 58	1100 ± 145	Rhiel et al. (1985)
" (-N)	7.5 et 10.0	9.2 ± 1.9		175 ± 60		"
Rhodomonas lens	12 / 13	12.57 ± 2.8	13.69 ± 3.1	159 ± 51	575 ± 97	Spear-Bernstein, Miller (1985)
Cryptomonas rufesc.	7.5 / 7.5 - 9.5			540 ± 50	950 ± 50	Lichtlé et al. (1986)
" (HL)	9.5 / 9.5 - 12.5			370 ± 50	525 ± 50	"
Gonyaulax polyedra	10 - 11			364 ± 26	1863 ± 81	Sweeney (1981)
Fucus serratus	8 - 9	8.3 ± 0.2	8.9 ± 0.2	479 ± 54	1531 ± 208	Berkaloff et al (1983)

ticles fall into the 10 nm size category (29).

The most surprising observation on freeze-fracture micrographs was the differentiation of two types of EF and PF faces in analogy to those of green algae which was possible mainly on the basis of differences in particle frequencies. In green algae and higher plants these fracture faces could be correlated with either stacked or unstacked thylakoids indicated by indices s or u. These terms cannot be applied for chromophytan algae as they do not have grana. Their thylakoids run in parallel in fairly tight bands of two or three. Adjacent thylakoids are separated by a gap of 4 to 6 nm with occasional bifurcation of single thylakoids but without forming partitions. Stacking is not completely understood at present, however, the involvement of the chl a/b LHC with cations is well established (30-32). It has been mimicked in proteoliposomes containing purified LHC II (33, 34). Moreover, membrane adhesion is increased in the non-phosphorylated state of chl a/b LHCII particles, whereby one or two threonine residues of a surface exposed segment of the chl a/b LHC are involved (4). Neither a protein phosphorylation/dephosphorylation as a regulatory mechanism of a lateral movement of chl a/c LHC nor any homology of the amino acid sequences of chl a/c LHC and chl a/b LHCII have been shown. So the prerequisites for membrane stacking in Cryptophytes are lacking at present and the unequal distribution of particles within EF and PF faces of cryptophytan thylakoids warrants further investigations.

Thylakoid Membrane Particle Analysis

General introduction. The main aim of the analysis of the photosynthetic membranes on the supramolecular level is to correlate different particle categories with functional units of the photosynthetic apparatus. Three strategies in particular, have been used for this purpose whereby indirect and direct methods can be distinguished. The indirect method makes use of the variation of antennae and photosystems under different conditions of light and nutrition in situ with a subsequent evaluation of freeze-fractured faces of the thylakoid membranes in comparison with functional efficiency. A direct method is the immunogold-labeling of specific constituents of the photosynthetic apparatus of cryptophytes in Lowicryl sections for unequivocal localization. The second direct method starts with the isolation and biochemical and functional characterization of constituents of the thylakoid membranes with subsequent electron microscopic analysis directly or after incorporation into liposomes where particle size evaluation allows a correlation with one of the particle size classes found in freeze fracture faces of thylakoid membranes in situ. In the following some results of these methods are presented.

Indirect method. Almost 90 % of the intrathylakoidal phycoerythrin was lost after culturing of C. maculata for 10 to 14 days under energy fluence rates of 4.4 Wm^{-2} whereas the controls under 1.28 Wm^{-2} kept their phycoerythrin content. When these membranes with empty intrathylakoidal lumen were freeze-cleaved only one PF and one EF face could be distinguished.

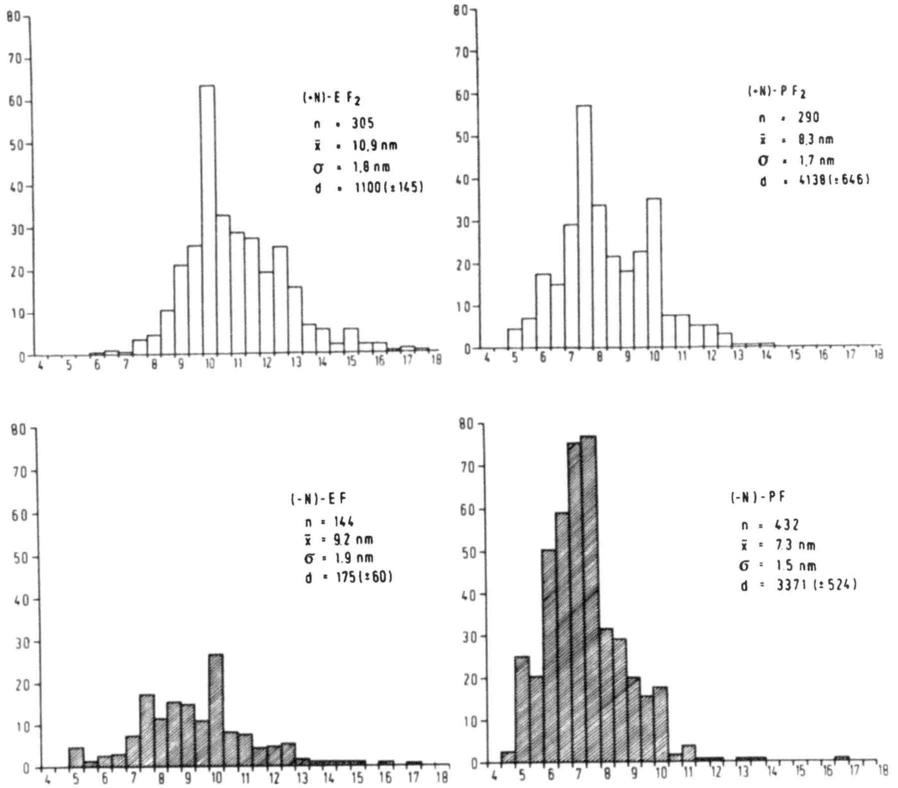

Fig. 1. Particle size distribution in EF and PF faces of thylakoid membranes of C. maculata chloroplasts after culture under 1.28 Wm^{-2} (above) and 4.4 Wm^{-2} (below). Details (18)

Fig. 1 shows the changes within the particle frequencies and in the particle size distribution in the PF and EF face. The most remarkable change was found within the EF face of high light thylakoids. EF faces were almost devoid of particles; only 10 to 15 % of the original number of the low light control were still present. The 12.5 nm particle population was almost lacking whereas some of the 10 nm size class were preserved and a new one at 7.5 appeared. Cells exhibiting this significantly decreased population of 12.5 nm particles in EF faces never reached a positive value in oxygen evolution after illumination with increasing energy fluence rates up to 8 Wm^{-2}; control cultures, on the other hand, were already saturated at 6.4 Wm^{-2} producing about 100 umol O_2 mg chlorophyll$^{-1}\cdot h^{-1}$. During fluorescence induction experiments at room temperature the initial fluorescence F_o and the maximal fluorescence F_{max} were significantly increased in these cells whereas the ratio of F_{var}/F_{max} decreased from 0.6 to 0.1. As the ratio of F_{var}/F_{max} reflects the amount of chlorophyll active in transferring energy to PS II reaction centers, this low value indicates a uncoupling of antennae components belonging to PS II complex. That this indeed is the case rather than a degradation of chl a/c LHC is substantiated by SDS gel electrophoresis. Triton x-100 solubilized membrane proteins of 4.4 Wm^{-2} and 1.28 Wm^{-2} treated cells show an almost similar banding pattern of the LHCP.That indeed the photosystem II complex was affected is further indicated by the 77 K fluorescence emission spectrum; its maximum shifted from 693$\pm$2 nm of the control cells to 684$\pm$2 nm in high light treated cells after excitation at 440nm (35).

These experimental results of an affected photosystem II complex with uncoupling of at least a part of chl a/c LHC and the lack of 12.5 nm size particles in the EF face of freeze-cleaved thylakoid membranes of Cryptomonas with some preservation of the 10 nm size particle population can be summarized most reasonably by correlating the 12.5 nm particle population with the PS II complexes. These can probably give rise to 10nm or even 7.5 nm particles after successive uncoupling of chl a/c light harvesting complexes.

Direct method 1. The direct method of membrane analysis by immunogold labeling was applied for the localization of phycoerythrin 545 (see above) and the chl a/c LHC. Here the results of only the latter will be presented. The heterogeneous distribution of chl a/b LHC in grana and stroma thylakoids of chloroplasts of green plants seems to be well established (4, 36). It is, however, an open question whether this fits also for the chl c-containing algal chloroplasts. In an immunocytochemical investigation the protein A gold label specific for chl a/c LHC was equally distributed all over the thylakoid membranes and neither on the outside nor on the inside of adjacent membranes of paired thylakoids could a concentration of gold label be seen (19). From this it can be concluded that the spatial destribution of photosystems and hence the chl a/c LHC may be different from that in green plants.

Direct method 2. The isolation of chl a/c LHC and PS I from solubilized thylakoid membranes of C. maculata was performed to correlate particle size classes of freeze-cleaved thylakoid membranes with functional units of the photosynthetic apparatus.

A chl a/c LHC fraction with absorption maxima at 463, 589 and 640, indicating chl c, and a maximal fluorescence emission at 682 nm was obtained. The unheated LHC protein with a M_r of 80000 was composed of three polypeptides banding in SDS gel electrophoresis with M_rs at 22000, 19000 and 18000 (37). This is in line with some other chl a/c LHC preparations of diatoms and brown algae (Table 2). The different LHCP-forms may arise from different members of a LHCP gene family as was shown for chl a/b LHC proteins (47, 48); they could however also be derived from a single gene by posttranslational processing into several polypeptides (49).

A non-spherical shape of the LHC protein was seen by electron microscopy after negative staining. This was confirmed by an electron microscopic study of freeze-fractured liposomes after extensive incorporation of chl a/c LHC (37). The average size of the particles was 6 to 7 nm (Fig. 2).

Table 2. Properties of Chl a/c Light Harvesting Protein Complexes of Various Algal Classes.

ALGAL CLASS	SPECIES	TYPE OF CHL C	CHL A/C RATIO	SUBUNIT(S) KDA	ABSORBANCE MAXIMA (NM)	FLUORESCENCE EMISSION (NM)	REFERENCES
DINOPHYCEAE	GLENODINIUM	CHL c_2	0.21	20	453, 585, 636 672, 485	--	BOCZAR ET AL. (1980)
	GLENODINIUM	CHL c_2	4.2	37, 24, 22, 19	453, 585, 636 672, 485	675	BOCZAR & PRÉZELIN (1986)
	GONYAULAX POLYEDRA		0.32	38, 32 21, 16	452, 585, 635 677	675 640	BOCZAR & PRÉZELIN (1987)
CRYPTO-PHYCEAE	CHROOMONAS CS 24	CHL c_2	1.4-1.54	24.0, 20.0	436, 672 458, 586, 640 488	682	INGRAM & HILLER (1983)
	CRYPTOMONAS MACULATA	CHL c_2	1.7	22, 19.0 18.0	436, 675 463, 589, 640 490	682 740	RHIEL ET AL. (1987)
+BACILLARIO-PHYCEAE	PHAEODAC-TYLUM TRICORNUTUM	CHL c_1+c_2	2.7	17.5, 18.0	CHLA OMITTED 450, 635	675 725	FRIEDMAN & ALBERTE (1984)
	PHAEODAC-TYLUM TRICORNUTUM	CHL c_1+c_2	2.7	16.4, 16.9	418, 440, 465 635, 670/78 490	675/80 735/40	OWENS & WOLD (1986)
	PHAEODAC-TYLUM TRICORNUTUM	CHL c_1+c_2	3.0	19.5, 19.0 18.0		685	FAWLEY & GROSSMAN (1986)
+PHAEOPHYCEAE	ACROCARPIA PANICULATA	CHL c_1+c_2	3	13-16	438, 674 465, 584, 636	683	ANDERSON & BARRETT (1979) BARRETT & THORNE (1981)

+ ALGAL CLASSES WITH FUCOXANTHIN (ABSORBANCE 475-540 NM) AS ASSOCIATED CAROTENOID

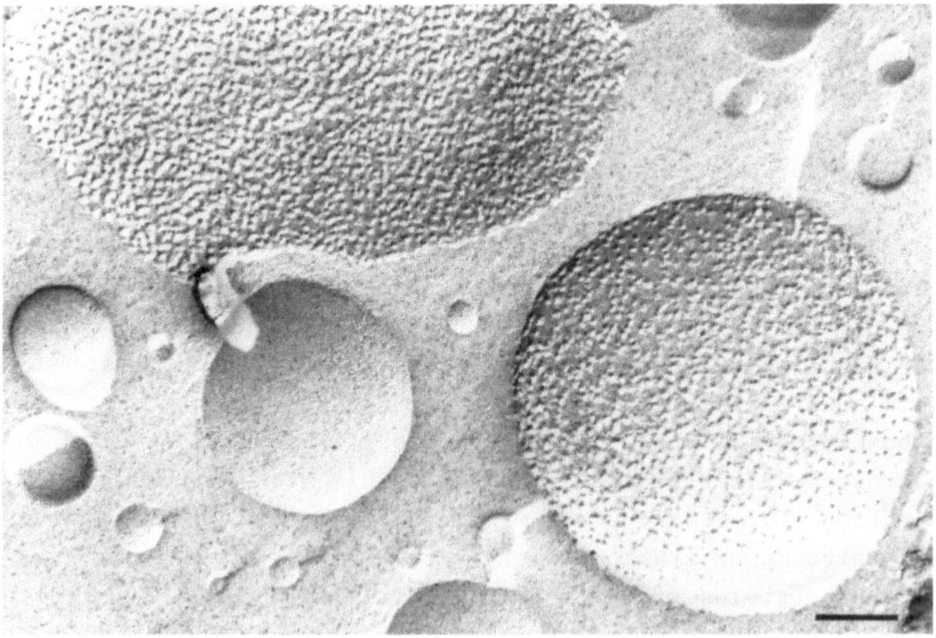

Fig. 2. Electron micrograph of freeze-fractured liposomes with densely incorporated particles of isolated chl a/c LHC of C. maculata (original 37). Bar 100 nm

Freeze-fracture micrographs of reconstituted chl a/b LHC vesicles suspended in low salt medium revealed 7.5±1.6 nm particles (33, 50). From these values it can be concluded that the chl a/c LHCs are slightly smaller than the chl a/b LHCs. In proteoliposomes, occasionally fractured perpendicular to the plane, the chl a/c LHCs were seen as transmembrane particles. This is the first clue that the chl a/c LHC have some similarities in the primary protein structure with the chl a/b LHC. For this complex three transmembrane helices have been postulated (51, 52).

The PS I enriched fraction showed a chlorophyll/P700 ratio of 100±20 and a 77 K fluorescence emission maximum at 680 nm after excitation at 460 nm. The lack of absorption at 460 nm and the small fluorescence emission at 635 nm indicated that there were only traces of some loosely associated or uncoupled chlorophyll c in the preparation. The undenatured PS I complex with a M_r of 140000 disappeared after heating at 100°C and generated a major band at M_r 61000. Under both conditions five smaller polypeptides in the range of M_r 13000 to 19000 were present (37). This is in agreement with observations for PS I fractions of cyanobacteria (53), diatoms (43) and green plants (54, 55). Negatively stained preparations of PS I complexes revealed an association to membraneous sheets with a particle size distribution around 10 (9.3±1.8) nm. This was confirmed by electron microscopy of freeze-fractured PS I complexes incorporated into liposomes. The banding pattern of SDS polyacrylamide gel electrophoresis fraction of C. maculata, the observed size of PS I particles and the lack of absorption at 460 nm suggest that the 10 nm particles represent PS I reaction center without additional LHC proteins.

Conclusion. The presence of two antenna pigment systems in Cryptophytes raises the question of their spatial and functional coordination with the photosystems. It has been shown earlier that there are multiple forms of the phycobiliproteins within the thylakoid lumen of Cryptophytes (56-58).

These can be traced back, in part, to different α and ß subunits, which are independent polypeptides as confirmed by amino acid sequence analysis (59, 60). One species is found to be in close association with the thylakoid membrane (61). This observation is substantiated by immunogold labeling techniques carried out independently by two groups (19, 62). Evidence for a specific energy transfer from phycoerythrin to PS II was obtained by Lichtlé et al. (63). Particle size distribution measurements with diameters for the EF particles representing PS II of 12.5, 10.0 and 7.5 with an increase of the smaller size categories under 4.4 Wm^{-2} show that the chl a/c LHC is in close connection with the PS II core, but is uncoupled under high light energy fluence rates. Excitation spectra of two Cryptomonas species showed an increased contribution by chl c_2 to PS II chl a fluorescence emission (64). This observation also supports the concept of a spatial and functional relationship. The correlation of chl a/c LHC with PS I at present remains open.

Acknowledgement

This research was supported by the Deutsche Forschungsgemeinschaft (Sonderforschungsbereich 305).

References

1. Gantt, E. 1980. In: Phytoflagellates (E. Cox, ed.). Elsevier, North Holland, Inc., Amsterdam. pp. 381-405.

2. Gantt, E. 1979. In: Biochemistry & Physiology of Protozoa, 2nd ed., Vol. 1 (M. Levandowsky, S.A. Hutner, eds.). Academic Press, New York. pp 121-137.

3. Anderson, J.M., J. Barrett. 1986. In: Photosynthetic Membranes and Light-Harvesting Systems (L.A. Staehelin, C.J. Arntzen, eds.). Springer, Berlin, New York, Tokio. pp. 269-285.

4. Staehelin, L.A. 1986. In: Photosynthetic Membranes and Light-Harvesting Systems (L.A. Staehelin, C.J. Arntzen, eds.). Springer, Berlin, New York, Tokio. pp. 1-84.

5. Zuber, H. 1985. Photochem. Photobiol. 42, 821.

6. Zuber, H., R. Brunisholz, W. Sidler. 1987. In: Photosynthesis (J. Amesz ed.). Elsevier Science Publ. B.V. Amsterdam. pp. 233-271.

7. Mörschel, E., E. Rhiel. 1987. In: Electron Microscopy of Proteins (J.R. Harris, R.W. Horne eds.), 6, Membranous Structures, Academic Press, London, New York, pp. 209-254.

8. Sepsenwol, S. 1973. Exp. Cell Res. 76, 395.

9. Hibberd, D.J. 1977. J. mar.biol.An. U.K. 57, 45.

10.Wilcox, L.W., G.J. Wedemayer. 1984. J. Phycol. 20, 236.

11.Dodge, J.D. 1969. Arch. Mikrobiol. 69, 266.

12.Wehrmeyer, W. 1970. Arch. Mikrobiol. 71, 367.

13.Gantt, E., M.R. Edwards, L. Provasoli. 1971. J.Cell Biol. 48, 280.

14.Lichtlé, C. 1978. Thèse de doctorat D'Etat, Paris.

15.Faust, M.A., E. Gantt. 1973. J. Phycol. 9, 489.

16.Lichtlé, Chr. 1979. Protoplasma 101, 283.

17.Thinh, L.V. 1983. Phycologia 22, 7.

18.Rhiel, E., E. Mörschel, W.Wehrmeyer. 1985. Protoplasma 129, 62.

19.Rhiel, E., J. Kunz, W. Wehrmeyer. 1988. (submitted)

20.Lichtlé, C., J.C. Thomas. 1976. Phycologia 15, 393.

21.Dwarte, D.M., M. Vesk. 1982. Micron 13, 325.

22.Dwarte, D.M., M. Vesk. 1983. Protoplasma 117. 130.

23.Lichtlé, C., J.C. Duval, N. Hauswirth, A. Spilar. 1986. Photobiochem. Photobiophys. 11, 159.

24.Spear-Bernstein, L., K.R. Miller. 1985. Protoplasma 129, 1.

25. Sweeney, B.M. 1981. J. Phycol. 17, 95.

26. Berkaloff, C., J.C. Duval, N. Hauswirth, B. Rousseau. 1983. J. Phycol. 19, 96.

27. Wollman, F.A., J. Olive, P. Bennoun, M. Recouvreur. 1980. J. Cell Biol. 87, 728.

28. Staehelin, L.A., C.L. Arntzen. 1983. J. Cell Biol. 97, 1327.

29. Staehelin, L.A., E.H. Giddings, P. Badami, W.W. Kryzmowski. 1978. In: Light Transducing Membranes, (D.W. Deamer, ed.), Academic Press. New York, pp. 335-355.

30. Mullet, J.E., C.J. Arntzen. 1980. Biochim. Biophys. Acta 589, 100.

31. Mullet, J.E., T.O. Baldwin, C.J. Arntzen. 1981. In: Photosynthesis III. Structure and Molecular Organization of the Photosynthetic Apparatus (G. Akoyunoglou ed.), Balaban Internat. Science Services, Philadelphia, pp. 577-582.

32. Steinback, K.E., J.J. Burke, C.J. Arntzen. 1979. Arch. Biochem. Biophys. 195, 546.

33. McDonnel, A., L.A. Staehelin. 1980. J. Cell Biol. 84, 40.

34. Ryrie, I.J., J.M. Anderson, D.J. Goodchild. 1980. Eur. J. Biochem. 107, 345.

35. Rhiel, E., K. Krupinska, W. Wehrmeyer. 1986. Planta 169,361.

36. Kyle, D.J.. L.A. Staehelin, C.J. Arntzen. 1983. Arch. Biochem. Biophys. 222, 527.

37. Rhiel, E., E. Mörschel, W. Wehrmeyer. 1987. Eur. J. Cell Biol. 43 (1), 82.

38. Boczar, B.A., B.B. Prézelin, J.B. Markwell, J.P. Thornber. 1980. FEBS-Lett. 120, 243.

39. Boczar, B.A., B.B. Prézelin. 1986. Biochim. Biophys. Acta 850, 300.

40. Boczar, B.A., B.B. Prézelin. 1987. Plant Physiol. 83, 805.

41. Ingram, K., R.G. Hiller. 1983. Biochim.Biophys.Acta 722,310.

42. Friedman, A.L., R.S. Alberte. 1984. Plant Physiol. 76, 483.

43. Owens, T.G., E.R. Wold. 1986. Plant Physiol. 80, 732.

44. Fawley, M.W., A.R. Grossman. 1986. Plant Physiol. 81, 149.

45. Anderson, J.M., J. Barrett. 1979. In: Chlorophyll Organization and Energy Transfer in Photosynthesis. Ciba Found. Symposium 61, Excerpta Medica, Amsterdam, pp. 81-104.

46. Barrett, J., S.W. Thorne. 1981. In: Photosynthesis II. Structure and Molecular Organisation of the Photosynthetic Apparatus (G. Akoyunoglou, ed.) Balaban Internat. Science Services, Philadelphia, pp. 347-356.

47. Cashmore, A.R. 1984. Proc.Natl.Acad.Sci. USA 81, 2960.

48. Dunsmuir, P. 1985. Nucleic Acid Res. 13, 2503.

49. Kohorn, B.D., E. Havel, P.R. Chitnis, J.P. Thornber, E.M. Tobin. 1986. J. Cell Biol. 102, 972.

50. Sprague, S.G., E.L. Camm, B.R. Green, L.A. Staehelin. 1985. J. Cell Biol. 100, 552.

51. Karlin-Neumann, G.A., B.D. Kohorn, J.P. Thornber, E.M. Tobin. 1985. J. Molee. Applied Geneties, 3, 45.

52. Bürgi, R., F. Suter, H. Zuber. 1987. Biochim. Biophys. Acta 890, 346.

53. Lundell, D.J., A.N. Glazer, A. Melis, R. Malkin. 1985. J. Biol. Chem. 260, 646.

54. Dunahay, T.G., L.A. Staehelin. 1985. Plant Physiol. 78, 606.

55. Ryrie, I.J. 1983. Eur. J. Biochem. 137, 205.

56. Mörschel, E., W. Wehrmeyer. 1975. Arch. Microbiol. 105, 153.

57. Mörschel, E., W. Wehrmeyer. 1977. Arch. Microbiol. 113, 83.

58. Hiller, R.G., C.D. Martin. 1987. Biochim. Biophys. Acta 923, 98.

59. Sidler, W., B. Kumpf, F. Suter, W. Morisset, W. Wehrmeyer, H. Zuber. 1985. Biol. Chem. Hoppe Seyler 366, 233.

60. Sidler, W., H. Nutt, B. Kumpf, H. Zuber, W. Wehrmeyer, A. Brenzel. 1987. In: Progress in Photosynthesis Research 2 (J. Biggins ed.) M. Nijhoff Publ. Dordrecht, 153-156

61. Brenzel, A., W. Wehrmeyer. in preparation

62. Spear-Bernstein, L., K.R. Miller. 1987. In: Progress in Photosynthesis Research 2 (J. Biggins ed.) M. Nijhoff Publ. Dordrecht, 4.309.4.312.

63. Lichtlé, C., H. Jupin, J.C. Duval. 1980. Biochim. Biophys. Acta 591, 104.

64. Bruce, D., J. Biggins, T. Steiner, M. Thewalt. 1986. Photochem. Photobiol. 44, 519.

STRUCTURAL AND PHYLOGENETIC RELATIONSHIPS OF PHYCOERYTHRINS FROM CYANOBACTERIA, RED ALGAE AND CRYPTOPHYCEAE

W. Sidler and H. Zuber
Institut für Molekularbiologie und Biophysik, ETH-Hönggerberg, CH-8093 Zürich, Schweiz.

Introduction

Phycoerythrins are red biliproteins, which absorb and transduce light energy. They occur in cyanobacteria, red algae and cryptomonads. C-phycoerythrin (C-PE), isolated from the cyanobacterium Calothrix (Fremyella diplosiphon), for example, has maximal absorption at 565 nm and a fluorescence maximum at about 590 nm. The chromphores responsible for the photobiological activities in the different phycoerythrin complexes are the open chain tetrapyrroles phycoerythrobilin (PEB) and phycourobilin (PUB). These pigments absorb at 555 nm and 495 nm, respectively.In cyanobacteria and red algae, phycoerythrins are located in the phycobilisome at the periphery of the rods, as was shown for the example of the hemidiscoidal phycobilisome from Calothrix. In cyanobacteria, which undergo complementary chromatic adaptation, abundant production of C-phycoerythrin is induced in these organisms under green light conditions. Porphyridium cruentum and some other red algae have a hemi-ellipsoidal type of phycobilisome, which has a double or threefold molecular weight (10 to 15 million daltons). Phycoerythrins are arranged in stacks of discs, forming the rods of the phycobilisome. Each disc contains six phycoerythrin-molecules, which are composed of an α- and a β-subunit (18 and 20 kDa, respectively.). In red algae, this $(\alpha\beta)_6$ complex contains a colored γ-subunit (30 kDa). In Calothrix, the linkers associated with C-PE are not yet fully

Photosynthetic Light-Harvesting Systems

characterized. Cryptomonad biliproteins are located in the intrathylakoidal lumen of the cryptomonad chloroplast and no supramolecular structure is discernable by electron microscopy. Various spectral types of cryptomonad phycocyanins and phycoerythrins have been characterized. Cryptomonad organisms contain only one biliprotein, either a phycocyanin-type or a phycoerythrin-type and no allophycocyanin. For a complete overview see ref. (1).

In this paper we present comparative studies of the complete amino acid sequences for the phycoerythrins from cyanobacteria (C-PE from Calothrix (formerly Fremyella diplosiphon)) (2), red algae (B-PE from Porphyridium cruentum) (3) and the biliproteins from cryptomonads (PC-645 from Chroomonas sp (4,5)), and a partial sequence of PE-545 from Cryptomonas maculata. Sequence homologies and the resulting phylogenetic relationships, the chromophore binding sites and functional aspects in comparison with the three-dimensional structure of C-phycocyanin (6,7) are discussed.

C - Phycoerythrin from the Cyanobacterium Calothrix

The first complete amino acid sequence of a phycoerythrin is that of C-phycoerythrin from Fremyella diplosiphon, now callled Calothrix (Fig. 1). C-PE consits of an α- and a β- subunit. The hexameric complex, as it exists in the phycobilisome, is not yet compleatly characterized. From its amino acid sequence, the α-subunit is seen to contain 164 and the β-subunit 184 amino acid residues. β-C-phycoerythrin is the largest biliprotein β-subunit, the others being 7-17 residues shorter. The amino acid sequence was established by N-terminal sequencer runs and by sequence analysis of large overlapping fragments generated by chemical and enzymic cleavage of C-PE (4).

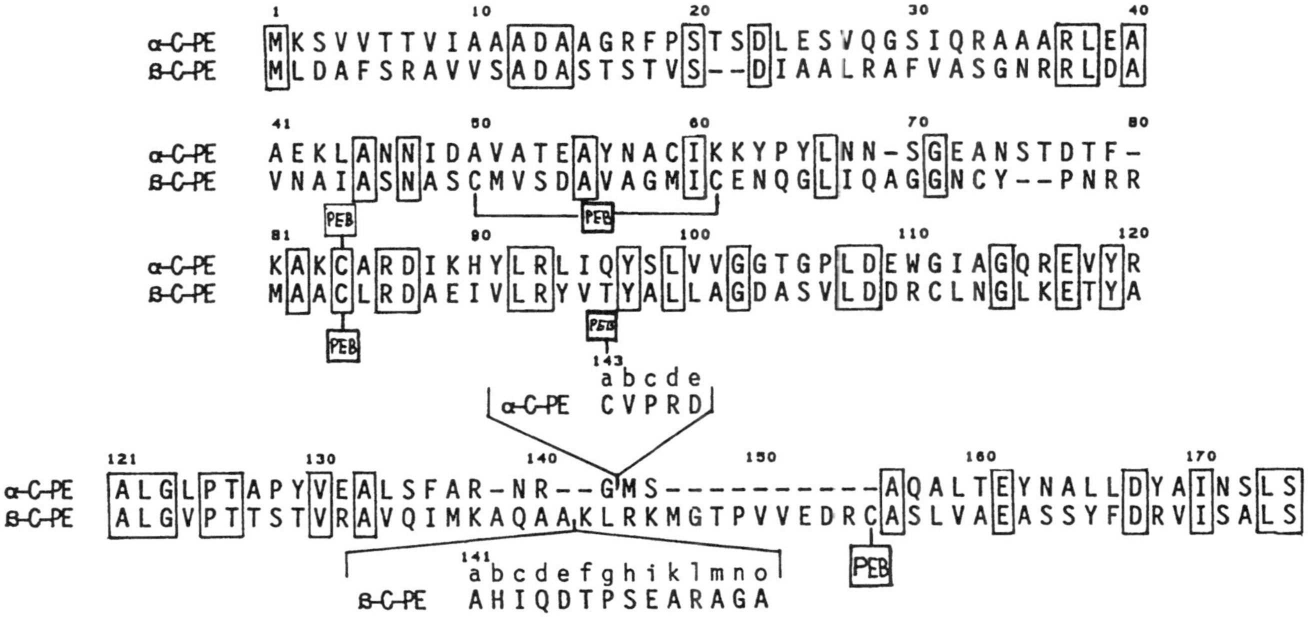

Fig.1. The Complete Primary Structure of C-Phycoerythrin from the Cyanobacterium Calothrix.
The homologous amino acid residues of the α- and β-subunit are in boxes. PEB = phycoerythrobilin chromophore. The position numbers of the amino acid residues correspond to those used in the C-phycocyanin X-ray structure.

The amino acid sequence homology between the α- and β-subunits is 23%. C-PE contains five chromophores, two more than phycocyanin (8). The PEB-binding sites at positions 84 in α- an β-C-PE are homologous to PEC, C-PC (8) and A-PC. One of the two additional chromophores is doubly bound to cys 50 and cys 61 in the β-subunit, and the fifth PEB-chromophore is inserted together with a pentapeptide in the α-subunit at position 143a (Fig. 1). A unique insertion of 10 residues in comparison with C-PC is found at postions β-141a-o.
In the nucleotide sequence (9) an asparagine instead of a serine was found at position β-72. This difference resulted from a methylated asparagin, as was found by Klotz et al. (10) in allophycocyanin and in B-PE at this position. In the amino acid sequence analysis of the corresponding tryptic β-C-PE fragment from Calothrix the PTH-derivate of the methylated asparagin appears at the serine-position in our HPLC-identification system (11, 12).

B - Phycoerythrin from the Red Algae Porphyridium cruentum

B-Phycoerythrin is a biliprotein analogous and homologous to C-phycoerythrin, but isolated from red algae. It was characterized as an $(\alpha\beta)_6\gamma$-complex (13). The γ-subunit with a molecular weight of about 30 KDa also contains chromophores (13). The X-ray diffraction analysis of B-PE crystals is in progress (14). We determined the complete amino acid sequence from α- and β-BP-E from the red algae Porphyridium cruentum (3). The B-PE α-subunit contains 164 amino acid residues and the B-PE β-subunit 177 amino acid residues, 7 less than C-PE. The amino acid sequence homology between the α- and β- subunit is 24%. In fact, despite a different phycobilisome type and a additional coloured γ-subunit, the similarity between red algal B-PE and cyanobacterial C-PE is very high: the sequence homology is 69% for the α-subunits of C-PE and B-PE and 65% for the β-subunits. The chromophore number, structure and binding type was determined earlier (16,17) as was the methylated asparagine (10). The number of PEB-chromophores and the binding sites are the same as in C-phycoerythrin (Fig.1). The most striking difference is the size of the ß-subunit, which is 7 residues shorter than the C-PE ß-subunit (Fig.2.).

Phycocyanin-645 from the Cryptomonad Chroomonas sp.

We established the complete amino acid sequence of Phycocyanin-645 from Chroomonas sp (4,5,26). In PC-645 three different subunits have been found by Wehrmeyer and Mörschel (18), two α-subunits of 9 kDa and 10 kDa, and a ß-subunit of 15 kDa. PC-645 has a unique spectrum, absorbing in a similar range as a whole phycobilisome. MacColl and coworkers identified three types of chromophores in PC-645, namely on the ß-subunit one purple cryptoviolin (CV) -, two blue phycocyanobilin (PCB) - and one green 697B - chromophores on each of the α-subunit (19). From the complete amino acid sequence analysis, the α_1-subunit contains 70 and the α_2-subunit 80 amino acid residues. The ß-subunit contains 177 residues and has the same size as the red algal B-PE ß-subunit. The existence of two different α-subunits is confirmed by the amino acid sequence. No homology to the ß-subunit can be recognized if the N-terminal sequences of the α-subunits are aligned with the N-terminal sequence of the ß-subunit. A low, but still significant homology results if the chromophore binding sites, which are at residue 18 in both α-subunits, are aligned with the PCB-binding site position 84 in the ß-subunit (4,5). With this alignment, a very low sequence homology of 11% (α_1-subunit) and of 15% (α_2-subunit) with the ß-subunit is achieved. The possibility that the cryptomonad α-subunits may also derive phylogenetically from a coloured γ-subunit or from a linker polypetide has to be considered unless we know the amino acid sequences for all proteins in the phycobilisome. The sequence homology between the α-subunits is considerably higher, namely 50%. The ß-subunit contains 1 CV-chromophore and two blue PCB-chromophores. The CV-chromophore was identified spectroscopically in the corresponding cyanogen bromide fragment (5). The CV-chromophore is probably doubly bound at cys 50 and cys 61. The two PCB-chromophores are bound to cystein residues at positions 84 and 155, homologous to the other biliproteins (5). In the protein sequence there is an insertion of 5 amino acid residues at homologous positions as in the phycoerythrins (Fig.2). Crystallographic studies are in progress (20).

Comparison of Cyanobacterial C-Phycoerythrin, Red Algal B-Phycoerythrin and Cryptophytan Phycocyanin-645 and PE-545

Homologies and phylogenetic relationships. PC-645 from the cryptomonad Chroomonas sp has the three chromophore binding sites at homologous postions as C-PE and B-PE. The amino acid sequence homology to C-PE and B-PE is hihger than to C-PC. If we compare all three sequences we find the highest homology between PC-645 ß-subunit and the red algal ß-B-PE, and the lowest to ß-C-PC (Table 1).

Table 1. Amino Acid Sequence Homologies.

	ß-PC-645
ß-C-PC	50%
ß-C-PE	65%
ß-B-PE	69%
ß-PE-545	86%

These amino acid homologies also support theories about phylogenetic development of the cryptomonad chloroplast from cyanobacteria via red algal chloroplasts (21,22,). Structural differences in the ß-subunits of C-PE, B-PE, PC-645 and PE-545 are found only in the region of the large insertion at positions ß 140-150 (Fig. 2).

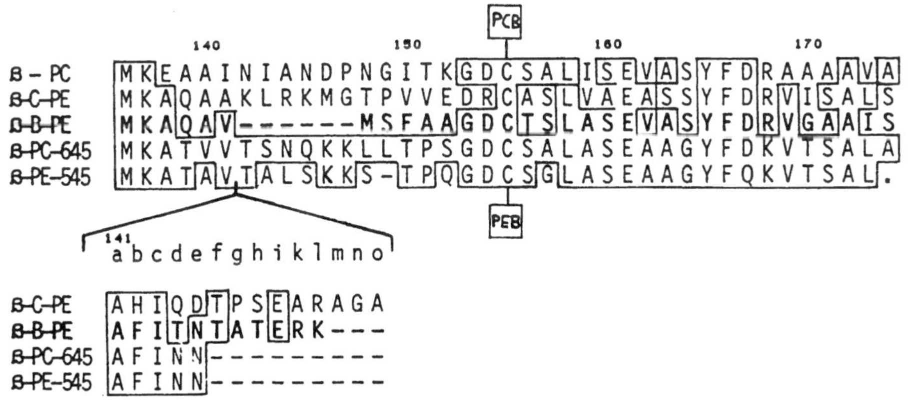

Fig. 2. Comparison of Cyanobacterial, Red Algal and Cryptophytan Phycoerythrins.
Amino acid residues homologous to PC-645 are in boxes. The position numbers correspond to those used in the X-ray structure of C-phycocyanin (6,7). PEB = phycoerythrobilin chromophore, PCB = phycocyanobilin chromophore.

All four ß-subunits exhibit parts of the insertion at position 141a-o, but in B-PE the insertion is 3 residues smaller and positions ß 142-147 are deleted. In ß PC-645 and in ß PE-545 the insertion (position 141a-e) is 9 residues smaller and positions 141f-o are deleted. The amino acid homology between cryptomonad phycocyanin and cryptomonad phycoerythrin at this position and at the positions sequenced up to now (4,5) is very high (86%). This demonstrates that cryptomonad phycocyanin-645 (and speculatively perhaps all cryptomonad biliproteins) are phylogenetically derived from red algal phycoerythrins (with respect to the apoprotein) rather than from a phycocyanin. Modifications with different chromophore types yields the different cryptomonad biliprotein types. PC-645 derives therefore from a red algal PE and not from a PC. Immunochemical studies by MacColl and coworkwers (23,24), who found cross-reactivity between cryptomonad biliproteins and red algal B-PE and also between cryptomonad PE and PC, are therefore supported by these amino acid sequence data.

<u>Cyanobacteria</u>		<u>Red Algae</u>		<u>Cryptomonads</u>
APC	→	APC		
C-PC	→	R-PC		
C-PE	→	B-PE, R-PE	→	PE I, PE II
			↘	PC-645, PC-612 etc.

<u>**Functional aspects.**</u> In the three-dimensional structure of C-PC from <u>Mastigocladus laminosus</u>, Schirmer and coworkers found that PCB chromophores arch around the aspartates A87, B87 and B39 and that the nitrogens of pyrroles B and C are within hydrogen-bonding distance of one of the carboxylate oxygens (7).In phycoerythrins we also find these aspartates highly conserved at the three homologous positions for the three homologous PEBs (Fig.3). Therefore we may speculate that the same principle of the PCB - protein interaction also may be applied to the PEB - protein interaction.

	30									40
β - PC	V	V	K	E	G	N	K	R	L	[D]	V
β-C-PE	F	V	A	S	G	N	R	R	L	[D]	A
β-PC-645	F	V	S	E	G	N	K	R	L	[D]	A
β-B-PE	**F**	**I**	**A**	**D**	**G**	**N**	**K**	**R**	**L**	**[D]**	**A**

	81			PCB						90
β - PC	M	A	A	C	L	R	[D]	M	E	I
β-C-PE	M	A	A	C	L	R	[D]	A	E	I
β-PC-645	M	A	A	C	L	R	[D]	A	E	I
β-B-PE	**M**	**A**	**A**	**C**	**L**	**R**	**[D]**	**G**	**E**	**I**
				PEB						

Fig.3. Comparison of Cyanobacterial, Red Algal and Cryptophytan Phycoerythrins and Phycocyanins.
The aspartates B39 (position β-39) and B87 (positions β-87), homologous to those which interact in β-C-phycocyanin with the PCB-chromophore, are in boxes.

If one assumes a similar tertiary structure of C-PC and phycoerythrins on the basis of sequence homology (2,8), analogous function and position in the phycobilisome rods, the aggregation state (hexameric structure) and similar shape of the hexamer observed by electron microscopy (25), then the positions of the α-84, β-84 and β-155 PEB- and PCB-chromophores in C-PE, B-PE and PC-645 may be similar to the PCB-positions found in the X-ray structure of C-PC. On the basis of the X-ray structure of C-PC one may estimate the possible positions of the fourth and fifth chromophore of C-PE and B-PE (Fig. 4). The second chromophore of the PE-α-subunits would be bound to the E helix (Fig. 4a), the other being at an homologous position (Fig. 4b). In the C-PE-, B-PE- and PC-645-β subunit the doubly bound chromophore would be attached to the B-helix (Fig. 4b), the other two being also at homologous positions. Applying this to the trimeric complex, we recognize, that the additional two chromophores of C-PE and B-PE may be near the periphery of the trimer (or hexamer) and therefore probably function as sensitizing chromophores (Fig. 4 c).

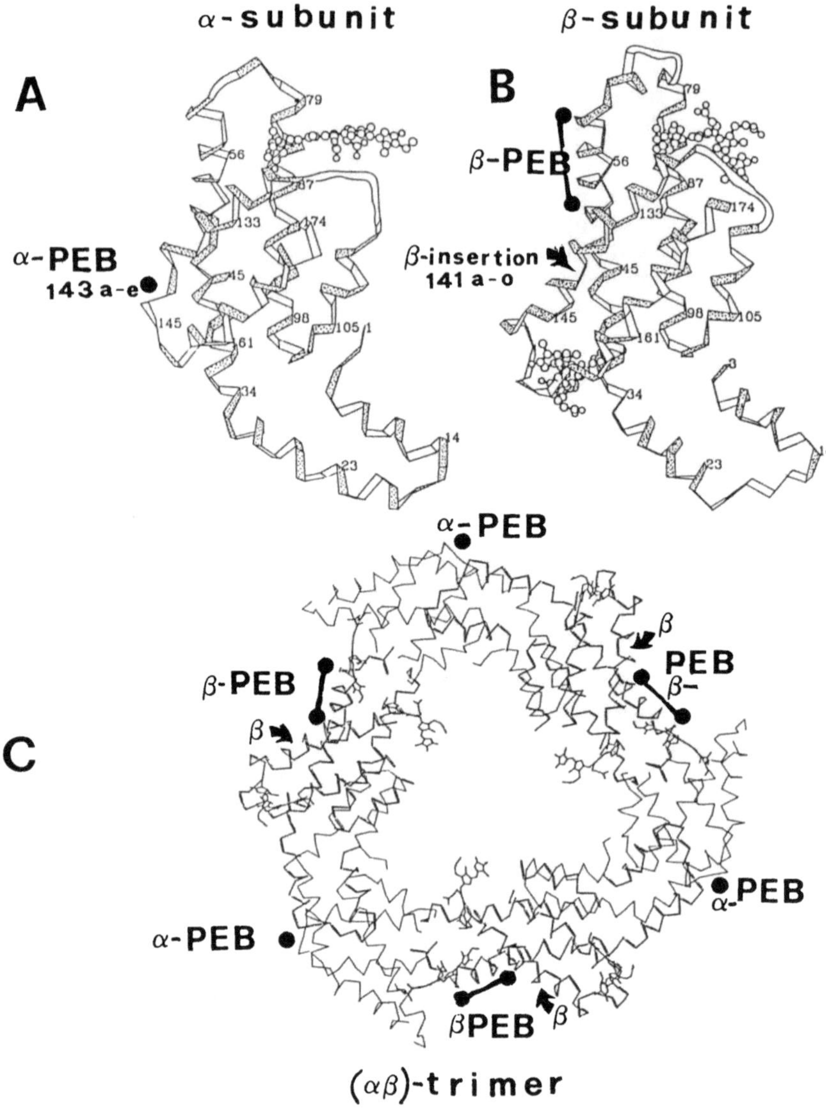

Fig.4. PEB-Chromophore Binding-Sites and Insertion.
The positions of the binding-sites of the two additional PEB-chromophores of C-PE and B-PE and the insertion in the ß-subunit (arrow) are implemented into the three-dimensional structure of C-phycocyanin from Mastigocladus laminosus (6).
a: location of PEB 143a in the α-subunit and b: of PEB 50-61 in the ß-subunbit; c: location of α-PEB 143a and ß-PEB 50-61 and of the insertion ß-141a-o (C-PE) and ß-141a-e (B-PE) in the trimeric C-PC aggregate.

This would account for 24 sensitizing chromopohres in the hexamer, twice the number found in the C-PC hexamer.The γ-methyl-asparagine may be located close to the B84 PEB-chromophore, as was reconstructed on a picture system using C-PC data (8,14,15). One may speculate γ-methyl asparagine interacts with this B-84 PEB and perhaps also with a linker in the central hole of the PE-Hexamer.

Cryptomonad biliproteins. The missing 60 residues in the PC-645 α-subunits correspond to the x and y helices in C-PC (5,6), which provide the α-β-subunit contact. This may explain why cryptomonad biliprotein do not form trimers, hexamers and higher aggregates, but form heterotetramers as in the case of PC-645 (5).

References

1. MacColl, R., Guard-Friar, D. 1987. Phycobiliproteins. CRC Press Inc. Boca Raton, Florida.

2. Sidler, W., Kumpf, B., Rüdiger, W., Zuber, H. 1986. Biol. Chem. Hoppe-Seyler 367, 627.

3. Sidler, W., Kumpf, B., Glazer, A.N., Zuber, H., in press.

4. Sidler, W., Kumpf, B., Suter, F., Morisset, W., Wehrmeyer, W., Zuber, H. 1985. Biol. Chem. Hoppe-Seyler 366, 233.

5. Sidler, W., Nutt, H., Kumpf, B., Zuber, H., Wehrmeyer, W., Brenzel, A. 1986. In: Progress in Photosynthesis Research. Vol. 2 (J. Biggins, ed.). Martinus Nijhoff Publishers, Dordrecht, The Netherlands. pp 153.

6. Schirmer, T., Bode W., Huber, R., Sidler, W., Zuber,H. 1985. J. Mol. Biol. 184, 257.

7. Schirmer, T., Bode, W. and Huber, R. 1987. J. Mol. Biol. 196, 677.

8. Frank, G., Sidler, W., Widmer, H., and Zuber, H. 1978. Hoppe-Seyler`s Z. Physiol. Chem. 359, 1491.

9. Mazel, D., Guglielmi, G., Houmard, J., Sidler, W., Bryant, D.A., Tandeau de Marsac, N. 1986. Nucleic Acids Research 14, 8279.

10. Klotz, A.V., Leary, J.A., Glazer, A.N. 1986. J. Biol. Chem. 261, 15891.

11. Rümbeli, R., Suter, F., Wirth, M., Sidler W., Zuber, H. 1987. FEBS Lett. 221, 1.

12. Rümbeli, R., Suter, F., Wirth, M., Sidler W., Zuber, H. 1987. Biol. Chem. Hoppe-Seyler 368, in press.

13. Glazer, A.N., Hixon, C.S. 1977. J. Biol. Chem. 252, 32.

14. Fisher, R.G., Woods, N.E., Fuchs, H. E., Sweet, R.M. 1980. J. Biol. Chem. 255, 5082.

16. Lundell, J.D., Glazer, A.N., DeLange, R.J., Brown, D.M. 1984. J. Biol. Chem. 259, 5472.

17. Klotz, A.V., Glazer, A.N., Bishop, J.B., Nagy, J.O., Rapoport, H. 1986. J. Biol. Chem. 261, 6797.

18. Mörschel, E., Wehrmeyer, W. 1975. Arch. Microbiol. 105, 153.

19. MacColl, R., Guard-Friar, D. (1983) J. Biol. Chem. 258, 14324.

20. Morisset, W., Wehrmeyer, W., Schirmer, T., Bode, W. 1984. Arch. Microbiol. 140, 202.

21. Staehelin, L.A. 1986. In: Photosynthesis III. (L.A. Staehelin and C.J. Arntzen, eds.).Springer-Verlag, Berlin, Heidelberg, New York, Tokyo. pp. 8.

22. Ludwig, M., Gibbs, S.P. 1985. Protoplasma 127, 9.

23. MacColl, R., Berns, D.S., Gibbons, O. 1976. Arch. Biochem. Biophys. 177, 265.

24. Guard-Friar, D., Eisenberg, B., Edwards, M.R., MacColl, R. 1986. Plant Physiol. 80, 38.

25. Rosinski, J., Hainfeld, J.F., Rigbi, M., Siegelman, H. W . 1981. Ann. Bot. 47, 1-12.

26. Sidler, W., Nutt, H., Kumpf, B., Zuber, H., Wehrmeyer, W., Brenzel, A. 1987. in press.

ISOLATION AND CHARACTERIZATION OF THE COMPONENTS OF THE PHYCOBILISOME FROM MASTIGOCLADUS LAMINOSUS AND CROSSLINKING EXPERIMENTS

R. Rümbeli and H. Zuber
Institut für Molekularbiologie und Biophysik, CH-8093 Zürich

a) Isolation and characterization of phycobiliprotein complexes and subunits

The primary structure of most phycobiliproteins and some linker polypeptides of the phycobilisome from *Mastigocladus laminosus* have been determined. The total sequences of α^{AP}, β^{AP} [1], α^{PC}, β^{PC} [2], α^{PEC}, β^{PEC} [3], $\beta^{16.2}$ [4], $L_R^{8.9}$ [5] and $L_C^{8.9}$ [6] are known as well as partial sequences of α^{APB} [7], $L_R^{34.5,PEC}$ [5] and $L_R^{34.5,PC}$ [5]. These components were isolated from well defined linker polypeptide - phycobiliprotein complexes of which the aggregation state and the spectroscopic behaviour were characterized. Fig. 1 represents the scheme of the isolation procedures and Tab. 1 shows some spectroscopic characteristics of phycobiliprotein - linker complexes.

Intact phycobilisomes were dissociated in 5mM potassium phosphate and a phosphate gradient was used to separate the resulting complexes on a Cellex-D anion exchange column [8]. All major phycobiliprotein subunits were isolated from the appropriate complexes on a Bio-Gel P60 gel filtration column in 63mM formic acid [6]. The β-subunits always eluted in front of the α-subunits due to differences in the shape of the molecule. The separated subunits could be renatured in phosphate buffer, whereas for example the small amount of contaminating β-subunits formed higher complexes with the α-subunits and therefore could be separated by gel filtration on a Sephadex G150 column from the pure α-subunits [4]. These subunits were available for sequence determination or spectro-

Photosynthetic Light-Harvesting Systems

Table 1. Spectroscopic Characteristics of Phycobiliprotein - Linker Polypeptide Complexes.

	λ_{Amax} nm	$\epsilon(\lambda_{Amax})$ $M^{-1}cm^{-1}$	λ_{Emax} nm	ref.
$(\alpha^{PEC}\beta^{PEC})_6L_R^{34.5,PEC}$	575	2076000	600	[8]
$(\alpha^{PC}\beta^{PC})_3$	615	883000	643	[10]
$(\alpha^{PC}\beta^{PC})_6L_R^{34.5,PC}$	631	1956000	647	[8]
$(\alpha^{AP}\beta^{AP})_3$	651	745000	660	[9]
$(\alpha^{AP}\beta^{AP})_3L_C^{8.9}$	652	1076000	661	[9]
$(\alpha^{APB}\alpha_2^{AP}\beta_3^{AP})L_C^{8.9}$	653	977000	679	[9]

scopic studies.

The PEC I fraction consisted of $(\alpha^{PEC}\beta^{PEC})_6L_R^{34.5}$ and the small linker polypeptide $L_R^{8.9}$ [8]. This peptide was not tightly bound to the complex and it was separated from the complex during gel filtration or sucrose density gradient centrifugation of the PEC I fraction. By gel filtration of PEC I in 50% formic acid the two linker polypeptides $L_R^{34.5,PEC}$ and $L_R^{8.9}$ were isolated and amino acid sequences thereof were determined [5].

The two complexes $(\alpha^{AP}\beta^{AP})_3L_C^{8.9}$ and $(\alpha^{APB}\alpha_2^{AP}\beta_3^{AP})L_C^{8.9}$ were isolated from the APC I fraction on a Bio-Gel hydroxyl apatite ion exchange column and were further purified by sucrose density gradient centrifugation [9]. From gel filtration of APC I in 63mM formic acid on a Bio-Gel P60 column the following components were obtained: α^{AP}, β^{AP}, $L_C^{8.9}$ and a fraction enriched with α^{APB} [9]. α^{APB} could not be completely purified by ion exchange chromatography in 8M urea but nevertheless the amino acid sequence thereof was partly determined [7]. After purification of $L_C^{8.9}$ on a Bio-Gel P30 column in 50% formic acid the primary structure determination of this linker polypeptide was completed [6].

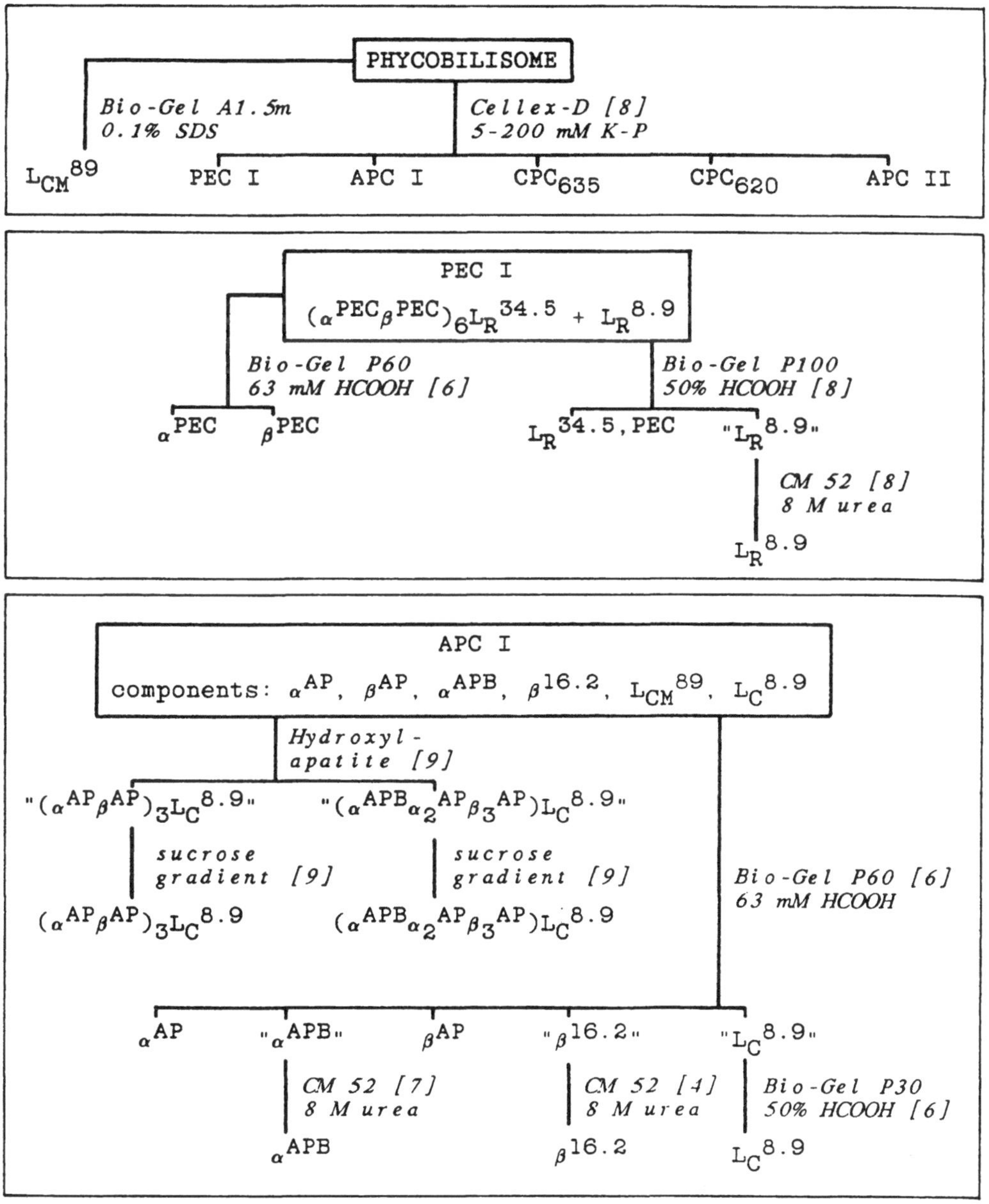

Fig. 1. Diagram of the isolation and purification steps to obtain pure phycobiliprotein complexes and subunits.
"X" : sample X contaminated with a minor component
K-P : potassium phosphate

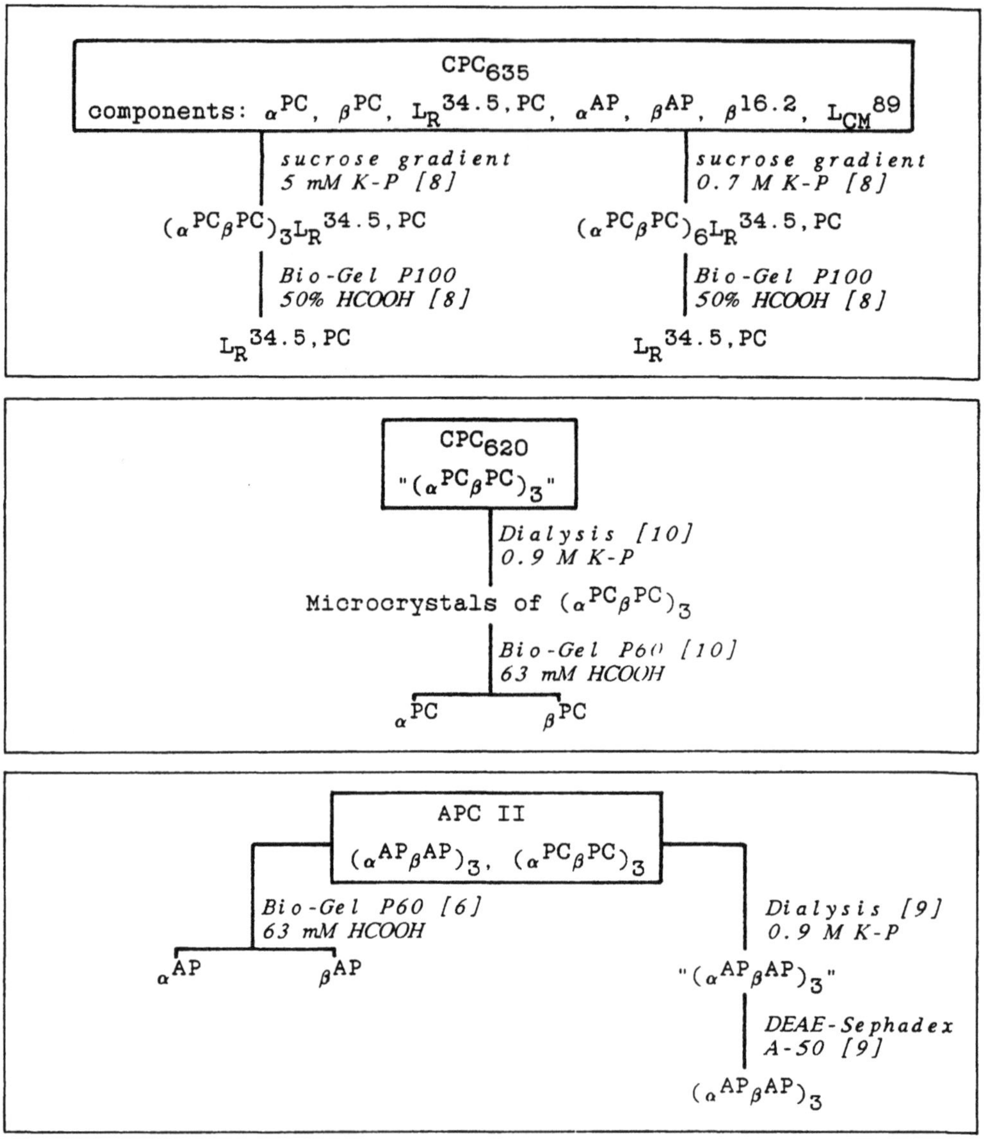

The minor phycobiliprotein $\beta^{16.2}$ is part of a complex containing also partly degraded L_{CM}^{89}. This complex eluted as

a contaminant of the APC I and the CPC_{635} fraction from the Cellex-D column. $\beta^{16.2}$ was well separated from the other components on the Bio-Gel P60 column in 63mM formic acid. It was further purified by ion exchange chromatography in 8M urea and the complete amino acid sequence was determined [4].
Sucrose density gradient centrifugation of the CPC_{635} fraction separated the PC complexes from the high molecular weight AP complexes. Depending on the phosphate concentration either $(\alpha^{PC}\beta^{PC})_3L_R^{34.5,PC}$ or $(\alpha^{PC}\beta^{PC})_6L_R^{34.5,PC}$ was obtained [8]. $L_R^{34.5,PC}$ was isolated by gel filtration of these complexes on a Bio-Gel P100 column in 50% formic acid [8].
The CPC_{620} fraction was purified by dialyzing it against 0.9M potassium phosphate. Under these conditions $(\alpha^{PC}\beta^{PC})_3$ formed microcrystals whereas the contaminating AP stayed in solution [10].
In the APC II fraction the PC was precipitated as microcrystals and the supernatant solution of $(\alpha^{AP}\beta^{AP})_3$ was further purified by anion exchange chromatography on DEAE-Sephadex A50 with a chloride gradient [9].
Isolation of a complex containing intact L_{CM}^{89} was difficult because this anchor polypeptide was rapidly degraded in a solution of dissociated phycobilisomes. Addition of PMSF and EDTA to all chromatography buffers slowed down but could not completely inhibit the protease activity. Therefore L_{CM}^{89} was isolated directly from intact phycobilisomes by denaturation in 2% SDS and gel filtration on a Bio-Gel A1.5m column in 0.1% SDS, 100mM NaCl and 1mM 2-mercaptoethanol. The fractions containing L_{CM}^{89} were dialyzed against 1mM NH_3 and lyophylized. Preliminary results of an Edman degradation of the intact protein showed that L_{CM}^{89} from *Mastigocladus laminosus* is completely homologous within the first seven N-terminal residues of L_{CM}^{94} of *Nostoc* [11]. Edman degradation of a mixture of intact L_{CM}^{89} and partly degraded polypeptide with an apparent molecular weight of 85kD yielded an inhomogenous sequence, indicating that the N-terminal part of L_{CM}^{89} is susceptible to degradation in the whole complex.
From the work of Minami et al. [12] and Klotz et al. [13] we were stimulated to reinvestigate the amino acid sequences of

the phycobiliproteins and some linker polypeptides from *Mastigocladus laminosus* for the occurrence of post-translationally modified asparagine residues. N^4-methylasparagine was not only found in β^{AP} but also in the homologous core component $\beta^{16.2}$ and in β^{PC} of the rod. No such residue was detected in the phycobiliprotein subunits α^{PEC}, β^{PEC}, α^{PC}, α^{AP} and in the linker polypeptides $L_R^{8.9}$ and $L_C^{8.9}$. The exact position of these modified residues was localized within the amino acid sequences by chemical and enzymatic fragmentations and Edman degradation of the corresponding peptides [14]. Residues $\beta^{AP}72$, $\beta^{16.2}72$ and $\beta^{PC}72$ were identified as N^4-methylasparagine. Although N^4-methylasparagine has not been verified yet in the x-ray structure of trimeric PC [15], one can see that the misinterpreted β^{PC}Ser 72 is pointing towards ring A of the chromophore β84 within a distance of 3.3 Å. Therefore this methylated residue might play a role in stabilizing the chromophore conformation and/or in modification of the spectroscopic characteristics of this chromophore. In the trimeric PC aggregate residue β72 is exposed to the central hole. It is also possible that this modified amino acid serves to position the linker polypeptides within the phycobiliprotein complexes.

b) Crosslinking experiments

Crosslinking experiments were performed on native phycobiliprotein complexes with several bis-imidates having various spacer lengths [16]. $(\alpha^{PC}\beta^{PC})_3$ was crosslinked with the homobifunctional amino specific reagent dimethyl pimelimidate having a maximal crosslinking distance of 10 Å. Two crosslinks were identified (Fig. 2a,b): an intramonomer intersubunit crosslink with a yield of 48 % and an intrasubunit crosslink within α^{PC} (57 %). These products were chemically and enzymatically fragmented and the crosslinks were localized within the amino acid sequence and the three-dimensional structure of $(\alpha^{PC}\beta^{PC})_3$ [16]. In the intermolecular crosslinked product the α-amino groups of α^{PC}Val 1 and β^{PC}Ala 1 were connected together.

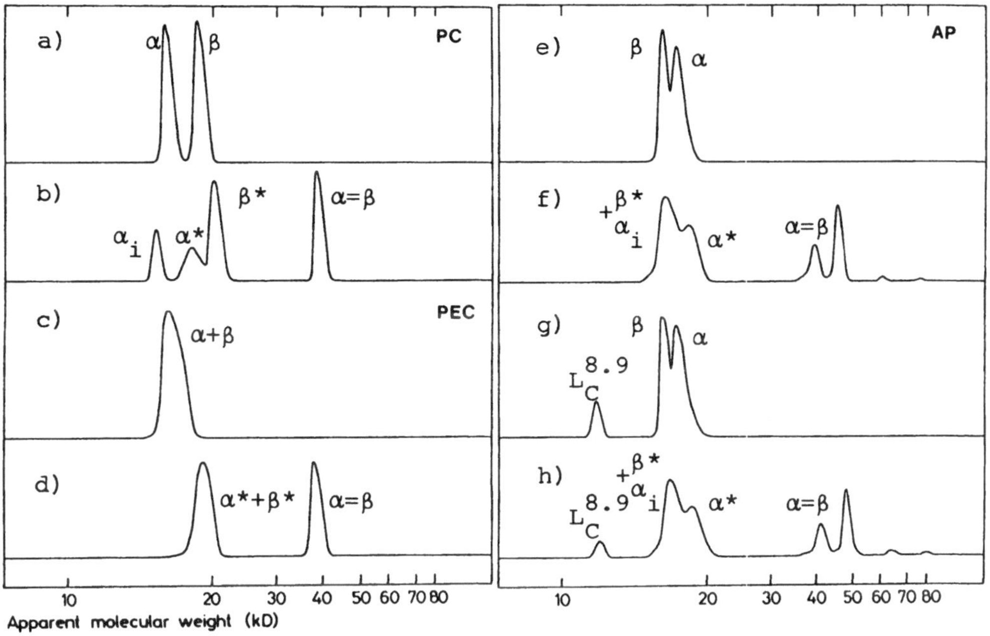

Fig. 2. SDS-Gel scans of the reaction products of phycobiliprotein complexes with dimethyl-pimelimidate (DPI). For experimental details see [16].

α_i : α with intrasubunit crosslink

α^* : product without crosslink

$\alpha=\beta$: intersubunit crosslink

a) $(\alpha^{PC}\beta^{PC})_3$, b) with DPI

c) $(\alpha^{PEC}\beta^{PEC})_3$, d) with DPI

e) $(\alpha^{AP}\beta^{AP})_3$, f) with DPI

g) $(\alpha^{AP}\beta^{AP})_3L_C^{8.9}$, h) with DPI

In the x-ray structure model these two residues point into the central hole of the complex and are 5.1 Å apart. The intramolecular crosslink was derived by connecting the ε-amino groups of α^{PC}Lys 62 and α^{PC}Lys 134, stabilizing the relative positions of helices B and G. The distance between these two residues was found to be 8.4 Å [15]. The same experiments were performed on $(\alpha^{PEC}\beta^{PEC})_3$ (Fig. 2c,d). No intramolecular crosslink was detectable. There is no Lys residue in the "G"-helix of α^{PEC} and this agrees well with the lack of an intrasubunit crosslink in PEC. As in PC there was also an intermolecular crosslink formed with comparable yield and apparent molecular weight, we therefore assume that in the case of PEC both N-termini were crosslinked by the reagent and that these residues have equivalent positions in the two complexes. After reaction of $(\alpha^{AP}\beta^{AP})_3$ with bis-imidates an intramolecular crosslink within α^{AP} and intermolecular crosslinks were found (Fig. 2e,f). As a hypothesis we propose that in α^{AP} the intrasubunit crosslink is between α^{AP}Lys 62 and α^{AP}Lys 137, giving about the same decrease in the apparent molecular weight as in α^{PC}. Two different intermolecular crosslinks were identified in trimeric AP. Taking into account the equal amount and relative positions of these crosslinked products on the SDS-gel compared with PC a crosslinking between the N-terminal residues of both subunits seems also probable in this case. As the N-terminal residues point into the central hole of the complex where the linker polypeptides are assumed to be located the question is to which extent the crosslinking reaction would be inhibited by the presence of a linker polypeptide in the central hole. In the reaction of the complex $(\alpha^{AP}\beta^{AP})_3L_C^{8.9}$ with dimethyl pimelimidate the yields of the crosslinked products were not significantly changed compared to the complex without linker polypeptide (Fig. 2g,h). As can be seen from Fig. 2h $L_C^{8.9}$ is not involved in the crosslinking reaction. Further crosslinking experiments on phycobiliprotein - linker polypeptide complexes will show if its possible to connect a linker polypeptide to a subunit within a complex and to locate the crosslinked residues within the amino acid sequence.

Abbreviations

AP = allophycocyanin, PC = phycocyanin, PEC = phycoerythrocyanin, APB = allophycocyanin B, α and β = subunits of phycobiliproteins. The phycobiliprotein complexes and polypeptides are abbreviated according to the system described by Glazer [17] : L = Linker polypeptide, R, C and M denominate the location of the linker polypeptide, i.e. rod, core and membrane, respectively. The superscript numbers indicate the apparent molecular weight of the polypeptide in kD.

Acknowledgement

This work was supported by the Swiss Federal Institute of Technology, Kredit Unterricht und Forschung, and the Swiss National Foundation, Grant number 3.286-0.82.

References

[1] Sidler, W., J. Gysi, E. Isker, H. Zuber. 1981. Hoppe Seyler's Z. Physiol. Chem. 362, 611-628.

[2] Frank, G., W. Sidler, H. Widmer, H. Zuber. 1978. Hoppe-Seyler's Z. Physiol. Chem. 359, 1491-1507.

[3] Füglistaller, P., F. Suter, H. Zuber. 1983. Hoppe-Seyler's Z. Physiol. Chem. 364, 691-712.

[4] Rümbeli, R., M. Wirth, F. Suter, H. Zuber. 1987. Biol. Chem. Hoppe-Seyler 368, 1-9.

[5] Füglistaller, F., F. Suter, H. Zuber. 1986. Biol. Chem. Hoppe-Seyler 367, 615-626.

[6] Füglistaller, P., R. Rümbeli, F. Suter, H. Zuber. 1984. Hoppe-Seyler's Z. Physiol. Chem. 365, 1085-1096.

[7] Suter, F., P. Füglistaller, D.J. Lundell, A.N. Glazer, H. Zuber. 1987. FEBS Letters 217, 279-282.

[8] Füglistaller, F., F. Suter, H. Zuber. 1986. Biol. Chem. Hoppe-Seyler 367, 601-614.

[9] Füglistaller, F., M. Mimuro, F. Suter, H. Zuber. 1987. Biol. Chem. Hoppe-Seyler 368, 353-367.

[10] Mimuro, M., F. Füglistaller, R. Rümbeli, H. Zuber. 1986. Biochim. Biophys. Acta 848, 155-166.

[11] Gantt, E., C. Cunningham, C.A. Lipschultz, M. Mimuro. 1987. EMBO Workshop on Oxygenic and Anoxygenic Electron Transport Systems in Cyanobacteria, Cape Sunion, Abstract p. 11.

[12] Minami, Y., F. Yamada, T. Hase, H. Matsubara, A. Murakami, Y. Fujita, T. Takao, Y. Shimonishi. 1985. FEBS Letters 191, 216-220.

[13] Klotz, A.V., J.A. Leary, A.N. Glazer. 1986. J. Biol. Chem. 261, 15891-15894.

[14] Rümbeli, R., F. Suter, M. Wirth, W. Sidler, H. Zuber. 1987. Biol. Chem. Hoppe-Seyler, in press.

[15] Schirmer, T., W. Bode, R. Huber, W. Sidler, H. Zuber. 1985. J. Mol. Biol. 184, 257-277.

[16] Rümbeli, R., M. Wirth, H. Zuber. 1987. Biol. Chem. Hoppe-Seyler, in press.

[17] Glazer, A.N. 1985. Annu. Rev. Biophys. Chem. 14, 47-77.

C-PHYCOCYANIN FROM *MASTIGOCLADUS LAMINOSUS*: CHROMOPHORE ASSIGNMENT IN HIGHER AGGREGATES BY CYSTEIN MODIFICATION

R. Fischer, S. Siebzehnrübl and H. Scheer

Botanisches Institut der Universität München
Menzinger Str. 67 8000 München 19 FRG

The phycobilisome (PBS) is a unique antenna system, because well-defined building blocs can be isolated from it, which allow principally to study the functional properties, e.g. light absorption and excitation energy transfer, as a function of aggregate size and functional complexity (1-4). Besides the so-called linker polypeptides, it is mainly composed of phycobiliproteins (phycoerythrin (PE) or phycoerythrocyanin (PEC), phycocyanin (PC) and allophycocyanin (APC)). Each phycobiliprotein is composed of two (or sometimes more) subunits (α,β) bearing 1-4 open-chain tetrapyrrolic chromophores, which are covalently bound to cystein residues of the apoprotein via thioether linkages (3-5). In any given phycobiliprotein the chromophores differ in their spectroscopic properties and their chemical reactivities. As shown in Table 1 for C-phycocyanin (PC) from *Mastigocladus laminosus* discussed here, this is even true for pigments bearing chromophores of identical molecular structure. These differences arise from the different proteinenvironment and distinct conformations of the individual chromophores in the native state.

The polypeptide and chromophore structures of tri- and hexamers of C-phycocyanins (PC) which are the basic building blocs of phycobilisomes, have been determined with high resolution by x-ray cristallography (5). For the analysis of energy transfer

Photosynthetic Light-Harvesting Systems

(6,7) and photochemical properties (8) it is necessary to know in addition the electronic structures (9) and component spectra (6,10) of the individual chromophores. The component spectrum for the single chromophore of PC (α-84, Fig.1) can be determined directly from the isolated α-subunit. The ones of the two β-chromophores (Fig. 1) have been obtained by resolution of the absorption spectrum of the β-subunit (6,10) and assignment of the β-84 chromophore by modification of the neighboring cys-111 with p-chloro-mercury benzenesulfonate (PCMS) (11). The same components can be used for the (α,β)-monomer (=heterodimer), because its spectrum is the sum of the subunit spectra.

Table 1 : Specific Attributes of Individual Chromophore Types of PC from <u>Mastigocladus laminosus</u> (from ref. 3,4,6,8,10,11)

Characteristics	α - 84	β - 84	β - 155
Reaction with PCMS	-	+	-
Absorption Maximum [nm]	616 - 618	622 - 624	598 - 600
Fluorescence	+	+	-
Optical Activity	+	-	+
Reversible Photochemistry	-	+	+/-(1)
Conformation According to X-Ray Crystal Structure	ZZZ	ZZZ	ZZX(2)

1) The contribution of this chromophore to photochemistry is still uncertain.
2) Ring D is almost perpendicular to ring C (5, see also 8)

The spectroscopic properties of larger aggregates can no longer be described as a sum of the same subunit component spectra (Table 2). The absorption maxima of higher aggregates show more or less pronounced red-shifts which have to be accounted for in the component spectra. This is in particular necessary with linker-containing trimers and phycobilisomes.

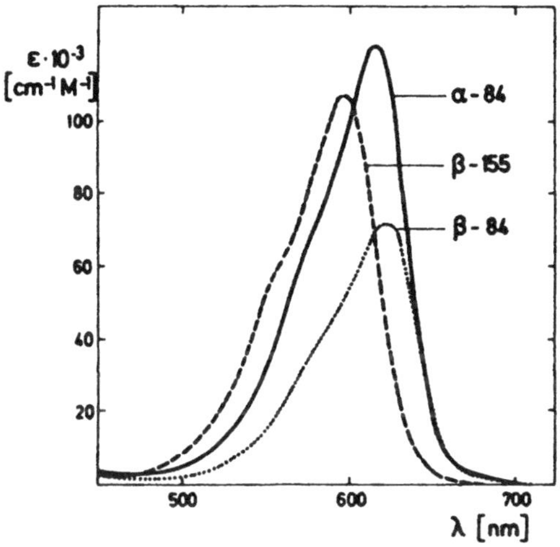

Fig. 1: Single chromophore spectra of PC from M. laminosus. The specra of chromophores β-84 and β-155 were obtained by deconvolution of the β-subunit spectrum assuming a similar lineshape as the α-84 chromophore.

Table 2 : Some Characteristics of PC in Different Aggregation States and its Two Subunits

Unit	Aggregation state of PC	n[1]	Absorption λ_{max} [nm]	[$cm^{-1}mM^{-1}$]	Molecular weight [kDa]
α-Subunit	dimer	2	616	244	36.0[2]
β-Subunit	dimer	4	604	328	38.8[2]
PC (α,β)	monomer	3	612	286	37.4[2]
PC $(\alpha,\beta)_3$	trimer	9	619	896	112.2[2]
PC+linker	trimer	9	630	nd	≈140
PBS	≥dodecamer[3]	≥36	≈632	-	≈560

1) n = number of chromophores
2) according to amino acid sequence (12)
3) calculated per rod, each phycobilisome contains six rods

Component spectra thus have to be derived separately in such aggregates. It has been suggested, that interactions of the linkers with chromophore β-84 may be an important factor in spectral modifications, because it is close to the central cavity of the doughnut-shaped trimers and hexamers in which the linkers are probably located (5,13).

To test the hypothesis, we have extended the PCMS method to tag this chromophore in phycocyanins from different origins and of different aggregation states. The results (Table 3) show that it is possible by this method to obtain perturbation spectra related to chromophore β-84 of PC in aggregates ranging from the β-subunit to PBS (APC is unreactive, and the PEC difference spectrum occurs at much shorter wavelengths than that of PC).

Table 3 : Treatment of PC-Aggregates and -Subunits with PCMS

Pigment	Absorption λ_{max} [nm]	Absorption difference λ_{min} [nm]	% 1)	λ_{max} [nm]	% 1)	$t_{½}$ 2) [min]
M. laminosus						
PC (monomer)	612	613	14.3	654	9.1	15
PC (trimer)	618	626	17.6	658	16.1	120
α-Subunit	616	-	-	-	-	-
β-Subunit	604	617	18.2	655	19.3	15
PC + linker	630	634	25.5	662	10.5	300
PEC	573	573/598 3)	25.9	645	13.1	30
APC	653	-	-	-	-	-
PBS	632	625	4.6	660	2.1	1440
Tolipothrix distorta PC (trimer)	613	623	37.2	657	10.2	180
Spirulina platensis PC (trimer)	620	621	26.4	657	9.7	60

1) Relative change in % of maximum absorbance before reaction
2) Time required for reaching 50% of final absorption change
3) double maximum

The absorption maximum shows a monotonous red-shift with in-the difference spectrum shifts first to the red with increasing aggregate size, but then again to the blue in PBS. However, the amplitude of the difference spectrum in PBS in much decreased as compared to smaller aggregates. Possibly, part of the β-84 chromophores are then inaccessible to the reagent, and the difference spectrum reflects only the accessible fraction. The much slower kinetics of the reaction, and the comparably high energy of the modified chromophores would be compatible with this interpretation.

Acknowledgement

This work was supported by the Deutsche Forschungsgemeinschaft. We thank G.Reng (Gesellschaft für Biotechnologische Forschung, D-3301 Stöckheim) for mass culture of M.laminosus.

References

1. Gantt, E., C.A. Lipschultz, T. Redlinger. 1985. In: Molecular Biology of the Photosynthetic Apparatus (K.E. Steinbach et al. eds). Cold Spring Harbor, New York, p. 223

2 Wehrmeyer, W. 1983. In: Proteins and Nucleic Acids in Plant Systematics (U. Jensen and D.E. Fairbrothers eds). Springer Verlag, Berlin, p. 143

3. Scheer, H. 1982. In: Light Reaction Path of Photosynthesis (F.K. Fong ed.). Springer Verlag, Berlin, p. 8

4 Glazer, A.N. 1985. Ann. Rev. Biophys. Chem. 14, 47

5. Schirmer T., W. Bode, R. Huber. 1987. J. Mol. Biol. 196, 677-695

6. Sauer K., H. Scheer, P. Sauer. 1987. Photochem. Photobiol. Vol. 46/3, 427-440

7. Sauer K., H. Scheer. 1988. Contribution in this vol.

8. Schmidt G., S. Siebzehnrübl, R. Fischer, H. Scheer. 1988. Contribution in this vol.

9. Schneider S., C. Scharnagl, M. Duerring, T. Schirmer, and W. Bode. 1988. Contribution in this vol.

10. Mimuro, M., P. Füglistaller, R. Rümbeli and H. Zuber. 1986. Biochim. Biophys. Acta 848, 155-166

11. Siebzehnrübl S., R. Fischer, H. Scheer. 1987. Z. Naturforsch. 42c, 258-262

12. Frank G., W. Sidler, H. Widmer, H. Zuber . 1978. Hoppe-Seyler's Z. Physiol. Chem. 359, 1491-1507

13. Mörschel, E., W. Wehrmeyer. 1979. Ber. Deutsch. Bot. Ges. 92, 393-402

14. John W., R. Fischer, S. Siebzehnrübl, H. Scheer. 1985. In: Antennas and Reaction Centers of Photosynthetic Bacteria: Structure, Interaction and Dynamics (M. E. Michel-Beyerle ed.) Springer Verlag Berlin, p. 17-35

PHOTOCHROMIC PROPERTIES OF C-PHYCOCYANIN

G.Schmidt, S.Siebzehnrübl, R.Fischer, and H.Scheer
Botanisches Institut der Universität, Menzinger Str. 67,
D-8000 München 19, F.R.G.

Summary

This contribution describes work aimed at understanding the different properties of the structurally very similar chromophores of the antenna pigment, C-phycocyanin (PC) from Mastigocladus laminosus, and of the photomorphogenetic reaction center pigment, phytochrome. In particular, the ZZZ-ZZE photoconversion of PC chromophores was explored. Native PC is unreactive except for a slow, irreversible bleaching. It attains photoreversibly photochromic properties upon partial uncoupling of the chromophores by certain perturbation of the protein structure. Two treatments are particulary effective: a) Dehydration and b) partial denaturation with mercaptoethanol. The results show that relatively small changes of the apoprotein structure are sufficient to induce an efficient photochemistry in PC, which is reminescent of the phytochrome primary reactions.

Introduction

The use of certain cofactors for a variety of functions is a common mechanism of adaptation on the molecular level for which the term "molecular ecology" seems appropriate. The plant biliproteins are an example for such adaptations. Phycobiliproteins, the photosynthetic antenna pigments of cyanobacteria, red and cryptophyte algae and the photomorphogenetic receptor pigments, phytochrome and phycochrome(s), bear very

Photosynthetic Light-Harvesting Systems

similar open-chain tetrapyrrolic chromophores (1). I.e, the chromophores of phycocyanins (PC) and allophycocyanins (APC) and the chromophore of phytochrome in the P_r-form, differ only by the substituent at C-18. In addition to these similarities in molecular structure, their covalent bonds to and their non-covalent interactions with the apoproteins are very similar as evidenced by their spectroscopic properties (2).
None the less are the functions of these pigments in vivo and many properties in the isolated states very similar for the two pigment classes. Phycobiliproteins like PC are chemically and photochemically inert. They are highly fluorescent in the isolated state and transfer excitation energy efficiently to the reaction centers in the native environment (1). Phytochrome is by contrast photochemically very reactive, and shows accordingly only little fluorescence. Its primary photochemistry is the reversible ZZZ-ZZE interconversion of the chromophore (3), (Fig. 1). It should be noted, that the properties of both pigments in the native state are drastically different from those of the free chromophores or denatured biliproteins, which are photolabile, chemically reactive and have very different conformations.

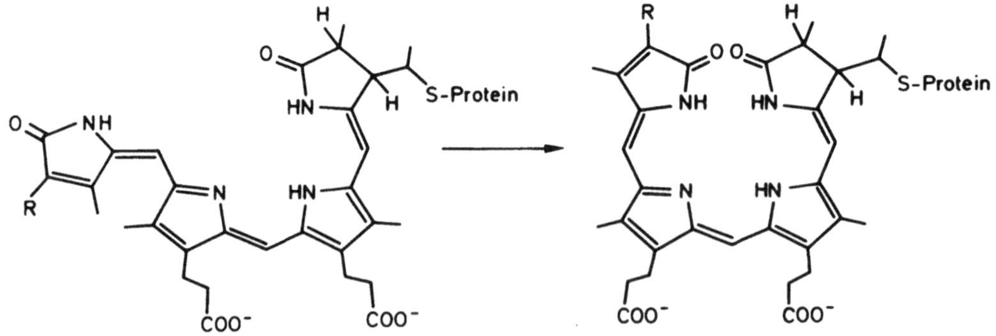

Phycocyanin R = C_2H_5
Phytochrom R = C_2H_3

In this contribution, we wish to present results which explore the possibility of photochromic reactions, similar to those of phytochrome, in the phycobiliproteins. There is a large body of data which relates the operationally defined phycochromes, e.g. photomorphogenetic receptor pigments of cyanobacteria, to subunits of the "common" antenna biliproteins (4,5). Partially denatured biliproteins are known to exhibit small, albeit distinct photoreversibly photochromic responses, which have in one case been related to a distinct chromophore (6). One of the chromophores in PC from two different species, e.g. ß-155, has also been shown by x-ray crystal structure analysis to be present in a strongly twisted conformation in which ring D is nearly perpendicular to the other rings (7). This has been related to the presence of small amounts of ZZE-chromophores being present in PC from Mastigocladus laminosus after denaturation (8). The large variations in the amount of ZZE chromophores prompted us to conduct a systematic search for the conditions under which ZZE chromophores are formed, and to study their reactions in the native and denatured state of the protein. Information on photoreactivity is also important with respect to applications of laser spectroscopic techniques to phycobiliproteins.

Materials and Methods

PC was prepared from frozen cells of Mastigocladus laminosus as previously described (9). Subunits were prepared by isoelectric focussing in a flat-bed aparatus on dextran gel. The separation was carried out in denatured state (7M urea) under argon and in the dark. After elution from the gel, the subunits were renatured by dialysis against 0.1M K-phosphate buffer pH=7, first at room temperature and then at 4C. Further purification was done by two steps of a ammonium sulphate precipitation (45%).

PC and subunits with ZZE-chromophores were prepared in 8M urea buffer, containing 0.1M K-phosphate, pH=7. The solution was treated in the dark with 15 vol% mercaptoethanol and then irradiated in an ice bath under nitrogen for ten minutes with white light. The following steps were carried out under green safety light. After elimination of the thiol at pH=2 (addition of hydrochloric acid), the equilibrium mixture of ZZZ/ZZE-products was cleaned up in two steps with a desalting gel to remove mercaptoethanol and to keep the photoisomerisation products in acidic denatured state, or in neutral K-phosphate buffer. The content of modified ZZE-chromophores was determined again by saturating irradiation with white light. Determinations of yield were based on the difference spectrum of ZZE/ZZZ interconversion of a pure ZZE PC-chromopeptide (3).

Absorption difference spectra were recorded with a ZWS II - spectrophotometer (Biochem, Puchheim) in split beam mode with thermostated (12C) cell holders. One of the cuvettes can be irradiated in the photometer by a Lumilux 150 Watt light source (Volpi, Denzlingen). The light guide from light source to the sample can be equipped with suitable interference filters. Analog data were digitized with a Technosystem (Darmstadt) model 1203A A- to -D converter and fed to a microcomputer (Apple IIe). Difference spectra against the unirradiated sample were recorded before and after irradiation.

Results and Discussion

Assay for presence of ZZE-configurated chromophores

PC and its subunits containing a part of the chromophores in the ZZE configuration (table 1) were prepared from native PC by a sequence of reactions which are well established. They involve denaturation of the protein, addition of thiols at C-10, irradiation, elimination of thiol and renaturation. The

15E-isomer is formed regioselectively, if this same sequence, with the omission of protein denaturation/renaturation steps, are applied to A-dihydrobilins (10). The ratio of ZZZ-to-ZZE in the product mixture was determined by denaturation of the modified pigment in acidic urea and illumination with white light to saturation. Under these conditions, the ZZE form is quantitatively converted to ZZZ (11), and the amplitude of the difference spectrum (δA, minimum - to - maximum, see fig.1) is proportional to the amount of ZZE in the original mixture [from the differential extinction coefficient δε=35100, taken from the spectra of PC-peptides (3) and the molar extinction coefficient of the denatured ZZZ-chromophore cation of PC ε=30300 (12); this amount can be calculated by the equation $Q_{ZZE} = 86 * \delta A / A$)]. The same assay has been used throughout the further studies, and the occurence of such light induced difference spectra has been taken as evidence for a ZZE-ZZZ conversion.

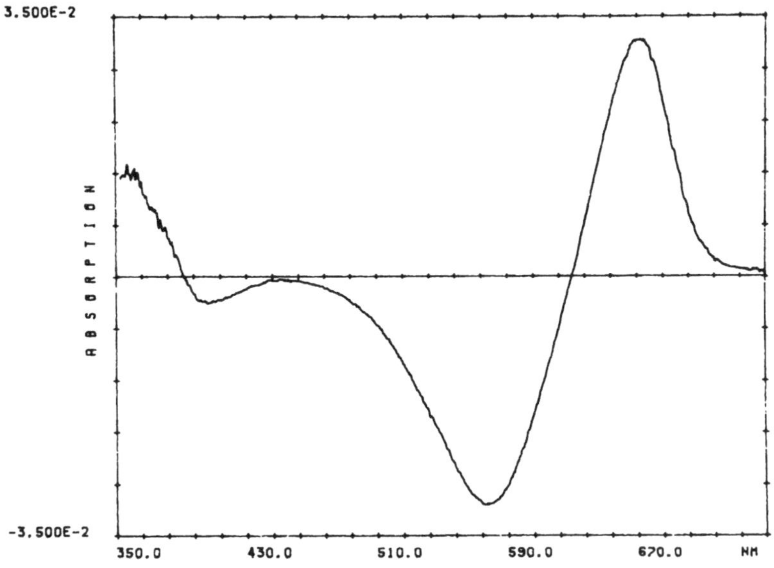

Fig. 1: ZZZ-ZZE difference spectrum of PC after denaturation (8M urea, 0.1M K-phosphate, pH=1.9) and irradiation with white light to saturation

Table 1: Photoisomerisation of ZZE-PC and its Subunits to ZZZ in Denatured State (8M Urea, 0.1M K-Phosphate, pH=1.9) by Irradiation with White Light (60/120sec).

ZZE- preparation-method	sample	content of ZZE-isomers (%)
denatured, via C-10 rubinoid addition product (see text)	PC	12.0
	α-subunit	15.5
	β-subunit	9.5
partially denatured (4M urea, 0.1M K-phosphate, pH=7.1, irradiation: 180sec at 630nm)	PC	2.4
	α-subunit	-
	β-subunit	5.9
partially denatured (0.1M K-phosphate, pH=7.4, 20vol% mercaptoethanol, irradiation: 300sec at 600nm)	α-subunit	13.8
	β-subunit	60.2

Photoconversion of ZZE-chromophores in native protein

To obtain the ZZZ/ZZE difference spectrum of a chromophore in a quasi-native environment, the PC modified as described above was irradiated in the native state (fig.2). It is dominated by a broad decrease in absorption at 575 nm, and a more intense, narrow-banded absorption increase around 630 nm. Even after saturating irradiation, only part (appx.30 - 40%) of the ZZE chromophores are phototransformed as evidenced by the further transformation after denaturation. This indicates a differential reactivity of the chromophores in the native protein, by steric hindrance or the like.

Fig. 2: ZZZ-ZZE difference spectrum of PC chromophores with native apoprotein (0.1M K-phosphate. pH=7). The spectrum has been obtained after saturating irradiation with white light.

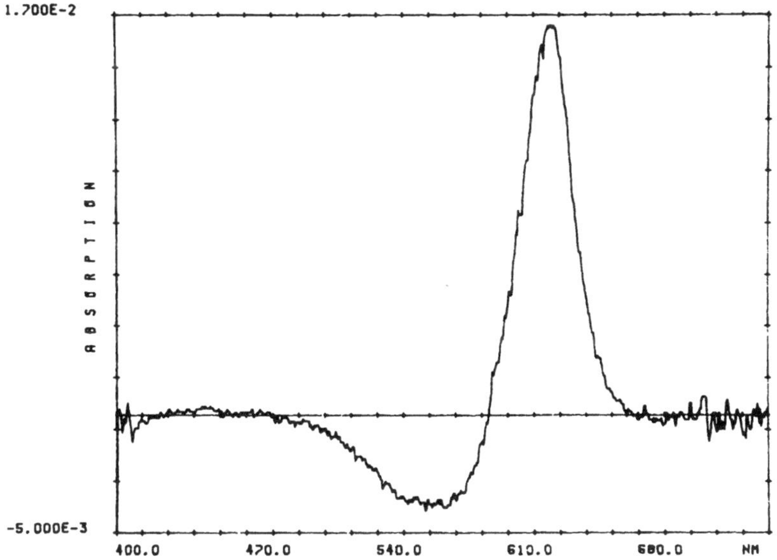

PC is photochromic in the dehydrated state

By measuring the light induced difference spectra obtained after acid urea denaturation of different PC preparations, it had been observed that many PC-samples, contained small and varying amounts of ZZE chromophores. A systematic investigation showed dehydration to be most effective in inducing the formation of ZZE chromophores, and that this formation is light dependent (table 2). ZZE chromophores were found in particular in samples which had undergone a lyophilisation process in the light. There is no detectable ZZE content if freshly isolated PC is lyophilised in the dark, whereas irradiation during lyophilisation or of the lyophilised material resulted in ZZE contents up to 5%. One possibility for photo-

isomerisation would be the formation of rubinoid addition products at C-10, which are known to photoisomerise (see previous section). Several amino acid side chains would principally allow such a reaction. However, this possibility is ruled out by monochromatic irradiation. Red light absorbed by the PC chromophore is more effective than blue light (table 2). Together with the results of the previous section, this indicates that PC becomes photochromic in the dehydrated state.
Small amounts of ZZE chromophores were also present in the isolated, not lyophilised, subunits. The factors leading to their formation were not further investigated, but they are formed most likely in the denaturation-renaturation cycle at intermediate urea concentrations. A complete shielding from light during the preparation is very difficult, and a light-

Table 2: Photoisomerisation of PC and its β-Subunit during Dehydration (Lyophilisation)

irradiation of PC during lyophilisation at nm:	content of ZZE-isomers %
dark control	0.3
370	-
430	2.4
530	3.4
630	4.6
660	3.8
white light (400-700)	5.3
irradiation of the β-subunit during lyophilisation at nm:	
dark control	1.7
white light (400-700)	5.2

induced isomerisation similar to the one described for partially denatured PC (6) may well be responsible.

PC is photochromic upon partial denaturation of the apoprotein

Various denaturing agents including acid (13), guanidinium chloride (14) and ethylene glycol (15) have been shown earlier to induce photochromic properties of PC chromophores. In a previous study with urea (6), it could be shown that "tickling" with urea resulted in a specific activation of only one of its three chromophores, e.g. β-84. The maximum amount of ZZE remained well below 10% in all previous studies, however. In an extension of the work with ethylene glycol, the thiol analogue mercaptoethanol now proved much more efficient (table 1).
This was originally a chance - observation during attempts to omit urea from the preparation of thiol adducts at C-10. If large amounts of mercaptoethanol (≥25 vol%) are added to native PC, the rubinoid addition products are formed nearly quantitatively even in the absence of any urea. Titration with the reagent clearly indicated that this addition proceeds in two steps. The first one is a denaturation of the protein, as evidenced by the decrease of the 620nm absorption of the chromophores, and an increase around 350nm. This is followed by the nucleophilic addition of mercaptoethanol at C-10 and the concomitant absorption increase around 415nm. It was thus not surprising to find ZZE chromophores in preparations of PC, subjected to treatment with mercaptoethanol and subsequent irradiation, since the photoreactivity of bilirubins is well documented (11,16). The amounts of ZZE chromophores in such preparations were unusually large, however, and they peaked under conditions where part of the phycocyanobilin chromophores were still unreacted with the reagent.
To further investigate this reaction, mercaptoethanol was added to the β-subunit of PC in an amount sufficient to trans-

form appx.50% of the chromophores to the yellow addition product. This mixture was then subjected to irradiation with light. The absorption at 604nm decreased rapidly with only little change in other regions of the spectrum above 300nm (fig. 3). An analysis of the mixture gave a ZZE yield of 60%,

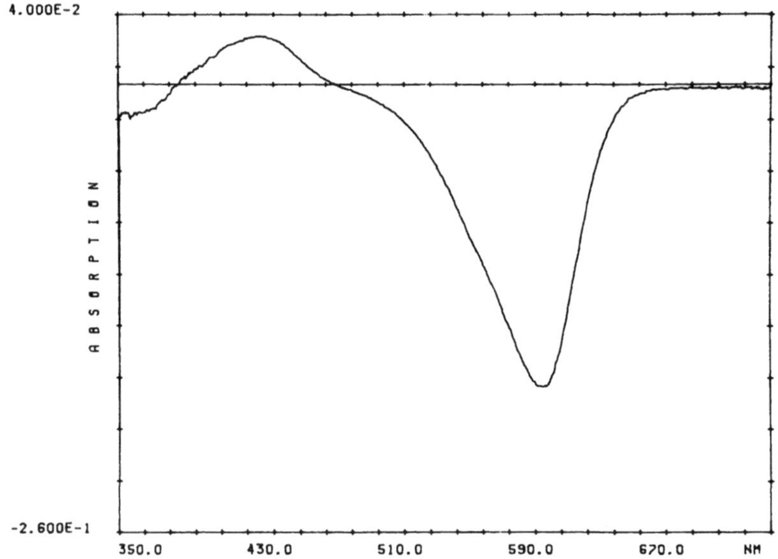

Fig. 3: Photoreaction of PC in the presence of mercaptoethanol (20%) with white light. The absorption at 620nm corresponds to the phycocyanobilin chromophore partially uncoupled from the apoprotein, that around 425nm to its addition product with mercaptoethanol.

which is four times the amount found after irradiation of the rubinoid pigment. Monochromatic irradiation furthermore gave maximum yields of ZZE chromophores with red light, absorbed by the PC chromophore, rather than with blue light absorbed by the thiol addition product (table 1). This shows clearly, that the photoreaction is not due to the well known isomerization of the yellow addition product, but rather to a very efficient photochemistry of the phycocyanobilin chomophore itself. It is noteworhy, that the proportion of ZZE is much larger in thr β- than in the α-subunit. This is similar to the reactivity in the present of 4M urea (6) and indicates a selective activity of (one of) the β-chromophores.

Conclusions

Photoreversibly photochromic properties can be induced in PC by several kinds of perturbation of the protein structure. The reactions involve most likely Z-E interconversions of the chromophores at the 15,16 double-bond between rings C and D. This reactivity is reminescent of the primary photochemistry of phytochrome. There is a differential reactivity of the chromophores.

Acknowledgement

This work was supported by the Deutsche Forschungsgemeinschaft, Bonn (SFB 143 and Forschergruppe "Pflanzliche Tetrapyrrole").
We thank G.Reng (Gesellschaft für Biotechnologische Forschung, D-3301 Stöckheim) for mass culture of M.laminosus.

References

1. Scheer, H.. 1982. In: Molecular Biology, Biochemistry and Biophysics, Vol 35 (F.K. Fong, ed.). Springer-Verlag Berlin, pp. 7-45.

2. Rüdiger, W., H. Scheer. 1983. In: Encyclopedia of Plant Physiology, New Series, Vol. 16 (W. Shropshire, Jr., H. Mohr, eds.). Springer-Verlag Berlin.

3. Thümmler, F., W. Rüdiger. 1981. Tetrahedron 39, 1943-1951.

4. Björn, L.O. 1979. Quart. Rev. Biophys. 12, 1-23.

5. Kufer, W. this book.

6. John, W., R. Fischer, S. Siebzehnrübl, H. Scheer. 1985. In: Antennas and Reaction Centers of Photosynthetic Bacteria: Structure, Interaction and Dynamics (M.E. Michel-Beyerle, ed.), Springer Verlag Berlin, pp.17-35.

7. Schirmer, T. 1987. J. Mol.Biol. 196, 677-695.

8. Schmidt, G., S. Siebzehnrübl, R. Fischer, H. Scheer, T. Schirmer, W. Bode, R. Huber. 1987. Z. Naturforsch. 42c, 845-848.

9. Füglistaller, P., H. Widmer, G. Frank, H. Zuber. 1981. Arch. Microbiol. 129, 268-274.

10. Kufer, W., H. Scheer. 1982. Z. Naturforsch. 37c, 179-192.

11. Falk, H., K. Grubmayr. 1979. Monatshefte für Chemie 110, 1237-1242.

12. Kufer, W., H. Scheer. 1981. Hoppe-Seyler's Z. Physiol. Chem. 360, 935-956.

13. Ohad, I., H.-J. Schneider, S. Gendel, L. Bogorad. 1980. Plant Physiol. 65, 6-12.

14. Ohki, K., Y. Fujita. 1979. Plant Cell Physiol. 20, 483-490.

15. de Kok, J. 1983. Private publication.

16. Mc Donagh, A., D. Lightner, T. Wooldridge. 1979. Journ. of the Chem. Soc., 109-112.

CONCERNING THE RELATIONSHIP OF LIGHT HARVESTING BILIPROTEINS TO PHYCOCHROMES IN CYANOBACTERIA

W. Kufer
Universität München, Botanisches Institut
Menzingerstr. 67
D-8000 München 19, FRG

Introduction

Besides being capable of oxygenic photosynthesis,cyanobacteria also show a broad range of photomorphogenetic responses. Among the best studied are the adaptation of the photosynthetic biliprotein antenna to the light quality received (complementary chromatic adaptation) and the influence of the light on the algal morphology (1). In these cases, the photomorphogenetic responses are ruled by a green red antagonism marked by photoreversibility. In this respect, cyanobacterial photomorphogenesis resembles classical phytochrome responses in higher plants with their red far-red antagonism. Consequently, it was proposed that a photoreversibly photochromic pigment similarly served as the photoreceptor in cyanobacteria. Björn and Björn (2,3) reported the spectral characteristics of four distinct photoreversibly photochromic pigments (phycochromes a, b, c and d) obtained from several cyanobacterial strains.

Results and Discussion

Phycoerythrocyanin is a light harvesting biliprotein occuring in certain genera of filamentous cyanobacteria (4). It was isolated from Mastigocladus laminosus and separated into its subunits by preparative scale isoelectric focussing in a

Photosynthetic Light-Harvesting Systems

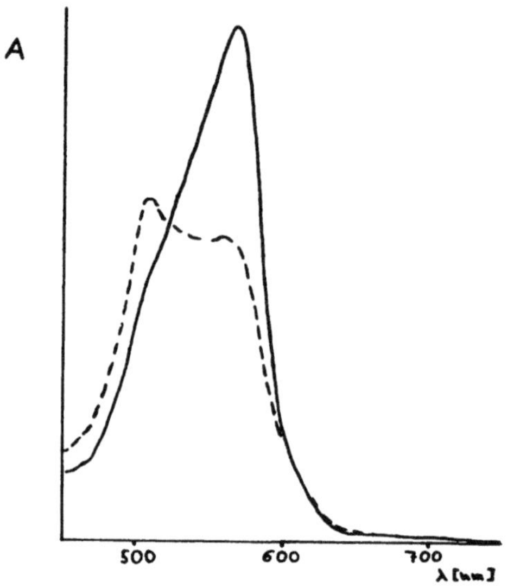

fig.1: absorption spectra of the phycoerythrocyanin α-subunit after irradiation with 510 (——) or 570 (- - -) nm light

granulated gel in the presence of ≈8M urea. The spectra of the subunits after exhaustive dialysis against 5mM K-phosphate buffer, pH 7.1, exhibited maxima at 564, 508 (sh), ≈370 (sh) and 328nm (α-subunit), and 597 and 344nm (β-subunit). Saturating irradiations with actinic light of 510 and 570nm produced the spectral forms shown in fig.1. Photoreversibility was observed for all 4 cycles measured. The spectral changes found are identical with those described for phycochrome b (5). A relationship between phycoerythrocyanin and phycochrome b had already been suggested, because the latter can only be found in algae containing this antenna pigment (6). Several authors demonstrated that photochromicity can be artificially induced in biliproteins, e.g. by chemical treatment (7,8). Such an artefact seems unlikely, as the phycochrome b activity was also found, when the α-subunit was prepared according to different methods (W.Kufer and G.S.Björn, unpublished results). Moreover, phycochrome b activity had also been demonstrated in the living cell (5).

It is concluded that the native subunit of a biliprotein known for its light harvesting function also exhibits photochromicity. Besides the direct evidence presented here, other findings additionally indicate that phycochromes are identical with common biliproteins or subpopulations thereof. Ohad et al. (9) found a loss of photochromicity after passing a purified cell extract over a column with bound antibodies against allophycocyanin. Z,E-isomerism has been suggested as a possible mechanism for photochromic absorption changes in phycocyanin (see the contribution of G.Schmidt et al. in this volume). However, it remains uncertain, whether cyanobacteria exploit the photochromicity inherent to certain light harvesting biliproteins for regulatory purposes. An action spectrum for a physiological response matching that of the photoconversion of phycochrome b has not yet been reported. An extensive study of several cyanobacterial strains containing phycoerythrocyanin revealed that the level of this biliprotein is regulated by light intensity rather than light quality (4). One must therefore consider, if the photochromicity of the phycoerythrocyanin α-subunit is without any physiological meaning and only a reflection of a "phytochrome analogous" environment of its chromophore. However, it may be a useful chemical model for those studying the phytochrome interconversions.
Most likely, the answer to the question, which pigments serve as photoreceptors in cyanobacterial photomorphogenesis, will be derived from further research on light inducible genes in these organisms.

Acknowledgements

Work on the α-subunit of phycoerythrocyanin originated during a stay at the laboratory of G.S.Björn and L.O.Björn in Lund, Sweden. The author is especially grateful for the support and encouragement he received during this time. The skilful technical assistance of Ms. G.Hopper is appreciated. This work was supported by a grant awarded by the Deutsche Forschungsgemeinschaft.

References

1. Tandeau de Marsac, N. 1983.
Bull. Institut Pasteur 81, 201.

2. Björn, G.S., L.O. Björn. 1976.
Physiol. Plant. 36, 297.

3. Björn, G.S. 1978.
Physiol. Plant. 43, 195.

4. Bryant, D.A. 1982.
J. Gen. Microbiol. 128, 835.

5. Björn, G.S. 1979.
Physiol. Plant. 46, 281.

6. Björn, G.S. 1980.
Physiol. Plant. 48, 483.

7. de Kok, J., S.E. Braslavsky, C.J. Spruit. 1981.
Photochem. Photobiol. 34, 705.

8. Chki, K., Y. Fujita. 1979.
Plant Cell Physiol. 20, 483.

9. Ohad, J., H.-J.A.W. Schneider, S. Gendel, L. Bogorad. 1980.
Plant Physiol. 65, 6.

SUBUNIT STRUCTURE AND REASSEMBLY OF THE LIGHT-HARVESTING COMPLEX FROM *RHODOSPIRILLUM RUBRUM G9+*

Robin Ghosh, Thomas Rosatzin and Reinhard Bachofen

Institute of Plant Biology, University, Zollikerstr. 107,
CH-8008 Zürich, Switzerland

Introduction

We are presently examining the light-harvesting complex (B873) from the carotenoidless mutant of *Rhodospirillum rubrum*, with the intention of defining its subunit structure and the biochemical conditions required for its functional assembly into membranes.

Results and discussion

(a) Isolation of B873 and characterisation of its physical state

Isolation of B873 was performed essentially according to the procedure of Picorel et al. [1] with the modification that the non-ionic detergent β-octyl glucoside (βOG) was used to elute the pigmented fractions from a first DEAE-cellulose column. Generally a second DEAE column was used to obtain pure B873, the complex was eluted with a combination of βOG and LDAO. Removal of the detergents from the fraction containing the light-harvesting complex (LH-complex) by extensive dialysis yielded a B873 complex with its characteristic absorption at 873nm and a near-IR CD spectrum identical to that observed for native chromatophores. The protein complex was pure as judged by silver-stained SDS-PAGE and contained 2 moles of phospholipid per αβ dimer. The dialyzed B873 was shown by freeze-fracture electron microscopy to exist as large vesicular-like structures containing regularly spaced

100 Å particles. This structure is practically identical to that observed in chromatophores (Fig. 1).

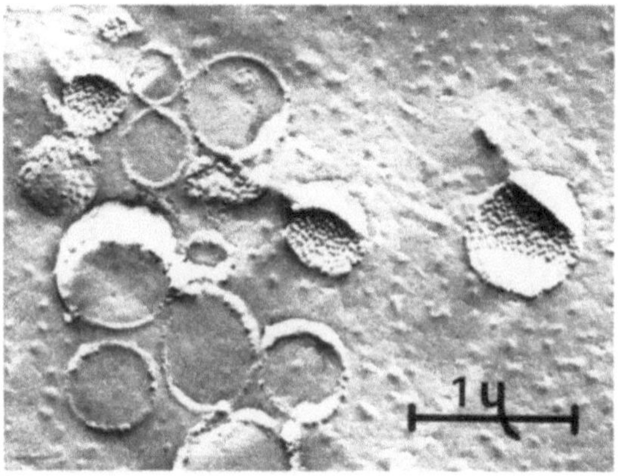

Fig. 1: Freeze fracture electron micrographs of isolated B873 after DE52 chromatography and extensive dialysis.

(b) Reversible dissociation of the B873 in detergent

The B873 complex was readily dissociated by the addition of LDAO. This led to a rapid blue shift of the absorption maximum to 775nm. Corresponding changes in the CD spectrum were also observed (Fig. 2). This shift was reversed completely upon dilution or dialysis. Using the non-ionic detergent octyl-pentaoxyethylene (O-POE), which has a density equal to that of water, it has been shown by analytical ultracentrifugation that the 873nm to 775nm transition is correlated with the formation of monomers (α-BChl or β-BChl) and dimers ($\alpha\beta \cdot (BChl)_2$). The addition of 1% Na-cholate to the B775 complex in 0.3% LDAO leads to a partial reversal of the blue shift, yielding finally a species absorbing at 860nm. Using spin-labelled complexes and ESR the cholate-induced transition was shown to involve association of the polypeptide components. No large change in secondary structure, as measured by UV-CD, was observed upon the addition of detergents [2,3]. Partial dissociation of the B873 complex by LDAO often leads to the appearance of a small absorption maximum at 820nm, possibly a discrete subunit form of the LH-complex [4].

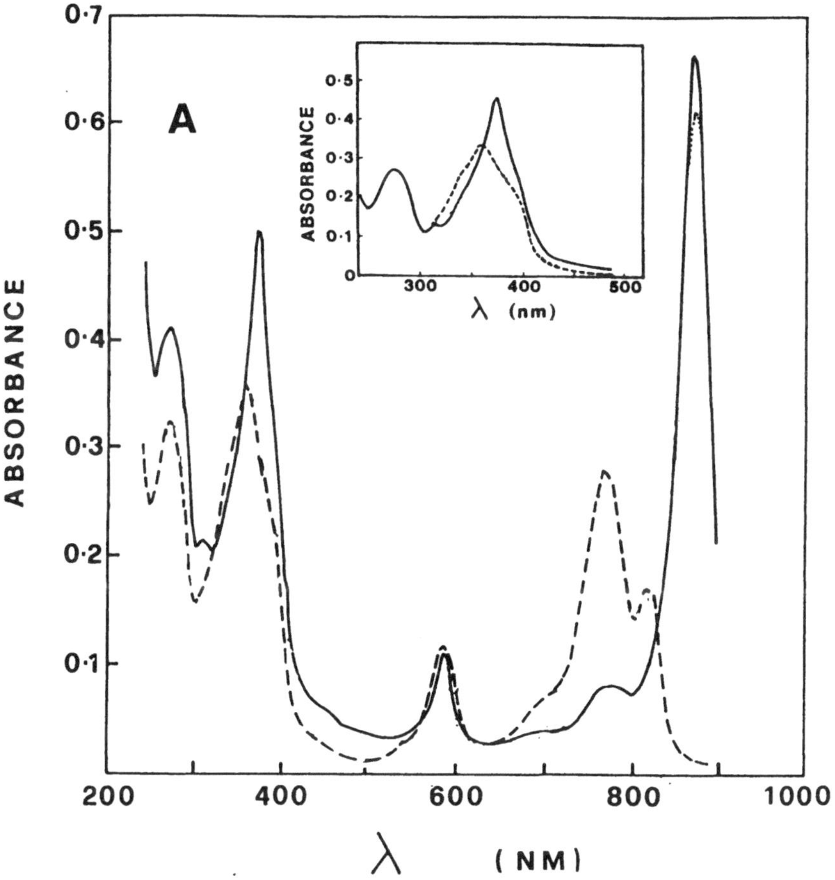

Fig. 2: (A) Absorption spectra of detergent-free B873 before (———) and after (– — — – –) solubilisation with 0.1% LDAO. The insert shows the Soret region after correction for residual light scattering.

(c) Subunit structure of the B820 oligomer

Loach and co-workers [4] have recently examined this B820 subunit form obtained from membranes of *R. rubrum* at high concentrations of βOG. They found that dilution or dialysis leads to a *partial* reversal of the blue shift to 865nm and, using column chromatography they estimated the size of the complex to be 45-50K. Starting from isolated B873 we also found that, at a defined βOG/B873 ratio, the B820 form is almost exclusively produced. The 873nm to 820nm transition is

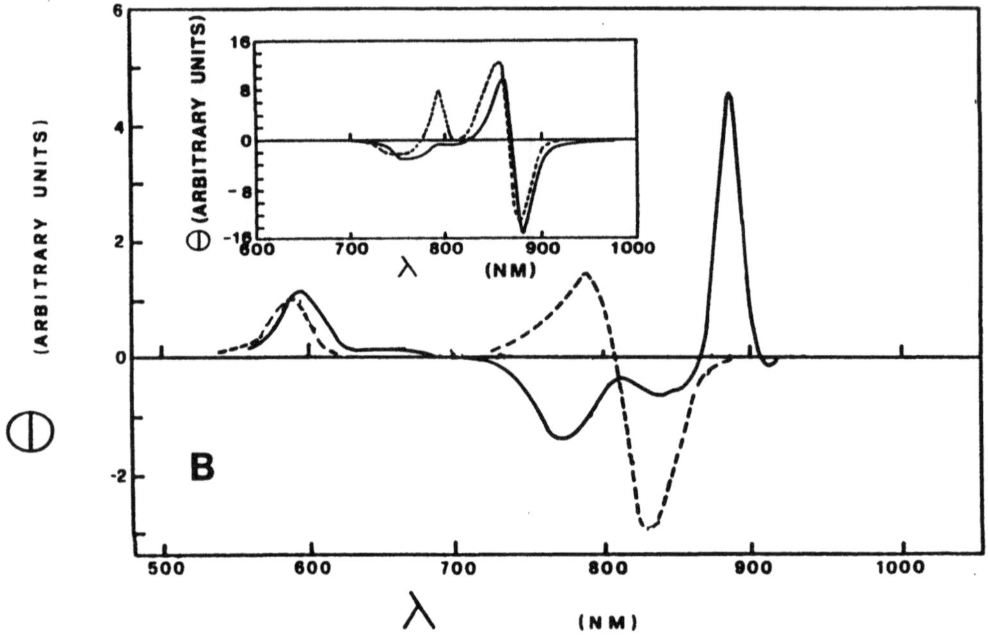

Fig. 2:(B) Near IR CD spectra of the detergent-free B873 (resuspended high speed pellet) before (———) and after (— — — —·) solubilisation with LDAO. The loss of the trough at 880nm (see insert) was due to scattering and could be recovered by sonication. In the insert the near IR CD spectrum of the detergent-free B873 (———) is superimposed upon the near IR CD spectrum of the isolated chromatophores (— — — —).

governed not only by the βOG/B873 ratio but also by the *absolute* concentration of βOG, since B820 formation occurs in the region of the critical micellar concentration (17.4mM ~0.5% (w/w)) (Fig. 3). Dilution of B820 at 1.0% βOG to 0.2% leads to the rapid reformation of B873. Preincubation of the B820 complex in 1% βOG for 1 min, however, yields a B865 complex, as reported by Miller et al. [4]. The molecular weight of B820 has been estimated using analytical ultracentrifugation in O-POE. Unfortunately, with this detergent, B820 cannot be generated exclusively without the simultaneous appearance of B775. Co-sedimentational analysis reveals that these forms are of low molecular weight (13K and 24-27k) and appear to consist of dimers

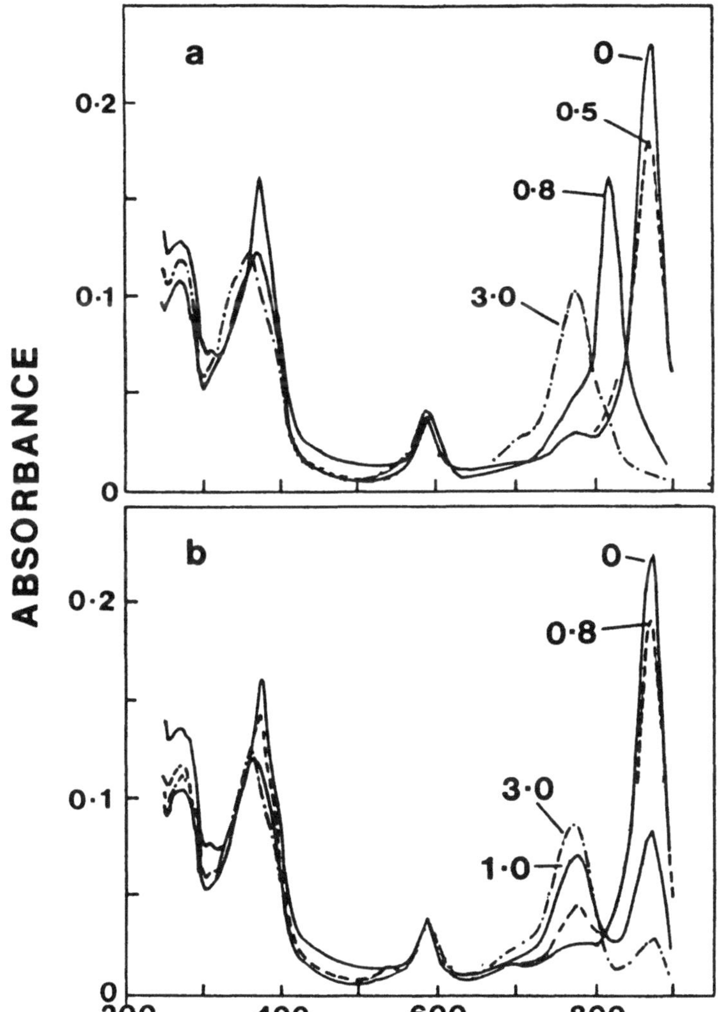

Fig. 3: Titration of absorption with increasing concentrations of βOG. Isolated detergent-free B873 was added to buffer containing the given concentration of βOG, allowed to equilibrate for 2 min at 26°C in(a) 50 mM NH_4HCO_3, pH 8.0 and (b) 50 mM Na-acetate, pH 4.0 before the measurement. The concentrations of βOG used are given in % w/w.

and tetramers, the monomer may also be present but is not observable for technical reasons. Loss of B820 by increasing the concentration of βOG, appears to be correlated with the loss of the tetramer, which is replaced by dimers and monomers. We therefore conclude that B820 corresponds to a tetramer $(\alpha\beta \cdot (BChl)_2)_2$.

The B820 in 0.8% βOG exhibits a near-IR CD spectrum with a maximum at 775nm and a minimum at 820nm (Fig. 3). Complete dissolution of the complex in 0.3% LDAO leads to the disappearance of the CD spectrum. It is not possible to conclude at present whether the dimer exhibits a CD spectrum.
A titration of the absorption spectrum of B873 at pH 7.8 in 50mM NH_4HCO_3 buffer with βOG to alter the βOG/B873 ratio shows two distinct phases. Up to 0.8% βOG the B820 form is preferentially produced, whereas at higher detergent concentrations B775 is mainly observed and becomes almost the exclusive form at 3% βOG (Fig. 4).

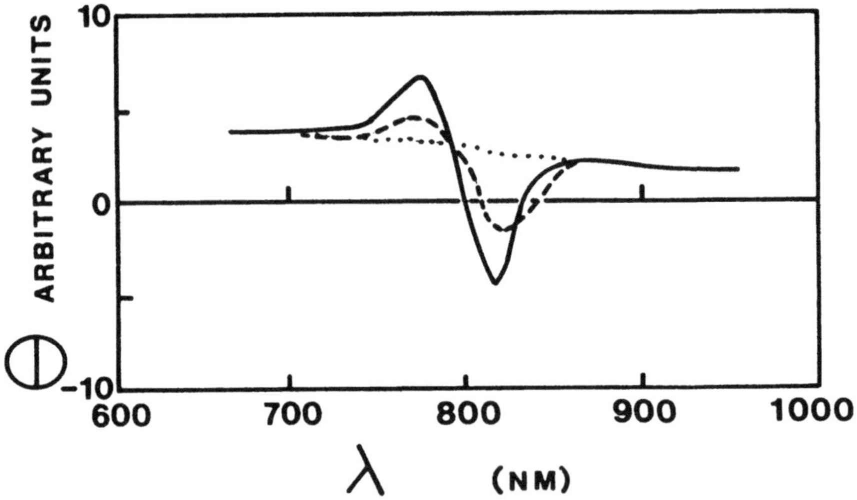

Fig. 4: Near-IR-CD spectra of B820 in 0.8% βOG (————) and B850 formed by dissolving B873 in 0.3% LDAO followed by addition of Na-cholate to 1% [2] equilibrated at 4°C for 3 h before the measurement (room temperature) (— — — —). The baseline (· · · · · · ·) is given separately.

At concentrations of 0.6% and 1% two crossover points are observed where either the forms B873/B820 or B820/B775 are present simultaneously. The addition of high salt (>200mM NaCl) at these βOG-concentrations preferentially enhances the lower wavelength form, corresponding to subunit dissociation. At low pH (pH 4.0) no B820 is observed as the βOG/B873 ratio is increased, whereas the appearance of B775 is almost unaffected.

This titration has been performed also in the opposite direction by dilution. In the most dilute case at a final βOG concentration of 0.2%, the complex is in the B873 state and for increasing concentrations it is shifted first to the B820 and then to the B775 state as expected. Interestingly, dilution to the final concentration of βOG shown in Fig. 3 yields the same spectral forms as when the reaction is performed in the original direction as shown above. This demonstrates that subunit dissociation is truly reversible and corresponds to an equilibrium state.

From these data a strategy has been developed for the total reassembly of the light-harvesting complex from its isolated components and for its reconstitution into phospholipid membranes.

(d) Total reassembly of a B862 complex

α- and β-polypeptides were purified using Sephadex LH60 chromatography in CH_2Cl_2/MeOH/NH_4OAc and CH_2Cl_2/MeOH/NH_4OAc/HOAC, respectively, after differential extraction of pure B873, as described [1,5] starting from pure B873. All steps were performed rapidly in the dark under N_2. The column eluate for β was neutralized with NH_4OH immediately after elution. Both polypeptide fractions were dried first under N_2 and followed by high vacuum. NH_4OH was then removed by washing four times with distilled H_2O, the pellets lyophilized and solubilized in 1% βOG. BChl was extracted from total chromatophores using dioxane-ethanol [6] and analysed by mass spectroscopy and NMR to prove that the isolated BChl was structurally intact and protein-free. Dried BChl was dissolved in 1% βOG and mixed separately with α and β polypeptides in βOG. The α-BChl and β-BChl complexes were mixed together and diluted immediately to 0.2% βOG. By this reassembly procedere a B862 complex (Fig. 5) was obtained after purification by sucrose gradient centrifugation and shown to contain *both* α and β polypeptides as detected by silver-stained SDS-PAGE. The absorption spectrum of the complex was shown to contain 3 components absorbing at 862nm, 824nm, and 775nm, respectively. The

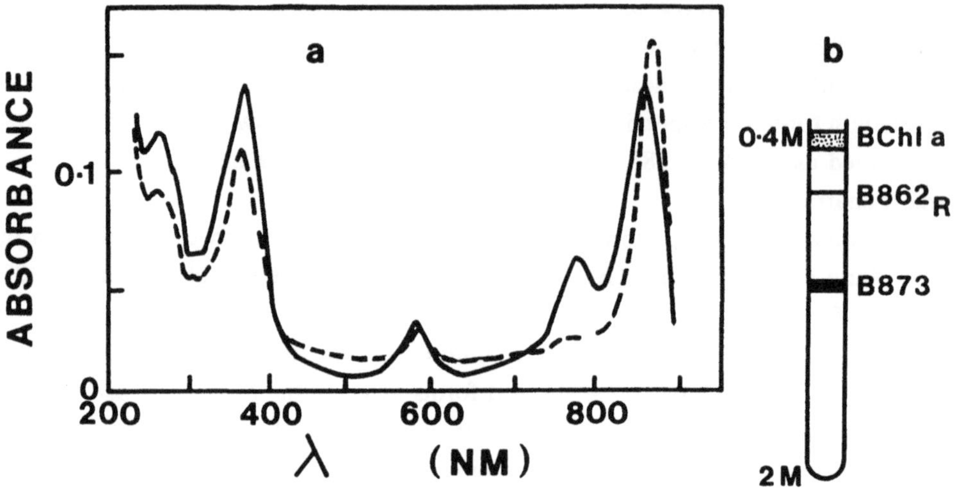

Fig. 5: (a) Absorption spectra of the reassembled B862 complex (———) in 50 mM NH_4HCO_3, pH 7.8, together with that of B873 prepared by dissolving B873 in 1% βOG and then diluting to 0.2% βOG in the same buffer (– – – –).
(b) Relative migration of BChla-detergent complexes (BChl a), reassembled B862 ($B862_R$) and isolated B873 (B873) on a sucrose gradient (0.4 to 2.0M) in 50 mM NH_4HCO_3, pH 7.8, containing 0.2% βOG.

B862 complex also exhibited a conservative CD spectrum in the near-IR region, with a crossover point at 862nm but with the reverse polarity to that observed for native B873. However, CD spectra for B873 reformed from B820 were essentially identical to that of the B862 complex (Fig. 5).

Several points concerning the reassembled B862 complex should be noted:
(1) in the presented experiments the ratio of BChl to protein was purposely adjusted to be less than that of original B873, hoping that the aggregation of BChl monomers in detergent would be less favourable than the formation of a BChl-protein complex. This strategy appeared to be effective, no free BChl was observed in sucrose gradients containing the reassembled complexes;
(2) the order of mixing is important, if α and β were premixed before the addition of BChl, no high wavelength species at 862nm was observed;

(3) the formation of an α-BChl or β-BChl *per se* was not sufficient to generate the red shift to 862 nm and thus to form the B862 complex.
(4) When BChl was allowed to equilibrate in micelles of βOG, some red-shifted species, corresponding to BChl aggregates were observed but never as high as 862 nm. At present, it is not clear why the reassembled complex does not absorb at 873nm. Possibly the non-stoichiometric BChl/protein ratio prevents the formation of a large aggregate, and/or the presence of phospholipids is necessary for B873 formation.

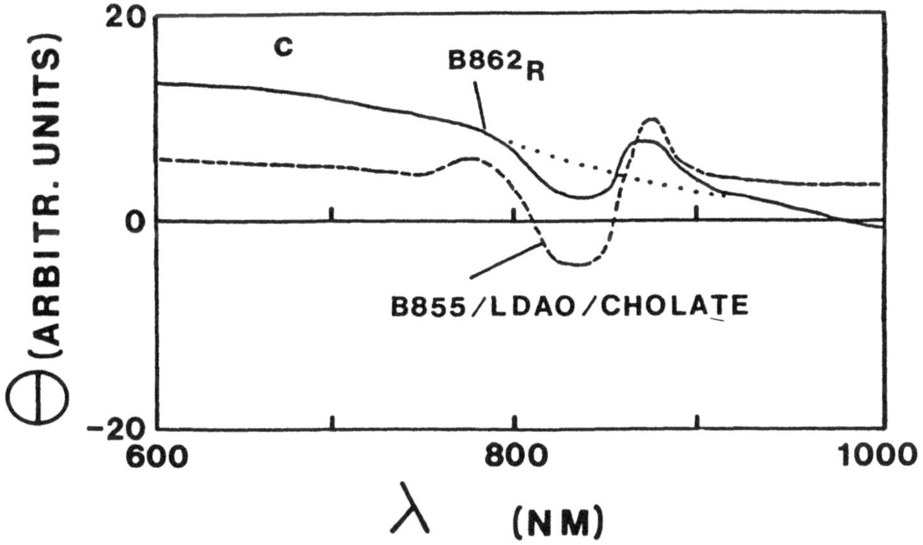

Fig. 5: (c) Near-IR-CD spectrum of the reassembled complex (———) (same conditions as in (a)) and the B850/LDAO/cholate complex described in fig. 4 (– – – – -). Baseline (· · · · · · · · ·).

Summary

The physical mechanism leading to the red-shift and hyperchromism of the antenna-bound BChl *in vivo* is subject of intense discussion. According to our observations the absorption of the pigment-protein complex at 873nm corresponds to a higher but stoichiometric aggregate of α- and β-polypeptides containing bound bacteriochlorophyll. Dissociation of the complex to yield lower molecular weight forms is *reversible* under suitable conditions. This can be used to examine the spectral properties of defined subspecies of the B873 complex and to design strategies for the reassembly of complex from its isolated components. Similar strategies may also be developed for the incorporation of light-harvesting complexes into natural and artificial membranes.

Acknowledgement

The financial support for this research from the Swiss National Foundation for Scientific Research is gratefully acknowledged (Grant 3.243-0.85). We thank A. Lustig (Biocentre, Basel) for performing the analytical ultracentrifugation experiments and Dr. M. Höchli (Dept. of Physiology, University of Zürich) for electron micrographs.

References

[1] Picorel, R., G. Belanger and G. Gingras. 1983. Biochemistry 22, 2491-2497.

[2] Bachofen, R., R. Ghosh and R. Schwerzmann. (1987). Progress in Photosynthesis Research II, 21-24.

[3] Ghosh, R., H. Hauser and R. Bachofen. 1987. Biochemistry, in press.

[4] Miller, J. F., S. B. Hinchigeri, P. S. Parkes-Loach, P. M. Callahan, J. R. Sprinkle, J. R. Riccobono and P. A. Loach. 1987. Biochemistry 26, 5055-5062.

[5] Brunisholz, R. A., F. Suter and H. Zuber. 1984. Hoppe-Seyler's Z. Physiol. Chem. 365, 675-688.

[6] Iriyama, K., N. Ogura and A. Takamiya. 1974. J. Biochem. 76, 901-904.

PRIMARY STRUCTURE ANALYSES OF BACTERIAL ANTENNA POLYPEPTIDES :
- CORRELATION OF AROMATIC AMINO ACIDS WITH SPECTRAL PROPERTIES
- STRUCTURAL SIMILARITIES WITH REACTION CENTER POLYPEPTIDES

R.A. Brunisholz and H. Zuber

Institut für Molekularbiologie und Biophysik, ETH-Hönggerberg
CH-8093 Zürich, Schweiz

Introduction

The basis for heterogeneous directed energy transfer between antenna complexes to the reaction center are well-ordered pigment molecules. Fundamental for specific pigment clusters are the structure and organization of the antenna polypeptides. In the case of purple bacteria the intra-membrane antenna pigment-protein complexes consist of relatively small hydrophobic polypeptides with specifically, non-covalently bound pigments. The basis for the three-dimensional structure and thus for the function of the antenna complexes is a specific aggregation of α/β pairs of antenna polypeptides to larger aggregates (1,2). An efficient directional flow of excitation energy to the reaction center is possible by a combination of chemically identical pigments absorbing at different wavelengths. Thus, the absorption characteristics of the pigments are determined mainly by their specific microenvironment (pigment-pigment -and pigment-protein interactions). i.e. in general, upon binding to their specific antenna polypeptides the pigments shift batho -and hyperchromically. Within the last years a large number of publications dealt with the question of the origin of the red-shifted pigment species (3-12). In order to explain the large spectral variety of e.g. bacteriochlorophyll (BChl)model systems with pigments have preferentially been used. Accordingly, the anomalous

Photosynthetic Light-Harvesting Systems

spectral shifts have not solely been attributed to interactions of inter-pigment electronic transitions moments. Although there were early suggestions that the spectral shifts are the result of complexing the pigments with the proteins (13,14) the nature of interactions remained unclear. Some authors called attention to polar/charged amino acids which, by perturbing the pigments would induce the spectral shifts (point-charge spectral shift hypothesis). However, up to this date no structural data of antenna polypeptides were available except for the BChl a binding protein of Prosthecochloris aestuarii. Our approach to study structure-spectra relationships was i) to determine the primary structures of the antenna specific light-harvesting polypeptides from purple and green bacteria (15-24) ii) to determine the organisation and arrangement of the trans-membrane antenna polypeptides within the photosynthetic membrane (25,26). The results obtained by this means offered a promising situation regarding detailed structure-function analyses since i) most of the antenna polypeptides studied exhibit only one membrane-spanning segment of about 21-23 amino acids flanked by rather short and polar N -and C-terminal domains ii) the dimeric BChl species (e.g. the pigments absorbing at 850, 880 nm) ligate most probably to a conserved histidine located in the bilayer half exposed to the periplasmic side. As a result, the structural model of the smallest compositional unit of e.g. a B880 antenna we proposed draw attention to invariant aromatic amino acids (16,17). By looking carefully at other chromophore binding polypeptides like heme binding proteins, rhodopsins, bacterio-rhodopsin and the BChl a binding protein of Prosthecochloris aestuarii we postulated that, as a general principle aromatic residues in the vicinity of the chromophore play the decisive factors for the specific spectral properties of chromoproteins (16). In this context it was interesting to note that the extremely red-shifted antennae of BChl b containing bacteria appear to have a third very small peptide component with additional aromatic residues (17,27). Recently, the "aromatic residue spectral shift" hypothesis has been substantiated by progress in crystallizing pigment-protein complexes and determining their structure to almost atomic resolution. The special pair of the reaction center of Rp. viridis interacts with eleven aromatic residues (28). Each of the seven BChl a of the soluble antenna protein of Prosthecochloris aestuarii interacts at least with one aromatic residue (29 reference therein). Here we focus more thoroughly to the correlation of specific amino acid side chain aromaticity with the spectral properties of the three antenna complexes B880, B1015 (core antenna), B800-850 and B800-820 (peripheral antennae). A careful comparison of the primary structures of the reaction center polypeptides L and M with the core antenna polypeptides α/β have corrobareted these findings: 9 out of 11 aromatic residues forming the special pair binding site (28) are distributed in 5 distinct clusters which appear as structural elements in the core antenna polypeptides as well. On the basis of this study, for the first time considerable evidence is presented that the interactions between special pair and reaction center polypeptides as well as between (core) antenna BChl and light-harvesting polypeptides are of the same type.

Correlation of aromatic residues with antenna absorption characteristics

Since in 1981 when for the first time the primary structure of an intramembrane-bound antenna polypeptide became available(15) a large number of equivalent light-harvesting polypeptides from other strains of Rhodospirillaceae has been sequenced (16-24). As compiled in the table most of the three spectrally different antenna complexes contain a distinct α/β -pair of polypeptides ranging from 40 to 70 amino acid residues.

Table: Compilation of type of intra-membrane antenna complex, type of antenna polypeptide α, β, γ (corresponding number of amino acids in parantheses) synthesized in seven purple non-sulfur bacteria and in Chloroflexus aurantiacus.

	Core antenna B880 (B1015)	Peripheral antenna B800-850	B800-820
Rs. rubrum	α/β(52/54)	—	—
Rp. marina	α/β(52,56)	—	—
Rp. viridis	α/β/γ(58,55,36)	—	—
Rb. sphaeroides	α/β(58,48)	α/β(54,50)	—
Rb. capsulatus	α/β(58,48)	α/β(60,49)γ?	—
Rc. gelatinosus	α/β(52,47)	α/β(71,50)	—
Rp. acidophila 7050	α/β(50,54)	α/β(53,48)	α/β(53,42)
7750	α/β(" ")	α/β(53,41)	$\alpha/\beta_1\beta_2$(53,41,43)
10050	α/β(" ")	α/β(53,41)	—
Chloroflexus aurant.	α/β(44,53)*)	—	—

*) B806-866

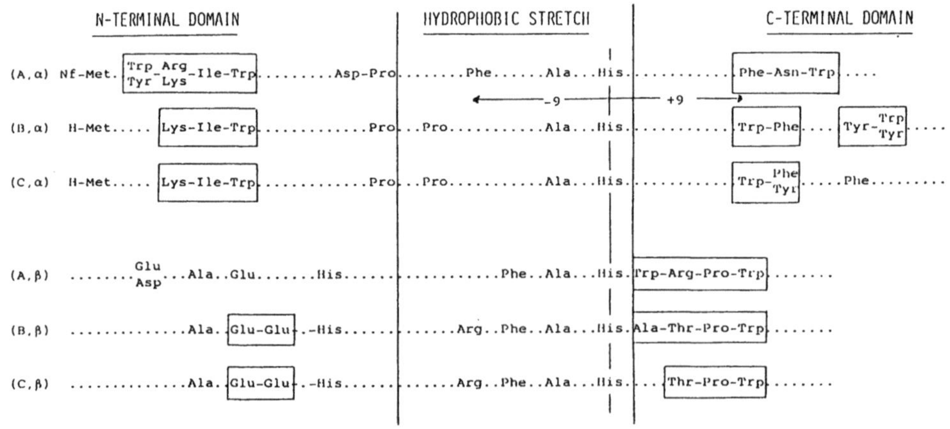

Fig. 1: Specific structural elements (amino acid residues boxed) of antenna polypeptides of the B880 (A), B800-850 (B) and B800-820 (C) light-harvesting complexes from purple bacteria.

Our approach to understanding the role of pigment-protein interactions is to carefully select and work out decisive structural features of the antenna specific light-harvesting polypeptides. The structures determined so far did provide considerable insight into pigment-protein binding. As depicted in Fig.1 besides the most significant structural feature, the overall conserved trans-membrane histidine (most probable ligand for BChl) the antenna specific structural elements are clustered into groups of 2 to 4 amino acids. Most of these clusters contain 1 - 2 aromatic residues. Interestingly, an ample number of these antenna specific aromatic residues are located within the hydrophobic stretches and in the C-terminal domains of which the latter have been shown to be exposed to the periplasmic side (25,26). In Fig.4 the specific location of the aromatic clusters is illustrated in more details for the core antenna polypeptides from seven purple non-sulfur bacteria. In (17) it has been postulated that for the core antenna B880 from Rs. rubrum the substantial redshift is mainly due to coupled oscillator interaction of the pigment with the $\pi - \pi^*$ transitions of the nearby located aromatic residues. The suggestion that aromatic residues are the crucial structural factor for the individual spectral properties of the antenna complexes gained more interest with the structural studies of the antenna polypeptides from Rp. acidophila (19): Comparing the polypeptide structures a substantial mutation has only been observed in the C-terminal domains of the α-polypeptides (see Fig. 2). The aromatic cluster Tyr_{44}-Trp_{45} in the B(800-850)-α polypeptides is replaced by Phe-Leu (strain 7050) and Phe-Thr (strain 7750) in the B(800-820)-α polypeptides. In conclusion, further evidence is provided that the multiplicity of infra-red absorption bands of BChl in vivo may arise from conjugation of the pigment with antenna polypeptide specific aromatic amino acid residues (batho -and hyperchromicity).

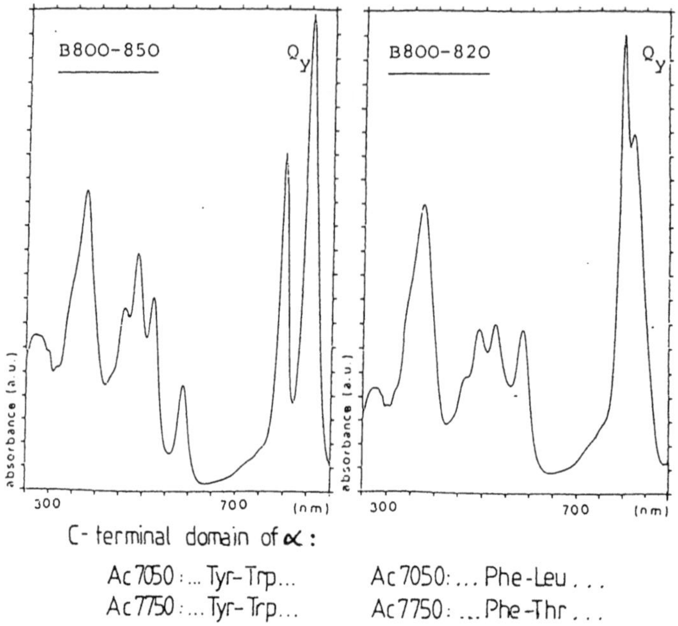

Fig. 2:
Absorption spectra (250-900 nm) of the light-harvesting complexes B800-850 and B800-820 from Rp. acidophila strain 7050. Correlation of a specific aromatic cluster in the α-polypeptides with the individual spectral absorption properties. (Q_y-band: 850/820)

Correlation of aromatic residues with antenna circular dichroism characteristics

The application of circular dichroism spectroscopy to photosynthetic membranes,purified antenna and reaction center complexes promises to provide important information about structure and organization of pigment-protein complexes (30). Near-infrared CD-spectroscopy applied to bacterial antenna complexes suggested strong coupling among the chromophores (30 and references therein). Accordingly, the strong CD-signals observed have mainly been ascribed to pigment-pigment interactions without taking the protein structure into account. However, a few reports did not ignore the effects of the protein when analyzing model pigment-protein complexes (11). They focused to the protein environment as a source for induced CD-signals as the following will emphasize. The core antennae of purple bacteria have been classified into two subdivisions according to their NIR-CD intensities and the exact positions of their red-most absorption bands (31). One core antenna class exhibits representatives (*Rs. rubrum*, *Rp. viridis*(?), *Rp. acidophila*) with intense biphasic CD-signals and absorption maxima around 880/890 nm whilst the CD-signals of the antenna complexes of the other group are much weaker. In addition, their Q_y-bands are slightly shifted to the blue, centered at around 870 nm (*Rb. sphaeroides*, *Rb. capsulatus*, *Rc. gelatinosus*). Interestingly, these subtle spectral differences can be correlated with the differences in the primary structures of the corresponding antenna polypeptides as illustrated in Fig. 3A and 3B. The structural features of the polypeptides of core antenna complexes with intense CD-signals are i) in the α-polypeptides: the adjacent amino acid of the conserved histidine is a phenylalanine. ii) in the C-terminal domain of the β-polypeptides a tyrosine/tryptophan residue is present adjacent to an extreme turn constellation (in Fig. 3B, circled letters). These specific structural elements are not present in the polypeptides of the core antenna complexes with weak CD-intensities indicative of a possible structural and functional role for the strongly interacting core antenna pigments.

Fig. 3A: (text see next page)

B880-α

						*	↓					
1	...	A	L	L	I	H	F	I	L	L	...	STRONG NIR-CD SIGNAL
2	...	A	L	L	I	H	F	G	L	L	...	
3	...	A	L	I	I	H	F	I	A	L	...	
4	...	A	V	L	I	H	L	M	L	L	...	WEAK NIR-CD SIGNAL
5	...	A	V	M	I	H	L	I	L	L	...	
6	...	A	V	L	I	H	L	I	L	L	...	
							↑					

B 880-β C-end

Fig. 3B :
1. . . R P W V (P) G (P) N G Y S
2. . . R P W I A (P) I (P) K G W V
3. . . R P W I (P) G (P) K G W A
— STRONG NIR-CD · SIGNAL

4. . . R P W L
5. . . R P W F
6. . . R P W F
— WEAK NIR-CD · SIGNAL

Fig. 3 : Correlation of the NIR-CD intensities of the core antenna with specific structural features of the α - and β-antenna polypeptides:
A. Sequence constellation in the vicinity of the conserved transmembrane histidine * of the α-polypeptide*.
B. A comparison of the C-terminal domains of the β-polypeptides of the core antennae.
1. Rs. rubrum 2. Rp. viridis 3. Rp. acidophila
4. Rc. gelatinosus 5. Rb. sphaeroides 6. Rb. capsulatus

Structural similarities between the core antenna -and reaction center polypeptides

Since the electronic absorption characteristics of the special pair pigments of the reaction center (RC) in general match the ones of the corresponding core antenna pigments (e.g. P870, B870 in Rs. rubrum) one would expect at least a few similar structural features in the core antenna -and the RC polypeptides. However, by comparing the linear sequences in the vicinity of the special pair histidines (28) (17 amino acids up -and downstreams) and the conserved "antenna histidines" only poor homologies have been observed: α/L (17%), α/M (9%), β/L (6%) and β/M (3%). Proteolytic digestion studies in vivo revealed (25) opposite directions of the helices bearing the special pair (D-helix in (28)) and the helices bearing the conserved "antenna histidines". If one compares the sequences in opposite direction the overall homology within this specific

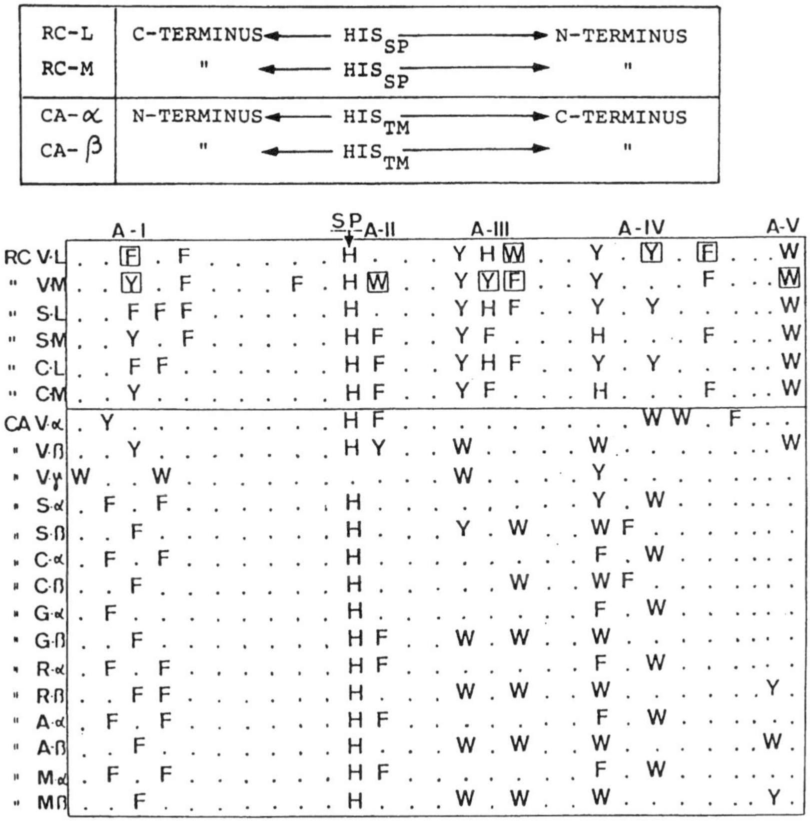

Fig. 4: Structural relatednes in the vicinity of the RC-polypeptides L and M histidines (His-L173, His-M200) liganding the special pair BChl molecules and the conserved trans-membrane core antenna polypeptide histidines liganding the dimeric antenna BChl molecules.
Top: scheme of how the sequences are compared.
Boxed letters: aromatic residues forming part of the special pair binding site in the reaction center of <u>Rp. viridis</u> (28).
Primary structure data: see references in the text (+24)
RC: reaction center, CA: core antenna, SP: special pair, TM: trans-membrane, V: Rp. viridis, S: Rb. sphaeroides, C: Rb. capsulatus, G: Rc. gelatinosus, R: Rs. rubrum, A: Rp. acidophila, M: Rp. marina, F: phenylalanine, Y: tyrosine, W: tryptophan, H: histidine.
AI -AV : aromatic amino acid clusters AI -AV.

domains **doesn't** change dramatically. However, in connection with the preceeding chapters an intriguing observation has been made (see Fig.4). Nine out of eleven aromatic residues forming the special pair binding site are located in five distinct clusters (termed AI - AV) in the vicinity of the special pair histidines which are present in the core antenna in equivalent positions as well. Interestingly, the third peptide component of the extremely red-shifted antenna B1015 from Rp. viridis does contain cluster AI, AIII, and AIV as well. In conclusion, it is conceivable that similarities regarding protein environment of special pair and (core) antenna pigments are not only restricted to histidines serving as 5th ligands.Moreover, it is striking to note that side chain aromaticity may provide the major interaction sites of the spectrally (absorption) equal but functionally different pigment types.

References

1. Zuber, H., R. Brunisholz, W. Sidler. 1987. In: Photosynthesis (J. Amesz, ed.) Elsevier Science Publishers p. 233-271
2. Zuber, H. 1985. Photochem. Photobiol. 42, 821-844
3. Pearlstein, R.M., R.C. Davis, S.L. Ditson. 1982. Proc. Natl. Acad. Sci. USA. 79, 400-402
4. Hanson, L.K., Hofrichter,J. 1985. Photochem. Photobiol. 41, 247-249
5. Gottstein, J., Scheer, H. 1983. Proc. Acad. Natl. Sci. USA. 80, 2231-2234
6. Warshel, A. 1979. J. Americ. Chem. soc. 101/3, 744-746
7. Kuki, A., S.G. Boxer. 1983. Biochemistry. 22, 2923-2933
8. Davis, R.C., R.M. Pearlstein. 1979. Nature. 280, 413-415
9. Mukherjee, T., A.V. Sapre, J.P. Mittal. 1978. Photochem. Photobiol. 928, 95-96
10. Davis, R.C., S.L. Ditson, A.I. Fentiman, R.M. Pearlstein. 1982. J. Americ. Chem. Soc. 103, 6823-6826

11. Wright, K.A., S.G. Boxer. 1981. Biochemistry. 20, 7546-7556

12. Cotton, T.M., P.A. Loach, J.J. Katz, K. Ballschmiter. 1978. Photochem. Photobiol. 27, 735-749

13. Wassink. E.C., E. Katz, R. Dorrestein. 1939. Enzymologia. 7, 113-129

14. Steffen, H., M. Calvin. 1970. Biochim. Biophys. Res. Comm. 41/2, 282-286

15. Brunisholz, R.A., Cuendet, P.A., R. Theiler, H. Zuber. 1981. FEBS letters. 129, 150-154

16. Brunisholz, R.A., F. Suter, H. Zuber. 1984. Hoppe Seyler Z. Physiol. Chem. 365, 675-688

17. Brunisholz, R.A., F. Jay, F.Suter, H. Zuber. 1985. Biol. Chem. Hoppe Seyler. 366, 87-98

18. Theiler, R., Wiemken, V., F.Suter, H. Zuber. 1984. Hoppe Seyler Z. Physiol. Chem. 365, 703-719

19. Brunisholz, R.A., I. Bissig, E. Niederer, F. Suter, H. Zuber. 1987. In: Progress in Photosyntesis Research (J. Biggins, ed.) Martinus Nijhoff Publishers II.1. 13-16

20. Tadros, M., Suter, F., H.H. Seydewitz, I. Witt, H.Zuber, Drews, G. 1984. Eur. J. Biochem. 138, 669-678

21. Youvan, D.C., S. Ismail. 1985. Proc. Natl. Acad. Sci. USA. 82, 58-62

22. Wechsler, T.D., R.A. Brunisholz, G. Frank, F. Suter, H. Zuber. 1987. FEBS letters. 210/1, 189-194

23. Wechsler, T.D., Brunisholz, R.A., F. Suter, R.C. Fuller, H. Zuber. 1985. FEBS letters. 191/1,34-38

24. Youvan, D.C., E.J. Bylina, M.Alberti, H. Begush, J.E. Hearst. 1984. Cell. 37, 949-957

25. Brunisholz, R.A., V. Wiemken, F. Suter, H. Zuber. 1984. Hoppe Seyler Z. Physiol. Chem. 365, 689-701

26. Brunisholz, R.A., H. Zuber, J. Valentine, J.G. Lindsay, K. J. Wolley, R.C. Cogdell. 1986. Biochim. Biophys. Acta. 849, 295-303

27. Brunisholz, R.A., R. Steiner, H. Scheer, H. Zuber. 1987. Workshop on Structure, Function and Formation of Membrane bound Complexes in Phototrophic Bacteria. Abstract I-P-1 Freiburg, West-Germany

28. Michel, H., D. Epp, J. Deisenhofer. 1986. The EMBO Journal 5/10, 2445-2451

29. Pearlstein, R.M. 1987. In: Photosynthesis. (J. Amesz, ed.) Elsevier Science Publishers. pp. 299-315

30. Cogdell, Rj., H. Scheer. 1985. Photochem. Photobiol. 42, 669-678

31. Thornber,J.P., R.J. Cogdell, B.K. Pierson, Seftor, R.E.B. 1983. J. Of Cellular Biochem. 23, 159-169

THE STRUCTURE OF THE 'CORE' OF THE PURPLE BACTERIAL PHOTOSYNTHETIC UNIT

Deborah J. Dawkins, Linda A. Ferguson and Richard Cogdell
Plant Molecular Biology and Biochemistry Group,
Department of Botany,
University of Glasgow,
Glasgow, G12 8QQ, U.K.

Introduction

The light-absorbing pigments in purple bacteria are organised into two main types of complexes; the well-defined reaction centres (RC) and discreet light-harvesting complexes. An antenna complex typically consists of a small number of bacteriochlorophyll (2-3) and carotenoid (1-2) molecules which are non-covalently bound to two low molecular weight (5-7KD) hydrophobic proteins (the α-and β-apoproteins) (1). Early studies by Aagaard and Sistrom (2), on two Bchl<u>a</u>-containing species <u>Rhodobacter sphaeroides</u> and <u>Rhodospirillum rubrum</u>, clearly demonstrated the existence of two different types of antenna complex. The first class, called the B875-(or B880-) type of complex, is intimately associated with the RC and represents, together with the RC, the 'fixed' portion of the photosynthetic unit. The second class, called the B800-850-complex, is present in variable amounts relative to the concentration of RC's present. This type of complex represents the 'variable' component of the photosynthetic unit.

In this study we set out to begin to investigate the structure of the fixed 'core' of the photosynthetic unit of Bchl<u>a</u>-containing purple bacteria. Examination of the structure of the photosynthetic membrane of Bchl<u>b</u>-containing species, such as <u>Rhodopseudomonas viridis</u> and <u>Ectothiorhodospira halochloris</u>, with high-resolution electron microscopy and image reconstruction techniques, has suggested that the B880-type of antenna complexes (actually B1012 in <u>Rps. viridis</u>) form a 'well-defined' core complex with the RC (3-5). In these species it has been suggested that 12 heterodimers of the antenna complex surround the RC, rather like

Photosynthetic Light-Harvesting Systems

a 'ring-doughnut' structure. We set out to see whether a similar fixed 'core' structure exists in Bchl*a*-containing species or whether the apparent fixed ratio of Bchl*a*:RC is purely statistical.

Materials and Methods

Rhodopseudomonas acidophila strains 7750 and 7050 were grown in Pfennig's medium (6). All other species were grown in Bose's medium (7) except *Chromatium vinosum* strain D, which was grown in the heterotrophic synthetic medium of Fuller (cited in (7)). All the bacteria were grown anaerobically in the light at 30°C under either 'high light' (~42Wm^{-2}) or 'low light' (~9Wm^{-2}). Following growth, the cells were harvested by centrifugation, washed and resuspended in 20mM MES pH 6.8, 100mM KCl. They were then either used immediately or stored at -20°C until required. The cells were broken by passage through a 'French' pressure cell at 10 Tons inch $^{-2}$, in the presence of a little DNase and $MgCl_2$. The broken cells were then centrifuged at 15,000 xg at 4°C for 10 mins. In those bacterial species which contain vesicular-shaped intracytoplasmic membranes i.e *Rh.sphaeroides*, *Chr.vinosum* and *R.rubrum*, the low speed pellet, containing unbroken cells and cell wall debris, was discarded. Chromatophores were then obtained from the subsequent centrifugation of the low speed supernatant at 243,000 x g at 4°C for 60 mins. In those species which contain lamellar intracytoplasmic membranes i.e. *Rps. acidophila*, *Rhodopseudomonas blastica*, *Rhodocyclus gelatinosus* and *Rhodopseudomonas palustris*, both the low and high speed pellets were retained and combined. The pelleted membranes were resuspended in 20mM Tris HCl pH 8.0.

Firsow and Drews(8), using *Rps. palustris*, described a method of fractionating the photosynthetic unit into its 'variable' component (the B800-850-complexes) and its 'fixed' component (the RC:antenna conjugate). Their method involved solubilisation of the photosynthetic membrane with a mild detergent, followed by fractionation by sucrose density-gradient centrifugation. We have adapted their basic method in our present study as follows. The concentration of photosynthetic membranes was adjusted

Table 1

Summary of the optimal solubisation conditions for each of the species and strains of bacteria used in the present study.

Species	Preferred Detergent	Detergent concentration (v/v)
Rps. acidophila strain 7050	LDAO	1.0%
Rps. acidophila strain 7750	"	1.0%
Rps. gelatinosus strain 149	"	0.5 - 1.0%
Rc. gelatinosus strain 1512	"	0.5 - 1.0%
Rps. palustris strn.DSM 8252	"	0.75%
Rps. palustris strn.'French'	"	0.75%
Rps. blastica	"	0.5%
Chr. vinosum strain D	"	0.5 - 1.0%
Rh. sphaeroides strain R26	Dodecyl-β-D-maltoside	1.0%
Rh. sphaeroides strain 2.4.1	N-octyl-B-D-glucopyranoside	1.0%
R. rubrum strain S1	LDAO	0.5%

to give an OD of 50 cm^{-1} at the strongest NIR Bchl<u>a</u> absorption band. 1cm^3 of this sample was then mixed with 2cm^3 of 10mM Tris HCl pH 8.0 containing detergent. This mixture was stirred at 4^oC for 10 mins. and then any unsolubilised material was removed by a low speed centrifugation at 15,000 x g for 10 mins. The clear pigmented supernatant from this spin was then loaded onto the sucrose gradients. The gradients were made up with three sucrose concentrations, 1.2, 0.6 and 0.3M sucrose in 10mM Tris HCl pH 8.0 and usually containing 0.2% (v/v) of the same detergent that was used in the solubilisation step. The gradients were then centrifuged at 197,000 x g for 16 hours at 4^oC in a fixed angle rotor. After centrifugation two, well separated, highly pigmented bands are typically seen (see Fig. 1. below). These two bands were removed from the gradient and investigated.

The type and concentration of detergent used for the membrane solubilisation was dependent upon the species of bacteria used, and to some extent on the sample batch. We attempted to optimise the solubilistion conditions for each species used so as to allow the isolation of their pigment-protein complexes in a form, as far as possible, still representative of their <u>in situ</u> state. The following detergents were tested: sodium and lithium dodecylsulphate, lauryl-dimethylamine-N-oxide (LDAO), triton X-100, dodecyl-β-D-maltoside, cholic acid and N-octyl-β-D-glucopyranoside. The optimal solubilisation conditions for each species are summarised in Table 1.

The polypeptide composition of the RC:antenna conjugates was determined by SDS-polyacrylamide gradient (11.5-16.5%) gel electrophoresis, using the method of Laemmli (9) as modified by Demartini et al. (10). The acrylamide gradient was stabilised by a 15-70% sucrose gradient. Often samples removed from the sucrose density-gradient centrifugation runs were too dilute for analysis by gel electrophoresis and so they were concentrated on DE52-cellulose mini-columns poured in pasteur pipettes.

The concentrations of reaction centres in the RC:antenna conjugates was determined by chemical oxidation with potassium ferricyanide (11). Oxidation of reaction centres induces characteristic spectral changes in the NIR; i.e. bleaching around 870nm and a blue shift of the 800nm RC absorption band (11). In the RC:antenna conjugates there is a problem with monitoring at 870nm since oxidation of the antenna Bchl a (which is irreversible) also causes bleaching at this wavelength. We therefore estimated the RC concentration from the blue shift at 800nm, using a differential extinction coefficient of $133 mM^{-1}cm^{-1}$ for $^{A}789$-808nm. Reversibility of these changes was confirmed by addition of sodium asorbate, once the oxidation was complete.

The total Bchla content of the RC:antenna conjugates was determined by measuring the absorbance at the NIR maximum i.e. between 870 and 880nm depending upon the species, and using an average *in vivo* extinction coefficient of 120 $mM^{-1}cm^{-1}$ (12).

The presence of fluorescence induction kinetics can be used as a test for the functional integrity of the RC:antenna conjugate. If the antenna complexes can still transfer energy to the RC, then when they are initially excited by light their fluorescence yield will be low. Then as the period of illumination proceeds energy transfer to the RCs will 'close' them and so the yield of antenna fluorescence will rise. If on the other hand the antenna complexes are not functionally connected to the RCs then no variable fluorescence will be detected. The fluorescence emitted by the antenna complexes in the RC:antenna conjugates was detected with a home-made fluorimeter (13). Broad-band blue excitation light was provided by passing light from a 150W Quartz Halogen lamp through the following combination of filters: Corning 4-97, Schott BG18 and Balzer's Calflex C. The fluorescence was collected at right angles to the direction of excitation, passed through an RG 715 cut-off filter and a Balzer's 913nm interference filter, and detected by an EMI photomultiplier (Type 9659 QB). The signal from the photo-tube was amplified, passed through an A to D converter and stored in a mini-computer (DEC PDP-11/23). An electronic timer triggered the shutter over the excitation beam and data storage device.

We attempted to assess whether the RC:antenna conjugates had a homogenous, well-defined structure by crystallising them. We used the general method of vapour diffusion, with the presence of small amphiphiles (14,15). The RC:antenna conjugates for the crystallisation trials were given a further purification by passing them over a Sephacryl S-200 molecular sieve column. This removed the minor contaminants that were present following the initial fractionation by sucrose density-gradient centrifugation. The successful crystallisation conditions will be described below.

Results and Discussion

Using sucrose density-gradient centrifugation we have been able to fractionate detergent-solubilised photosynthetic membranes from a range of Bchl*a*-containing species of purple bacteria into RC:antenna conjugates and B800-850-complexes. Fig. 1. shows a typical sucrose gradient after the centrifuge run. The upper band is the B800-850-complex and the lower band is the RC:antenna conjugate. Fig. 2 shows the absorption spectra of these two bands, in this case isolated from *Rps. palustris* strain French. Although this method of fractionating the photosynthetic membrane is also good for producing B800-850-complexes, in this study we are only going to report on the properties of the RC:antenna conjugate.

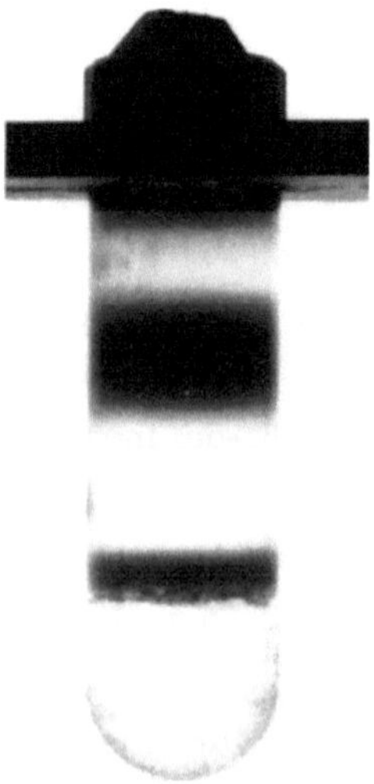

Fig. 1. A picture of a typical fractionation of the photosynthetic membrane of Rps. palustris strain French.

The top band is the B800-850-complex and the bottom band is the RC: antenna conjugate.

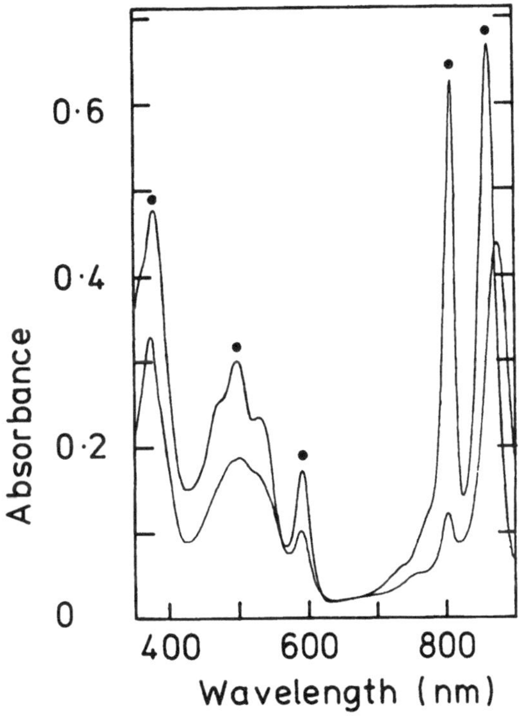

Fig. 2. The absorption spectra of the two bands removed from the gradient shown in Fig. 1.

The spectra were recorded with a Pye-Unicam SP8-500 spectrophotometer. Top band marked peaks. Bottom band unmarked peaks.

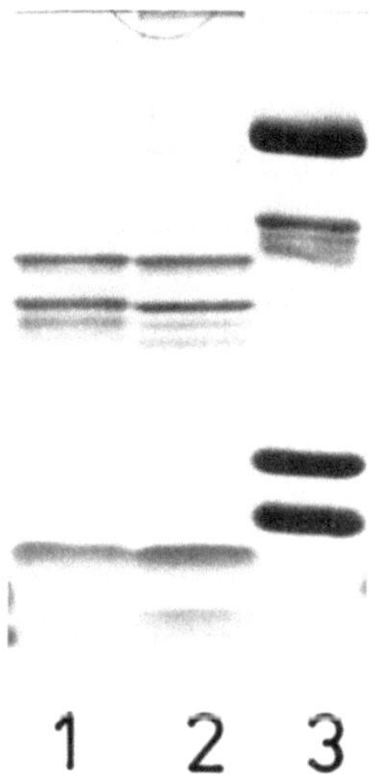

Fig. 3. An SDS -polyacrylamide gradient gel of the RC:antenna conjugate isolated from Rps. palustris strain French.

Lanes 1 & 2:
the RC:antenna conjugate
Lane 3
standard proteins - Bovine serum albumen-68KD,
Alcohol dehydrogenase - 41KD,
Myoglobin - 17.2 KD and Cytochrome c - 12.1 KD.

Fig. 3. shows the typical polypeptide composition of these RC:antenna conjugates. Two low molecular weight antenna apoproteins are seen at the bottom of the SDS gel, together with the L,M and H subunits of the reaction centre. In this gell the 'L' subunit has 'split' into two bands.

Table 2.

The ratio of RC: Bchl<u>a</u> determined for the RC: antenna conjugates isolated from a range of Bchl<u>a</u> - containing purple bacterial species.

Species	RC: Bchl<u>a</u> ratio mole:mole ± 1SD	number of different isolates analysed
<u>Rps. blastica</u>	1:21 ± 4	18
<u>Rps. palustris</u> French	1:23 ± 3	30
<u>Rps. palustris</u> DSM 8252	1:27 ± 7	13
<u>Rc. gelatinosus</u> 149	1:35 ± 9	52
<u>Rc. gelatinosus</u> 151	1:29 ± 7	26
<u>Rps. acidophila</u> 7050	1:41 ± 7	66
<u>Rps. acidophila</u> 7750	1:39 ± 7	27
<u>Rh. sphaeroides</u> R26	1:33 ± 13	10
<u>R. rubrum</u> S1	1:33 ± 9	48

This happens occasionally and we do not know the reason for it.
Table 2 summarises the results of our determination of the RC:Bchl*a* ratio in the various RC: antenna conjugates that we were able to prepare. In each case these determinations have been made on many separate preparations.
It can be seen from this table that within a given species the ratio of RC:Bchl*a* in these RC:antenna conjugates is rather constant. Some of the variation seen from species to species, undoubtedly reflects the lack of species-specific extinction coefficients. However allowing for this it is apparent that there does seem to be a constant stochiometry of about 1RC per 30 Bchl*a* molecules, irrespective of the species studied. Two species *Rps. acidophila* and *Rps. blastica* deviate significantly from this ratio. We have investigated why this should be so and the answer seems to lie in the sensitivity of these two conjugates to detergents. The B875-complexes present in the RC: antenna conjugate from *Rps. blastica* are rather unstable, even in the presence of the mild detergent triton X-100. The RC: Bchl*a* ratio in the *Rps. blastica* conjugate steadily declines upon storage after the initial centrifugation. This decline is clearly seen as a loss of the absorption of 875 nm. In contrast the antenna portion of the conjugate from *Rps. acidophila* is quite stable but the RC content decreases upon storage. In this case the RC is rather susceptible to destruction by the detergents. We feel therefore that these two species represent special cases and that their RC: Bchl*a* ratios deviate strongly from ~1:30 for reasons of instability. RC:Bch*a* ratios consistent with those presented here have also been reported by Ueda et al. (16) for RC:antenna conjugates isolated from *Rh. sphaeroides*, *R. rubrum* and *Chr. vinosum*.

Table 3 shows the results of some experiments where we investigated whether the RC:Bchl*a* ratio remained to same in the RC:antenna conjugates when the cells were grown at different light intensities. Under these different growth conditions the cells radically alter their specific pigment content (2).

Table 3

A comparison of the RC: Bchl*a* ratios for RC: antenna conjugates isolated from 'high' and 'low' light grown cells.

Species	Ratio of RC:Bchl*a* mole:mole (± 1SD)		No. of different experiments.
	High Light	Low Light	
Rps. palustris - French	1:23 ± 4	1:24 ± 3.5	7
Rps. acidophila - 7050	1:41 ±10.5	1:38 ± 9.5	27
Rc. gelatinosus - 149	1:34 ± 7	1:31 ± 10	13
R. rubrum - S1	1:28 ± 10	1:29 ± 5.5	10

It is quite clear that in all cases there is no significant difference in the RC:Bchl*a* ratio in the conjugates isolated from either 'high' or 'low' light grown cells. These results confirm the hypothesis put forward by Aagaard and Sistrom (2) and provide support for the idea that the 'core' complexes represent definite fixed structures.

Fig. 4 illustrates a typical trace of the fluorescence induction kinetics obtained with our RC: antenna conjugates. In this case the data are for the conjugate isolated from *R.rubrum*.

Upon illumination the level of fluorescence emitted by the antenna complexes in the conjugate gradually rises from F_o to F_{max}. The variable component of this fluorescence is removed by chemical oxidation of the RC and by 'incoupling' the RC from the antenna complexes by incubation with higher detergent concentrations. Only in the case of *Rps. acidophila* have we consistently failed to see any variable fluorescence. We

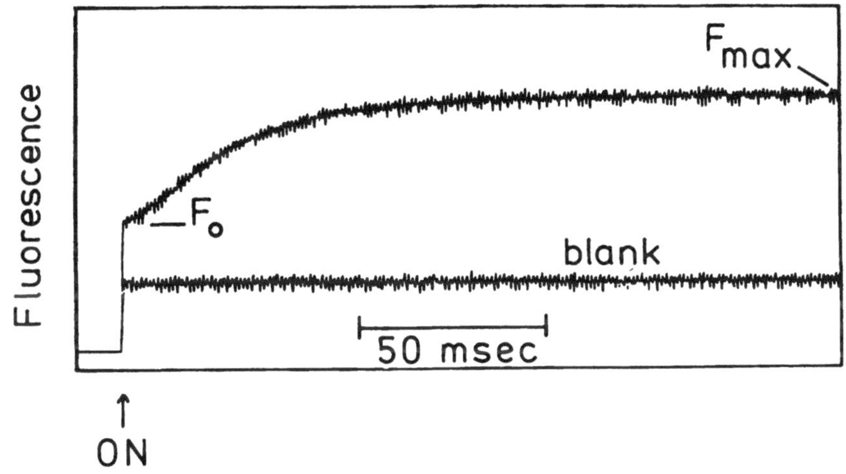

Fig. 4. The fluorescence induction kinetics seen upon excitation of the RC:antenna conjugate from R. rubrum

F_o - The instantaneous level of fluorescence reached immediately following the onset of illumination.

F_{max} - The final maximum level of fluorescence.

therefore conclude that in all the other cases, the antenna complexes are functionally connected to the RCs in the RC:antenna conjugates.

So far we have only really tried to crystallise the RC:antenna conjugate from R. rubrum. We have been successful with this conjugate using PEG 2000 as the precipitant and heptane-1,2,3-triol as the small amphiphile. Some typical crystals are shown in Fig. 5. It is worth noting in this case that the crystals were quite dichroic. Wacker et al. (17) have also crystallised the RC:antenna conjugate isolated from Rps. palustris. We feel that since it is clearly possible to crystallise these RC: antenna conjugates, it lends strong support to the idea that they have a fixed structure. In other words the constant RC: Bchl*a* ratio reflects the presence of a well-defined complex rather than a statistical average of a heterogenous population of complexes. It appears therefore that Bchl*a*-containing species have a similar 'core' structure in their photosynthetic units to the Bchl*b*-containing species.

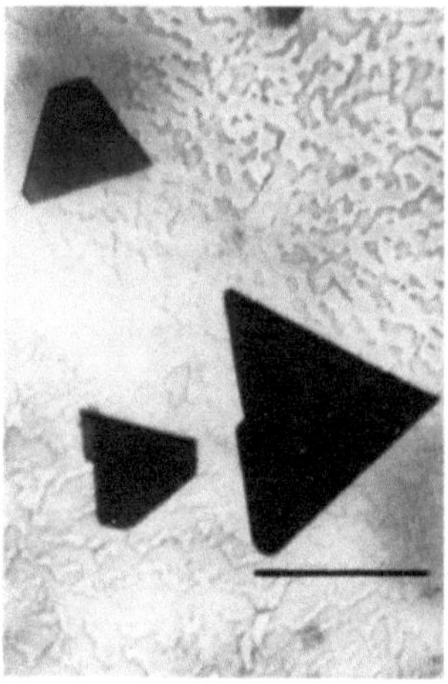

Fig. 5. Crystals of the RC: antenna conjugate from R. rubrum
The RC:antenna conjugate was made in a solution containing 8.5% PEG 2000 (w/v), 20mM $MgCl_2$, 10mM- Tris HCl pH8.0 and 0.04% LDAO and then concentrated by vapour diffusion against 18% PEG 2000, 20mM $MgCl_2$ and 10mM Tris-HCl pH 8.0 at 20°C. The crystals grew slowly over a period of 1-2 weeks. The bar represent 100µ.

Acknowledgements.

D.J.D. thanks the SERC for a postgraduate studentship.

References

1. Thornber, J.P., R.J. Cogdell, B.K. Pierson and R.E.B. Seftor. 1983. J. Cell. Biochem. 23, 159.

2. Aagaard, J. and W.R. Sistrom. 1972. Photochem. Photbiol. 15, 209.

3. Miller, K.R. 1982. Nature 300, 53.

4. Stark, W., W. Kühlbrandt, I. Wildhaber and K. Mühlethaler. 1984. EMBO J. 3, 777.

5. Engelhardt, H., R. Gukenberger, R. Hegerl and W. Baumeister. 1985. Ultramicroscopy 16, 395.

6. Pfennig, N. 1969. J. Bacteriol. 99, 597.

7. Bose, S.K. 1963. In: Bacterial Photosynthesis (H. Gest, A. San Pietro and L.P. Vernon, eds.) Antioch Press, Yellow Springs, Ohio, p. 501.

8. Firsow, N.N. and G. Drews. 1977. Arch. Microbiol. 115, 299.

9. Laemmli, U.K. 1970. Nature 227, 680.

10. DeMartini, M. and M. Inouye. 1978. J. Bacteriol. 133, 329.

11. Clayton, R.K. 1980. Photosynthesis: physical mechanisms and chemical patterns. Cambridge University Press. pp. 94

12. Clayton, R.K. 1963. In : Bacterial Photosynthesis (H. Gest, A. San Pietro and L.P. Vernon, eds.) Antioch Press, Yellow Springs. Ohio, p. 377.

13. Bell, D.H. and M.F. Hipkins 1984. In: Advances in Photosynthesis Research (C. Sybesma, ed.) Martinus Nijhoff/Dr. W.Junk, The Hague 2 p.113.

14. Michel, H. and D. Oesterhelt, 1980. Proc. Natl. Acad. Sci. USA. 77, 1283.

15. Garavito, R.M. and J.P. Rosenbusch. 1980. J. Cell. Biol. 86, 327.

16. Ueda, T.Y. Morimoto, M. Sato, T. Kakuno, J. Yamashita and T. Horio. 1985. J. Biochem. 98, 1487.

17. Wacker, T., N. Grad'on, A. Becker, W. Mantele, W. Kreutz, G. Drews and W. Welte. 1985. FEBS Lett. 197, 267.

A COMPARISON OF THE BACTERIOCHLOROPHYLL C--BINDING PROTEINS OF CHLOROBIUM AND CHLOROFLEXUS

P. D. Gerola

Department of Ecology, University of Calabria
I-87030 Arcavacata di Rende (Cosenza), Italy

P. Højrup

Institute of Molecular Biology, Odense University
Campusvej 55, DK-5230 Odense M, Denmark

J. M. Olson

Institute of Biochemistry, Odense University
Campusvej 55, DK-5230 Odense M, Denmark

Introduction

In green bacteria the main light-harvesting pigment (BChl c, d, or, e) organized inside oblong bodies (chlorosomes) lying in the cytoplasm and adhering to the cytoplasmic membrane, where the reaction centers are localized.

In Chlorobiaceae the BChl c molecules appear to be associated with 15-kDa proteinaceous subunits organized hexagonally in 10-nm diameter rod elements running the length of each chlorosome (1). The path of excitation energy transfer is from BChl c to Bchl a inside the chlorosome(2), then to BChl a-protein between the chlorosome and the cytoplasmic membrane and finally to the BChl a antenna and/or the relation centers embedded in the membrane.

When chlorosomes were characterized by polyacrylamide gel electrophoresis (PAGE), there was one major band at approx.

Photosynthetic Light-Harvesting Systems

4-5 kDa (3). It was the only band observed after Coomassie brilliant blue staining. Silver staining revealed the presence of six minor bands at about 9, 20, 26, 27, 32 and 38 kDa (M_r). Since the BChl c-binding protein was the major protein of the chlorosome, it was assumed to be a constituent of the major band (4-5 kDa, M_r) seen after PAGE.

Recently Wechsler et al. (4) determined the amino acid (aa) sequence of the BChl c-binding protein from chlorosomes of the green filamentous bacterium Chloroflexus aurantiacus, which is only distantly related to Chlorobium limicola (S_{AB} = 0.2) (5,6). In Chloroflexus the BChl c molecules have stearyl tails, whereas in Chlorobium the tails are farnesyl. Furthermore the rod elements in Chloroflexus are only 5.2 nm in diameter compared to 10 nm in Chlorobium. Nevertheless both kinds of chlorosomes, in addition to BChl c, contain small amounts of a special BChl a (792 nm in Chloroflexus and 794 nm in Chlorobium) probably localized in the lipid monolayer covering the chlorosome. The BChl c absorption spectra in the two kinds of chlorosomes are quite similar, but differences are observed in the circular dichroism spectra (7,8). We wanted to compare the primary and secondary structures of the two BChl c-binding proteins in view of the large evolutionary distance between Chlorobiaceae and Chloroflexaceae and the apparent similarity between the BChl c antenna systems in these two types of organism.

Materials and Methods

Cb. limicola f. thiosulfatophilum 6230 (Tassajara) cells were broken and chlorosomes were prepared as described previously (3, 9).

The chlorosomes were extracted with 80% acetone, and the residue was dissolved in 0.1% trifluoroacetic acid in isopropanol-water (1:1). The dissolved residue was loaded on a Lichrosorb RP8 column which was equilibrated with 0.1% (v/v) trifluoroacetic acid and 50% (v/v) isopropanol in water. The protein was eluted by increasing the percentage of isopropanol from 50% to 82%. The BChl c-binding protein was eluted from the column at about 77% isopropanol.

The protein was dissolved in 0.1 M NH_4HCO_3 and 1 mM $CaCl_2$, pH 8.0. Trypsin or chymotrypsin was added, and the digestion was allowed to proceed for two hours at 37°C. Separation of the peptides was carried out on a reverse phase ODS column. The column was equilibrated with 98% buffer A (0.1% trifluoroacetic acid in water) and 2% buffer B (0.1% trifluoroacetic acid in 96% ethanol). The peptides were eluted with a gradient of 1.5% buffer B/min at 50°C.

For sequencing the intact protein, 30 nmole was applied to a Beckman 890C spinning-cup sequencer. This determined the first 54 residues with a few ambiguities. Tryptic and chymotryptic peptides were sequenced on an Applied Biosystems gas-phase sequencer using the ZNVAC program supplied by the manufacturer. For all samples the molecular weight was determined by plasma desorption mass spectrometry.

The molecular weight of the intact protein was determined on a Bio-Ion 10K plasma-desorption time-of-flight mass spectrometer. The sample was dissolved in 5 µl of 0.1% trifluoroacetic acid in water and applied to an aluminium Mylar foil coated with nitrocellulose. After a brief wash with 0.1% trifluoroacetic acid the sample was inserted in the mass spectrometer and was bombarded with fission fragments from a 10 µCi ^{252}Cf source.

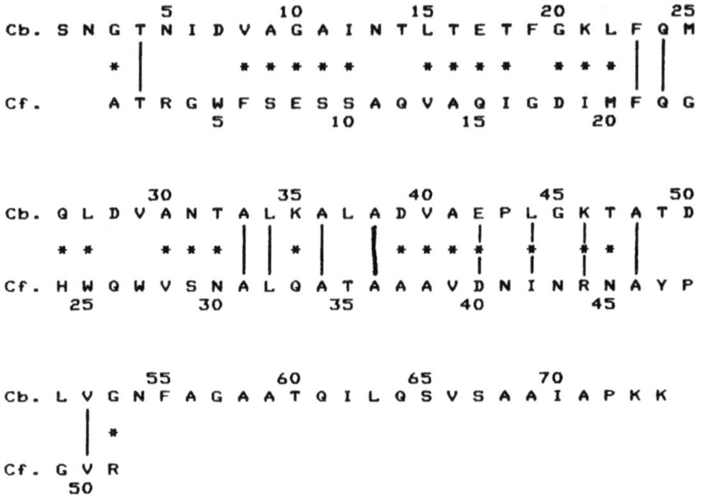

Fig. 1. Amino acid sequences of BChl *c*-binding proteins from Chlorobium (Cb) and Chloroflexus (Cf). The homologies are indicated by solid vertical lines. Functional similarities are indicated by broken lines. Amino acid differences which can be explained by single mutation events are indicated by asterisks.

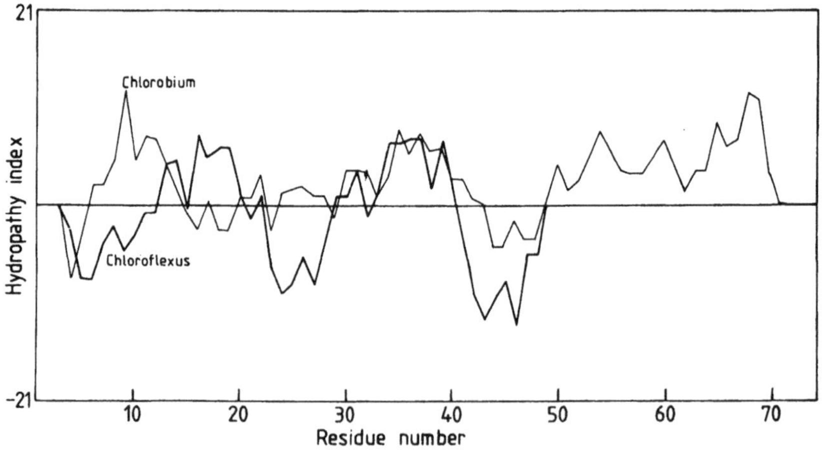

Fig. 2. Hydropathy plots (10) for the BChl *c*-binding proteins from Chlorobium (light line) and Chloroflexus (heavy line). The average hydropathy was determined within a 7-aa running segment.

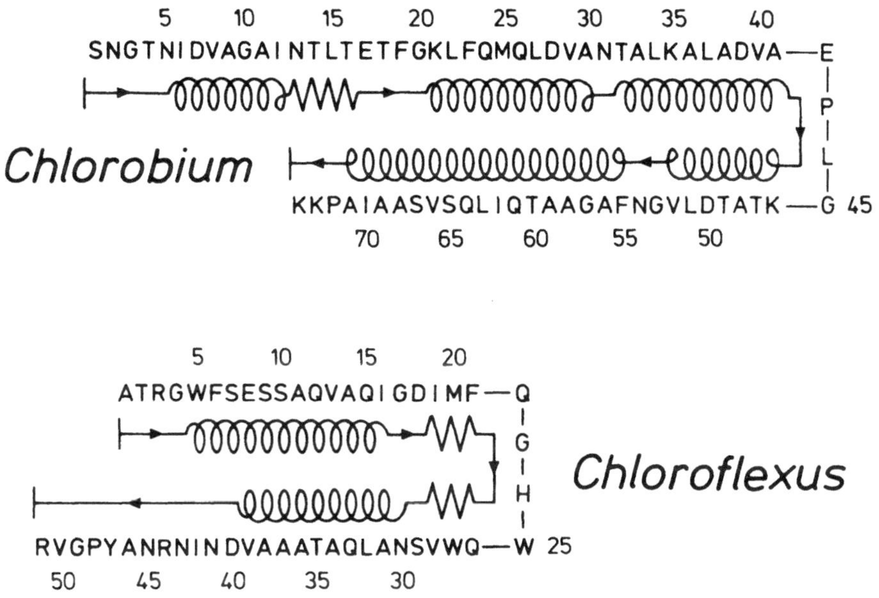

Fig. 3. Secondary structures predicted according to Chou and Fasman (11,12) for the BChl c-binding proteins from Chlorobium and Chloroflexus.

Results and Discussion

The molecular mass of the BChl c-binding protein was found to be 7474 ± 3 compared to 5592 for the Chloroflexus protein (4). In Fig. 1 the sequences of the Chlorobium (9) and Chloroflexus (4) BChl c-binding proteins are aligned for maximum homology. In Fig. 2 the hydropathy plots (10) of the two proteins are shown, and in Fig. 3 the secondary structures for the two proteins have been tentatively predicted according to the method of Chou and Fasman (11,12).

The BChl c-binding proteins from Chlorobium and Chloroflexus are quite different in length (74 and 51 aa respectively), and that might explain the difference in the rod element diameters in the chlorosomes of the two organisms. Both proteins are characterized by a large number of amide-containing residues and both lack Cys; in the Chlorobium protein Tyr, Trp, His and Arg are also missing, while Lys is missing in the Chloroflexus protein. The number of aromatic residues is higher in the Chloroflexus protein than in the Chlorobium protein. The differences in protein length and aa sequence might be expected because of the large evolutionary distance between the two organisms (5, 6). However most of the aa sequence differences can be explained by single mutation events (asterisks in Fig. 1). In light of so little overall homology between the two proteins, the high homology (30%) in the region between Chlorobium residues 31 and 43 is particularly interesting from a functional point of view. This aspect is reinforced by the hydropathy plots (7-aa running segment) for the two proteins shown in Fig. 2. The two profiles are very similar from Chlorobium residues 31 through 49, but are completely different from Chlorobium residues 4 through 29. (If the hydropathy indices for individual aa residues are considered, the two proteins are quite similar from Chlorobium residue 30 to residue 50.)

Wechsler et al. (4) have proposed a model for the secondary structure of the BChl c-binding protein from Chloroflexus which was stated to be based on the method of Chou and Fasman (11, 12). The model consists of an α-helix between Trp 5 and Ile 42 with 7 amide-containing residues proposed as BChl c-binding sites: Gln 12, 15, 22, 26, 33 and Asn 30 and 41. However the Chou-and-Fasman method for predicting secondary structures from aa sequences does not support the model proposed by Wechsler et al. (4). The analysis of Gerola et al. (9) indicates the presence of 11 tetrapeptides which are α-helix breakers ($P\alpha < 1$) in the Trp 5 - Ile 42 region. Moreover the 6-aa segments 21-26, 22-27, 23-28, 25-30, have $P_{\alpha} < 1$, and all the 6-aa segments from 12-17 up to 27-32 have $P_{\alpha} < P_{\beta}$. It is not possible according to the method of Chou and Fasman (11,12) to predict an unbroken α-helix between Trp 5 and Ile 42. A similar conclusion is reached when the method of Garnier et al. (13) is used to predict the secondary structure.

Secondary structure predictive methods based only on the primary structure are not very reliable when applied to proteins like the BChl c-binding proteins. The predictive methods are based on known secondary structures of water-soluble globular proteins, while the BChl c-binding proteins are in hydrophobic environments, and moreover the BChl c molecules bound to the proteins probably influence their secondary structures. However if one tries to predict the secondary structure for the BChl c-binding protein of Chloroflexus according to the method of Chou and Fasman, an α-helix region is clearly predicted from Ala 31 to Val 39 and possibly from Trp 5 or Ala 11 to Ile 16 as shown in Fig. 3. A β-sheet region should be present from Ile 19 to Val 28, with a possible turn in the region 22-25 (9).

More difficult is the prediction for the Chlorobium protein, as aa sequences with high P_α and P_β values overlap in several regions. As shown in Fig. 3 an α-helix region is clearly predicted from Thr 32 to Glu 42, while other α-helix regions might be present between Ile 6 and Ile 12, Lys 21 and Ala 30, Lys 46 and Val 52 and between Phe 55 and Ala 71. A β-sheet region might be predicted from Asn 13 to Thr 16, and a turn between Glu 42 and Gly 45 (9). In Fig. 4 helical wheels are drawn according to Schiffer and Edmundson (14) representing the α-helix regions from Ala 31 to Val 39 of the Chloroflexus protein and from Lys 21 to Ala 30 and from Thr 32 to Glu 42 for the Chlorobium protein. On each wheel all the hydrophobic residues lie on one side, while almost all polar and charged residues are localized on the other side. There are two exceptions: Gln 33in the Chloroflexus protein and Gln 26 in the Chlorobium protein. Each of these charged residues is on

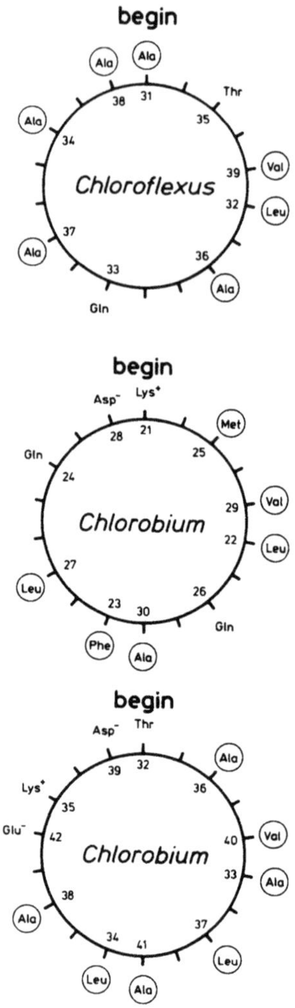

Fig. 4. Helical wheels drawn according to Schiffer and Edmundson (14) representing the α-helix regions from Ala 31 to Val 39 in the protein from Chloroflexus and from Lys 21 to Ala 30 and from Thr 32 to Glu 42 in the protein from Chlorobium.

the hydrophobic side of its wheel. Because of their locations these two amide-containing residues are likely candidates for BChl c-binding sites. Also the other α-helix regions predicted by the Chou-and-Fasman method, when represented as helical wheels, show distinct hydrophobic arcs, and the hydrophobic arcs of the two α-helix regions (Lys 21-Ala 30 and Thr 32-Glu 42) in Chlorobium lie on the same side of the protein, as can be seen by constructing an unbroken α-helix wheel from Lys 21 to Glu 42. According to the Chou-and-Fasman method α-helix is predicted for both proteins from Chlorobium residues 33 to 41. This is the only region in the two proteins with similar profiles of α-helix potential values of 6-aa running segments (9). It is also the protein region showing the highest homology (Fig. 1) and similar hydropathy profiles (Fig. 2). These facts suggest a functional role for that protein region, i.e., involvement in binding BChl c molecules. In their model for the Chloroflexus protein Wechsler et al. (4) proposed a rod-like protein with 75% of the residues forming a long (57 Å) α-helix in which 7 Gln/Asn residues were proposed as binding sites for BChl c molecules. However, taking into account the P_β values and the presence of tetrapeptide α-helix breakers, a long α-helix does not agree with our analysis (9) according to the method of Chou and Fasman (11, 12). The presence of turns and bends in both the Chlorofelxus and Chlorobium proteins suggests that both proteins may have somewhat globular tertiary structures.

Wechsler et al. (4) proposed a chlorosome subunit composed of 6 dimers of BChl c-binding proteins (12 polypeptide chains) binding 84 BChl c molecules. The molecular mass (7.5 kDa) of the BChl c-binding polypeptides in Chlorobium might permit a dimeric organization of that protein also. According to the model proposed by Olson (1) the six subunits forming each ring in the rod element are made up of the BChl c-binding protein. Unfortunately the size estimation of ca. 15 kDa (1) of these subunits based on electron microscopy is not precise enough to distinguish between 1 or 2 polypeptides in each subunit. If each subunit consists of 1 polypeptide, it should bind 6 ± 1 BChl c molecules, but if there are 2 polypeptides in each subunit, it should bind 12 ± 2 BChl c molecules. It is questionable whether there is enough room for 12 ± 2 BChl c molecules per subunit in the rod element of Chlorobium. Experiments with crosslinking agents are needed to resolve this matter.

References

1. Olson, J.M. 1980. Biochim. Biophys. Acta 594, 33.

2. van Dorssen, R., P.D. Gerola, J.M. Olson, J. Amesz. 1986. Biochim. Biophys. Acta 848, 77.

3. Gerola, P.D., J.M. Olson. 1986. Biochim. Biophys. Acta 848, 69.

4. Wechsler, T., F. Suter, R.C. Fuller, H. Zuber. 1985. FEBS Lett. 181, 173.

5. Gibson,J., W. Ludwig, E. Stackebrandt, C.R. Woese. 1985. System. Appl. Microbiol. 6, 152.

6. Woese, C.R. 1987. Microbiol. Rev. 51, 211.

7. Betti, J.A., R.E. Blankenship, L.V. Natarajan, L.C. Dickinson, R.C. Fuller. 1982. Biochim. Biophys. Acta 680, 194.

8. Olson, J.M., P.D. Gerola, G.H. van Brakel, R.F. Meiburg, H. Vasmel. 1985. In: Antennas and Reaction Centers of Photosynthetic bacteria (M.E. Michel-Beyerle, ed.). Springer-Verlag, p. 67.

9. Gerola, P.D., P. Højrup, J. Knudsen, P. Roepstorff, J.M. Olson. 1988. In: Green Photosynthetic Bacteria (J.M. Olson, ed.) Plenum Press (in press).

10. Kyte, J., R.F. Doolittle. 1982. J. Mol. Biol. 157, 105.

11. Chou, P.Y., G.D. Fasman. 1978. Adv. Enzymol. Relat. Areas Mol. Biol. 47, 45.

12. Chou, P.Y., G.D. Fasman,. 1979. Biophys. J. 26, 367.

13. Garnier, J., D.V. Osgathorpe, B. Robson. 1978. J. Mol. Biol. 120, 97.

14. Schiffer, M., A.B. Edmundson. 1967. Biophys. J. 7, 121.

INTERACTIONS BETWEEN BACTERIOCHLOROPHYLL C MOLECULES IN OLIGOMERS AND IN CHLOROSOMES OF GREEN PHOTOSYNTHETIC BACTERIA

Daniel C. Brune, George H. King and Robert E. Blankenship
Department of Chemistry, Arizona State University, Tempe, AZ 85287-1604 USA

Introduction

Bacteriochlorophyll (BChl) c is the major pigment in chlorosomes of the thermophilic green gliding bacterium Chloroflexus aurantiacus. Chlorosomes are membrane-attached antenna structures approximately 30 nm wide, 110 nm long, and 12 nm thick (1). BChl c in chlorosomes is highly ordered, with its Q_y electronic transition dipole making an angle of about 37° relative to the long axis of the chlorosomes (2,3). The red and Soret maxima of BChl c in chlorosomes occur at 740 and 460 nm rather than at about 665 and 435 nm, the values observed for monomeric BChl c in polar organic solvents.

The structure of BChl c is particularly suited for self-association (4-6). Each molecule has a central Mg atom able to accept 1 or 2 ligands on either or both sides of the chlorin macrocycle in addition to the 4 nitrogens within the macrocycle. Each molecule also contains a 2a-hydroxyl group and a 9-keto group, either or both of which may ligate the Mg of an adjacent molecule. Unlike most other chlorophylls (but like the related chlorosome pigments BChl d and BChl e and like pyrochlorophyll a), BChl c lacks a 10-carbomethoxy group that could sterically interfere with interactions between the 9-keto group and adjacent molecules.

Several previous reports have shown that BChl c from Chlorobium limicola or the related green sulfur bacterium Prosthecochloris aestuarii spontaneously self-associates in dry, nonpolar solvents to form oligomers with spectral features very similar to those of BChl c in chlorosomes from

Photosynthetic Light-Harvesting Systems

those organisms (5-9). Similar results have been reported for BChl c from Chloroflexus aurantiacus (10,11). The purpose of the work presented here is to examine the self-association of BChl c in different organic solvents.

In addition to monomeric BChl c, which occurs in polar organic solvents, self-associated forms with absorption maxima at 680 nm, 710 nm, and 740 nm are observed in solvents of low polarity. Pyrochlorophyll (PyChl) a, which differs from BChl c most significantly in having a vinyl instead of an α-hydroxyethyl group at the 2 position, forms only a 680 nm oligomer. Visible and Fourier transform infrared (FTIR) spectral data on these oligomers are presented and previously proposed structures for the 740 nm oligomer are discussed. In addition, two possible new structures for this oligomer are presented for consideration.

Materials and Methods

Chloroflexus aurantiacus J-10-fl was cultivated at 55°C in the medium of Pierson and Castenholz (12). Chlorosomes were prepared using the procedure of Feick and Fuller (13). BChl c was extracted and purified as described by Brune et al. (10). Chl a was extracted from spinach and converted to PyChl a by standard methods (14). Absorption spectra were recorded on a Cary 219 spectrophotometer or on an instrument constructed as described previously (15). The solvents used in some experiments were dried by storing over molecular sieves (EM Science, Type 4A), and, in the case of CH_2Cl_2, $NaHCO_3$ was also added to remove any HCl liberated by solvent decomposition. FTIR spectra were obtained using a Nicolet MX-1 spectrometer. The chlorosome sample was dried on the inner surface of a CaF_2 plate that formed one face of a demountable liquid sample cell. Samples in organic solvents were placed inside a 0.1 mm pathlength KBr cuvette (Aldrich Chemical Co.).

Results

Figure 1 (a-d) shows the absorption spectra of monomeric BChl c and of the species absorbing maximally at 680 nm, 710 nm, and 740 nm. Increasingly red-shifted aggregates were formed as the solvent polarity decreased. Thus the 680 species (probably a dimer) formed best in dry CH_2Cl_2. The 710 nm aggregate formed in slightly less polar solvents (CCl_4 and CS_2) and the extent of its formation increased as hexane was mixed with either of these solvents. The mixture of 60% hexane-40% CS_2 used to obtain the spectrum shown in Figure 1b was optimal for formation of the 710 nm species at the BChl c concentration used in these experiments (about 10 µM). Increasing the amount of hexane in the mixture led to formation of the 740 oligomer, with essentially all of the BChl c incorporated into this oligomer in pure hexane.

Increasing pigment concentration had an effect similar to decreasing solvent polarity. Thus at a concentration 100 x that used in Fig. 1b, substantial amounts of the 710 species as well as the 680 nm dimer formed in dry CH_2Cl_2. Likewise, at a 100 x higher BChl c concentration, most of the BChl c in pure CCl_4 existed as the 740 nm oligomer.

PyChl a, in contrast to BChl c, self-associated only to form a species absorbing maximally at 680 nm, even when dissolved at high concentrations in hexane or heptane (Figure 1f). Previous reports have noted that, like Chl a, PyChl a can form a hydrated oligomer absorbing maximally at 740 nm in wet nonpolar solvents (16). However, even without taking specific precautions to dry the hexane or heptane solvents, this 740 nm oligomer was never observed. This result implies a substantial difference between the oligomerization reactions of BChl c and PyChl a, and it seems reasonable to attribute this difference to the presence of the 2-hydroxyethyl group in BChl c in place of the 2-vinyl group of PyChl a. Several other groups have also noted that BChl c forms 740-750 nm oligomers in dry nonpolar solvents while Chl a (and PyChl a) do not (5-7,17).

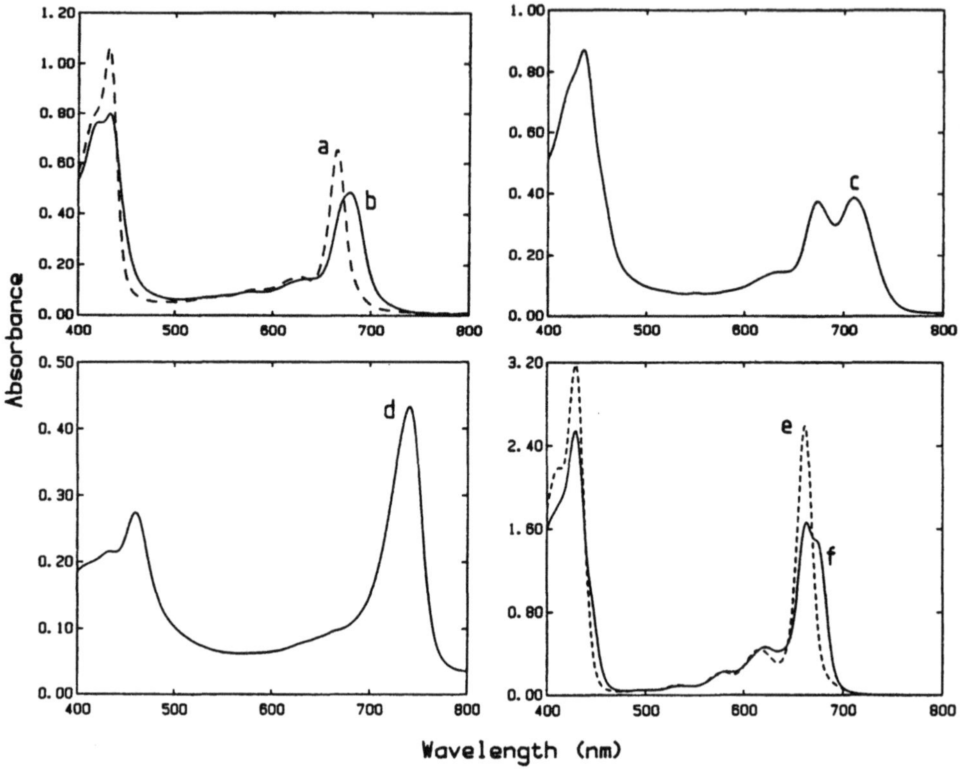

Figure 1. Absorption spectra of BChl c (a-d) and PyChl a (e, f). a. Monomeric BChl c in CH_2Cl_2 + 0.3% CH_3OH. b. 680 nm (dimeric) species of BChl c in dry CH_2Cl_2. c. 710 nm BChl c species in 60% hexane, 40% CS_2. d. 740 nm BChl c oligomer in hexane. e. Monomeric PyChl a in heptane + 0.3% CH_3OH. f. 680 nm species of PyChl a in heptane.

The coincidence in the absorption maxima of PyChl a in hexane and BChl c in CH_2Cl_2 does not necessarily imply that the aggregates formed are structurally identical. Smith and coworkers (18) concluded that the methyl bacteriochlorophyllide d species analogous to our BChl c oligomer in CH_2Cl_2 is a dimer in which the 2a-hydroxyl group of each molecule ligates the Mg atom of the other molecule. As PyChl a lacks this hydroxyl group, that structure is not possible for PyChl a. Katz et al. (17) have presented a detailed discussion of the structure of the analogous 680 nm-absorbing Chl a species that forms in dry nonpolar solvents.

An attempt was made to examine the 680 nm dimer of BChl c in CH_2Cl_2 by FTIR spectroscopy. Unfortunately, the BChl c concentration required for this experiment was sufficiently high that an absorption spectrum revealed the presence of both the 680 nm and the 710 nm species as well as some monomeric BChl c. The FTIR spectrum of this sample exhibited a small peak at 1700 cm^{-1}, a shoulder at 1680 cm^{-1}, and larger maxima at 1735 and 1650 cm^{-1} (not shown). The peak at 1700 cm^{-1} suggests participation of the 7-propionic ester carbonyl group in pigment-pigment interaction in either the 680 or the 710 nm oligomer (or in both) while the peak at 1650 cm^{-1} implicates the 9-keto group. After converting all of the BChl c to the monomeric form by adding 0.2% pyridine, the carbonyl absorbances were at 1735 and 1680 cm^{-1}, values characteristic of the free ester and 9-keto groups, respectively. While these results are promising, the relative amounts of the 680 and 710 nm species of BChl c in the mixture was unknown, thus precluding any conclusions about the structure of either.

An FTIR difference spectrum of a concentrated PyChl a sample in hexane before and after adding pyridine revealed only a 1654 to 1685 cm^{-1} band shift (Figure 2). This type of shift has been observed before for PyChl a (4), and was attributed to disruption of 9-keto-Mg interactions. Absorption spectra taken before and after adding pyridine were similar to Figure 1 f and e, respectively.

Figure 3 shows an FTIR difference spectrum obtained when 0.1% pyridine was added to a concentrated solution of BChl c in CCl_4, converting the oligomeric form absorbing at 740 nm to a monomeric form absorbing at 670 nm. The most pronounced effect of this addition was the shift of a 1650 cm^{-1} peak to 1687 cm^{-1}, implicating the 9-keto group in oligomer formation. Also shown in Figure 3 is a difference spectrum obtained by exposing a dried chlorosome film to pyridine vapor. This treatment also converted the 740 nm-absorbing species of BChl c to a 670 nm-absorbing (monomeric) form and produced a shift in the frequency of the 9-keto group from 1650 to 1680 cm^{-1}. The slight difference between the positions of the 9-keto absorbances in CCl_4 and in the dried chlorosome film after adding pyridine probably reflects a slight difference in the environment

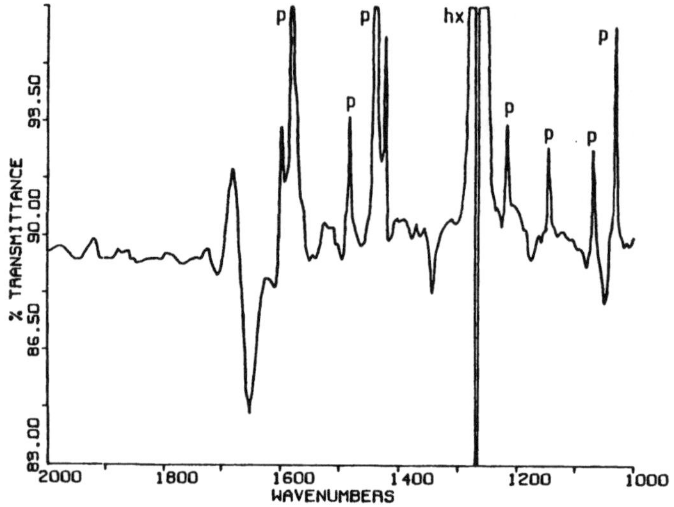

Figure 2. FTIR difference spectrum for PyChl a in hexane. The spectrum of the sample after adding 0.2% pyridine was substracted from that of the sample without pyridine. Peaks marked "p" are due to pyridine; "hx" denotes a strong hexane solvent peak that was incompletely subtracted out.

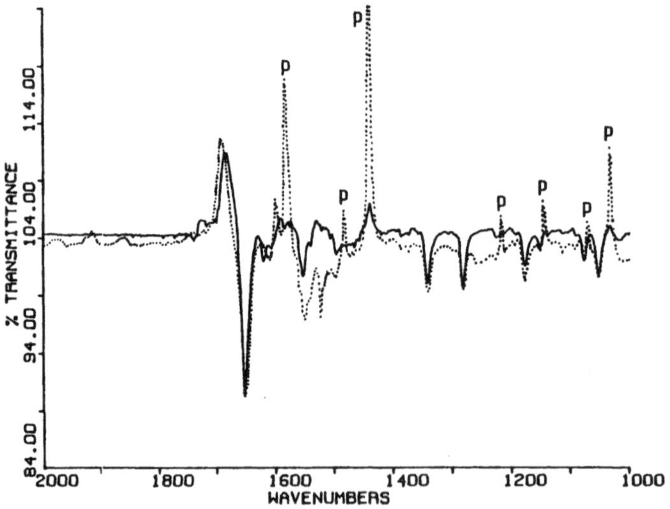

Figure 3. FTIR difference spectra for dried chlorosomes (solid line) and for BChl c in CCl_4 (dotted line). In both cases, the spectrum of the pyridine-treated sample was subtracted from that of the untreated sample. Peaks marked "p" are due to pyridine.

of monomeric BChl c in the two samples. Similar difference spectra comparing a dried film of oligomeric BChl c with a dried chlorosome film were presented previously by Blankenship et al. (11), and in those spectra, both samples exhibited 9-keto absorbances at 1680 cm^{-1} after exposure to pyridine vapor. In addition to the shift in position of the intense 9-keto absorbance, spectral bands at 1553, 1342, 1281, 1176, and 1048 cm^{-1} exhibited decreases in absorbance upon oligomer disruption that remain to be interpreted.

Resonance Raman (RR) data presented by Lutz and van Brakel (9) also demonstrated that the 9-keto vibrational frequency for BChl c in dry CCl_4 or in chlorosomes (1638-1640 cm^{-1}) was different from that for BChl c in polar solvents such as pyridine and tetrahydrofuran (1685-1690 cm^{-1}), and Nozawa and Tasumi (personal communication) have obtained similar results. The positions for the keto carbonyl resonances, particularly for BChl c in the 740 nm oligomer and in chlorosomes, differ somewhat from the values observed in our FTIR experiments. The reason for this discrepancy is unknown, although it may be related to the fact that the RR samples were cooled (to 20 K in the experiments of Lutz and van Brakel) while ours were at room temperature.

Discussion

It is clear from the results presented here that formation of the 740 nm BChl c oligomer both requires the 2-hydroxyethyl group and involves the 9-keto group. To account for both of these factors, Blankenship et al. (11) advocated an asymmetric 6-coordinate model for the 740 nm oligomer in which the Mg atom of each BChl c is ligated on opposite sides by the 9-keto and the 2-hydroxyethyl groups of adjacent molecules (Fig. 4a). In this model, which is essentially the same as that proposed earlier by Bystrova et al. (5), the out-of-plane position of the Mg atom indicated by the RR data (Nozawa and Tasumi, pers. commun., and ref. 9) is accounted for by the inequality of the ligands on opposite sides of the chlorin ring. Lutz and van Brakel (9) have argued that the close agreement

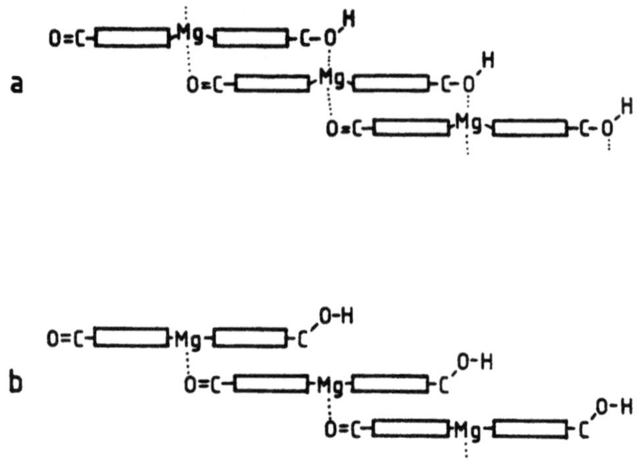

Figure 4. a. Asymmetric 6-coordinate model for the 740 nm oligomer of BChl *c* suggested by Blankenship *et al.* (11). b. 5-coordinate model involving only 9-keto coordination proposed by Lutz and van Brakel (9).

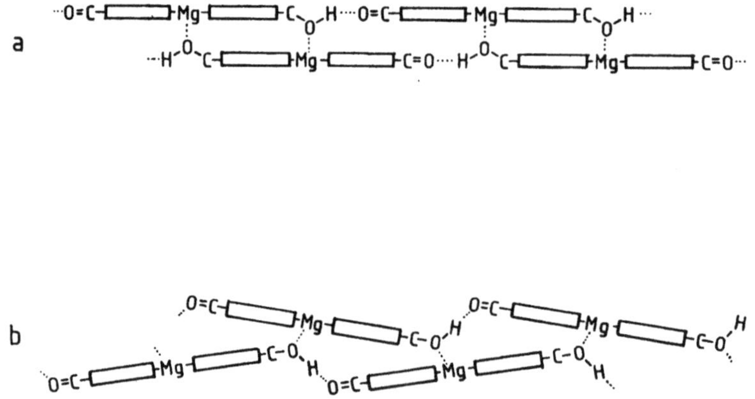

Figure 5. Structural models for the 740 nm BChl *c* oligomer with 5-coordinate Mg but involving both the 2-hydroxyethyl and 9-keto groups. a. Antiparallel chain model. b. Stepped, parallel chain model.

between the vibrational modes in the chlorin macrocycle of 5-coordinate BChl c in solution and of BChl c in the 740 nm oligomer and in chlorosomes implies that oligomeric and chlorosomal BChl c must be 5-coordinate. They proposed a structure for the 740 nm oligomer in which the BChl c monomers are linked only by ligation of Mg by the 9-keto group (Figure 4b).

Even accepting their argument that Mg atom of BChl c is 5-coordinate, we feel that their model is inadequate because it fails to account for the observed requirement for the 2-hydroxyethyl group. Thus their model implies that PyChl a should also form a 740 nm oligomer in dry nonpolar solvents, which it clearly does not. For this reason we would like to suggest two new structural models incorporating both the 2-hydroxyethyl group and the 9-keto group. In the first of these (Figure 5a), the BChl c molecules are arranged in antiparallel chains, with the molecules in each chain linked by hydrogen bonds between the 2a-hydroxyl and the 9-keto groups. BChl c molecules in opposite chains are attached by hydroxyl ligation of the central Mg atoms. This structure can also be derived by linking up dimers having the structure proposed by Smith et al. (18) via hydroxyl-to-keto hydrogen bonds. Hydroxyl groups on either (or both) end(s) might be capped by hydrogen bonding to a folded back 7-propionic ester group. In the second structural model (Figure 5b), the interacting BChl c molecules are arranged in parallel, stepped chains. Each BChl c molecule is linked to the next by hydroxyl-to-Mg ligation and to the one after the next by hydroxyl-to-9-keto hydrogen bonding.

While it is premature to speculate which of these structural models is more appropriate, both are consistent with the available data. Lutz and van Brakel (9) noted that BChl c dissolved in alcohol solvents (e.g. phenyl-1-ethanol) exhibited a smaller shift in the position of the 9-keto resonance (to 1657 cm^{-1}) than was observed in the 740 nm oligomer and in chlorosomes (to 1640 cm^{-1}) and used this as an argument against hydrogen bonding to the 9-keto group being involved in oligomer formation. However, it should be pointed out that a Mg-ligated hydroxyl group is not the same as a bulk solvent hydroxyl group. Thus Ballschmiter and Katz (19) found that the 9-keto vibration was shifted to 1640 cm^{-1} in the hydrated 740 nm Chl *a* oligomer in which crystallographic studies have shown that the Chl *a* molecules are joined by hydrogen bonding between the 9-keto group of one molecule and a Mg-ligated water molecule on the next (16). This shift is in fact larger than that (to 1655 cm^{-1}) caused by direct binding of the keto group of one molecule to the Mg of another (20). A similar, but smaller, shift was observed in Chl *a* dimers joined by hydrogen bonding of the 9-keto group to a ligated ethanol molecule (21). In this case, the vibrational frequency of the hydrogen bonded keto group was shifted to either a slightly higher or slightly lower value than that of a Mg-ligated keto group, depending on the solvent used.

In conclusion, the results presented here support the suggestion that BChl *c* in chlorosomes exists as an oligomer similar in structure to the 740 nm oligomer that forms spontaneously in hexane. Both the 2-hydroxyethyl group and the 9-keto group of BChl *c* are involved in forming that oligomer. The two new structural models presented here account for the participation of both groups and also allow the BChl *c* Mg atom to be pentacoordinated, as suggested by the RR data.

Acknowledgements

We would like to thank Gilbert Seely for donating the PyChl *a* used in these experiments, Marc Lutz for sending us a preprint of reference 9, and Tsunenori Nozawa for sending us preliminary RR data on BChl *c* and chlorosomes. This work was supported by a grant from the United States Department of Energy Biological Energy Storage Program (DE-FG02-85ER1388).

References

1. Staehelin, L.A., J.R. Golecki, R.C. Fuller, G. Drews. 1978. Arch. Microbiol. 119, 269.
2. Betti, J.A., R.E. Blankenship, L.V. Natarajan, L.C. Dickinson, R.C. Fuller. 1982. Biochim. Biophys. Acta 680, 194.
3. van Dorssen, R.J., H. Vasmel, J. Amesz. 1986. Photosynth. Res. 9, 33.

4. Katz, J.J., R.C. Dougherty, L.J. Boucher. 1966. In: The Chlorophylls (L.P. Vernon and G.R. Seely, eds.), Academic Press, New York, p. 185.

5. Bystrova, M.I., I.N. Mal'gosheva, A.A. Krasnovskii. 1979. Mol. Bio. 13 582.

6. Smith, K.M., L.A. Kehres, J. Fajer. 1983. J. Am. Chem. Soc. 105, 1387.

7. Olson, J.M., P.D. Gerola, G.H. van Brakel, R.F. Meiburg, H. Vasmel. 1985. In: Antennas and Reaction Centers of Photosynthetic Bacteria (M.E. Michel-Beyerle, ed.) Springer Verlag, Berlin, p. 67.

8. Worcester, D.L., T.J. Michalski, J.J. Katz. 1986. Proc. Natl. Acad. Sci. USA 83, 3791.

9. Lutz, M., G. van Brakel. 1987. In: Proc. EMBO Workshop on Green Photosynthetic Bacteria (J.M. Olson, ed.) Plenum Publishing Co. (in press).

10. Brune, D.C., T. Nozawa, R.E. Blankenship. 1987. Biochem. (in press).

11. Blankenship, R.E., D.C. Brune, J.M. Freeman, G.H. King, J.D. McManus, T. Nozawa, J.T. Trost, B.P. Wittmershaus. 1987. In: Proc. EMBO Workshop on Green Photosynthetic Bacteria (J.M. Olson, ed.) Plenum Publishing Co. (in press).

12. Pierson, B.K., R.W. Castenholz. 1974. Arch. Microbiol. 100, 283.

13. Feick, R.G., R.C. Fuller. 1984. Biochem. 23, 3693.

14. Pennington, F.C., H.H. Strain, W.A. Svec, J.J. Katz. 1964. J. Am. Chem. Soc. 86, 1418.

15. Brune, D.C., G.H. King, A.I. Infosino, T. Steiner, M.L.W. Thewalt, R.E. Blankenship. 1987. Biochem. (in press).

16. Kratky, C., J.D. Dunitz. 1977. J. Mol. Biol. 113, 431.

17. Katz, J.J., L. L. Shipman, T.M. Cotton, T.R. Janson. 1978. In: The Porphyrins, Vol. 5 (D. Dolphin, ed.) Academic Press, New York, p. 401.

18. Smith, K.M., F.W. Bobe, D.A. Goff, R.J. Abraham. 1986. J. Am. Chem. Soc. 108, 1111.

19. Ballschmiter, K., J.J. Katz. 1969. J. Am. Chem. Soc. 91, 2661.

20. Lutz, M. 1984. Adv. Infrared Raman Spectrosc. 11, 211.

21. Cotton, T.M., P.A. Loach, J.J. Katz, K. Ballschmiter. 1978. Photochem. Photobiol. 27, 735.

LIGHT-HARVESTING COMPLEXES OF CHLOROPHYLL C-CONTAINING ALGAE

A.W.D. Larkum

School of Biological Sciences, University of Sydney, NSW 2006, Australia.

R.G. Hiller

School of Biological Sciences, Macquarie University, N.Ryde, NSW 2113, Australia.

Introduction

Chl c-containing algae, representing the most diverse group of eukaryotic algae, make up the bulk of the phytoplankton in the world's oceans and are extremely important for marine productivity. Some authorities have grouped these algae into one division, the Chromophyta, with some 9 classes (1), not all of which contain Chl c. Others prefer to treat each major group as a division or phylum (2). We adopt the latter approach here, considering the Chromophyta as a superdivision. We also consider the Prymnesiophyta as a division separate from the Chrysophyta. Recently a small group of algae within the Prasinophyceae (Chlorophyta), has been shown to contain a chlorophyll c-like pigment and a light-harvesting chlorophyll-protein (LHC) different to the LHCs of Chlorophyta (3,4). Although this new group is probably allied with the green algae (Chlorophyta), its affinities with the Chromophyta need to be considered.

The pigmentation of the Chromophyta has been well established for many years (see ref.5) but the pigment-protein complexes involved are only now beginning to be resolved. This contribution attempts to summarise the very limited set of information so far available.

Photosynthetic Light-Harvesting Systems

The major pigments of Chromophyta

Chl $\underline{c}_1$ and Chl $\underline{c}_2$ are not true Chls, but magnesium complexes of porphyrins which have the isocyclic ring found in Chls *a* and *b* but lack the phytol tails of the latter (6). Ring IV is not reduced (Fig 1) and has an acrylic acid side chain. In comparison with Chl *a*, Chl $\underline{c}_1$ and Chl $\underline{c}_2$ have 5 or 6 less H atoms, respectively (Fig 1). Both Chl $\underline{c}_1$ and $\underline{c}_2$ have very low red absorbance peaks compared to Chl *a* and *b*. Chl $\underline{c}_2$ has relatively greater absorbance than Chl $\underline{c}_1$ in the violet region: in ether the peaks are at 448 and 444 nm respectively (Fig 1) but when complexed to protein in light-harvesting cholorphyll-protein complexes (LHCs) the peak is closer to 450 nm.

Two carotenoids, fucoxanthin and peridin, both polyoxygenated xanthophylls are major components of all Chl *a*-*c* complexes so far investigated. When

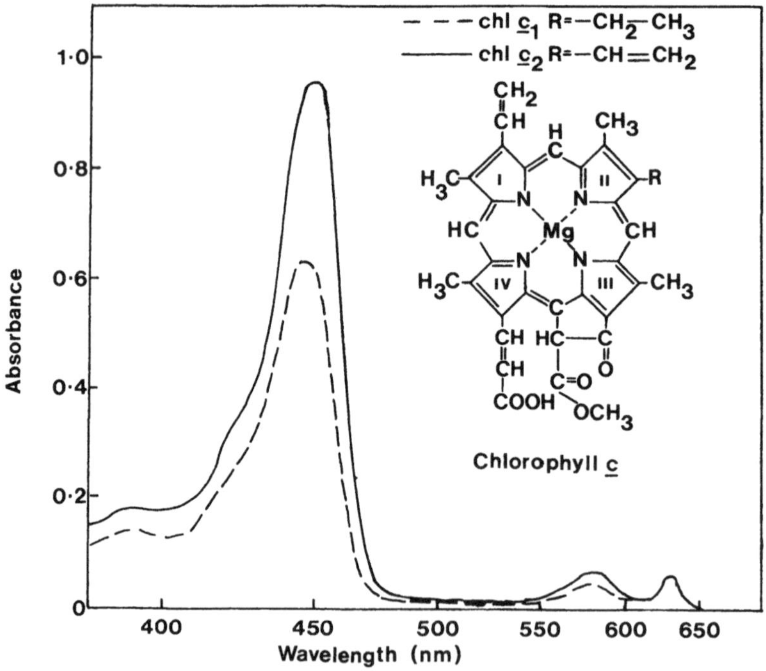

Fig. 1. The absorption spectra of Chl $\underline{c}_1$ and Chl $\underline{c}_2$ in ether (after ref. 6.) and their chemical formulae.

Table 1. A summary of current information on light-harvesting Chl *a* - *c* protein complexes.

Algal Group	Accessory pigments	Organism	Chl *a*/*c* ratio	Mr polypeptide(s) (kDa)	Chl *a* A_{max} (nm)	Chl *c* A_{max} (nm)	Fluorescence Emission (77K)(nm)	Ref
Bacillariophyta	Chl c_1 + c_2 Fucoxanthin	*Phaeodactylum tricornutum*	2.4	17.5, 18.0	441, 450	450	675	9
			11, 3.6	16.4, 16.9	436	460	678	11
			3.0	18, 19, 19.5	433	460	680, 690	10
			1.8 - 3.2	15 - 20	440, 460	460	670	12
Phaeophyta	Chlc_1 + c_2 Fucoxanthin	*Fucus* *Acrocarpia*	3.0	17 - 20	438, 671	450	680, 683	13, 14 15
Prymnesiophyta	Chlc_1 + c_2	*Pavlova*		16.5, 17, 18, 20.5	438, 675	460	-	18
	Fucoxanthin	*Pavlova*	4.7	21	440, 670	460	676	19
Chrysophyta	Chlc_1 + c_2 Fucoxanthin	*Synura*	-	<32?	439, 670	?	-	22
Xanthophyta	Chl c ? Carotenoids	*Tribonema*	No Chl *c*?	20?	443, 669	-	-	22
Eustigmatophyta	Violaxanthin Vaucheriaxanthin	*Polyedriella*	No Chl *c*	22	438, 678	-	673	24
Cryptophyta	Chl c_2 diadinoxanthin	*Chroomonas*	1.4	20, 24	436, 672	460	682	20
	phycobiliprotein	*Cryptomonas*	-	18.5, 19.5,24	438	460	681	21
Pyrrophyta	Chl c_2 Peridinin	*Glenodinium* *Gonyaulax*	1.0 - 4.0	17, 19, 24?	438, 672	452, 460	-	16, 17
Prasinophyceae	Chl b Chl c_1 ?	*Mantoniella* *Micromonas*	5.7	20.5, 22	438, (643), 672	Chl *b* (465)	-	3,4

these carotenoids are complexed to protein they show large red shifts, so that the principle absorption band at 480 nm in polar solvents is moved to 490 nm which, together with a large shoulder extending to 540 nm, enables these complexes to harvest blue and green light which predominates at depth in the world's oceans (see ref 5).

Light-Harvesting Chlorophyll Complexes of Chromophyta

The present information on LHCs of Chromophyta is summarised in Table 1. A typical absorption spectrum of an light-harvesting chlorophyll a-c-fucoxanthin complex (for the diatom Phaeodactylum tricornutum) is shown in Fig 2.

In terms of LHC the diatoms are the most studied of all the Chromophyta. The earlier studies of Alberte et al (7) and Gugliemelli (8) have now been superseded by no less than 4 studies of LHC in Phaeodactylum tricornutum

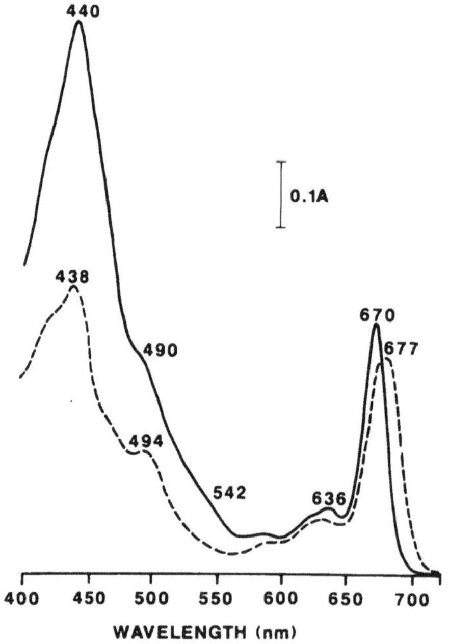

Fig. 2. Absorption spectrum of the light-harvesting Chl a-c-fucoxanthin complex (——) and photosystem I complex (- - -) of Phaeodactylum tricornutum prepared as in ref. 35.

(9-12). In the earlier study of Skeletonema costatum (and brown algae), Alberte et al,(7) reported, for non-denaturing SDS-PAGE, two complexes one of which contained Chl a+c plus fucoxanthin and another which lacked fucoxanthin. In the later paper Friedman and Alberte (9) suggested 'that the earlier protocol, using very low SDS levels to preserve energy transfer, resulted in two populations of pigment proteins which were separated upon electrophoresis based on aggregation state - - and were subject to selective pigment rempoval during electrophoresis'. This point illustrates a common problem. In recent work gentle detergent extraction has been used in combination with sucrose density gradient centrifugation. and a reasonably consistent set of characteristics has emerged for diatoms (Table 1), suggesting a complex with ratios of Chl a: Chl c:fucoxanthin of 1:0.3:2.5 (indicating a molar ratio of 2Chl a, 1 Chl c and 5 fucoxanthin). Only Owens and Wold (1986) have analysed for Chl c_1 + c_2 and arrived at a ratio of 1Chl a_1: 3.1 Chl c_2.

In Phaeophyta (brown algae) the Chl a-c-fucoxanthin complexes is still not fully resolved despite a number of attempts (13-15). The problem is similar to that discussed for diatoms of not knowing whether pure complexes have been isolated or partial mixtures or complexes stripped of some of their pigments. Similar difficulties have arisen with the attempts to isolate LHCs from Pyrrophyta (dinoflagellates)(16,17).

In the prymnesiophyta, an LHC from Pavlova gyrans and P.lutherii has been prepared by detergent solubilisation and sucrose density gradient centrifugation (18,19). This is a typical light-harvesting Chl a-c-fucoxanthin complex In Cryptophyta in addition to phycobiliprotein, there is an LHC with Chl a+c_2 plus a carotenoid (diadinoxanthin?), which comprises 40% of the total Chl a (20,21).

In the Chrysophyta and Xanthophyta, only one alga from each of these Divisions has been studied, Synura petersenii (Chrysophyta) and Tribonema aequale (22) and separation of LHC has been achieved only by SDS-PAGE. In the Eustigmatophyta, a division of algae that does not contain Chl c but structurally is very similar to the Xanthophyta from which it was separated (23), it is of interest that an LHC has been isolated from

Polyedriella helvetica by SDS-PAGE that has a polypeptide of 22 kDa (24). Chl a is the only Chl in this group and the major light-harvesting accessory pigment is violaxanthin (24,25).

In all those LHCs isolated so far coupling of energy transfer between Chl c and fucoxanthin or peridinin to Chl a has been demonstrated, but in harsher treatments these pigments may be partially uncoupled (17,18).

Prasinophyceae

This class of algae, usually allied with the Chlorophyta (but separated on the presence of scales and other characters) has recently attracted considerable attention by the discovery in only two of its members, Mantoniella squamata and Micromonas pusilla, of a LHC distinct from the LHC typical of other prasinophytes and green algae (3) and by the claim that these two species contain Chl c_1 (3,26,27). It has been reported previously that these two species contained Mg- 2,4,divinylpheoporphyrin a_5 monomethyl ester in addition to Chl a and b (28). This pigment is concentrated in the LHC and is coupled to Chl a (26,29). There is now good evidence that this pigment is Chl c_1 (3,27).

Mattox and Stewart (30) suggested that Mantoniella should be placed in the Micromonadophyceae, a group of primitive green algae, up till then placed in the Chlorophyta. However if their recent findings can be substantiated for other algae, this group might well deserve division-level recognition (3).

Chromophyte groups needing further investigation

LHCs in only a very few algae have been investigated throughout the large number of chromophyte groups. For, at least one division Chloromonadophyta, there is no information at all, and in the Xanthophyta and Chrysophyta only one species of each has been investigated. In the

Xanthophyta only the genus Vaucheria has been shown definitely to possess Chl c (31) but its LHC(s) has not so far been investigated.

The apoproteins and their antigenic properties

Despite some previous confusion, a set of consistent values is now emerging for the M_r of the constituent polypeptides. In most cases the polypeptides lie in the region of 17 to 22 kDa (Table 1). In the Bacillariophyta there is a clear doublet, or pair of polypeptides, separated by about 1 kDa (in the region 16-19 kDa) and the heavier of the pair may in fact show heterogeneity (10). Such a doublet may also occur in other groups especially the Phaeophyta where it is found at about 17 - 20 kDa (J.M. Anderson, pers. comm.). In the Cryptophyta two polypeptides are separated to a much greater degree (21 and 24 kDa) but the lighter twin shows heterogeneity (21). In Prymnesiophyta polypeptides have been found at 16.5, 17, 18 and 20.5 kDa (18)and at 21 kDa (19).

Few immunological studies have been carried out to compare cross-reactivity of apoprotein polypeptides in chromophyte algae Friedman and Alberte (9,32) found cross-reactivity in a number of diatoms. Hiller et al.(19) found cross-reactivity between the major LHC polypetide of the prymnesiophyte apoprotein and two other groups (diatoms and higher plant LHC) but little or no cross-reactivity with dinoflagellates or cryptophytes. In Fig. 3 we present a new Western blot analysis in which these findings are demonstrated and cross-reactivity to Colpomenia sinuosa (Phaeophyta) is added to the previous list. These results show a clear homology across a range but not all LHCs of chromophyte algae and surprisingly with higher plant LHC. Antibodies have been raised to both the 20 and 24 kDa polypeptides of Chroomonas sp. (Cryptophyta) but these showed no cross-reactivity to LHCs of any other algal group or to LHC of higher plants and the reverse is also true (Hiller, unpublished results).

Structural aspects of the light-harvesting system of chromophytes

Little is known of the detailed molecular structure of the protein

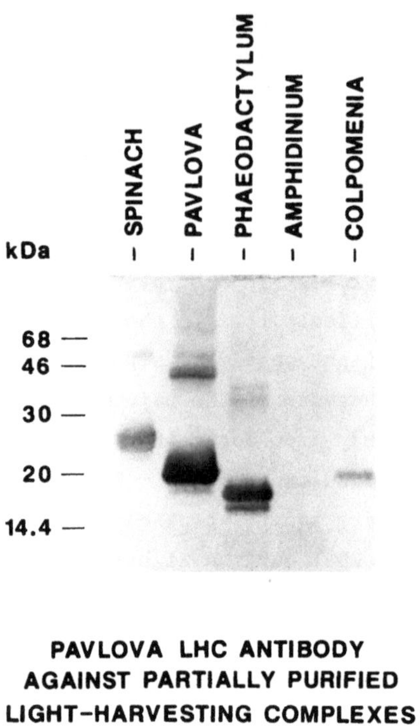

Fig. 3. Western blot showing cross-reactivity of antibody raised to the 21 kDalton polypeptide of the light-harvesting Chl a-c-fucoxanthin complex of Pavlova lutherii to polypeptides of partially purified LHCs of other algae, separated by SDS-PAGE (as in ref. 19).

Fig. 4. The absorbance and linear dichroism spectra of LHC from Chroomonas sp., prepared as in ref. 20 (R. G. Hiller and J. Breton, unpublished results)

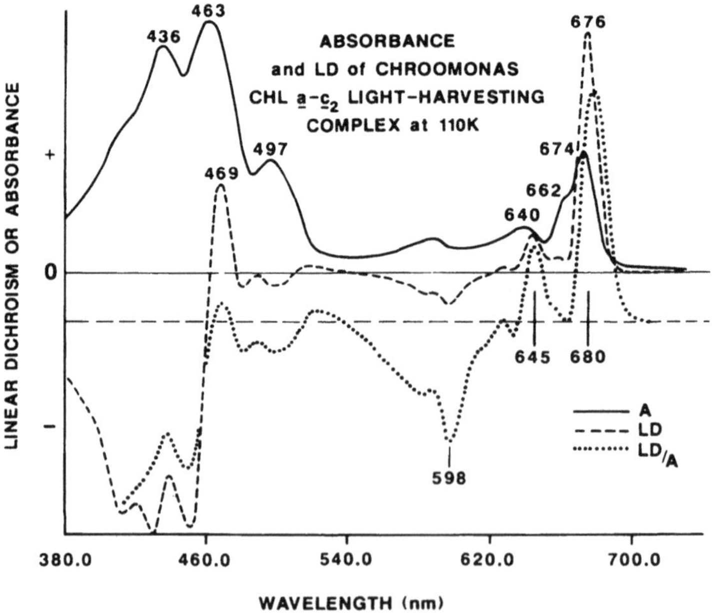

complexes of chromophyte algae or their arrangement in the thylakoid membrane. The thylakoids of these algae are arranged typically in groups of three, but whether the thylakoids show the same degree of appression as green algae and higher plants is a matter of debate (5). There is some evidence to suggest that the arrangement of Chl-protein complexes in the thylakoid membrane of chromophytes is different from that in other algae and higher plants, based on fluorescence characteristics and state transitions (5,33,34).

Hiller et al (35) have used UV circular dichroism to study the α-helical content of LHC and PSI complex from Pavlova lutherii (Prymnesiophyta) and Phaeodactylum tricornutum (Bacillariophyta). For PSI they found approx 55% α-helix content and 7% β-structure for both algae. For LHC they found approx 40% α-helix content and 14% β-structure. Thus all antenna Chl-protein complexes studied so far have an α-helix-dominant secondary structure. However until sequence data are available it will not be possible to match the UVCD data against models as can be done for LHCP II (36-38) where the evidence supports four membrane-spanning α-helices. Work on CD and linear dichroism of Chroomonas sp. LHC, has indicated comparability between that and LHC of higher plants (Hiller and Breton, unpublished results). They found strong CD signals at 645 and 680 nm, due to Chl c_2 and Chl a, at wavelengths longer than their respective absorbance maxima. A comparison of the absorbance spectrum with the linear dichroism spectrum (Fig 4) taken at 110 k in a squeezed poly-acrylamide gel showed that, as in LHCP II of higher plants, a large part of the long-wavelength absorbance (Q_y transition) of Chl a has its transition dipole oriented preferentially in the plane of the membrane. There is similar orientation of part of the Chl c_2 centred at 645 nm although this component cannot be seen in the absorbance spectrum. A puzzling feature of the LD spectrum is the strong negative component centred at 597 nm. Assignment of this transition to Q_y-1 transition of Chl a or c_2 is difficult but it suggests a strong perpendicular orientation of a minor component. It may be speculated that this phenomenon plays a role in energy transfer from phycoerythrin to Chl. A precedent for this is the perpendicular orientation of the minor chlorosomal BChl a component which is postulated to be an intermediary in energy transfer between BChl c of the chlorosome and BChl a of the cytoplasmic membrane in Chloroflexus aurantiacus (39).

Phylogeny of Chromophytes

The discovery of Prochloron (40) established the origin of Chl b at the prokaryotic level but because there is no antigenic cross-reactivity between LHC II from higher plants and LHC from Prochloron it is still not clear whether Prochloron is an ancestor of green algal and higher plant chloroplasts (41). On the other hand the cross-reactivity of some chromophyte LHCs with higher plant LHC does establish a link between the protein that binds the Chl c and the Chl b, respectively, in these two groups. However, it sheds no light on whether the origin of Chl c is prokaryotic or eukaryotic, although it does suggest a greater evolutionary distance between Prochoron and chromophytes than between chromophytes and higher plant chloroplasts.

The lack of cross-reactivity between the LHCs of cryptomonads and dinoflagellates and the LHCs of other chromophytes so far tested suggests a long evolutionary history amongst this diverse group and indicates that Chl c and its binding protein is of very ancient origin. The presence therefore of a Chl c-like pigment, together with Chl b, on a protein similar to the LHC of chromophytes (3) in a primitive green alga is of very great interest. This might suggest that Chl c arose after Chl b and became attached to the same apoprotein. However the M_r of the major polypetide of Prochloron LHC at 32 kDa is very much greater than that of green algae and higher plants (24-28 kDa) and of chromophytes (17-24 kDa). Until sequence data are available for the LHCs of Prochloron, chromophytes and Mantoniella it will not become clear which of the several evolutionary explanations is correct. If the Chl c-like pigment of Mantoniella is confirmed as Chl c_1 (4,27) this will be the only case where Chl c_1 occurs separately from Chl c_2 and would raise the possibility that Chl c_1 arose before Chl c_2. However the recent discovery of Chl c_3 in several members of Chrysophyta and Prymnesiophyta (42), where Chl c_3 replaces Chl c_1, indicates that there is inherent plasticity in the expression of Chl c.

Ecological Considerations

Chl c, complexed to protein in LHC, has absorption peaks at 450 and 630 nm (6). Since the absorption at 630 nm is comparatively small the possession of Chl c in addition to Chl a and carotenoids does not appear to be advantageous in terms of light harvesting capacity. There is some cost-saving because Chl c does not have a phytol chain (5). However a much more important fact seems to be that Chl c occurs, in almost all instances, in combination with fucoxanthin or peridinin. It is tempting to suggest that Chl c evolved together with these polyoxygenated xanthophylls to facilitate energy transfer and to take advantage of the special properties of these particular xanthophylls, which have efficient light-harvesting properties up to 540 nm when complexed with protein. The presence of this light-harvesting mechanism provides the chromophytes with a very competitive system for capturing light in the deeper parts of the ocean where the maximum transmission of light is at 480-520 nm (5). Thus the saving of a long hydrocarbon chain on each light-harvesting Chl and its diversion to a more efficient light-harvesting xanthophyll molecule is of great adaptive significance.

Summary

Relatively few chromophyte algae have so far been studied. However in those that have been studied a relatively consistent picture is emerging of a light-harvesting Chl protein complex (LHC) that binds Chl a and Chl c and carotenoid and has polypeptides in the region of 17-24 kDa. Even in Eustigmatophyta, which lack Chl c but have been placed in the Chromophyta on structural grounds, the LHC has a polypeptide in this region at 22 kDa. Thus on apoprotein size and pigment content chromophyte LHCs appear to be different to the LHCs of green algae and higher plants but on antigenic evidence some groups (Bacillariophyta, Phaeophyta, Prymnesiophyta) show affinity. The discovery of a primitive group of green algae (Mantoniella and Micromonas) that possess a Chl c-like pigment and a chromophyte LHC also points to affinities between the two groups but until protein sequence data become available the exact situation will remain unclear.

The evolution of fucoxanthin or peridinin together with Chl c in this diverse group of algae appears to have been of great adaptive and ecological significance.

References

1. Lee,R.E. 1980. Phycology. Cambridge University Press, Cambridge.

2. Whittaker,R.H. and L.Margulis. 1978. Biosystems 10,3-18.

3. Fawley,M.W., K.D.Stewart and K.R.Mattox. 1986. J.Mol.Evol. 23,168-176.

4. Wilhelm,C. 1987. Biochim.Biophys.Acta in press.

5. Larkum,A.W.D. and J.Barrett. 1983. Adv.Bot.Res. 10,1-219.

6. Jeffrey,S.W. 1980. In: Primary productivity in the sea. (P.G. Falkowski, ed.). Plenum Publ., New York. pp.33-58.

7. Alberte,R.S.,A.L.Friedman,D.L.Gustafson,M.S.Rudnick and H.Lyman. 1981. Biochim.Biophys.Acta 635,304-316.

8. Gugliemelli,L.A. 1984. Biochim.Biophys.Acta 766,45-50.

9. Friedman,A.L. and R.S.Alberte. 1984. Plant Physiol. 76,483-489.

10. Fawley,M.W. and A.R.Grossman. 1986. Plant Physiol. 81,149-155.

11. Owens,T.G. and E.R.Wold. 1986. Plant Physiol. 80,732-738.

12. Caron,L. and Brown,J. 1987. Plant Cell Physiol. 28,775-786.

13. Barrett,J. and J.M.Anderson. 1980. Biochim.Biophys.Acta 590, 309-323.

14. Berkaloff,C.,J.C.Duval,H.Jupin,F.Chrissovergis and L. Caron. Photosynthesis III (G. Akoyonoglou, ed.). Balaban Intern. Sci.Serv., Philadelphia,P.A. pp.485-494.

15. Caron,L.,J.P.Dubacq,C.Berkaloff and H.Jupin. Plant Cell Physiol. 26, 131-139.

16. Boczar,B.A.,B.B.Prezelin,J.P.Markwell and J.P.Thornber. 1980. FEBS Lett. 120,243-247.

17. Boczar,B.A. and B.B.Prezelin. 1987. Plant Physiol. 83,805-812.

18. Fawley,M.W., S.J.Morton,K.D.Stewart and K.R.Mattox. 1987. J.Phycol. 23,377-381.

19. Hiller,R.G.,A.W.D.Larkum and P.Wrench. Biochim.Biophys.Acta in press.

20. Ingram,K. and R.G.Hiller. 1983. Biochim.Biophys.Acta 722,310-319.

21.Rhiel,E.,K.Krupinska and W.Wehrmeyer. 1986. Planta 169,361-369.

22.Wiedemann,I., C.Wilhelm and A.Wild. 1983. Photosynthesis Res. 4,317-329.

23.Hibberd,D.J. and G.F.Leedale. 1970. Nature (Lond.) 225,750-760.

24.Chrystal,J. and A.W.D.Larkum. 1987. Progr. Photosynthesis Res. Vol.II.(J. Biggins,ed.). Martinus Nijhoff Pub., Dordrecht. pp.189-192.

25.Owens,T.,J.C.Gallagher and R.S.Alberte. 1987. J.Phycol. 23,79-85.

26.Wilhelm,C., I.Lenartz-Weiler,I.Wiedemann and A.Wild. Phycologia 25,304-312.

27.Wilhelm,C. 1987. Botanica Acta in press.

28.Ricketts,T.R. 1970. Phytochemistry 9,1835-1842.

29.Brown,J. 1985. Biochim.Biophys.Acta 807,1143-1146.

30.Mattox,K.R. and K.D.Stewart. 1984. In: Systematics of the green algae (D.E.G. Irvine and D.M. John, eds.). Academic Press, London Orlando. pp.29-72.

31.Caubergs,S.R and J.de Greef. 1971. Bot. Mag.,Tokyo 84,222-230.

32.Friedman,A.L. and R.S.Alberte. 1986. Plant Physiol. 80,43-51.

33.Owens,T.G. 1986. Plant Physiol. 80, 739-746.

34.Chrystal,J. and A.W.D.Larkum. Biochim.Biophys.Acta in review

35.Hiller,R.G.,A.-M.Bardin and E.Nabedryk. Biochim.Biophys.Acta in press.

36.Nabedryk,E.,S.Andrianambinintsoa and J.Breton. Biochim.Biophys.Acta 765,380-387.

37.Karlin-Neumann,G.A.,B.D.Kohon,J.P.Thornber and E.M.Tobin. 1985. J.Mol.Appl.Genet. 3,45-61.

38.Anderson,J.M. and D.J.Goodchild. 1987. FEBS Lett. 213,29-33.

39.van Dorssen,R.J.,H.Vasmel and J.Amesz. 1986. Photosynthesis Res. 9,33-45.

40.Lewin,R.A. 1976. Nature (Lond.) 261,697-698.

41.Hiller,R.G. and A.W.D.Larkum. 1985. Biochim.Biophys.Acta 806, 107-115.

42.Jeffrey,S.W. and S.W.Wright. Biochim.Biophys.Acta in press.

ISOLATION AND CHARACTERIZATION OF A CHLOROPHYLL A/C-HETEROXANTHIN/DIADINOXANTHIN LIGHT-HARVESTING COMPLEX FROM PLEUROCHLORIS MEIRINGENSIS (XANTHOPHYCEAE)

C.Wilhelm, C. Büchel
Institut für Allgemeine Botanik, Johannes Gutenberg Universität, D-6500 Mainz, FRG

B. Rousseau
Laboratoire des Biomembranes, Ecole Normale Superieure, F-75231 Paris (France)

Introduction

All photosynthetically active carotenoids and chlorophylls are non-convalently bound in so-called chlorophyll-protein complexes. In higher plants and eukaryotic algae the reaction center complexes were found to be uniform with respect to their pigment and polypeptide composition, whereas the light-harvesting systems differ widely in their pigmentation and protein structure. It was stated that the different light-harvesting systems reflect both taxanomic relations and evolutionary vestiges. Three different lines of light-harvesting have been postulated: the phycobiliprotein containing "biliphyta", the chlorophyll c-containing "chromophyta" and the chlorophyll b-containing "chlorophyta" (1,2). Intensive studies of the last years have elucidated a multiplicity of light-capturing carotenoids in the antenna systems not only in chlorophyll c- but also in chlorophyll b-containing algae (3,4).

Nevertheless, there are some algal groups, whose molecular organization of the photosynthetic pigments is not well known. Especially, the antenna systems of yellow-green algae, including the Eustigmatophytes and the Xanthophytes are less understood. These two different groups belonging to the "chromo-

Photosynthetic Light-Harvesting Systems

phytes" do not contain fucoxanthin and chlorophyll c is much less concentrated than in the brown diatoms, chrysophytes or brown algae. On the basis of different ultrastructural and biochemical features the Xanthophytes and Eustigmatophytes were separated in two different taxa (5). In the context of light-harvesting it is important to note that the xanthophytes contain heteroxanthin and diadinoxanthin as main xanthophylls instead of violaxanthin, which is common in the Eustigmatophytes (6, 7).
Till today it was unclear if these xanthophylls really act as antenna pigments. Because of a very strong instability of their chlorophyll-proteins against detergent-based solubilization, the isolation of intact antenna complexes is strongly hindered. There are only two reports about the chlorophyll-proteins of Eustigmatophytes (8,9) and one about Xanthophytes (10). Because of the lack of chlorophyll b and the little content of chlorophyll c the xanthophylls are expected to play the most important role in light-harvesting. For the present, there is no reliable information about the nature and the stoichiometry of the xanthophylls in the chlorophyll-proteins of Xanthophytes.

The present study continues our work about xanthophycean algae giving more detailed information about the stoichiometry of pigments in the chlorophyll-protein complexes, their polypeptid composition and the molecular basis of energy transfer.

Materials and Methods

Pleurochloris meiringensis (Culture Collection of Göttingen, No.860-3) was cultivated under low light conditions (3,5 W/m^2) as previously published for xanthophycean algae (10). For the isolation of the chlorophyll-protein complexes the thylakoids were solubilized with digitonin:Chl a=40:1 and separated by a sucrose density centrifugation adopted from Berkaloff et al.

(11). The gelelectrophoresis of the denaturated polypeptides was done according to Laemmli (12). The absorption spectra of the eluted protein complexes were determined by a Shimadzu MPL 2000 photometer. The fluorescence spectra were recorded at 77K with a Hitachi F-3000 fluorometer at chlorophyll concentrations of 0.5 µg/ml. The excitation spectra were performed with a band pass of 3 nm for the excitation and 20 nm for the emission. The emission spectra were recorded with a band pass of 20 nm for the excitation and of 3 nm for the emission. The wavelength of excitation and emission are given in the figure legends. The pigment separations were performed by an HPLC system which was earlier described in detail (13).

Results

Using sucrose density centrifugation three pigmented bands can be obtained (Fig. 1). The two heaviest fractions (number 1 and 2 in figure 1) do not show any differences with respect to their absorption or fluorescence spectra or to their polypeptide pattern. The following characterizations of the PS I proteins were shown only for the fraction 2. The third band which contains more than 50 % of total chlorophyll is considered to be the LHC. Figure 2 gives the absorption spectra of total cells, the LHC and the PS I complex. Comparing the three absorption spectra the LHC exhibits a stronger absorption at 487 nm, whereas in the PS I fraction this maximum is less pronounced than in total thylakoids. Obviously, the LHC contains a higher amount of accessory pigments per chlorophyll a than the thylakoids and all the more the PS I complex. The fluorescence excitation spectra at 77K (figure 3) give evidence that the LHC-bound carotenoids (absorbing at 490 nm) transfer the captured energy efficiently to chlorophyll a. In the PS I complex (dotted line) the small amounts of accessory pigments revealed by absorption spectroscopy are found to be well connected with chlorophyll a, too. The fluorescence emission spectra indicate the existence of an light-harvesting antenna emitting at 685 nm and of an PS I-antenna-complex at 720 nm,respectively (Fig. 4). There is no indication of some deconnected pigments. Table 1 gives the

Table 1
Pigments of 1000 Chl a in thylakoids, PS I and LHC isolated by sucrose density centrifugation.

$\frac{\text{mol Pigment}}{\text{1000 chl a}}$	total thylakoids	LHC	PS I
Chl c	53	224	10,4
Heteroxanthin	40	148	34
Diadinoxanthin	117	264	98
Vaucher-ester	63	129	51
ß-Carotene	15	-	23

molar amounts of the pigments from total thylakoids, LHC and PS I. Comparing the LHC and the PS I complex with total thylakoids it must be emphazised that the LHC is strongly enriched in chlorophyll c, diadinoxanthin and vaucheriaxanthin-ester; its molar ratio of xanthophyll per chlorophyll a is double as high as in thylakoids. By contrast, the PS I complex contains less of accessory pigments in comparison to the whole thylakoid fraction. Because the chlorophyll proteins isolated by sucrose density centrifugation are usually not as pure as done by gelelectrophoresis, the protein fractions were run on a denaturating gel system to check their purity and to measure the molecular weights of their apoproteins. Figure 5 shows that the LHC contains only two subunits in the range of 18 and 23 kD. The polypeptide pattern of the PS I fraction is more complex. Beside the PS I subunits of 60 and 43 kD two bands at lower molecular weights are found, relating to a small antenna.

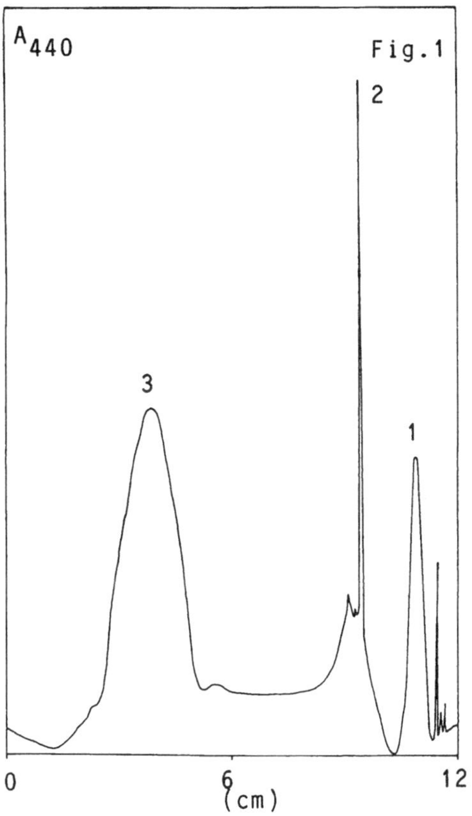

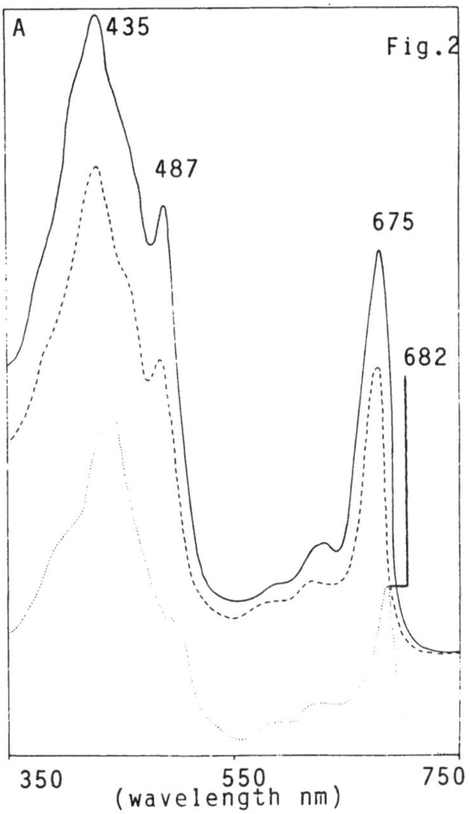

Legends of figures.

Figure 1: Densitogram of sucrose density centrifugation

Figure 2: Absorption spectra of total cells (solid line), the light-harvesting complex (broken line) and the photosystem I protein (dotted line)

Figure 3: Fluorescence excitation spectra of the light-harvesting complex (broken line) and the photosystem I protein (dotted line). Emission wavelength: 685 nm

Figure 4: Fluorescence emission spectra of the light-harvesting complex (broken line) and the photosystem I protein (dotted line). Excitation wavelength: 490 nm

Figure 5: Polypeptide composition of the denaturated light-harvesting and photosystem I complex.

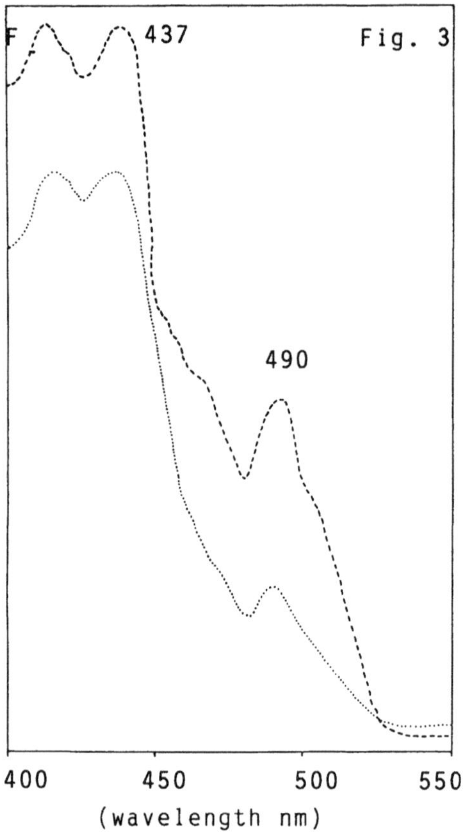

Fig. 3

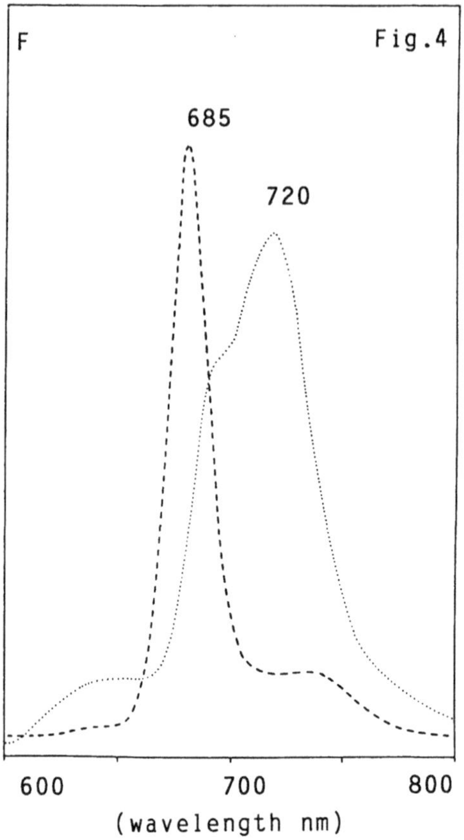

Fig.4

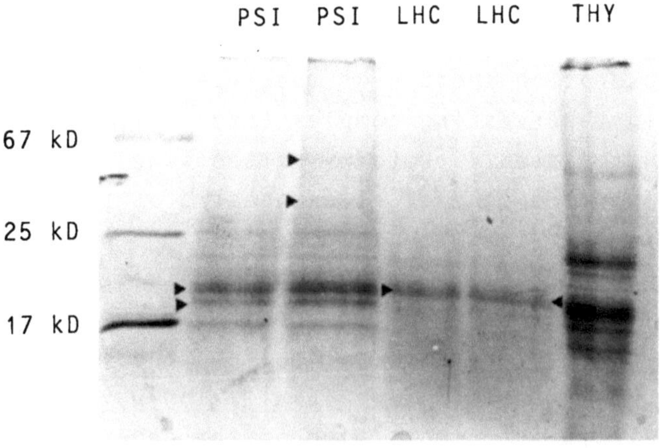

Fig.5

Discussion

The light-harvesting apparatus of Xanthophyceae differs from those of Eustigmatophyceae by the existence of heteroxanthin and diadinoxanthin instead of violaxanthin. The present work shows that under mild separation conditions at least two different chlorophyll proteins can be isolated without any losses of pigments. The resulting LHC is considerably pure, whereas the PS I complex contains some non specific polypeptides. The features of the PS I complex from Pleurochloris are very similar to those of chlorophycean algae (13,14,15). It exhibits a long wavelength absorption at 682 nm and a long wavelength emission at 720 nm in liquid nitrogen. Our results clearly show that this long wavelength emission is strongly correlated with PS I. There is no variety observed in fluorescence spectra as reported from other chlorophyll c-containing algae (16). The PS I complex is associated with a light-harvesting antenna system, whose specificity for photosystem I is not yet clear.

The pigment composition of the LHC shows mainly the xanthophylls, heteroxanthin, diadinoxanthin and vaucheriaxanthin-ester, which amount in total to 0,5 molecules per 1 chlorophyll a. The ratio chlorophyll a/chlorophyll c is found to be 8:1, which probably reflects the in-vivo conditions because no pigment was lost. Brown (9) reports similar xanthophyll /chlorophyll a ratios in the yellow-green alga Nannochloropsis, but a member of the Eustigmatophytes, that does not contain comparable amounts of chlorophyll c. Our results give clear evidence that the three mentioned xanthophylls act as true light-harvesting antenna pigments. Partial deconnection of the xanthophylls in the LHC provoke a strong chlorophyll c fluorescence but not an enhancement of chlorophyll a emission (data not shown). Therefore, in xanthophytes chlorophyll c functions mainly in the mediation of energy transport from xanthophylls to chlorophyll a but less as light capturing pigment.

Acknowledgements

This research was supported by the Deutsche Forschungsgemeinschaft grant no. Wi 764/1-4

References:

1: Cavalier-Smith, T. 1982. Biol. J. Linn. Soc. 17, 289.

2: Larkum, A.W.D., J. Barret. 1983. Adv. Bot. Res. 10, 1.

3: Anderson, J.M. and J. Barret. 1986. In: Encyclopedia of Plant Physiology (Staehelin, L.A. and Arntzen, C.J., eds.) Springer, Heidelberg, New York, p. 269-285.

4: Wilhelm, C., I. Lenartz-Weiler, I. Wiedemann, A. Wild. 1986. Phycologia 25, 304.

5: Hibberd, D., G.F. Leedale. 1970. Nature 225, 758.

6: Wittle, S., P. Casselton. 1975 a. Br. Phycol.J. 10, 179.

7: Wittle, S., P. Casselton. 1975 b. Br. Phycol.J. 10, 192.

8: Chrystal, J., A.W.D. Larkum. 1986. In: Progress in Photosynthesis Research, Vol. 2 (Biggins, J., ed.). Martinus Nijhoff Publishers, Dordrecht, p. 189.

9: Brown, J. 1987. Plant Physiol. 83, 434.

10: Wiedemann, I., C. Wilhelm, A. Wild. 1983. Photosynth. Res. 4, 317.

11: Berkaloff, C., J.C. Duval, B. Rousseau. 1983. In: Advances in Photosynthesis Research (Sysbesma, C. ed.) Martinus Nijhoff Publishers, The Hague p. 449.

12: Laemmli, U.K. 1970. Nature 227, 680.

13: Wilhelm, C., I. Lenartz-Weiler: Photosynth. Res.(in press)

14: Wollmann, F.-A., P. Bennoun. 1982. Biochim. Biophys. Acta 680, 352.

15: Senger, H., K. Krupinska. 1986. Plant Cell Physiol. 27, 1127.

16: Duval, J.C., H. Jupin, H., C. Berkaloff. 1983. Physiol. Veg. 21, 1145.

THE ANTENNA COMPONENTS OF PHOTOSYSTEM II WITH EMPHASIS ON THE MAJOR PIGMENT-PROTEIN, LHC II*b*

Gary F. Peter and J. Philip Thornber

Department of Biology, University of California, Los Angeles, California 90024, USA.

The pigmented portion of photosystems I and II of green plants can be conveniently thought of as consisting of two distinct parts: a core complex (CC) and a light-harvesting complex (LHC) [1]. The core complex contains those polypeptides and cofactors, mainly chlorophyll (chl) and carotenoid molecules, that are required for a plant to have a stable and functional photochemical reaction center in vivo, whereas the light-harvesting complex houses most of a photosystem's antenna pigment molecules but is not essential for the structure and function of its associated core complex. Photosystem I has a simpler pigment-protein composition than photosystem II: CC I consists of a P700-beta-carotene-chlorophyll a-protein, having two similar subunits of ~70kDa apparent sizes. LHC I contains a xanthophyll-chl a/b-protein which has a 20-21kDa protein subunit; a possibility exists that a second LHC I chl-protein occurs of a somewhat similar size. Photosystem II, on the other hand, has at least seven different caroteno-chl-proteins. Its core complex (CC II) has an 85kDa P680-containing pigment-protein complex which bears considerable biochemical and biophysical similarities to the purple bacterial photochemical reaction center, i.e. it has two ~30kDa polypeptides associated with chlorophyll and pheophytin molecules [e.g. 2]. In addition, CC II has two beta-carotene-chla-proteins (having the trivial names of CP 47 and CP 43), which account for ~10% of the total chl in a green plant. There are at least four xanthophyll-chlorophyll-proteins that constitute LHC II. This article aims to describe the status of our biochemical and structural knowledge of the LHC II pigment-proteins.

Photosynthetic Light-Harvesting Systems

1. CHARACTERISTICS OF THE LHC II PIGMENT-PROTEINS.

1a) Fractionation of thylakoid membranes into pigment-proteins

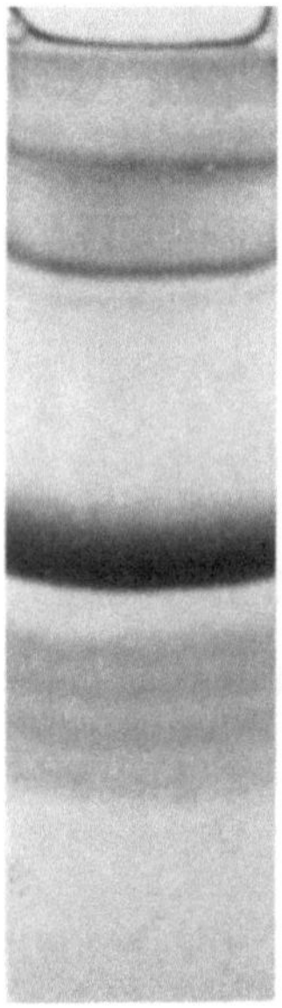

Fig. 1. *Non-denaturing gel electrophoresis (see text) of a photosystem II preparation [4]. The positions of the LHC IIa,b,c and d pigment-proteins are marked; FP is free pigment, essentially carotenoid only. The uppermost bands are representations of CC II and a ~300kDa LHC IIb component (see text).*

A recently devised electrophoretic system [3] resolves the multiple pigment-proteins in thylakoids without displacing any of the chl, and probably carotenoid, molecules from their association with protein. In this system thylakoid membranes are solubilized by a glycosidic surfactant (e.g. octyl- or nonyl-glucoside, dodecyl-maltoside) and then fractionated by polyacrylamide gel electrophoresis (PAGE) using an upper reservoir buffer containing the surfactant, monosodium-N-lauryl beta-iminodiproprionate (Deriphat 160). Thus, for the first time, it is possible to describe unequivocally the pigment content of each green plant pigment-protein. Since there is a relatively large number of higher plant pigment-proteins, it is easier to examine the pigment-proteins separated from a part of the photosynthetic apparatus, e.g. photosystem I or photosystem II fractions, rather than from the whole thylakoid. Isolated Photosystem II particles [4] treated with heptylthio-glucoside and decyl-maltoside, can be separated into seven green bands on Deriphat-PAGE (Fig.1). Four of them are LHC II pigment-proteins: LHC II*a*, also known as CP29 [5]; LHC II*b*, the major pigment-protein of green plants [cf. 1]; LHC II*c*

(CP27)[6,7]; LHC II*d* (CP 24)[8]. Their apparent sizes on non-denaturing PAGE are 35, 72,30 and 24kDa, respectively, in extracts of pea thylakoids or photosystem II particles.

1b) Absorbance spectra

LHC II*b* and LHC II*d* contain approximately equal amounts of chls a and b. All LHC II pigment-proteins have substanial absorbance in the 400-500nm region due to carotenoid. LHC II*a* contains a peak at ~645nm which is characteristic of this chl-protein [9].

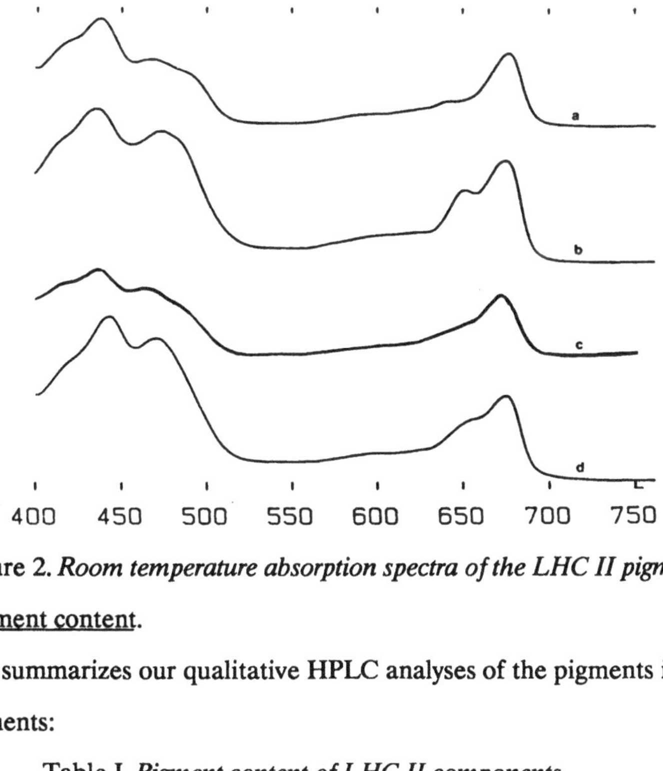

Figure 2. *Room temperature absorption spectra of the LHC II pigment-proteins.*

1c) Pigment content.

Table I summarizes our qualitative HPLC analyses of the pigments in the LHC II components:

Table I. *Pigment content of LHC II components*

	Alternative name	*Chlorophyll a/b*	*Xanthophyll content*
LHC IIa	*CP 29*	*a only*	*Lutein, violoxanthin*
LHC IIb	*CP II*	*1.15*	*Lutein, neoxanthin, violaxanthin*
LHC IIc	*CP 27*	*a + b*	*Lutein, neoxanthin, violaxanthin*
LHC IId	*CP 24*	*1.25*	*Neoxanthin, violaxanthin, lutein*

1d) Subunit composition

The polypeptides associated with each LHC II component have been difficult to assign unequivocally because not only can all four pigment-proteins migrate very near each other on non-denaturing PAGE, but so also can their apoproteins on fully denaturing PAGE. All LHC II apoproteins are within 10kDa in apparent size of each other. A two-dimensional electrophoretogram (Fig. 3) demonstrates the problem. After separating

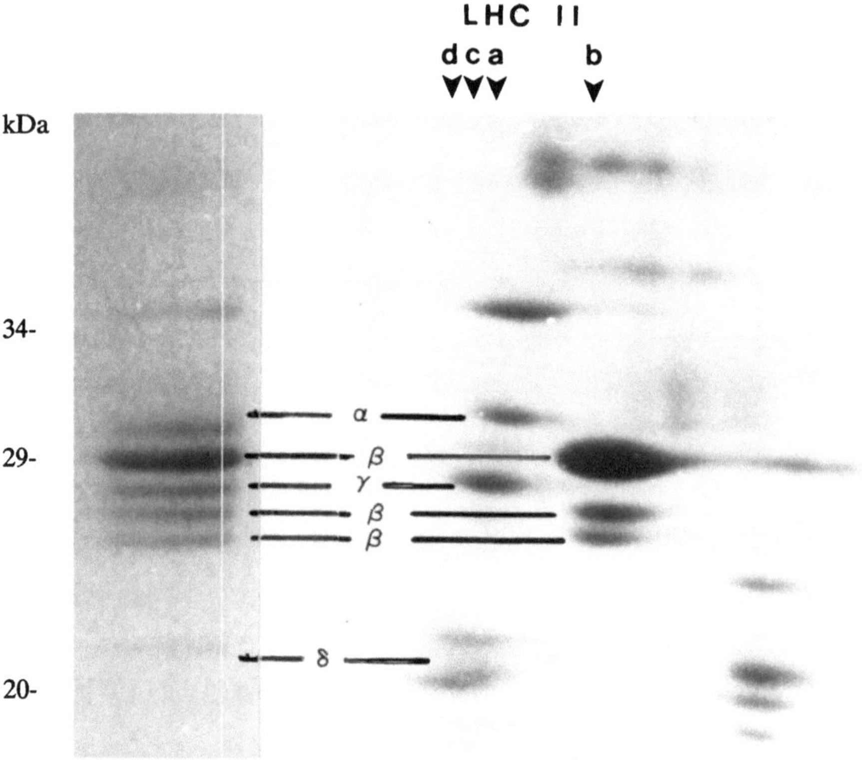

Figure 3. *Left, fully denaturing PAGE of a solution enriched in LHC II pigment-proteins. Right, two dimensional PAGE (see text) of solubilized thylakoids. At the top of the figure the positions of the various LHC II pigment-proteins are shown after the first dimension PAGE; note that LHC IIb runs as an oligomer of 72kDa and LHC IIa, c and d run as pigment-proteins of 35, 30 and 24 kDa, respectively. The apoproteins associated with each pigment protein are indicated by their respective Greek letters.*

the pigmented complexes of thylakoids by Deriphat-PAGE (cp. Fig.1), a gel strip is excised, incubated with 4% SDS-100mM dithiothreitol-Tris pH6.8, and placed along the top of a second polyacrylamide gel so that the original direction of electrophoretic migration is now at right angles to that to be used for the second electrophoresis. This second system is strongly denaturing so that the pigment-proteins are dissociated into their individual constituents during electrophoresis. The region of the gel after electrophoresis where polypeptides of 20-40kDa migrate is shown (Fig.3) with the chl-protein origin of each 20-30kDa polypeptide identified. Note that LHC II*b* migrates only as an oligomeric form (72kDa) under the conditions used here, thus permitting its apoproteins to be well separated in the second dimension from those of LHC II*a*, *c* and *d*, and simplifying unequivocal assignments. Two-dimensional electrophoretograms as well as re-electrophoresis of each dissected pigmented band on denaturing SDS-PAGE, provide the following assignment of polypeptide subunits [3]:

Table II. *Subunit composition of LHC II pigment-proteins of pea and barley.*

	LHC IIa	*LHC IIb*	*LHC IIc*	*LHC IId*
Pea	*29*	*28,26,25*	*27*	*21*
Barley	*29*	*28,27,25*	*31*	*21*

1e) Protein phosphorylation

Phosphorylation of LHC II polypeptides is believed to be involved in controlling State 1-State 2 transitions in green plants [10]. We used standard conditions (2 min. in light with 200 μM ATP) to label thylakoid proteins with ^{32}P[11] which resulted in the 28 and 27kDa polypeptides of LHC II*b* being relatively strongly labeled. LHC II*d*'s apoprotein was also labeled but to a much lesser extent [cp.7]. All other LHC II apoproteins were not phosphorylated, including the 25kDa LHC II*b* polypeptide.

1f) LHC II content of a chlorophyll b-less mutant of barley

It was originally proposed that LHC II was lacking in this mutant [e.g. 12]. Later, others provided evidence for the presence of some LHC II apoproteins in the mutant [e.g. 13-16]. Using our improved fractionation technique (Fig.1) we examined which LHC II pigment-proteins are present in the mutant: LHC II*a* is present in slightly reduced amounts compared to those in the wild type; only chl a and xanthophylls are present in this complex, even in the wild-type plant. The 25kDa polypeptide of LHC II*b* is also present but we are uncertain that pigment is associated with it. We do not detect the in the mutant the two major polypeptides of 28 and 27kDa that are associated with wild-type LHC II*b*, and LHC II*d* is much reduced in quantity (see [3] for details).

2. THE MAJOR CHLOROPHYLL a/b-PROTEIN OF GREEN PLANTS (LHC II*b*)

Many research groups study LHC II*b*, which comprises the major thylakoid proteins, because it is a convenient marker to use for this cellular compartment. Thus, LHC II*b* has been used to study the intricacies of synthesis and assembly of a nuclear gene family coding for a membrane polypeptide: LHC II*b* apoproteins are synthesized on cytoplasmic ribosomes as larger precursor polypeptides, imported into plastids, and then processed to the final size and inserted into thylakoid membranes (see [17] for review). As a consequence, characterization of genes for LHC II*b* apoproteins and production of antibodies that react with these proteins are relatively well advanced. LHC II*b* has also been an attractive component to study biochemically and biophysically since it is the depository for large amounts of the chls and xanthophylls synthesized by green plants, and because it provides a readily available water-insoluble pigment-protein in which to examine, the molecular arrangement of pigment molecules and energy transfer in a defined micro-environment. Some caution should be exercised in automatically accepting some of the older data on this chl-protein: the sections above indicate how easily LHC II*b* preparations might contain other LHC II pigment-proteins; furthermore, older preparative procedures often removed substantial quantities of the pigment molecules originally bound to the complex.

2a) The relationship between the different LHC II*b* apoproteins

Three apoproteins (Table II) are readily separated after denaturing PAGE of the isolated pigment-protein. If urea (4M) is omitted from the separating gel and/or if the electrophoretic separation is done at low temperatures, then the smallest apoprotein co-migrates with the largest, i.e. only two apoproteins are readily seen (cf. 18). It is possible that further changes in the SDS-PAGE conditions might fractionate more than three apoproteins. Regardless of the exact number, we do not yet know how the apoproteins are related. The explanation that each apoprotein is solely a different gene product now seems less probable in light of the gene studies in Arabidopsis [19]. Only three members of this plant's LHC II*b* gene family have been isolated and all three gene products are predicted to have identical amino acid sequences. Yet, LHC II*b* in Arabidopsis has a similar multiplicity of apoproteins [20] as occurs in other plants that have as many as 15 cross-hybridizing genes [21]. The notion cannot be entirely abandoned, however, since there is a weakly cross-hybridizing fourth gene in Arabidopsis which has yet to be isolated and sequenced [19]; nevertheless, it is more likely that the apoproteins differ in their post-translational modifications. Certainly the two slowest migrating forms are phosphorylatable whereas the fastest is not (see above). This difference, however, is probably not the complete explanation of LHC II*b* apoprotein multiplicity, since whether or not an LHC II*b* polypeptide can be phosphorylated may be indicative of whether or not proteolytic processing has occurred at the N-terminus [cf. 28]. Also, LHC II*b* has recently been found to have a palmitylated polypeptide(s) [22]. A combination of modifications may, therefore, explain the variation in mobility of the apoproteins on PAGE.

2b) Models for the structure of LHC II*b*

Although crystals have been made of this hydrophobic chl-protein [23], none have yet yielded the high resolution X-ray data needed to have a detailed three-dimensional structure of this pigment-protein. In the absence of such data, we and others have resorted to modelling the complex on the basis of its predicted amino acid sequence [24-26]. More than twenty LHC II*b* genes from ten different species have now been sequenced (summarized in [17]). There are certainly two classes of genes based on the predicted amino acid sequences: one type codes for a slightly smaller polypeptide with obviously different yet similar transit and N-terminal sequences than the other type. Interestingly, all genes coding for the smaller polypeptide have, so far, contained an intron. Based on comparisons of the upstream sequences of the genes, it has been proposed that five gene classes occur in petunia and tomato [21,27]. The deduced primary sequences of all the mature polypeptides are, nevertheless, very similar [cp.17]. Using one of the deduced sequences, a model for the folding of this polypeptide with respect to the lipid bilayer (Fig. 4) has been predicted [24] and tested [28]. Three membrane-spanning alpha-helical segments of ~23 residues were suggested from computer analyses of the sequence. Outside of the bilayer little other alpha-helical structure was predicted except for two short amphiphilic helical stretches that would be located just prior to the first and just after the third putative membrane helices (cf. Fig. 4). Several closely spaced beta-turns were predicted in the segment of protein connecting the putative second and third helices [cf. 24,28]. The percentages of alpha-helical and beta-structures, measured spectrally for this pigment-protein [29], agree well with the predicted amounts. The helices in the membrane are not aligned exactly perpendicular to the bilayer as shown in Fig. 4, but rather they make a net angle of 80° with respect to it [29]. We proposed [24] that the charged residues within the membrane may form salt bridges either between helices of the same polypeptide, which would stabilize the folding of the monomer, or between monomers which would yield an oligomeric structure (see section 2c and [30]). Since the N-terminus of the protein is firmly established to be in the stroma [31], and if the polypeptide chain spans the bilayer three times, its C-terminus will be in the lumen (Fig. 4). Support for the C-terminus being located in the lumen, and hence for a model in which the polypeptide spans the membrane three times, has come from tyrosine-labelling studies of the protein in right-side-out and inside-out thylakoid vesicles [25]. Data from work with deletion mutants [28] also is consistent with three spans. Alternatively, work with a synthetic peptide antibody to the C-terminal residues agglutinated thylakoids but not inside-out vesicles which suggests that the C-terminus is in the stroma and, therefore, that four membrane spans occur [26].

Electron microscope data on two-dimensional arrays of LHC II*b* [32,33] indicate that the polypeptide chain extends ~20Å on one side of the membrane and ~7Å on the other. If the three membrane-spanning helices are the only portions of the protein that are located in the lipid bilayer, as portrayed in Fig. 4, then the folding of the remainder of the protein would be consistent with such protrusions.

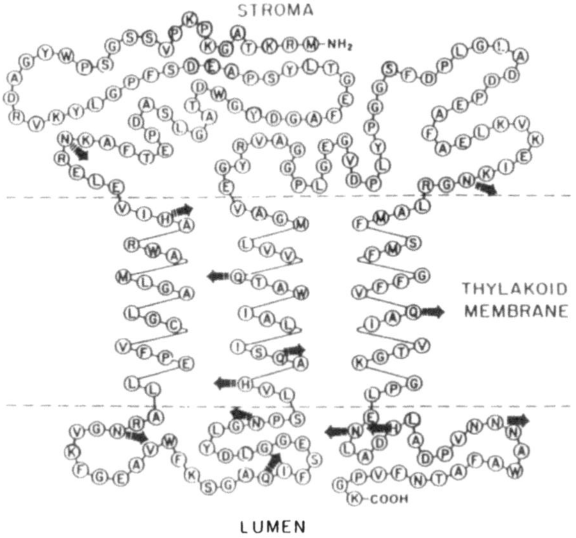

Figure 4. *Proposed model for the folding of the LHC IIb polypeptide chain [cf. 24] The chlorophyll molecules are proposed (see text) to be co-ordinated to the polypeptide chain via liganding to the conserved histidine, glutamine and asparagine residues (marked by* ⇛*)*

The location of the pigment molecules with respect to the folded polypeptide is of major interest. Direct determinations of their number per polypeptide have given equivocal results [cf.1], and hence the best estimate of the ratio, at present, is probably that deduced from the ratio of chl a/b in the complex, which has been repeatedly determined to be 1.15 $\pm$ 0.02. The most reasonable numbers that fit this ratio are 7 chl a and 6 chl b, and hence, from the known ratio of chl/carotenoid in LHC II*b*, 3 xanthophyll molecules per polypeptide. Histidine residues have been firmly established to coordinate chl molecules to polypeptides in purple and green bacteria [cf. 2]; however, there are far too few of them in LHC II*b* to accommodate 13 chl molecules. Glutamine and asparagine residues are the next mostly likely candidates and, interestingly, the number of conserved His + Gln + Asn in the LHC II*b* sequence [cf. 17] equals the proposed number of chl molecules; furthermore, those His + Gln + Asn residues that do not occur in membrane-spanning helices are almost invariably located close to the end of a helix (see Fig. 4), and thereby would be close to the lipid bilayer. It can be envisaged how either a chl a or chl b molecule can be selected to coordinate to a particular his, gln or asn residues so that when the helices are arranged in three dimensions the chl b molecules would be surrounded by chl a molecules to give the relative positions indicated by physical chemical data [34]. The porphyrin rings of the chl molecules are oriented at right angles to the plane of the membrane [cf. 35].

2c) The form of LHC II*b* in situ

In the model (Fig. 4) we have considered the structure of a monomer of LHC II*b*. Much evidence exists, however, that it occurs in situ in an oligomeric form(s): all of the complex can migrate on non-denaturing PAGE with a considerably larger apparent size than that of the supposedly monomeric form (e.g. Figs. 1 and 3, see also [36]), and electron diffraction data [32,33] are interpreted to indicate that LHC II*b* occurs as a trimer. We observe two oligomeric forms of LHC II*b* on Deriphat-PAGE of thylakoids or photosystem II particles, one having an apparent size of 72kDa and the other of 300kDa [3]. Whether the two forms represent two different arrangments of LHC II*b* in situ or whether the smaller form is derived from dissociation of the larger one, has yet to be elucidated. Several studies have indicated that the basic structure of LHC II*b* is trimeric; however, our preliminary protein cross-linking studies on LHC II*b* reveal that it may be tetrameric: the major linked form we observe has the size of a dimer of the apoproteins but trimeric and tetrameric sizes of the cross-linked proteins are also obtained. Incidentally, no unequivocal evidence has yet been presented to show that the smallest pigmented form (30-35kDa) of LHC II*b* observed on PAGE is indeed monomeric. The interaction between subunits in the oligomer may involve one or more of the three charged residues in the putative membrane-spanning helices (Fig. 4) [30]; salt bridges would permit more specific interactions than hydrophobic interactions alone.

Concluding remarks

Much remains to be elucidated about the four LHC II pigment-proteins. Thus, even though predicted amino acid sequences abound for LHC II*b* polypeptides [17] and those for the other LHC II components will appear soon, we are still a long way from knowing unequivocally the folding of these polypeptides with respect to the lipid bilayer, how the pigments are associated with the polypeptide, what are the oligomeric forms of each LHC II component and how they interact with each other and other photosystem II components to yield the entire structure. Some of the required information will come when their structures are determined by X-ray crystallography but in the interim, and to obtain other unequivocal data, more protein chemical studies are required to advance our knowledge of these important antenna components.

Acknowledgments

The authors wish to thank Drs. Jacques Breton and Don Bryant for particularly helpful discussions about the proposed model for the pigment-protein. Our research described in this paper was supported by grants from NSF and USDA (to JPT); GFP received graduate fellowships from the Shell and McKnight Foundations. We thank Dr. Elaine Tobin for very useful suggestions for improving our manuscript, and Ms. Camille Peterson for help in preparing it.

References

1. Thornber, J.P. 1986. Encyl. Plant Physiol., New Ser., 19, 98-142.
2. Diesenhofer, J.,O. Epp, K. Miki, R. Huber and H. Michel. 1985. Nature (Lond.) 318, 618-624.
3. Peter, G.F. and J.P.Thornber. 1988. J. Cell Biol., submitted
4. Ghanotakis, D.F. and G.T.Babcock. 1983. FEBS. Lett. 153, 231-234.
5. Camm, E.L. and B.R.Green. 1980. Plant Physiol. 66, 428-432.
6. Bassi, R and D. Simpson. 1987. Progr. Photosyn. Res. 2, 81-88.
7. Dunahay, T.G., G. Schuster, and L.A. Staehelin. 1987. FEBS. Lett. 215, 25-30.
8. Dunahay, T.G. and L.A.Staehelin. 1986. Plant Physiol. 80, 429-434.
9. Machold, O. and A. Meister. 1979. Biochim. Biophys. Acta 546, 472-480.
10. Barber, J. 1982. Annu. Rev. Plant Physiol. 33, 261-295.
11. Markwell, J.P., N.R. Baker, M. Bradbury, and J.P.Thornber. 1984. Plant Physiol. 74,

12. Thornber, J.P. and H.R. Highkin. 1974. Eur. J. Biochem. 41, 109-116.

13. Burke J.J., K.E. Steinback and C.J.Arntzen. 1979. Plant Physiol. 63, 237-243.

14. Darr, S.C., S.C. Somerville, and C.J.Arntzen. 1986. J. Cell Biol. 103, 733-740.

15. White, M.J. and B.R. Green. 1987. Eur. J. Biochem. 165, 531-535.

16. Ryrie, I.J. 1983. Eur. J. Biochem. 131, 149-155.

17. Chitnis, P.R. and J.P.Thornber. 1988. Photosynth. Res., in press.

18. Thornber, J.P., G.F. Peter, R. Nechushtai, P.R. Chitnis, F.A. Hunter, and E.M.Tobin. 1986. In: Regulation of Chloroplast Differentiation. Alan Liss, N.Y., pp.249-258.

19. Leutwiler, L.S., E.M. Meyerowitz, and E.M.Tobin. 1986. Nucl. Acid Res. 14, 4051-4064.

20. Tobin, E.M., J. Silverthorne, S. Flores, L.S. Leutwiler and G.A. Karlin-Neumann. 1987. In: Molecular Biology of Plant Growth Control, Alan Liss, NY. pp 401-411.

21. Dunsmuir, P., S.M. Smith, and J.Bedbrook. 1983. J. Mol. Appl. Genet. 2, 285-300.

22. Mattoo, A.K. and M. Edelman. 1987. Proc. Natl. Acad. Sci. USA. 84, 1497-1501.

23. Kuehlbrandt, W. 1987. J. Mol. Biol. 194, 757-762.

24. Karlin-Neumann, G.A., B.D. Kohorn, J.P. Thornber and E.M. Tobin. 1985. J. Mol. Appl. Genet. 3, 45-61.

25. Buergi, R., F. Suter and H. Zuber. 1987. Biochim. Biophys. Acta 890, 346-351.

26. Anderson, J.M. and D.J.Goodchild. 1987. FEBS. Lett. 213, 29-33.

27. Pichersky, E., R. Bernatzky, S.D. Tanksley, B.R. Briedenbach, A.P. Kausch and A.R.Cashmore. 1985. Gene 40, 247-258.

28. Kohorn, B.D., E. Harel, P.R. Chitnis, J.P. Thornber and E.M. Tobin. 1986. J.Cell. Biol. 102, 972-981.

29. Nebedryk, E., S. Andriaambinintsoa and J.Breton. 1984. Biochim. Biophys. Acta 765, 380-387.

30. Kohorn, B.D. and E.M.Tobin. 1987. J. Biol. Chem. 262, 12897-12899.

31. Steinback, K.E., J.J. Burke and C.J.Arntzen. 1979. Arch.Biochem.Biophys. 195,972-981

32. Kuehlbrandt, W. 1984. Nature (London) 307, 478-480.

33. Li, J. 1985. Proc. Natl. Acad. Sci. USA. 82, 386-390.

34. Knox, R.S. and R.L. Van Metter. 1979. CIBA Found. Symp. 61, 177-190.

35. Breton, J. 1986. Encyl. Plant Physiol., New Ser. 19, 319-36.

Section II

Organization: Molecular Genetics and Crystallography

MOLECULAR BIOLOGY OF ANTENNAS

Gerhart Drews
Insitute of Biology, Microbiology, Albert-Ludwigs-University, D-7800 Freiburg, Federal Republic of Germany

During the long lasting process of evolution a great variety of light-harvesting systems originated in photosynthetic organisms. As detailed in the introductory chapter of Section I, four main pigments are involved in the light-harvesting function (chlorophylls, bacteriochlorophylls, carotenoids and phycobilins), which are bound to polypeptides and are organized either in core complexes (reaction center and surrounding invariable antenna) or in variable peripheral complexes, which are either membrane-attached (chlorosomes and phycobilisomes) or integral membrane-bound antenna complexes. Thus light-harvesting pigment-protein complexes show a larger variability of their composition and organization in the various photosynthetic apparatuses as the reaction centers of the respective organisms. Some antenna complexes like chlorosomes and phycobilisomes are present only in distinct groups of organisms and are considered as characteristic taxonomical features of these organisms. Since the member of these groups live in different ecological habitats the antenna systems must be present in the organisms before they adapted to the ecological niche wherethe resprective organism live. So the great variety of antenna types might be originated very early in evolution.

In some eukaryotic and prokaryotic organisms the amino acid sequence of pigment-binding polypeptides have been determined recently by sequencing proteins or DNA. The results are not only important for structural investigations but they help

Photosynthetic Light-Harvesting Systems

also to detect highly conserved domains in these proteins. This type of analysis canbe used to explore different structure principles. Thus in bacterial membrane-bound antenna complexes two different short polypeptides with one central hydrophobic alpha helical domain bind the pigments while in chloroplast antenna one large polypeptide, possibly spanning the membrane more times, has the function of pigment-binding (see Ref. 1 and 2 and articles of Brunisholz, Drews and Thornber in this volume).

The most powerful method to study the structure and organization of antenna complexes is the high-resolution X-ray diffraction measurement of crystallized complexes. From the combination of this method with the analysis of the primary structure of the pigment-binding proteins the secondary and tertiary structure of the proteins can be deduced and the exact localization and orientation of pigment molecules and their neighboring and binding amino acyl residues determined (3). Phycobiliproteins are the only light-harvesting systems which have been crystallized and analyzed by high resolution X-ray spectroscopy (article of Schirmer et al.). Some bacterial light-harvesting complexes have been crystallized (4 - 7). Linear dichroism spectroscopy has revealed high ordered bacteriochlorophyll molecules in the crystallized complexes (4, 5, 7 and article of Welte et al.). Core complexes of Rhodopseudomonas viridis which are regularly arranged in a two-dimensional fashion and crystals of chlorophyll a/b protein complexes, have been studied combining electron diffraction with image processing of electron micrographs (article of Kühlbrandt). It will be a great progress if the crystallized core complexes are good enough for high resolution X-ray spectroscopy in order to learn more about the arrangement of those antenna molecules which donate excitation energy directly to reaction centers.

The research on antenna systems was strongly stimulated by the progress in molecular biology and genetics. The structure and organization of genes coding for pigment-binding proteins and tetrapyrrol and carotenoid synthesis have been determined in some bacteria, cyanobacteria, algae and plants. From the gene structure the amino acid sequence of polypeptides was deduced and many genetic studies on function-structure relationship were initiated.

In purple bacteria the genes coding for the biosynthesis of pigments and the pigment-binding proteins have been mapped, cloned and partially sequenced (8 - 12). Genes for bacteriochlorophyll, carotenoid synthesis and for the reaction center and B870 light-harvesting complexes are clustered on the genome of Rhodobacter capsulatus and Rb. sphaeroides (8-10, 12). Genes for the variable antenna complex are localized outside of this cluster (see article of Drews and Ref. 11).

In cyanobacteria groups of genes coding for proteins of phycobilisomes and photosystem I and II have been cloned and partially sequenced (see article of Bryant).

The arrangement of genes for antenna system in plants is complex because genes coding for constituents of the photosynthetic apparatus are localized on the nuclear and the chloroplast genome as well). Many plastid genes, coding for reaction center and antenna complexes, are organized in polycistronic transcription units of heterogenous composition and size (13).

Although similarities in DNA sequences, for example between subunits D-1 and D-2 of photosystem II and the L and M subunits of the bacterial reaction center or between sequences for biliproteins in cyanobacteria and red algae have been found, these data are not sufficient to deduce a homology in the meaning of evolutionary relationship among the photosynthetic organisms at present time (14).

Sequence information is important for studies on structure - function relationships (see introduction of section I). Molecular genetics has developed methods to test structural predictions on the function of specific amino acid positions in pigment-binding polypeptides. Amino acyl residues in specific positions of polypeptides can be replaced by other amino acids using site-specific mutagenesis(15). By this method the role of specific amino acyl residues for binding and positioning of pigment molecules can be determined and the secondary structure and orientation of proteins in the antenna complexes deduced, provided that enough data on amino sequence and pigment localization are available. Amino acyl residues, which are predicted from X-ray studies to bind bacteriochlorophylls and quinones, have been replaced by other residues (16). The results confirm the prediction that amino acids which can replace amino acyl residues in the neighborhood of the pigment molecules must have a very similar atomic structure and charge distribution.

Interposon mutagenesis, i.e. the introduction of marker-bearing insertions into points or deletions of cloned segments of the respective genes, is also a valuable technique for knock-out mutations or introduction of modified genes (17, 18). Randomly generated mutations by chemical agents or transposon mutagenesis and selection of markers by screening (for example lost of a specific antenna complex) is still the most efficient tool for obtaining functionally and structurally changed mutant strains (18). The transposon can be introduced into the chromosom or into genes cloned in plasmids and kept in Escherichia coli. The effect of mutations on structure and function of antennae of different organisms can help to find general principles of organization for light-absorption and efficient transfer of excitation energy to the photochemical reaction centers.

Cloning of genes and determination of DNA sequences has also considerably stimulated research on the biogenesis of the photosynthetic apparatus. The formation of constituents, their transport and translocation through membranes and the final assembly in the target thylakoid membrane are the major steps in the biogenesis. Regulation has been observed on all levels of this multistep-process. The influence of oxygen tension on messenger RNA formation for pigment-binding proteins has been studied in purple bacteria (article of Drews) or the role of light quality in mRNA formation during differentiation of chloroplasts (article of Westhoff et al.). Regulation on the level of transcription was investigated i) by hybridization of mRNA with specific DNA probes or ii) by measuring of promoter activity in cloned genes and iii) by mutation or partial deletion of promoter or enhancer elements and testing of promoter activity in expression vectors. The regulation on the level of translation has been studied by comparative measurements of the formation of specific mRNAs and their products. The assembly process in or on the membrane depends from pigments, proteins, ATP, electro-chemical proton gradient across the membrane and possibly other proteins essential for posttranslational processes. The multistep process of formation of light-harvesting complexes is presumably regulated by a cascade of regulatory factors which have to be studied in the future. We are at the very beginning to understand how photosynthetic organisms build the photosynthetic apparatus under the directive of the genome and specific environmental factors. In some systems parts of the signal chain between the external factor and the gene regulation habe been identified, so the phytochrome system which controls the chloroplast differentiation in plants under the regime of light.

The ideas and methods of molecular genetics stimulate and force the investigations of structure - function relationship but determine the progress in the research on biogenetics and differentiation of the antenna systems in all photosynthetic organisms. For all organisms in principle the experimental approach has been developed.

References

1. Zuber, H. 1986. Trends Biochem. Sci. 11, 414

2. Drews, G. 1985. Microbiol. Rev. 49, 59.

3. Deisenhofer, J., H. Michel, R. Huber, 1985. Trends Biochem. Sci. 10, 243.

4. Allen, J.P., R. Theiler, G. Feher. 1985. In: Antennas and Reaction Centers of Photosynthetic Bacteria (M.F. Michel-Beyerle, ed.) Springer Verlag, Berlin, Heidelberg, New York pp. 82-84

5. Cogdell, R.J., K. Woolley, R.C. Mackenzie, J.G. Lindsy, H. Michel, J. Dobler, W. Zinth. 1985. In: Antennas and Reaction Centers of Photosynthetic Bacteria (M.E. Michel-Beyerle, ed.) Springer Verlag, Berlin, Heidelberg, New York. pp. 85-87.

6. Wacker, T., N. Gad'on, A. Becker, W. Mäntele, W. Kreutz, G. Drews, W. Welte. 1986. FEBS Lett. 197, 267.

7. Welte , W., T. Wacker, M. Leis, W. Kreutz, J. Shiozawa, N. Gad'on, G. Drews. 1985. FEBS Lett. 182, 260.

8. Yen, H.C., B. Marrs, 1977. J. Bacteriol. 126, 619.

9. Zsebo, K.M., J.E. Hearst. 1984. Cell 37, 937.

10. Youvan, D.C., R.J. Bylina, M. Alberti, H. Begusch, J. Hearst. 1984. Cell 37, 949.

11. Ashby, M.K., S.A. Coomber, C.N. Hunter. 1987. FEBS Lett. 213, 245.

12. Kiley, P.J., T.J. Donohue, W.A. Havelka, S. Kaplan. 1987. J. Bacteriol. 169, 742.

13. Hagemann, R., M.M. Lindenhahn, M. Metzlaff. 1985. Progr. Bot. 47, 208.

14. Doolittle, W.F. 1982. In: The Biology of Cyanobacteria (N.G. Carr and B.A. Whittong, eds.) Blackwell Sci. Publ. pp. 307-331.

15. Zoller, M.J., M. Smith. 1984. DNA 3, 479.

16. Bylina, E.J., S. Ismail, D.C. Youvan. 1986. In: Microbial Energy Transduction (D.C. Youvan, F. Daldal., eds.) Cold Spring Harbor Laboratory, New York, pp. 63-70

17. Scolnik, P., R.H. Haselkorn. 1984. Nature (London) 397, 289.

18. Scolnik, P.A., B.L. Marrs. 1987. Ann. Rev. Microbiol. 41, 703.

HIGH-RESOLUTION CRYSTAL STRUCTURE OF C-PHYCOCYANIN AND POLARIZED OPTICAL SPECTRA OF SINGLE CRYSTALS

T.Schirmer*), W.Bode, R.Huber

Max-Planck-Institut für Biochemie, Am Klopferspitz, D-8033 Martinsried, F.R.G.

Several biliprotein crystals are composed of oligomeric aggregates which most probably closely resemble the trimeric and hexameric discs of native phycobilisomes (1-3). Therefore, X-ray analysis of these crystals yields not only information about the monomeric structure, but about the interaction of the monomers and relative geometry of their chromophores as well. Particularly, the structure of C-phycocyanin (C-PC) hexamers from *Agmenellum quadruplicatum* (2, 4) shows that there is a multitude of paths for intra-hexameric energy transfer. Chromophore α84 and β84 of the neighbouring monomer (same trimer) have the most favourable relative position, whereas α84 - α84 and β155 - β155 seem to represent the main paths for inter-trimeric energy transfer. Energy transfer across the interface of hexamers in the phycobilisome rods most probably proceeds from β84 to β84.

The crystal structure of C-PC from *Mastigocladus laminosus* has recently been refined (4). The high resolution (2.1 Å) of this structure allowed an accurate determination of the geometry of all three chromophores. All chromophores are defined in their entirety and exhibit very similar geometry in their tetrapyrrole part. The structure of a phycocyanobilin chromophore resembles a cleaved porphyrin which has been unfolded by a twist of roughly 180° around the C5-C6 and C14-C15 bonds. Accordingly, the configuration / conformation of

----

*) present address: Laboratory of Molecular Biology, M.R.C., Cambridge CB2 2QH, U.K.

Photosynthetic Light-Harvesting Systems

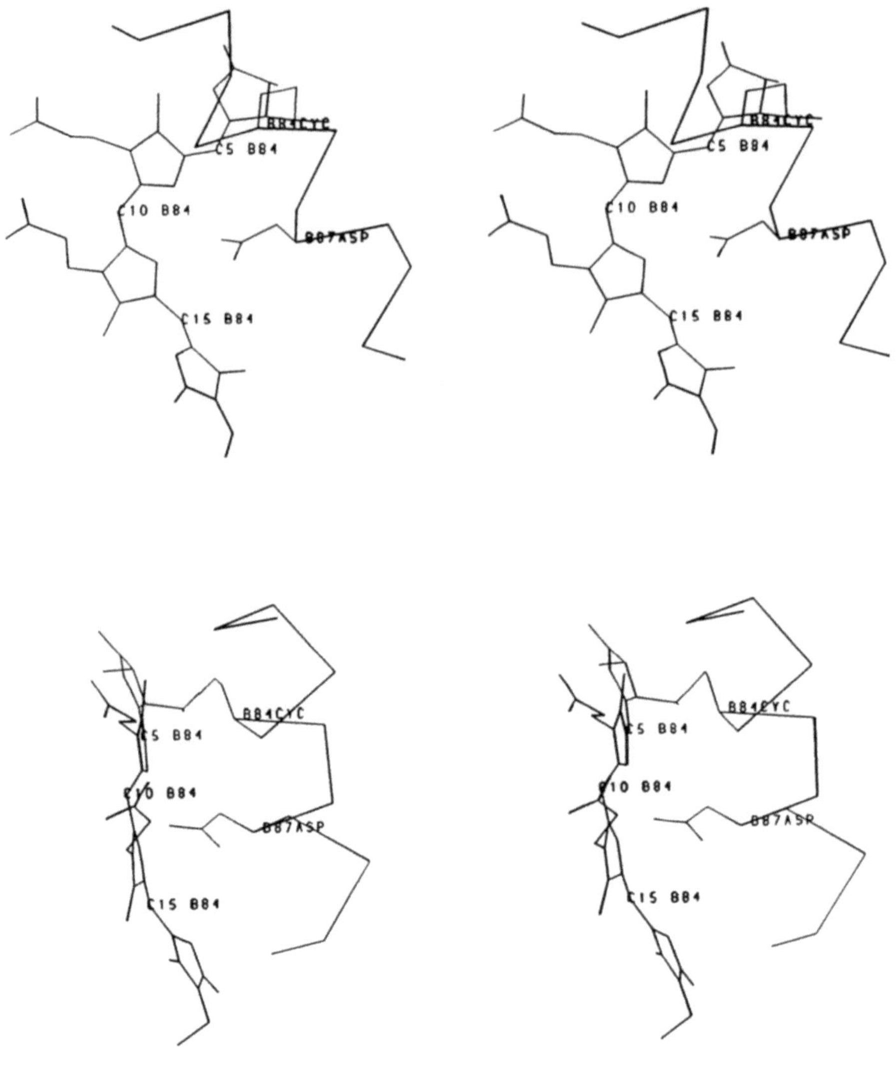

Fig. 1:
a) Stereoscopic view of chromophore β84, the side-chain of aspartate β87 and the Cα-backbone (part of helix E).
b) Same model, rotated by approximately 90 degree around the vertical direction.

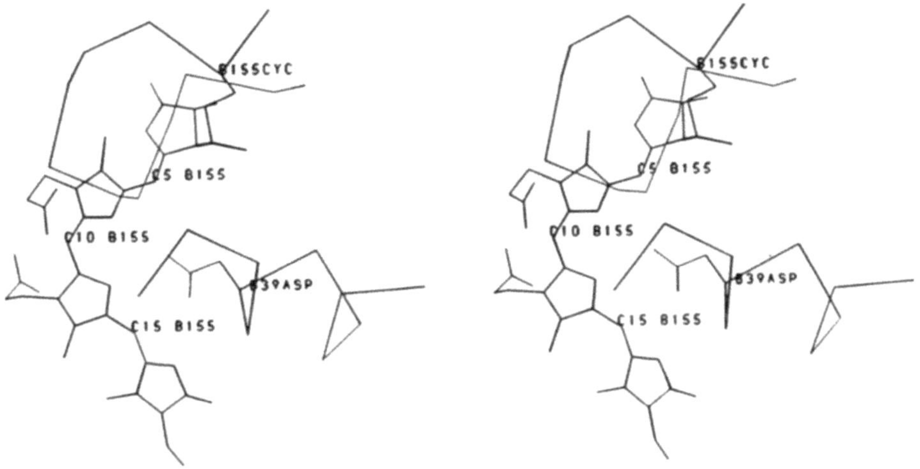

Fig. 2:
a) Stereoscopic view of chromophore β155, the side-chain of aspartate β39 and the Cα-backbone (G-H loop and part of helix A).

the chromophores is Z-anti, Z-syn, Z-anti (with the exception of the "configuration" of C14-C15 of chromophore β155, which is almost midway between Z and E).

All three chromophores show a common principle of interaction with the protein. They arch around aspartate residues (α87, β87, β39), the nitrogen atoms of pyrroles B and C being within hydrogen-bonding distance to one of the carboxylate oxygens (Figs. 1 and 2). Most of the propionic side-chains of the chromophores form salt-bridges with arginine and lysine residues. In view of the local symmetry between the α- and β-subunit (1), the similarity of the structures of α84 and β84 is not surprising, although it has to be considered that they occupy quite different positions in the trimeric aggregate. α84 is shielded by the neighbouring monomer, whereas β84 is located at the inner side of the central channel. It is striking, however, that also β155 exhibits a very similar geometry, despite its distinct position in the subunit at the G-H loop. Moreover, the same chromophore-aspartate interaction is observed, the interacting aspartate β39 being provided by helix B (Fig. 2).

The three chromophores of C-PC differ in their spectral properties (see e.g. ref. 5). To assign the different spectra to the individual chromophores, polarized optical spectroscopy on single crystals together with the knowledge of the chromophore positions and orientations provides a powerful method.

Polarized absorption and fluorescence spectra of single C-PC crystals from *Mastigocladus laminosus* have been recorded (6). The crystals show dichroism. The position of the absorption maximum shifts from 612 nm (E parallel to the optical axis c) to 626 nm (E perpendicular to c). The polarized absorption spectra can be simulated well only - using the deconvoluted solution spectra of the individual chromophores (5) and the orientations of the chromophores (4) - if the short and long wavelength component spectra of the β-subunit (β_s and β_f, respectively) are assigned to chromophores β155 and β84, respectively.

The polarization component of the fluorescence from single crystals is about 2.9 times more intense perpendicular to the three-fold trimer axis than parallel to it. From the known inclination of the chromophores towards the trimer axis intensity ratios of 11.0, 1.6 and 0.25 would be expected, if one of the chromophores α84, β84 and β155, respectively, would fluoresce solely. Fluorescense originating from both chromophores α84 and β84, however, would yield an intensity ratio of 3.0, comparing favourable with the observed ratio. It can be excluded that β155 contributes a substantial part to the observed fluorescense. This gives further direct evidence that β155 is the β_s-chromophore. It can be deduced, therefore, that the net energy flow in the trimers (and hexamers) is directed from the peripheral β155 chromophores to the α84 - β84 pair.

References

1. Schirmer, T., W. Bode, R. Huber, W. Sidler, H. Zuber. 1985. J.Mol.Biol. 184, 257.

2. Schirmer, T., R. Huber, M. Schneider, W. Bode, M. Miller, M.L. Hackert. 1986. J.Mol.Biol. 188, 651.

3. Rümbeli, R., T. Schirmer, W. Bode, W. Sidler, H. Zuber. 1985. J.Mol.Biol. 186, 197.

4. Schirmer, T., W. Bode, R. Huber. 1987. J.Mol.Biol. 195. 677.

5. Mimuro M., P. Füglistaller, R. Rümbeli, H. Zuber. 1985. Biochim. Biophys. Acta. 848. 155.

6. Schirmer. T., M.G. Vincent. 1987. Biochim. Biophys. Acta. 893. 379.

CRYSTALLIZATION AND SPECTROSCOPIC INVESTIGATION OF PURPLE BACTERIAL B800-850 AND RC-B875 COMPLEXES

W.Welte, T.Wacker, A.Becker

Inst.f.Biophysik&Strahlenbiologie der Universität
Albertstr. 23, 78 Freiburg, FRG

1. Introduction

In this chapter we want to give a report of our work with crystallized pigment-protein complexes and a introduction to the evaluation of spectra obtained with polarized light from crystals together with a brief discussion of structural results obtained with other techniques.

2. Purification and Crystallization of B800-850 and RC-B875 complexes

The B800-850 light harvesting complex and the reaction center (RC)-B875 complex are both intrinsic membrane protein complexes of the intracytoplasmic membranes of purple bacteria. Crystallization requires homogenous solutions of purified protein complexes. All proteins present in the solution thus must be precise copies to atomic resolution of one macromolecular structure. Consequently , the protein containing membranes first must be solubilized with suitable detergents. Subsequently the mixed detergent-protein micelle which is to be crystallized (1), must be purified by separation from other protein-detergent micelles as well as detergent-lipid mixed micelles. Lauryldimethylamine-oxide (**LDAO**) often is used as a detergent for solubilizing and purifying the purple bacterial Bchl-protein complexes. A bulk separation of the RC-B875 and of the B800-850 complex was achieved by loading the solubilized intracytoplasmic membrane on top of a sucrose gradient, typically a step gradient with layers of 0.3, 0.6 and 1.2 M sucrose containing 0.6 % of LDAO and centrifuging at 45000 rpm for 15 h (2).

The two upper pigmented bands of the gradient were enriched in the B800-850 complex and the RC-B875 complex respectively. The purity of

Photosynthetic Light-Harvesting Systems

these preparations could be improved by anion exchange chromatography in the presence of Tris buffer, pH 8 and 0.05% LDAO (3) .
Nevertheless, the result of the crystallization of both complexes depended on a final purification by chromatofocusing (4). By this method, the proteins are separated according to their isoelectric point (IEP) by a pH gradient which is formed and migrates downwards in the column. As the isoelectric point is determined by the distribution of charged amino acids located on the hydrophilic surface of the protein, the procedure has a capacity of sorting out denatured proteins as well as protein complexes containing undesired polypeptides. Some proteins may be labile at the isoelectric point. This holds for instance for the B800-850 complexes which are stabile at alkaline pH around 8., but possess IEP-s around pH 7.. The RC-B875 complex in general is more labile at all conditions than the B800-850 complex. In order to stabilize the complexes during chromatofocusing, we run the columns at 4° C and add 15% glycerol to the elution buffers. Glycerol, as well as sugars are known to stabilize proteins by repulsive interaction with the protein surface (5).
If we crystallized the B800-850 prepared without chromatofocusing according to (6) we got a mixture of triclinic and orthorhombic crystals. However, when we included chromatofocusing as final step of the purification, the crystals were exclusively orthorhombic. We think that some lipid bound to the protein may be responsible for the triclinic crystals.
If we omitted glycerol from the elution buffers, the crystals of the B800-850 complex frequently changed their color from red to green during crystallization, indicating weakly bound carotenoid. The color change was accompanied with a lower yield of crystals. Some preparations lacking carotenoids did not crystallize at all.

The RC-B875 preparation eluting from DEAE chromatography was separateded into three bands by chromatofocusing, which eluted at pH 7.2 , pH 6.9 and pH 6.4. The first two bands were spectrally similar. The difference of the spectra of the third band from either the spectrum of the first or the second band gave a spectrum typical for B800-850. We think that the first two bands represent aggregates of both complexes possibly held together by lipids. The best yield of crystals was obtained from the third band. As an alternative to LDAO, we also chromatofocused the RC-B875 complex in 1-dodecyl propanediol-3-phosphorylcholine (7). The crystals obtained in this detergent grew to a bigger size as compared to those grown in LDAO. It was however difficult to handle these crystals as they had a tendency to dissolve when taken out of the original crystallization chamber.

2. X-ray and spectroscopic studies with crystals

So far we were unable to get high resolution X-ray diffraction from either the crystals of the B800-850 complex or the crystals of the RC-B875 complex. The outermost diffraction spots were at 6 Å for the former and at 30 Å for the latter crystals.

We therefore tried to get structural information from the crystals by optical spectroscopy. By spreading the presaturated protein solution on a cover slide, which is equilibrated with a reservoir solution in a closed box, thin crystals of suitable optical density can be grown.

When light of a wavelength λ and polarized along a unit vector $\vec{p}$ passes through a unit cell containing the transition moments $\vec{t}_i$:(t_{1i}, t_{2i}, t_{3i}) it is absorbed according to the formula for the absorption A :

$$A = \text{const.} \sum (\vec{p} \cdot \vec{t}_i)^2 \qquad [1]$$

Several conclusions can be drawn form this formula :

1) If the $\vec{t}_i$ possess a three-or higher fold crystallographic symmetry axis the absorption is constant when $\vec{p}$ rotates within the plane orthogonal to the axis.

2) If the $\vec{t}_i$ are all inverted, the absorption does not change. The absorption process thus adds an inversion symmetry to the point symmetry group of the $\vec{t}_i$.

3) It is possible to simplify the formula [1] by using some tensor mathematics (8).

$$(\vec{p} \cdot \vec{t}_i)^2 = \sum_{kl} (p_k p_l \cdot t_{ki} t_{li}) = \vec{p}\, T_i\, \vec{p} \qquad \textbf{[2]}$$

with the tensor T_i :

$$\begin{matrix} t_{1i}^2 & t_{1i}\, t_{2i} & t_{1i}\, t_{3i} \\ t_{2i}\, t_{1i} & t_{2i}^2 & t_{2i}\, t_{3i} \\ t_{3i}\, t_{1i} & t_{3i}\, t_{2i} & t_{3i}^2 \end{matrix} \qquad \textbf{[3]}$$

As this matrix of the tensor T_i is real and symmetric, it can be orthogonalized by rotating the coordinate system to the principal axes. The formula then assumes the form :

$$A = \text{const.} \, (p_1^2 \cdot T_{11}^2 + p_2^2 \cdot T_{22}^2 + p_3^2 \cdot T_{33}^2) \qquad \textbf{[4]}$$

For polar coordinates the direction cosines are :

$$\begin{aligned} p_1 &= \sin(\vartheta) \cdot \cos(\varphi) \\ p_2 &= \sin(\vartheta) \cdot \sin(\varphi) \\ p_3 &= \cos(\vartheta) \end{aligned} \qquad \textbf{[5]}$$

We can introduce a vector $\vec{v} = (T_{11} \cdot p_1 , T_{22} \cdot p_2 , T_{33} \cdot p_3)$ the length of which is equal to the square root of the absorption A (up to a constant), which satisfies the equation of an ellipsoid :

$$\frac{v_1^2}{T_{11}^2} + \frac{v_2^2}{T_{22}^2} + \frac{v_3^2}{T_{33}^2} = 1 \qquad \textbf{[6]}$$

with half axes T_{11} , T_{22} and T_{33} . The advantage of this definition is, that we can describe the result of the measurement of $\sqrt{A}$ as a function of the polarization vector by three half axes which are proportional to the diagonal elements of the orthogonalized tensor T_i .We must however note, that the new vector $\vec{v}$ is not parallel to the polarization vector $\vec{p}$. However, both are parallel when pointing into the direction of the ellipsoid half axes.

A pigment-protein complex contains several transition moments absorbing at a fixed wavelength. In a crystal, several of these complexes are arranged to form the asymmetric unit (9). Finally all point symmetry elements form images of the asymmetric unit. The superposition of all these images gives the transition moments of the unit cell. Intuitively, when proceeding along these steps to the larger aggregates, the dichroism, i.e. the differences between the half axes a,b and c, gets smaller or remains at best equal. The lower the symmetry of a crystal, the more information can be gained about the geometry of the pigments of the single complex from the crystal spectra obtained with polarized light.

We would like to discuss two cases here, spectra from crystals with high- and with low dichroism :

2.1. *Spectra from crystals with high dichroism :*

2.1.1. Spectra from rectangular crystal platelets of B800-850 complexes of R.acidophila, obtained with light polarized along the long and the short edge of the crystals (Fig. 6 in (10)).

The Q_y transition moments of the B800 and the B850 chlorophylls absorb preferentially along the short edge of the crystals, while the Q_x transition moments of both types of Chlorophylls and the carotenoid band at approximately 525 nm absorb preferentially along the short edge. We can thus conclude, that the B800 and B850 Q_y transition moments of the complex are confined roughly to a plane orthogonal to the long edge of the crystal, while the Q_x transition moments as well as the carotenoid transition moment of the complex are roughly aligned and parallel to the long edge of the crystal. These results from the crystals are in agreement with linear dichroism spectra from uniaxially oriented B800-850 complexes(11). These authors used membranes containing the B800-850 complex which was oriented uniaxially by air-drying of the membranes. A similar uniaxial orientation can be obtained by stretching polyvinyl alcohol films into which detergent solubilized B800-850 complexes are embedded(12). Spectroscopic results as given in (11-13) were used to propose a model for the structure of the B800-850 complex (13). In a recent study it was shown, that the α-helices are tilted (14).

2.1.2. Spectra from crystals of the RC-B875 complex (corecomplex) from R. palustris published in (3).

These crystals have the habit of a thin, rectangular platelet with a long edge called x-edge , a small edge called y-edge and an axis normal to the surface of the platelet, called z-edge.

The spectra show, that the Q_x transition moments of the B875 chlorophylls absorbing at 590 nm, are preferentially aligned along the x-edge, the complexes thus are arranged roughly parallel or antiparallel to each other in the unit cell.

The Q_y -transition moments absorbing at 875 nm are distributed in the y-z-plane. However, upon tilting the polarization vector from the y-edge to the z-edge, it is seen, that the Q_y transition moments are distributed in an dichroic manner around the RC. Our results are consistent with a one-or two fold symmetry axis in the distribution of the Q_y -transition moments. We think, that this symmetry of the arrangement of the B875 antennae complexes associated with the RC would fit reasonably with the observed crossection of the RC (15). Such an arrangement could be of importance for a quick and efficient transfer of excitation energy to the special pair of the RC.

In the thylakoids of *Rps .viridis* and of several other BChl b containg purple bacteria, the corecomplex forms hexagonal arrays. Electron micrographs have been subjected to image filtering- and reconstruction(16 -18). On the basis of the results, a model with a hexagonally symmetric arrangement of B875 complexes around the RC has been proposed(16 ,17) . These models do seem to be conflicting with our results. *Rps.palustris* is a BChl-a containing organism. It might be that this class has a different arrangement of B875 complexes. The difference in the wavelength of the far infrared absorption could be a consequence not only of the different type of BChl but also of a structural difference. With respect to the size, both types of corecomplexes are similar and possess some 30 BChl molecules or some 12 B875 units per RC .

2.2. *Spectra from crystals with low dichroism*

If the dichroism is not of the order of magnitude of 90% or more, it is likely, that crystal symmetry has diminished the dichroism of the single complex. This is the case in the orthorhombic crystals from B800-850 complexes from *Rps.palustris* (19).

In this case, in order to gain structural information from the crystal spectra, it is necessary to know the symmetry elements of the crystal. In the case of crystals from the B800-850 complex from *Rps.palustris* the space group could be determined by X-ray diffraction to be orthorhombic, C222. The point group thus consists of three mutually orthogonal 2-fold axes. These axes thus have to be applied to every transition moment of the single B800-850 complex.

Let t_{1i}, t_{2i} and t_{3i} be the direction cosines of a single transition moment $\vec{t}_i$ of the B800-850 complex along the unit cell axes. The images of this transition moment by the two-fold axes then are :

$$\begin{array}{lll} t_{1i}, \ t_{2i}, \ t_{3i} & \text{identity operation} \\ t_{1i}, \ -t_{2i}, \ -t_{3i} & \text{two fold axis along x-axis} \\ -t_{1i}, \ t_{2i}, \ -t_{3i} & \text{two fold axis along y-axis} \\ -t_{1i}, \ -t_{2i}, \ t_{3i} & \text{two fold axis along z-axis} \end{array} \qquad [7]$$

If we use [3] to calculate T_i from these vectors, the result is :

$$\begin{matrix} 4t_{1i}^2 & 0 & 0 \\ 0 & 4t_{2i}^2 & 0 \\ 0 & 0 & 4t_{3i}^2 \end{matrix} \qquad [8]$$

As is seen, the matrix is orthogonal in the coordinate system formed by the three unit cell vectors of the orthorhombic space group. This is a handsome consequence of the orhogonality of these axes and the two-fold axes. The aforementioned ellipsoid of the images of every transition moment of the B800-850 complex, i.e. of the unit cell, thus has its half axes parallel to the unit cell axes of the orthorhombic unit cell.

By inserting [8] and [5] into [4], we can calculate A as a function of angles with respect to the unit cell axes. The ellipsoid half axes of the transition moment $\vec{t}_i$ thus are $a_i = 4\,t_{1i}^2$, $b_i = 4\,t_{2i}^2$, $c_i = 4\,t_{3i}^2$. In order to get the absorption of all transition moments, we have to calculate the sum over all transition moments i. The resulting half axes : $a = \sum a_i$, $b = \sum b_i$, $c = \sum c_i$, can be directly compared with those determined experimentally by measuring the absorptions with the polarization vector along the three unit cell vectors and taking the square roots from them. Thus there is a set of equations for each absorption band, connecting the corresponding transition moments of a single complex with the half axes determined experimentally.

These relations can be used in different ways : If we know the arrangement of one type of transition moments in the single complex by other experiments, we can use the ellipsoid equations to calculate the orientation of this set of transition moments with respect to the unit cell axes. The Q_y-transition moments of the BChl 800 and 850 for instance are known from other studies to be parallel to a common plane but distributed in a nondichroic manner within that plane(11,13). The ellipsoid equations then can be used to calculate the unknown direction cosines of the normal to that plane with the unit cell axes.

After knowing this orientation, we can attend to the ellipsoid of the absorption at 595 nm of the Q_x transitions. As the Q_x transitions of the B800 and of the B850 chlorophylls absorb at the same wavelength, there will be two unknown ellipsoids and only one experimentally known set of half axes. If the tilting angle of Q_x with respect to the aforementioned normal is unknown for each of the two sets of Bchl-s, we can expect to get a relation connecting both tilting angles with each other.

References :

(1) Michel,H. 1983. Trends in the Biochem. Sci. 8 ,56-59.

(2) Firsow,N.J. and Drews,G. 1977. Arch.Microbiol. 115,299-306.

(3) Wacker,T., Gad'on,N., Becker,A., Mäntele,W., Kreutz,W., Drews,G. and Welte,W. 1986. FEBS Lett. 197 , 267-273.

(4) Chromatofocusing, Pharmacia Fine Chemicals AB, Uppsala, Sweden.

(5) Arakawa,T. and Timasheff,S.N. 1985. Biochemistry 24 ,6756-6762.

(6) Welte,W., Wacker,T. ,Leis,M., Kreutz,W., Shiozawa,J., Gad'on,N. and Drews,G. 1985. FEBS Lett. 182 , 260-264.

(7) Weltzien,H.U., Richter,G. and Ferber. 1979. Eur.J.Biol.Chem. 254 , 3652-3657.

(8) Arfken,G. 1970. Mathematical methods for physicists, Academic Press New York.

(9) Buerger,M.J. 1970. Contemporary crystallography. McGraw Hill, New York.

(10) Cogdell,R.J., Woolley,K., Mackenzie,R.C., Lindsay,J.G., Michel,H., Dobler,J. and Zinth,W. 1985. In : Antennas and Reaction Centers of Photosynthetic bacteria, (Michel-Beyerle,ed.) Springer Verlag Berlin, pp. 85-87.

(11) Breton, J., Vemeglio, A., Garrigos, M. and Paillotin. G., 1981. In : Proceedings of the Fifth Int. Photosynthesis Congress, Vol III (G. Akoyunoglou, ed.). Balaban Int. Science Services, Philadelphia, USA. pp 445-459.

(12) Bolt,J. and Sauer,K. 1979. Biochim.Biophys.Acta, 546, 54-63.

(13) Kramer, H.J.M., van Grondelle, R.,Hunter, C.N., Westerhuis, W.H.J. and Amez, J. 1984. Biochim. Biophys. Acta 765, 156-165.

(14) Breton,J. and Nabedryk,E. 1984. FEBS Lett. 176, 355-359.

(15) Deisenhofer,J., Epp,O., Miki,K., Huber,R. and Michel,H. 1985. Nature 318 , 618-624.

(16) Miller,K.R.. 1982. Nature 300,53-55.

(17) Stark,W., Kühlbrandt,W., Wildhaber,I., Wehrli,E. and Mühlethaler,K. 1984. EMBO J. 3,777-783.

(18) Engelhardt,H., Baumeister,W. and Saxton,W.O. 1983. Arch.Microbiol. 135,169-175.

(19) Wacker,T., Steck,K., Becker,A., Drews,G., Gad'on,N., Kreutz,W., Mäntele,W. and Welte,W.1986 In : Prog. in Photosynthesis Research (J.Biggins,ed.) Martinus Nijhoff Publ., Dordrecht,NL. I.4.383

STRUCTURE OF THE LIGHT-HARVESTING CHLOROPHYLL a/b-PROTEIN COMPLEX FROM CHLOROPLAST MEMBRANES

W. Kühlbrandt

Biophysics Section, Department of Physics and AFRC Photosynthesis Group, Department of Pure and Applied Biology, Imperial College of Science and Technology, Prince Consort Road, London SW7 2BZ, England

1. Introduction

The light-harvesting chlorophyll a/b-protein complex (LHC II) is the main antenna protein of green plants and algae. It is an integral membrane protein which is found predominantly in proximity to the reaction centre complex of photosystem II in the chloroplast grana. Apart from intercepting solar energy, LHC II has two other important functions in mediating cation induced membrane appression and hence grana formation, and in regulating the distribution of excitation energy to photosystems I and II (for reviews, see 1-3).

The LHC II polypeptide is the product of a nuclear multi-gene family (4,5). The mature protein isolated from pea chloroplasts contains at least three polypeptides (6) which seem to be very similar in terms of their peptide sequence (7). Two genes encoding the major polypeptide of pea LHC II have been sequenced (4,8), indicating a molecular weight of 25,000 (9). Each polypeptide binds chlorophyll a (Chl a) and Chl b at a molar ratio of 1.15 (6) as well as apparently non-stoichiometric amounts of carotenoids (10) and lipid.

Two-dimensional (11) and three-dimensional crystals (6) of LHC II have been obtained and used to determine the molecular structure of the complex by electron crystallography. The structure of LHC II was also investigated by other methods, including microspectrophotometry, analytical ultracentrifugation and time-resolved fluorescence spectroscopy.

Photosynthetic Light-Harvesting Systems

2. Crystal Structure of LHC II

Electron microscopy and image processing of negatively stained two−dimensional crystals revealed the overall three−dimensional shape of the complex at 16 Å resolution. The complex was a trimer with a three−fold axis of symmetry in z−direction, perpendicular to the plane of the membrane crystal. The trimer was asymmetrical with respect to the crystal plane (coincident with the plane of the chloroplast membrane) and spanned the membrane crystal, measuring 60 Å in thickness (11).

The structure of thin three−dimensional crystals growing in the shape of hexagonal plates was analyzed at 7 Å resolution in z projection. Specimens preserved in glucose or in tannic acid diffracted electrons to 3.7 Å resolution (6). The dimension of the unit cell (a= 127 Å) and the symmetry of the diffraction pattern was the same as had been found for two−dimensional crystals, indicating that the hexagonal crystals consisted of layers of highly ordered two−dimensional crystals which have a tendency to stack. Image analysis of low−dose, high−resolution micrographs of very thin specimens preserved in tannic acid revealed that the stacked layers were not precisely in register but translationally displaced with respect to one another by seemingly random amounts so that interference between layers perturbed the apparent crystal symmetry in projection. A method for locating the displaced layers by cross−correlation with a low−resolution projection map was developed. The effects of interference were reversed after least−squares refinement of displacement vectors. Corrected phases of structure factors from six crystals agreed with one another and were averaged and combined with electron diffraction amplitudes.

A 7 Å projection map was generated. The map confirmed that LHC II was a trimer, consisting of three monomers related by three−fold symmetry. Regions of density above average showed the distribution of protein. Within the monomer, several peaks were resolved. The dimensions of these peaks and the distance between them suggested α−helices in a direction roughly perpendicular to the membrane plane. The density of regions in the centre and on the periphery of the trimer was below average. A comparison with the projection map of negatively stained two−dimensional crystals showed that these regions were stain−excluding and therefore contained low−density matter, such as chlorophyll and lipid. An estimate of the volume enclosed by the three LHC II monomers indicated that most of the Chl could be accommodated in the central region of the trimer (12).

3. Optical Properties of LHC II Crystals

To find out more about the orientation of Chl molecules in the complex, the Chl dichroism of hexagonal LHC II crystals was investigated by microspectrophotometry. There was no detectable linear dichroism in the x,y−plane, as expected for an antenna complex with a three−fold axis of symmetry normal to the crystal plane. Due to this known orientation of the complex, the observed dichroism reflected the absorption characteristics of individual LHC II trimers in the crystals and therefore probably *in vivo*. The linear dichroism of Chl molecules in z direction was determined by extrapolation of measurements on tilted crystals. The absorbance in the Q_y region of Chl a and Chl b in z−direction was 35 % higher than in the crystal plane, indicating a partial orientation of the Q_y transition moments in this plane. The weaker Q_x transition moments, on the other hand, showed no linear dichroism in z direction and seemed to be wholly isotropic. The linear dichroism of carotenoids could not be measured due to the high optical density of the crystals in this region of the spectrum. Since the x,y−plane of the crystals coincided with the plane of the thylakoid membrane, it seemed that LHC II absorbed more light in a direction perpendicular to this plane, presumably because, on average, more light reaches the antenna from this direction owing to the proximity of other Chl protein complexes in the same membrane and the average, flattened shape of chloroplast grana.

4. Aggregate Size of LHC II in Solution

A simple, generally applicable technique was developed for determining the molecular weight, and hence the aggregate size of membrane protein complexes in detergent solution by sedimentation equilibrium ultracentrifugation of the protein at dialysis equilibrium with the detergent. Applying this method to LHC II solubilized in 35 mM n−octyl−β,D−glucopyranoside (OG) it was found that the complex was a trimer, dissociating into monomers and dimers at monomer concentrations below 2 μM. The accurate determination of protein concentration required for this method also allowed the molar Chl to protein ratio to be measured as 15. The Chl a to Chl b ratio of pure LHC II (6) was consistent with 8 Chl a and 7 Chl b per monomer (14).

5. Circular Dichroism and Time–Resolved Fluorescence of LHC II

Circular dichroism measurements indicated a Chl b exciton of three–fold symmetry and a Chl a–Chl b exciton that did not show this symmetry. Both were characteristic of LHC II trimers solubilized in OG. A third, "quenching" Chl b exciton became prominent upon precipitation of the complex. Picosecond time–resolved fluorescence revealed that efficient energy transfer between Chl molecules within the solubilized complex depended on the three–fold symmetry which was disrupted by high detergent to protein ratios (15).

6. General Conclusions

Taken together, these results are consistent in showing that LHC II is a trimer having three–fold symmetry, not only in the crystalline state but also in detergent solution. The trimer is thus the preferred, most stable form of LHC II. It seems more than likely that the complex has the same configuration *in vivo*, since circular dichroism and time–resolved fluorescence have shown that only the trimeric complex is functionally active in isolated LHC II.

Comparing the structure of LHC II to that of other light–absorbing protein complexes, the universal occurrence of three–fold symmetry is striking. Bacteriorhodopsin in the purple membrane of *Halobacterium halobium* (16), the water–soluble bacteriochlorophyll a–protein complex from *Prostecochloris aestuarii* (17), the c–phycocyanin from *Mastigocladus laminosus* (18) and the photosystem I reaction centre complex from *Synechococcus* (19) all show this symmetry. Even the light–harvesting complexes of the purple bacterium *Rhodopseudomonas viridis* and related species which surround the reaction centre in a ring of six–fold and therefore also three–fold symmetry (3,20) fit into this pattern. The widespread use of this symmetry in this particular class of pigment protein complexes may be due to the fact that an oligomer consisting of three monomers related by three–fold symmetry is the simplest, most economical macromolecular assembly to function as a rotationally isotropic antenna, absorbing light independent of the rotational orientation of the complex relative to the light source. Three–fold symmetry may be one of the organizing principles of the structure of antenna complexes in photosynthesis.

References

1. Barber,J. 1982. Ann.Rev.Plant Physiol. 33, 261–295

2. Staehelin, L.A. and Arntzen, C. 1983. J.Cell Bioil. 97, 1327–1337

3. Kühlbrandt, W. 1987. In: Electron Microscopy of Proteins, Vol.6 (J.R.Harris and R.W.Horne, eds.). Academic Press, London. pp 155–207

4. Coruzzi, G., Broglie, R., Cashmore, A. and Chua, N.H. 1983. J.Biol.Chem. 258, 1399–1402

5. Dunsmuir, P., Smith, S.M. and Bedbrook, J. 1983. J.Mol.Appl.Gen. 2, 285–300

6. Kühlbrandt, W. 1987. J.Mol.Biol. 194, 757–762

7. Bürgi, R., Suter, F. and Zuber, H. 1987. Biochim.Biophys. Acta 890, 346–351

8. Cashmore,A.R. 1984. Proc.NatlAcad.Sci.USA 81, 2960–2965

9. Li,J. 1985. Proc.NatlAcad.Sci.USA 82, 386–390

10. Ryrie,I.J., Anderson,J.M. and Goodchild,D.J. 1980. Eur.J.Biochem. 107, 345–354

11. Kühlbrandt,W. 1984. Nature 307, 478–480

12. Kühlbrandt,W. 1987. Manuscript submitted

13. Kühlbrandt,W., Becker, A, and Maentele, W. 1987. Manuscript submitted

14. Butler,P.J.G. and Kühlbrandt,W. 1987. Manuscript submitted

15. Ide,J.P., Klug,D.R., Kühlbrandt,W., Giorgi,L.B. and Porter,G. 1987. Biochim.Biophys Acta 893, 349–364

16. Henderson, R. and Unwin, P.N.T. 1975. Nature 257, 28–32

17. Matthews, B.W., Fenna, R.E., Bolognesi, M.C., Schmid, M.F. and Olson, J.M. 1979. J. Mol. Biol. 131, 259–285

18. Schirmer, T., Bode, W., Huber, R., Sidler, W. and Zuber, H. 1985. J.Mol.Biol. 186, 257–277

19. Boekema, E.J., Dekker, J.P., van Heel, M.G., Rögner, M., Saenger, W., Witt, I. and Witt, H.T. 1987 FEBS Lett. 217, 283–286

20. Stark, W., Kühlbrandt, W., Wildhaber, I., Wehrli, E. and Mühlethaler, K. 1984. EMBO J. 3, 777–783.

PHYCOBILISOMES OF *SYNECHOCOCCUS* SP. PCC 7002, *PSEUDANABAENA* SP. PCC 7409, AND *CYANOPHORA PARADOXA*: AN ANALYSIS BY MOLECULAR GENETICS

Donald A. Bryant

Department of Molecular and Cell Biology, Penn State University, University Park, Pennsylvania 16802 USA

Introduction

The cyanobacterial photosynthetic apparatus is remarkably similar to that observed in the chloroplasts of eucaryotic algae and higher plants. In fact, four of the major multiprotein complexes of the thylakoids--the Photosystem II complex = H_2O-plastoquinone photo-oxidoreductase; the cytochrome b_6/f complex = plastoquinol-plastocyanin (cytochrome c-553) oxidoreductase; the Photosystem I complex = plastocyanin (cytochrome c-553)-ferredoxin (flavodoxin) photo-oxidoreductase; and the ATP synthase complex-- are virtually identical, both structurally and functionally to their eucaryotic analogues (1). The major source of diversity among the various oxygenic photosynthetic organisms is found in the light-harvesting antennae complexes associated with Photosystem II (and to a lesser extent Photosystem I). The cyanobacteria contain phycobiliproteins, which assemble into large, supramolecular complexes known as phycobilisomes (PBS), which are structurally and functionally associated with the Photosystem II reaction centers (1,2). Although PBS are also found in chloroplasts of the eucaryotic red algae, all other eucaryotic algae, higher plants, and certain procaryotes such as *Prochloron* sp. and *Prochlorothrix hollandica* contain one or more proteins from a highly diverse ensemble of caroteno-chlorophyll proteins as the antennae for Photosystem II and/or Photosystem I (1,3). The PBS and phycobiliproteins, because they are extremely abundant

Photosynthetic Light-Harvesting Systems

water-soluble proteins, have been intensively studied by biologists, chemists, biochemists, biophysicists, X-ray crystallographers, and most recently molecular biologists. The purpose of this overview is to describe recent progress on a structural and functional analysis of PBS by the techniques of molecular genetics. A brief discussion of the current understanding of PBS structure as influenced by an analysis of the genes encoding the proteins of these structures will be presented. Finally, aspects of the chemical and physical parameters (nutrient availability, light wavelength, and light intensity) which govern the expression of the genes encoding these proteins will be discussed.

Analysis of the Phycobiliprotein Genes of *Synechococcus* sp. PCC 7002

Synechococcus sp. PCC 7002 (*Agmenellum quadruplicatum* PR-6)is a marine, unicellular cyanobacterium which is naturally competent for the uptake of DNA (i.e., is transformable (4)). This organism grows well at high light intensities (100-250 μE m^{-2} s^{-1}) and elevated temperatures (34-39° C). Under optimal conditions, doubling times of approximately 3.75-4.0 h can be obtained. The organism is a "photoheterotroph"and is capable of rapid growth (doubling time about 10 h) under conditions where Photosystem II is non-functional (e.g., 10 μM DCMU) if 10 mM glycerol is added to the medium (5).

The transformation phenomenon, for both linear and plasmid DNA molecules, is highly efficient and the process has been well characterized (4). The genetic complexity (5000 Kbp) of this cyanobacterium, as well as its moles G + C content (49%), are similar to these values for *Escherichia coli* (6,7). A wide variety of biphasic, shuttle-vector plasmids, have been developed for this organism (8,9,10). These vectors carry replicons for plasmid maintenance in both *E. coli* and *Synechococcus* sp. PCC 7002; carry a variety of drug resistance markers (Km^R, Ap^R, Em^R, Cm^R); and in some cases allow for the expression of inserted DNA from a cyanobacterial promoter with insertion-selection on appropriate media by lacZ (β-galactosidase) inactivation. Although recombination-deficient mutants for this cyanobacterium have not yet been reported, the recA gene of Synechococcus sp. PCC 7002 has been cloned and sequenced (11); efforts to inactivate the gene are in progress. Finally, there is some evidence that the promoters for some Synechococcus sp. PCC 7002 genes are functional in E. coli (11,12,13). Correspondingly, the promoters for some E. coli genes appear to be functional in the cyanobacterium (9). Thus, Synechococcus sp. PCC 7002 is a procaryote which possesses many physiological and genetic properties which could be advantageous in an analysis of the photosynthetic apparatus by recombinant DNA and molecular genetics approaches. With these properties in mind, an analysis of the PBS of Synechococcus sp. PCC 7002 was initiated in 1983.

The PBS of Synechococcus sp. PCC 7002 contain only phycocyanin (PC) and allophycocyanin (AP) as major biliproteins and are composed of 12 structural polypeptides. A list of these polypeptides, the gene loci which encode them, their locations in the PBS, and other pertinent information are shown in Table 1. The gene encoding the α subunit of PC was cloned by using oligonucleotides whose sequences were based upon the known amino acid sequence of this protein (14). Subsequent expression studies in E. coli demonstrated the presence of the gene (cpcB) encoding the β subunit of PC on the same DNA fragment (12). A deletion mutant of the cpcBA genes has been constructed by interposon mutagenesis. Although this mutant is totally devoid of PC, the PBS cores apparently are normally assembled and functional in this mutant. Additional nucleotide sequence analysis of the region surrounding the cpcBA genes (in total 7.428 Kbp) has revealed four open reading frames which are tightly linked to and apparently co-transcribed with the cpcBA

Table 1. Composition of the Phycobilisomes of *Synechococcus* sp. PCC 7002.

Subunit	Copies/PBS	Gene Locus	Cloned	Sequenced	Mutant[a]
PERIPHERAL RODS:					
α^{PC}	72	*cpcA*	+	+	D
β^{PC}	72	*cpcB*	+	+	D
L_R^{33}	6	*cpcC*	+	+	I
$L_R^{8.5}$	6	*cpcD*	+	+	D
L_{RC}^{29}	6	*cpcG*[b]	-	-	-
L^{50}	1-2	----	-	-	-
CORE:					
α^{AP}	32	*apcA*	+	+	D
β^{AP}	34	*apcB*	+	+	D
α^{AP-B}	2	*apcD*	+	+	I[c]
β^{18}	2	*apcF*[d]	-	-	-
L_C^{8}	6	*apcC*	+	+	I
L_{CM}^{94}	2	*apcE*[e]	-	-	-

[a] D = deletion; I = insertion

[b] The *cpcG* gene has been detected immediately downstream from the *cpcBACDEF* cluster in *Anabaena* sp. PCC 7120 by W. Belknap and R. Haselkorn (personal communication). The *cpcG* gene product shares sequence homology with the *cpcC* gene product. Hybridization of a *cpcG* probe to genomic DNA of *Synechococcus* sp. PCC 7002 has been demonstrated.

[c] Mutant construction is in progress.

[d] Hybridization signals have been detected by using gene-internal fragments of the *apcB* genes as hybridization probes.

[e] This highly conserved gene has been cloned from the cyanelle genome of *C. paradoxa* (see discussion below). Efforts to isolate the *Synechococcus* sp. PCC 7002 gene, which hybridizes to probes from the *C. paradoxa* gene, are in progress.

genes. These open reading frames have been mutagenized by interposon (cartridge mutagenesis), and the phenotypes of the resulting mutants have been analyzed. The cpcC gene encodes the 33 kDa PC-associated linker (L_R^{33}). This assignment has been confirmed by amino acid sequence analysis of the amino terinus of this protein. Mutants in this locus cannot attach the terminal hexamer of PC $(\alpha\beta)_6$ to the PBS which correspondingly are devoid of this protein as well as the $L_R^{8.5}$ protein (15). The PC which is not assembled (50% of the total) is highly fluorescent and apparently found free in the cytoplasm. Electron microscopic examination of the PBS of this mutant shows that the peripheral rods of this mutant contain only one hexameric $(\alpha\beta)_6$ equivalent of PC.

The cpcD gene encodes the $L_R^{8.5}$ PC-associated linker polypeptide. This assignment has also been confirmed by amino acid sequence analysis of the corresponding protein and by virtue of the high degree of homology exhibited by this protein to that of the analogous protein sequenced from Fischerella sp. PCC 7603 (16). A deletion of the cpcD gene was effectively created by interposon replacement of the cpcD gene. This mutant cannot attach the terminal PC trimer $(\alpha\beta)_3$ to the peripheral rods and is very fluorescent. Approximately 25% of the PC is found free in the cytoplasm while 75% of the PC is found attached to the PBS in the peripheral rods. PBS of this mutant are devoid of the cpcD gene product, and although more variable in length than those in the wild-type, contain peripheral rods which mostly end in "half-disc" trimers. These data suggest that the attachment of the terminal trimer of PC to the rods is normally dependent upon and stabilized by the $L_R^{8.5}$ linker polypeptide.

The cpcE gene lies downstream from the cpcD gene, as found in Anabaena sp. PCC 7120 (17), and consists of 269 codons. This gene does not encode any known structural component of the PBS, since no proteins of this size and/or amino acid sequence can be demonstrated to occur in the PBS of this organism. A mutation in this gene was constructed by inserting an interposon cartridge into the coding sequence (J.-H. Zhou and D. A. Bryant, unpublished results). These mutants appear to have a much lowered content of PC. However, analysis of the biliproteins and PBS revealed an unexpected phenotype: the complete absence of phycocyanobilin chromophores on the PC α subunit. Both the α and β subunits of PC are present in the cells, and the amounts of these subunits are similar to those observed in the wild-type. The β subunit probably carries both chromophores as usual (This point is under further investigation.). AP levels are normal, and all other phycobiliproteins apparently have properly attached chromophores. Some PC α and β subunits can be found associated with the core substructures of the PBS; this suggests that the α subunit can in part adopt a conformation compatible with its assembly onto the core substructure.

The cpcF gene is composed of 206 codons and also does not encode a known structural component of the PBS. A mutation in this gene was constructed by insertion of an interposon cartridge into the 3rd codon of the gene (R. de Lorimier, unpublished results). This mutant is characterized by a high reversion frequency, and proper determination of the phenotype associated with this locus has been difficult. Nonetheless, it is clear that this mutation causes greatly reduced PC levels in cells but has little, if any effect on the transcription of the cpc locus. This suggests that the cpcF gene product operates at the translational or post-translational level. A possible role in chromophore attachment is suggested by the observation that the cpcF product has limited but significant homology to the cpcE gene product (D. Mazel and N. Tandeau de Marsac, personal communication). A further analysis of this gene locus, including the construction of a second insertion mutation, is in progress.

The second major locus encoding biliproteins in Synechococcus sp. PCC 7002, the apc operon, was cloned using the previously cloned apcAB genes from Cyanophora paradoxa as probes (8; see below). The operon encodes the α(apcA) and β(apcB) subunits of

AP as well as the L_C^8 AP-associated linker polypeptide of the core complex. The genes are arranged in the order 5' apcABC 3' and their nucleotide sequence has been completely determined (V. L. Stirewalt and D. A. Bryant, unpublished results). A deletion mutation of the apcAB genes was created by interposon cartridge replacement. This mutant is totally devoid of AP but synthesizes normal levels of PC; as expected, this PC is highly fluorescent and aberrantly assembled in the cells. Secondary mutations have been isolated from this mutant which no longer have the capacity to produce PC. In fact, biliproteins are no longer detectable in extracts of these cells even after concentration. The nature of the secondary mutations has not yet been characterized; however, these mutations do not appear to involve large insertion/deletion events in the cpcBAC structural genes.

The apcC gene lies 226 bp downstream from the apcB gene and contains 68 codons. The product of this gene has been identified by comparison of its sequence to the highly homologous protein sequenced by Zuber and coworkers (19) from *Fischerella* sp. PCC 7603. An insertion mutant was constructed by the interposon technique (J.-H. Zhou and D. A. Bryant, unpublished results). This mutant synthesizes normal levels of AP and PC and assembles functional PBS; these PBS are somewhat less stable than those of the wild-type when isolated from the cells but appear to function normally *in vivo* since the cells are not highly fluorescent. Hence, an essential role, if any, of this component is not revealed by the phenotype of this mutant.

The apcD gene, which encodes the α^{AP-B} subunit, has very recently been cloned and sequenced (J.-H. Zhou and D. A. Bryant, unpublished results) using a hybridization probe derived from the highly homologous gene isolated from *Calothrix* sp. PCC 7601 (20). This subunit is a minor but probably important component of the core complex since it serves as one of the two terminal energy acceptors in the PBS. The role of this protein in energy transfer to the Photosystem II core may be revealed from a careful spectroscopic analysis of transfer kinetics in a mutant (which is presently being constructed).

Efforts to clone, sequence, and characterize the remaining genes encoding PBS components in *Synechococcus* sp. PCC 7002 are in progress. Hybridization signals with genomic DNA have been obtained for the *apcE* gene, encoding the L_{CM}^{94} linker phycobiliprotein by using probes from either the *C. paradoxa* cyanelle gene or from the *Calothrix* sp. PCC 7601 gene (J. Houmard, personal communication). Hybridization signals have also been detected for the *cpcG* gene using a probe from the *Anabaena* sp. PCC 7120 gene. The *apcF* gene has been detected using gene-internal fragments of the *apcB* gene as probes (N. Tandeau de Marsac and D. A. Bryant, unpublished results). For the moment no amino acid sequence information nor other information exists which would allow the gene encoding the L^{50} protein to be cloned.

The analysis of the *Synechococcus* sp. PCC 7002 PBS by genetic methods supports the models for tricylindrical PBS cores proposed by Anderson and Eiserling (21). Sequence comparisons of the L_C^8 and $L_R^{8.5}$ linker polypeptides indicate that these proteins share structural homology and are derived from a single ancestral gene (15). Hence, it is likely that the L_C^8 linker complexes occupy the terminal positions of the core cylinders as proposed by Anderson and Eiserling (21). The remainder of the two membrane-associated core cylinders shares homologies with the peripheral rods as well. It is suggested that the L_{CM}^{94} protein is composed of three domains which resemble a "cruciform." The "transept" domain would be homologous to the L_R^{33} protein and would serve to "link" the two basal AP hexamers in a fashion analagous to the function of the L_R^{33} protein in the peripheral rods (see 15). A domain within the 3' third of the *apcE* gene is indeed homologous to the *cpcC* gene product. The lower "vertical" domain would provide the interaction with the thylakoid membrane and Photosystem II and may provide the chromophore binding site. The upper "vertical" domain, consisting of some 20 kDa and which should be missing from the L_{CM}^{75} protein in Synechococcus sp. PCC 6301, should provide an interaction stabilizing the attachment of the third core cylinder. This domain could in part duplicate the function of the "transept" domain and facilitate the attachment of the upper two AP hexamers. A comparative analysis of apcE gene products may verify this model.

Speculations on the function of the L^{50} protein can also be proposed. Mutants largely but not completely devoid of PC contain this polypeptide, but mutants totally devoid of PC in Synechocystis sp. PCC 6803 do not contain this polypeptide (22,23). Hence, this polypeptide cannot be associated with the distal portions of the peripheral rods and is probably not internally contained within the core complex itself; it could be peripherally associated with the core, however. Dissociation studies (21,24) are consistent with the notion that, under some conditions, this polypeptide is also associated with PC fractions (either hexamers or dodecamers) which contain the L_{RC}^{29} linker. Hence, the L^{50} linker may also be rod-core linker and could be specifically associated with the attachment of 1-2 peripheral rod complexes. In Synechococcus sp. PCC 6301 these two rods would be those not accounted for by the "18 S" subassembly particle (24), which contains two peripheral rod attachment sites and occurs twice within one PBS. A role of the L^{50} linker in stabilization of the upper core cylinder in species containing tricylindrical cores might also be anticipated.

Control of Gene Expression in Synechococcus sp. PCC 7002.

It is well known that the specific biliprotein content and composition of cyanobacteria, as well as the ratio of their Photosystem I and Photosystem II reaction centers, are dependent on the growth conditions employed (1). Changes in nutrient availability (N, S, C, P, and Fe), light intensity, and light wavelength can cause major changes in the photosynthetic apparatus. The effects of light intensity and the availability of N and S on phycobiliprotein synthesis have been studied in Synechococcus sp. PCC 7002 (10; G. Gasparich and D. A. Bryant, unpublished results). Two approaches have been employed. Firstly, the transcription of the cpc and apc genes have been directly evaluated by hybridization analyses with total RNA extracted from cells grown under different conditions (Northern blotting). Secondly, the expression of the cpc operon under different conditions has been measured by using a cpcB-lacZ translational fusion and by an enzymic assay for β-galactosidase (10).

Nitrogen starvation of Synechococcus sp. PCC 7002 results in the specific repression (de-activation?) of transcription of the cpcBAC and apcABC genes. It apparently takes 2-3 h for the cells to deplete internal N pools and "perceive" that the combined N has been removed from the medium. Thereafter, transcription of these genes is arrested within approximately 1 h. The kinetics with which the apcABC and cpcBAC mRNAs disappear is similar to that observed upon the addition of 150 μM rifamycin to the medium. Under these conditions mRNA half-lives are less than 10 min for all genes tested. Starvation for S, as removal of sulfate from the medium, produces qualitatively similar effects but requires a somewhat longer period of starvation to deplete internal pools (about 12 h). These results suggest that there may be a global, nutrient-status-sensing system in Synechococcus sp. PCC 7002 which can respond quickly to chemical stimuli from the environment and adjust the biosynthesis of the biliprotein antennae. This system is also sensitive to light intensity. The cpc locus is more highly transcribed, as assayed by monitoring β-galactosidase levels, when cells are grown at low light intensity than at high light intensity (by a factor of 1.75 X; see 10). mRNA levels for components of Photosystem I (psaA and psaB) and Photosystem II (psbA, psbB, psbC, and psbD) are also somewhat affected by the nutrient status of the cells. The mRNAs for all of these genes decrease significantly (2-10 X) under conditions of N limitation (G. Gasparich and D. A. Bryant, unpublished results).

Analysis of Biliprotein Genes in Pseudanabaena sp. PCC 7409.

Pseudanabaena sp. PCC 7409 is a filamentous, non-heterocystous cyanobacterium which is a "Type III" chromatic adapter and hence exhibits light-wavelength-mediated control of expression of both PC and phycoerythrin (PE; 25, 26). By using gene-internal fragments of the cpcA and cpcB genes of Synechococcus sp. PCC 7002 as heterologous hybridization probes, the genes encoding the α and β subunits of PE (cpeA and cpeB, respectively)and the genes encoding the α and β subunits of the "constitutive" PC (cpcA-I and cpcB-I, respectively) and of the "red-light-inducible" PC (cpcA-II and cpcB-II) were cloned (27). By varying the hybridization stringency, it was possible to demonstrate that a 4.2 Kbp EcoRI fragment carried the cpe coding sequences while the cpc coding sequences were borne by EcoRI fragments of approximately 3.8 and 14.0 Kbp. Sequence analysis of the putative cpe-coding fragment confirmed its identity (27).

The complete nucleotide sequence of this cpe-coding fragment has now been determined on both strands by the chain termination method. The cpeB gene is found 73 bp upstream from the cpeA gene. A large open reading frame, consisting of 421, 434, or 445 codons (depending upon the actual start codon used) is found 134-206 nucleotides downstream from the cpeA termination codon. Two additional open reading frames of 121 and >107 codons are found still further downstream. These open reading frames do not appear to encode structural components of the PBS of Pseudanabaena sp. PCC 7409. It is not yet known whether these sequences are transcribed.

The transcription patterns of the cpeBA, cpcBA-I, and cpcBA-II loci of Pseudanabaena sp. PCC 7409 are similar to those which have been described for Calothrix sp. PCC 7601 (28,29). The cpeBA mRNA is only found in cells grown in green light and is not found in cells grown in red light. The cpcBA-I genes are encoded on the 14.0 Kbp EcoRI fragment; transcripts from these genes are found in cells grown in either red or green light. The major transcript is about 1500 bp in length. The cpcBA-II genes are encoded by the 3.8 Kbp EcoRI fragment. Transcripts of these genes are only found in cells grown in red light. Two transcripts are observed; an abundant 1450 bp transcript probably encodes the α and β subunits of PC-II only; a larger, less abundant transcript of about 3000 bp probably encodes the PC subunits as well as one or more linker polypeptides as found in Calothrix sp. PCC 7601 (29)and in Synechococcus sp. PCC 7002 (G. Gasparich and D. A. Bryant, unpublished results). The results described above confirm those found for Calothrix sp. PCC 7601 (28,29) and establish that the light-wavelength gene regulation phenomenon, complementary chromatic adaptation, is mediated at the level of transcription.

The sequences upstream from the cpeB translational start codons of Pseudanabaena sp. PCC 7409 and Calothrix sp. PCC 7601, which contain the "green-light-regulated" promoters for the cpe loci, have been compared. The transcription initiation site has been mapped for Calothrix sp. PCC 7601 (D. Mazel, personal communication). A similar analysis is in progress for Pseudanabaena sp. PCC 7409; however, the promoter regions are sufficiently conserved that some conclusions can immediately be drawn. The promoters are quite similar to the E. coli consensus promoter sequences. The -10 sequence (TTTTAT) is separated from the -35 sequence (TTGAGA) by 17 bp in Pseudanabaena sp. PCC 7409. The nucleotide sequence between the -10 and -35 sequences is not conserved. However, the nucleotide sequence upstream from the -35 sequence is nearly invariant in the two organisms for 32 bp. The sequences are also nearly invariant in the region surrounding the -10 sequence. The region between the conserved Shine-Dalgarno-type sequence and the -10 sequence is not conserved in length or nucleotide sequence. From this analysis it is not possible to determine, as yet, whether a positive or negative transcription regulatory factor, is responsible for the light-wavelength-regulated transcription of these genes.

Biliprotein Genes in the Cyanelle Genome of *Cyanophora paradoxa*

Cyanophora paradoxa is an enigmatic, eucaryotic alga which contains plastids referred to as cyanelles (30). Cyanelles resemble cyanobacteria in the organization of their photosynthetic apparatus and are enclosed in a wall layer of peptidoglycan but have a genome which is similar in size to those found in higher plant chloroplasts (30). The cyanelle genome of *C. paradoxa* was shown to encode several PBS components by Grossman *et al.* (31). These workers found that the synthesis of the subunits of PC and AP, as well as the core linker phycobiliprotein L_{CM}^{94}, were synthesized in the cyanelles while the major PC-associated linker polypeptides were synthesized in the cytoplasm of the eucaryotic host cell. Lemaux and Grossman (32,33) successfully isolated clones encoding the *cpcA* and *cpcB* genes from a plasmid expression library, and Bryant *et al.* (34) isolated the *apcA* and *apcB* genes by using oligonucleotide probes for the *apcB* gene. Attempts to localize the gene encoding L_{CM}^{94} by immunological screening of *E. coli* extracts, prepared from cells harboring cloned fragments of *C. paradoxa* cyanelle DNA, suggested that the gene was found at least in part on the same 7.8 Kbp *PstI* fragment which encodes the *apcAB* genes (J. M. Dubbs, E. Gantt, and D. A. Bryant, unpublished results). Zilinskas and Howell (35; personal communication) have recently shown that the *apcE* gene of *Nostoc* sp. PCC 8009 is located 5' to the gene encoding *apcA*. A similar arrangement apparently exists in *Calothrix* sp. PCC 7601 in which a large open reading frame, characterized by transcripts in the 3-6 Kb range, exists upstream from the *apcA* gene (J. Houmard, N. Tandeau de Marsac, and A. Grossman, personal communication). When a fragment from this open reading frame was employed as a hybridization probe against *C. paradoxa* cyanelle DNA, hybridization was noted upstream from the *apcA* gene. DNA sequence analysis of this region has indeed revealed a large open reading frame and further analysis is in progress. The deduced carboxyl-terminal sequence of this protein includes a domain which is homologous to the L_R^{33} PC linker (see above). Interestingly, the *apcC* gene, encoding the L_C^{8} AP-associated linker polypeptide, is not found in the 873 bp which lie 3' to the *apcB* gene. This result suggests that the *apcC* gene may be nuclear-encoded in *C. paradoxa*.

Acknowledgements

The author would like to thank his students, post-doctoral associates, and colleagues for the communication of results prior to publication. This work has been supported by NIH grant GM-31625, NSF grant INT-8514249, NSF grant DMB-8504294, and the Agriculture Experiment Station of Penn State University (Project 2612).

References

1. Bryant, D. A. 1987. In: Photosynthetic Picoplankton (T. Platt and W. K. W. Li, eds.). Can. Bull. Fisheries Aquatic Sci. Vol. 214, Dept. Fisheries Oceans, Ottawa, p. 423.

2. Glazer, A. N. 1985. Ann. Rev. Biophys. Biophys. Chem. 14, 47.

3. Glazer, A. N. 1983. Ann. Rev. Biochem. 52, 125.

4. Porter, R. D. 1986. CRC Crit. Rev. Microbiol. 13, 111.

5. Lambert, D. H. and S. E. Stevens, Jr. 1986. J. Bacteriol. 165, 654.

6. Herdman, M., J. Janvier, R. Rippka, and R. Y. Stanier. 1979. J. Gen. Microbiol. 111, 73.

7. Herdman, M., J. Janvier, J. B. Waterbury, R. Rippka, R. Y. Stanier, and M. Mandel. 1979. J. Gen. Microbiol. 111, 63.

8. Buzby, J. S., R. D. Porter, and S. E. Stevens, Jr. 1983. J. Bacteriol. 154, 1446.

9. Buzby, J. S., R. D. Porter, and S. E. Stevens, Jr. 1985. Science 230, 805.

10. Gasparich, G. J. Buzby, D. A. Bryant, R. D. Porter, and S. E. Stevens, Jr. 1987. In: Progress in Photosynthesis Research, Vol. IV (J. Biggins, ed.). Martinus Nijhoff/ Dr. W. Junk Publishers, p. 761.

11. Murphy, R. C., D. A. Bryant, R. D. Porter, and N. Tandeau de Marsac. 1987. J. Bacteriol. 169, 2739.

12. Bryant, D. A., J. M. Dubbs, P. I. Fields, R. D. Porter, and R. de Lorimier. 1985. FEMS Microbiol. Lett. 29, 343

13. Porter, R. D., J. S. Buzby, A. Pilon, P. I. Fields, J. M. Dubbs, S. E. Stevens, Jr. 1986. Gene 41, 249.

14. de Lorimier, R., D. A. Bryant, R. D. Porter, W.-Y. Liu, E. Jay, and S. E. Stevens, Jr. 1984. Proc. Natl. Acad. Sci. USA 81, 7946.

15. Bryant, D. A., R. de Lorimier, G. Guglielmi, V. L. Stirewalt, A. Cantrell, and S. E. Stevens, Jr. 1987. In: Progress in Photosynthesis Research, Vol. IV (J. Biggins, ed.). Martinus Nijhoff/Dr. W. Junk Publishers, p. 749.

16. Füglistaller, P., F. Suter, and H. Zuber. 1985. Biol. Chem. Hoppe-Seyler 366, 993.

17. Belknap, W. R. and R. Haselkorn. 1987. EMBO J. 6, 871.

18. Bryant, D. A., R. de Lorimier, D. H. Lambert, J. M. Dubbs, V. L. Stirewalt, S. E. Stevens, Jr., R. D. Porter, J. Tam, and E. Jay. 1985. Proc. Natl. Acad. Sci. USA 82, 3242.

19. Füglistaller, P., R. Rümbeli, F. Suter, and H. Zuber. 1984. Hoppe-Seyler's Z. Physiol. Chem. 365, 1085.

20. Houmard, J., V. Capuano, T. Coursin, and N. Tandeau de Marsac. 1987. Mol. Microbiol., in press.

21. Anderson, L. K. and F. A. Eiserling. 1986. J. Mol. Biol. 191, 441.

22. Lundell, D. J. and A. N. Glazer. 1983. J. Biol. Chem. 258, 894.

23. Elmorjani, K., J.-C. Thomas and P. Sebban. 1986. Arch. Microbiol. 146, 186.

24. Yamanaka, G., D. J. Lundell, and A. N. Glazer. 1982. J. Biol. Chem. 257, 4077.

25. Tandeau de Marsac, N. 1977. J. Bacteriol. 130, 82.

26. Bryant, D. A. and G. Cohen-Bazire. 1981. Eur. J. Biochem. 119, 415.

27. Dubbs, J. M. and D. A. Bryant. 1987. In: Progress in Photosynthesis Research, Vol. IV (J. Biggins, ed.). Martinus Nijhoff/Dr. W. Junk Publishers, p. 765.

28. Mazel, D., G. Guglielmi, J. Houmard, W. Sidler, D. A. Bryant, and N. Tandeau de Marsac. 1986. Nucl. Acids Res. 14, 8279.

29. Conley, P. B., P. G. Lemaux and A. Grossman. 1985. Science 230, 550.

30. Wasmann, C. C., W. Loeffelhardt, and H. J. Bohnert. 1987. In: The Cyanobacteria (P. Fay and C. Van Baalen, eds.). Elsevier, p. 303.

31. Grossman, A., L. Talbott, and T. Egelhoff. 1983. Carnegie Inst. Washington Yearbk. 82, 112.

32. Lemaux, P. G. and A. Grossman. 1984. Proc. Natl. Acad. Sci. USA 81, 4100.

33. Lemaux, P. G. and A. Grossman. 1985. EMBO J. 4, 1911.

34. Zilinskas, B. A. and D. A. Howell. 1987. In: Progress in Photosynthesis Research, Vol. II (J. Biggins, ed.). Martinus Nijhoff/Dr. W. Junk Publishers, p. 161.

ORGANIZATION AND ASSEMBLY OF BACTERIAL ANTENNA COMPLEXES

Gerhart Drews

Institute of Biology 2, Microbiology, Albert-Ludwigs-University, Schaenzlestrasse 1, D-7800 Freiburg, Federal Republic of Germany

1. Introduction

Phototrophic Prokaryotes have the capability to utilize the energy of light for growth and metabolism. Different light energy transducing systems have been developed during evolution. In this chapter only antenna complexes of Anoxyphotobacteriae will be discussed. These are Eubacteriae, having an anoxigenic type of photosynthesis (no production of oxygen in photosynthesis), are generally Gram-negative organisms and are facultative or strictly anaerobic. Photosynthetic metabolism depends from low oxygen partial pressure. The species and genera of Anoxyphotobacteriae have very little relationships between one another (1). Some of the phototrophic bacteria are even closer related to chemotrophic bacteria than to other phototrophic bacteria.

Our knowledge of the composition and genetic structure of the bacterial photosynthetic apparatus is restricted to few species. In this chapter the antenna system of *Rhodobacter capsulatus*, a facultative non-sulfur purple bacterium (Rhodospirillaceae), and of some other members of this family will be described.

Antenna or light-harvesting pigment-protein complexes have the function of light absorption and conversion of light energy to singlet excited states which migrate over the antennae by a random walk to the photochemical reaction centers where the singlet excited state energy is converted into electrochemical potential free energy. Antenna systems work with high efficiency and can adapt to varying incident light intensities. Theoretical considerations (2), spectroscopic mea-

Photosynthetic Light-Harvesting Systems

surements with polarized light in membrane (3) and crystal preparations (4) have shown that the molecular organization of pigment molecules is of high order. The pigment molecules in the membrane are oriented by pigment-binding polypeptides of the antenna complexes which span the membrane once by a central α-helical domain of hydrophobic amino acids (5). Structure, organization and function of bacterial antenna systems have been described in recent published reviews (5 - 10). The major results which are essential to understand the molecular mechanisms of assembly and its control will be briefly summarized.

2. The organization of the antennae complexes in facultative phototrophic purple bacteria

Most purple bacteria have two antennae complexes, one, which is the direct donor of excited singlet states to the reaction center (RC), surrounds the RC by a fixed number of units forming the core complex. This antenna complex is named by the farthest red absorption band "B870" in bacteria having bacteriochlorophyll (Bchl) a and B1020 in Bchl b containing bacteria. In *Rhodospirillum (R.) rubrum* B880 and in *Rhodopseudomas (Rps.) viridis* (Bchl b) B1020 are the only light-harvesting antenna complexes.

The second, variable complex is named B800-850 because of its two prominent NIR absorption bands. This complex interspaces and connects a variable number of core complexes. The synthesis of constitutents of the core complex (RC + B870) is coregulated, but the synthesis of the B800-850 complex is regulated independently by light intensity, oxygen tension and other factors of differentiation (11). Lowering of light intensity effects an increase of B800-850 synthesis and an enlargement of the area of the intracytoplasmic membrane. The higher rate of B800-850 synthesis raised the ratio (B800-850) /(RC+B870) and therewith the size of the photosynthetic unit (12, 13).

Oxygen partial pressure is the major external factor which regulates the synthesis of the pigment-protein complexes of the photosynthetic apparatus. Lowering of the oxygen tension below species specific threshold values induces the synthesis of the pigment-protein complexes. In cells of *Rb. capsulatus* RC and B870 were synthesized first, followed in a sequential process by the synthesis of B800-850 (14). Different types of the B800-850 complex have been studied and detailed (8).

All antenna complexes so far as have been studied contain two low-molecular weight (Mr below 10,000) pigment-binding polypeptides of different primary structure (5, 7). Non-pigment-binding proteins have been found in the B800-850 complex of *Rb. capsulatus* and in the B1020 complex of *Rps. viridis*. Two (B870) or three (B800-850) mol Bchl and one or two mol carotenoids are noncovalently bound to both pigment-binding polypeptides (5, 15, 16). The isolated B800-850 complexes (solubilized or crystallized) of *Rb. capsulatus*, *Rps. palustris* and *Rhodospirillum salexigens* form oligomers of the basic subunit. The most probable structure is the tetrameric state in *Rb. capsulatus* (W. Welte, in this volume).

Absorption spectroscopy with linear polarized light of crystals of antenna complexes has shown that the Q_x and the Q_y transition moments of the Bchl's are highly ordered relative to the plan of the membrane (see contribution of W. Welte).

The core structure containing RC + B870(B1020) has been studied in various purple bacteria, especially in crystals of *Rps. palustris* (4) and in membrane preparations of *Rps. viridis* and *Ectothiorhodospira halochloris* where regular arrays of particles were formed, which have been interpreted to be RCs surrounded by light-harvesting complexes (17, 18). Although we have at present time no high resolution data of X-ray studies on the organization of core complexes there is indirect evidence that the polypeptides of RC and B870(B1020) interact with each other to form supercomplexes between RC and B870(B1020) oligomers not only in isolated preparations but also in vivo in the membrane. Experimental evidence which supports the idea of strong protein-protein interactions between polypeptides of RC and B870(1020) complexes came from chemical cross-linking of RC and B870(1020) polypeptides in *Rhodospirillum rubrum*, *Rps. viridis* and *Rb. caspsulatus* (19 - 21) in conjunction with immunofractionation and two-dimensional mapping of products on SDS polyacrylamide gels (22, 23). The strong linear dichroism of Q_x and Q_y transition moments observed in crystals of RC-B870 of *Rps. palustris* is also in favor of an orientation of B870 light-harvesting pigment molecules relative to RC (4). A mutant of *Rb. capsulatus* which did not contain a B870 complex could not correctly insert the RC into the membrane. It was concluded that the B870 complex is essential to stabilize the RC and to facilitate the correct assembly of both complexes in the membrane (24, 25) and vice versa the B870 complex in *Rb. capsulatus* is not stabil if the RC is absent (26). In *Rb. sphaeroides* and *R. rubrum* the B870 complexes are more stable (16, 25). Kinetic studies on singlet-triplet fusion (27) and excitation spectroscopy (28, 29b) indicate a close contact between RC and B870 in normal membranes but a decrease of energy transfer after fusion of membranes with liposomes.

3. DNA structure of photosynthetic genes

Rb. capsulatus was the first phototrophic bacterium for which a useful genetic exchange system was discovered (29a). Most of the genes for the synthesis of bacteriochlorophyll *(bch)* and carotenoids *(crt)* and for the pigment-binding polypeptides were found to be clustered on the genome of *Rb. capsulatus* and mapped on a 46 kb fragment of the chromosome (30-32). Genes for the pigment-binding proteins of RC and B870 form the *puf* operon. The genes for the pigment-binding polypeptides alpha and beta of the B800-850 complex (*puc* operon) are located outside of the 46 kb photosynthetic gene cluster.The DNA sequence of *puf* and *puc* operons have been determined (33, 34). The gamma polypeptide of the B800-850 complex, which does not bind Bchl, is not in the *puc* operon. We have determined the complete sequence of the *puc* DNA and the DNA 700 bp upstream and downstream of B870 α,β in strain 37b4 and could not find the gamma protein in this region (H.V. Tichy, unpublished). The reconstitution of some B800-850-negative mutants with the crtI region of the BamHI fragment of pRPS404 and a hybridization of oligonucleotides prepared from a partial amino acid sequence of the B800-850 gamma protein with DNA in the *crt* cluster of pRPS404 supports the idea that genes for the gamma protein and one or more of the *crt* synthesizing enzymes are co-expressed and obligate for the formation of the B800-850 complex (H. Boos, B. Oberlé, M.H. Tadros, unpublished). Recently a mutant of *Rb. capsulatus*, which contains the B800-850 complex and all three proteins α, β and γ, but no carotenoids, has been isolated. This mutant GK2, however, forms the B800-850 complex only when grown anaerobically in the light and not in the dark (35).

The photosynthetic genes of *Rb. sphaeroides* are organized similarly (36 - 39). Genes for pigment binding polypeptides from *Rhodospirillum rubrum* (40), *Rps. viridis* (41) and *Rps. palustris* (M.H. Tadros et al., unpublished) have been cloned and sequenced.

4. Expression of genes coding for pigment-binding proteins of antenna complexes

The variation of the oxygen tension in the surroundings of facultative phototrophic bacteria is the major external signal to trigger a process of functional and structural membrane differentiation from a respiratory to a photosynthetic mode of energy transduction and vice versa. Lowering of the oxygen partial pressure induces a rapid synthesis of Bchl, pigment-binding proteins and other components of the photosynthetic apparatus. The process of oxygen sensing and signal transfer in the cell is unknown. Redox- or energy charge-controlled molecules have been proposed but not proved to be involved in regulation (12). The synthesis of pigment-binding polypeptides has been shown to be controlled on the level of transcription (42-45). After lowering of oxygen tension the level of mRNA for pigment-binding proteins of RC and antenna proteins increased strongly followed by the synthesis of these proteins and their assembly in the membrane with pigments (43). The *puf* operon (coding for RC-B870) in *Rb. capsulatus* is transcribed from more than one promoter. The most active of these promoters is located about 700 bp upstream from the *pufB* gene. This promoter is regulated by oxygen tension. In the region between *pufB* and the second promoter a new gene, called *pufQ*, has been found which is required for an early stage of Bchl synthesis (C.E. Bauer, M. Leuterio, D.A. Yopung and B.L. Marrs, unpublished). This gene product may be responsible for coupling of pigment and protein synthesis and assembly. The synthesis of Bchl is regulated on the level of the activity of key enzymes for tetrapyrrol synthesis (46) and by activation of genes for Bchl-synthesizing enzymes (47). The activity of the first enzyme, the δ-aminolevulininate synthetase, was proposed to be regulated by a feed back (46) or a switch-off/on mechanism (48). Inhibition of Bchl synthesis by the competitive inhibitor levulinic acid stops immediately not only Bchl formation but also the synthesis of pigment-binding polypeptides but not the synthesis of other membrane or cytoplasmic proteins (49). The induced synthesis of mRNA for the pigment-binding proteins is not impaired. Mutant strains, blocked in Bchl synthesis, synthesize mRNA for pigment-binding proteins but in reduced amounts or different kinetics (49, M. Young, unpublished). The pigment-binding polypeptides are only present in very low concentrations in membranes of Bchl-less mutant cells. It is speculated that the coordinated expression for genes of pigment synthesis and pigment-binding proteins is regulated by products of regulatory genes which are modified by effector molecules (Fig. 1). The following observation supports this idea: A transposon insertion downstream of the *puc* operon inhibits the formation of B800-850. In this region there is an open reading frame which might be of importance for gene expression (H.V. Tichy, unpublished). Another transposon insertion, which is located outside of all known photosynthetic operons and pigment clusters, inhibits the formation of the B800-850 complex after lowering of oxygen tension. B800-850 is formed in this mutant only in anaerobic light cultures (R. Liebetanz, unpublished).

Another important factor for regulation of gene expression is the half-life for mRNAs. In the *puf*operon between RC and B870 genes of *Rb. capsulatus* is an intercistronic stem and loop structure which is responsible for differences in transcript stability (50). The deletion of this structure decreased the stability of mRNA for B870 α,β polypeptides and consequently the amount of gene products formed in induced cells (51). The few examples may indicate that there are different mechanisms for the regulation of gene expression of the photosynthetic apparatus. We are just at the beginning to detect regulatory genes, to identify their products and to understand the oxygen regulation at the promoter and possible enhancer regions.

5. The assembly of antenna complexes and its regulation

The formation of pigment-protein complexes in membranes is a multi-step process controlled by different factors and regulatory processes. The transcription rates of genes for pigment synthesis, possibly for key enzymes, and genes for pigment-binding and complex-associated polypeptides are coordinately regulated.

The activation/inactivation or switch on/off of key enzymes of pigment synthesis and the regulation of translation of the mRNA for photosynthetic proteins are also unknown in their mechanisms. It has been proposed that tetrapyrrol compounds (precursors of Bchl) have an effect on enzyme activity (46) and translation of proteins (49). There is a mutual influence of pigment and protein synthesis in the regulation of the photosynthetic apparatus. As mentioned in chapter 4 levulinate inhibits specifically the synthesis of Bchl and of pigment-binding polypeptides (49) and vice versa inhibition of protein synthesis by mutation or translation blockers influences the rate of Bchl synthesis (49, 52, 53). Unbound, free Bchl and Bchl precursors are not only in mutant cells but also in wild type cells always present but in very low concentrations (54). Besides a positive control, for example by an oxygen sensing system, on Bchl and pigment-binding protein synthesis there are other control systems which may influence the mutual regulation of levels of constitutents of the antenna complexes.

Pigment-binding polypeptides are synthesized on free polysomes and bind afterwards to the membrane by an unknown mechanism (55). It has to befound out whether docking proteins play an essential role in that process. The translocation of the pigment-binding polypeptide through the lipid layer into a transmembrane orientation, was sensitive to carbonylcyanide m-chlorophenylhydrazone (CCCP), an uncoupler, which lowers the proton motive force (56). After inactivation of CCCP the accumulated protein continued translocation into the membrane (56). Protein which is attached to the membrane but inhibited in translocation is easily digested by protease K, showing that it is located on the membrane surface and not protected by the membrane lipid layer (57). It was suggested that proteins unable to assemble an antenna complex are quickly degraded by cytoplasmic proteases because they are not correctly inserted into the membrane (58). The fast turnover of non-assembled pigment-binding proteins is another mechanism to coordinate pigment and protein synthesis properly. The assembly of the antenna complex depends from all constituents. The mutant NK9 of *Rb. capsulatus* is unable to assemble the B800-850 complex although Bchl, carotenoids and both pigment-binding polypeptides α and β are synthesized (59).

Hypothetical scheme for regulation of the assembly of pigment-proteins of Rb. capsulatus

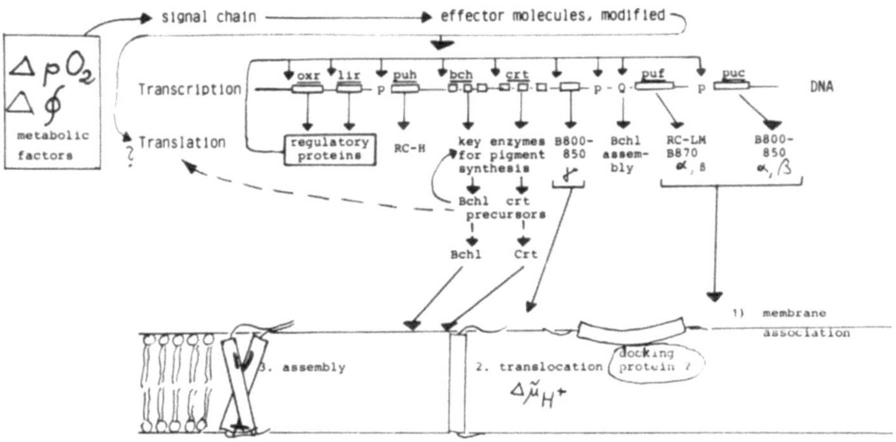

Fig. 1. For symbols and abbreviations see text.
ΔpO_2 = gradient of oxygen tension or light intensity($\Delta\phi$)

The pigment-binding polypeptides of the antenna complex B870 of *Rhodospirillum rubrum* (60) and of the B870 and B800-850 complexes of *Rb. capsulatus* and *Rb. sphaeroides* expose the N-terminal domain on the cytoplasmic membrane surface or point the N-terminus towards the cytoplasm (57, 61, 62, M.H. Tadros et al. unpublished). This observation suggests a similar mechanism of insertion of the antenna proteins into the membrane. An alignment of N-terminal domains of α and β polypeptides shows conservative amino acyl residues and a reasonably high degree of respective charge complementarity (M.H. Tadros, R. Frank, P. Argos, J.Y. Takemoto. G. Drews, submitted). It is possible that the basic residues of α proteins could interact with the acidic amino acids of B870β, providing a strong force for association of the two chains or that the hydrophobic, charged and other conservative amino acid residues of the N-termini play an important role during assembly and orientation of the transmembrane helices.

To investigate the role of charged amino acyl residues in the N-terminal domains of α and β polypeptides of B870, positively charged amino acyl residues were exchanged for negatively charged ones and vice versa by site-specific mutagenesis (63).

Preliminary results show that an exchange of four charged amino acids yields in a complete inhibition of the B870 formation (B. Dörge, G. Klug, unpublished).
Besides charged amino acids there are aromatic, hydroxylated and hydrophobic amino acyl residues in conservative positions in the N-terminal regions which are proposed to have an pivotal function in the assembly and stabilization of the antenna complexes in the membrane.
The different steps of synthesis and assembly of the antenna complex constituents and their regulation have been summarized in a hypothetical scheme (Fig. 1). It is an exciting task to elucidate the oxygen and light sensitive regulatory system of antenna formation.

Acknowledgement

This work was supported by a grant from the Deutsche Forschungsgemeinschaft.

References

(1) Trüper, H.G. 1987. Microbiologia SEM 3, 71-89.

(2) Knox, R.S. 1986, In: Photosynthesis III, Ecyclopedia of Plant Physiology, Vol. 19, (L.A. Staehelin and C.J. Arntzen, eds). Springer-Verlag, Berlin Heidelberg New York. pp. 286-298.

(3) Breton, J. 1986. In: Photosynthesis III, (L.A. Staehelin and C.J. Arntzen, eds.) Springer-Verlag, Berlin Heidelberg New York. pp. 319-326

(4) Wacker, T., N. Gad'on, A. Becker, W. Mäntele, W. Kreutz, G. Drews and W. Welte, 1986. FEBS Lett. 197, 267-273.

(5) Drews, G. 1985. Microbiol. Rev. 49, 59-70.

(6) Thornber, J.P. 1986. In: Photosynthesis III, (L.A. Staehelin and C.J. Arntzen, eds.) Springer-Verlag, Berlin Heidelberg New York. pp. 98-142.

(7) Zuber, H. 1986. Trends Biochem. Sci. 11, 414-419.

(8) Cogdell, R.S. 1986. In: Photosynthesis III, (L.A. Staehelin and C.J. Arntzen, eds.) Springer-Verlag, Berlin Heidelberg New York. pp. 252-259.

(9) Sprague, S.G. and A.R. Varga 1986. In: Photosynthesis III (L.A. Staehelin and C.J. Arntzen, eds.) Springer-Verlag, Berlin Heidelberg New York. pp. 603-619.

(10) Blankenship, R.E. and R.C. Fuller, 1986. In: Photosynthesis III. (L.A. Staehelin and C.J. Arntzen, eds.) Springer-Verlag, Berlin Heidelberg New York. pp. 390-399.

(11) Aagaard, J. and W.R. Sistrom. 1972. Photochem. Photobiol. 15, 209-225.

(12) Drews, G. and J. Oelze. 1981. Adv. Microb. Physiol. 22, 1-92.

(13) Donohue, T.J. and S. Kaplan. 1986. In: Photosynthesis III (L.A. Staehelin and C.J. Arntzen, eds.) Springer-Verlag, Berlin Heidelberg New York, pp. 632-639.

(14) Schumacher, A. and G. Drews. 1978. Biochim. Biophys.Acta 501, 183-194.

(15) Kramer, H.J.M., J.D. Pennoyer, R. van Grondelle, W.H.J. Westerhuijs, R.A. Niederman and J. Amesz. 1984. Biochim. Biophys. Acta 767, 335-344.

(16) Picorel, R., A. L'Ecuyer, M. Potier and G. Gingras. 1986. J.Biol.Chem. 261, 3020-3024.

(17) Stark, W., F. Jay and K. Mühlethaler. 1986. Arch.Microbiol. 146, 130-133.

(18) Engelhardt, H., A. Engel and W. Baumeister. 1986. Proc. Natl.Acad.Sci. USA 83, 8972-8976.

(19) Peters, J., W. Welte and G. Drews. 1984. FEBS Lett. 171, 267-270.

(20) Peters, J., J. Takemoto and G. Drews. 1983. Biochemistry 22, 5660-5667.

(21) Wiemken, V., R. Theiler and R. Bachofen, 1981. J. Bioenerg. Biomembr. 13, 181-193.

(22) Takemoto, J.Y., J. Peters and G. Drews. 1982. FEBS Lett. 142, 227-230.

(23) Broglie, R.M., C.N. Hunter, P. Delepelaire, R.A. Niederman, N.-H. Chua and R.K. Clayton. 1980. Proc. Natl. Acad. Sci. USA 77, 87-91.

(24) Jackson, W.J., R.C. Prince, G.J. Stewart and B.L. Marrs. 1986. Biochemistry 25, 8440-8446.

(25) Jackson, W.J., P.J. Kiley, C.E. Haith, S. Kaplan, R.C. Prince. 1987. FEBS Lett. 215, 171-174.

(26) Drews, G., P. Wevers and R. Dierstein. 1979. FEMS Microbiol. Lett. 5, 139-142.

(27) Monger, T.G. and W.W. Parson. 1977. Biochim.Biophys.Acta 460, 393-407.

(28) Pennoyer, J.D., H.J.M. Kramer, R. van Grondelle, W.H.J. Westerhuijs, J. Amesz and R.A. Niederman, 1985, FEBS Lett. 182, 145-150.

(29a) Marrs, B.L. 1974. Proc.Natl.Acad.Sci. USA 71, 971-973.

(29b) Takemoto, J.Y., T. Schonhardt, J.R. Golecki, and G. Drews. 1985. J. Bacteriol. 162, 1126-1134.

(30) Marrs, B.L. 1981. J. Bacteriol. 146, 1003-1012.

(31) Taylor, D.P., S.N. Cohen, W.G. Clark, and B.L. Marrs. 1983, J.Bact. 154, 580-590.

(32) Zsebo, K. and J.E. Hearst. 1984. Cell 37, 937-947.

(33) Youvan, D.C., E.J. Bylina, M. Alberti, H. Begusch, and J.E. Hearst. 1984. Cell 37, 949-557.

(34) Youvan, D.C. and S. Ismail. 1985. Proc.Natl.Acad.Sci. 82, 58-62.

(35) Dörge, B., G. Klug and G. Drews. 1987. Biochim.Biophys. Acta 892, 68-74.

(36) Ashby, M.K., S.A. Coomber and C.N. Hunter. 1987. FEBS Lett. 213, 245-248.

(37) Kiley, P.J. and S. Kaplan. 1987. J. Bacteriol. 169, 3268 -3275.

(38) Donohue, T.J., A.G. McEwan, and S. Kaplan. 1986. J. Bacteriol. 168, 962-972.

(39) Kiley, P.J., T.J. Donohue, W.A. Havelka, and S. Kaplan, 1987. J. Bacteriol. 169, 742-750.

(40) Bérard, J., G. Bélanger, P. Corriveau, and G. Gingras. 1986. J. Biol. Chem. 261, 82-87.

(41) Michel, H., K.A. Weyer, H. Gruenberg, I. Dunger, D. Oesterhelt, and F. Lottspeich. 1986. EMBO Jour. 5, 1149-1158.

(42) Clark, W.G., E. Davidson, and B.L. Marrs. 1984. J. Bacteriol. 157, 945-948.

(43) Klug, G., N. Kaufmann, and G. Drews. 1985. Proc.Natl. Acad.Sci. USA 82, 6485-6489.

(44) Zhu, Y.S., and S. Kaplan. 1985. J. Bacteriol. 162, 925-932.

(45) Zhu, Y.S., and J.E. Hearst. 1986. Proc.Natl.Acad.Sci. USA 83, 7613-7617.

(46) Lascelles, J. 1978. In: The Photosynthetic Bacteria (R.K. Clayton and W.R. Sistrom, eds.) Plenum Press, New York, pp. 795-808.

(47) Biel, A.J., and B.L. Marrs. 1983. J. Bacteriol. 156, 686 -694.

(48) Viale, A.A., E.A. Wider, and A.M. del C. Battle, 1987. Int. J. Biochem. 19, 379-383.

(49) Klug, G., R. Liebetanz, and G. Drews. 1986. Arch. Microbiol. 146, 284-291.

(50) Belasco, J.B., J.T. Beatty, C.W. Adams, A. von Gabain, and S.N. Cohen. 1985. Cell 40, 171-181.

(51) Klug, G., C.W. Adams, J. Belasco, B. Dörge and S.N. Cohen. 1988. EMBO J., in press.

(52) Biedermann, M., G. Drews, R. Marx, and J. Schröder. 1967. Arch. Microbiol. 56, 133-147.

(53) Reidl, H., J.R. Golecki, and G. Drews. 1985. Biochim.Biophys.Acta 808, 328-333.

(54) Beck, J. and G. Drews. 1982. Z. Naturforsch. 37c, 199-204.

(55) Dierstein, R., 1984. Eur. J. Biochem. 138, 509-518.

(56) Dierstein, R. and G. Drews. 1986. J. Bacteriol. 168, 167-172.

(57) Tadros, M.H., R. Frank, and G. Drews. 1986. FEBS Lett. 196, 233-273.

(58) Dierstein, R., M.H. Tadros, and G. Drews. 1984. FEMS Microbiol. Lett. 24, 219-223.

(59) Kaufmann, N., H. Hüdig, and G. Drews. 1984. Mol. Gen. Genet. 198, 153-158.

(60) Brunisholz, R.A., F. Suter, and H. Zuber, 1984. Hoppe-Seyler's Z. Physiol. Chem. 365, 675-688.

(61) Tadros, M.H., R. Frank, B. Dörge, N. Gad'on, J.Y. Takemoto, and G. Drews. 1987. Biochemistry, in press.

(62) Takemoto, J., R.L. Peterson, M.H. Tadros, and G. Drews. 1987. J. Bacteriol., in press.

(63) Zoller, M.S. and M. Smith. 1984. DNA3, 479-488.

THE USE OF MUTANTS TO INVESTIGATE THE ORGANISATION OF THE PHOTOSYNTHETIC APPARATUS OF RHODOBACTER SPHAEROIDES

C.N. Hunter

Department of Pure and Applied Biology, Imperial College, London. SW7 2BB U.K.

R. van Grondelle

Department of Biophysics, Physics Laboratory of the Free University, 1081 HV Amsterdam, The Netherlands.

Introduction

The photosynthetic apparatus of the purple bacterium Rhodobacter sphaeroides is housed within a system of intracytoplasmic membranes, and is composed of two light harvesting complexes, LH1 and LH2, which surround and interconnect photochemical reaction centres. This bacterium provides a relatively simple system for the study of light harvesting and energy transfer phenomena. In order to obtain a full description of these complexes it is desirable to study them in isolation in order to avoid complications that arise from the crossover of absorption and emission bands. Detergent solubilisation of membranes followed by a variety of fractionation techniques will produce LH1 and LH2 complexes either singly (1,2) or in the combinations LH2/LH1, LH1/RC (1,3). Much of our present understanding of isolated antennae has relied on this strategy.

The approach described here relies on the use of chemical and transposon mutagenesis techniques to produce strains which contain the combinations LH2 (mutant NF57) LH1+RC (M21) and LH1 (M2192). In this way the use of detergents has been avoided. Northern blot analysis has been used to examine transcripts of the relevant genes. The composition of mutant membranes has been analysed by SDS-

Photosynthetic Light-Harvesting Systems

polyacrylamide gel electrophoresis. All three mutants have been analysed by low temperature fluorescence and absorption spectroscopy; the exciton annihilation properties have enabled conclusions to be drawn on the state of aggregation of these antennae in vivo.

Materials and Methods

Mutagenesis of Rb. sphaeroides Mutants M21 and NF57 were obtained from Rb. sphaeroides wild-type strain 8253 following chemical mutagenesis with N-nitrosoguanidine. M2192 was derived from M21 by inserting transposon Tn5 into puf L encoding the reaction centre L subunit (4).

Northern Blots These were performed according to methods described in 5.

SDS-polyacrylamide gel electrophoresis This was performed according to Laemmli (6).

Cell fractionation Preparation of pigmented membrane fraction from wild-type and mutant strains was carried out according to methods in 7.

Spectroscopy Absorption and fluorescence emission spectroscopy was carried out using methods described in 8. Singlet-singlet annihilation was carried out using apparatus described by Vos (9).

Results

Northern blot analysis of total RNA from wild-type and mutant strains (Figure 1) shows that the 2.6 kb transcript of the puf operon (10) which encodes subunits of reaction centre and LH1 complexes is absent in mutants M2192 and NF57, both of which are incapable of

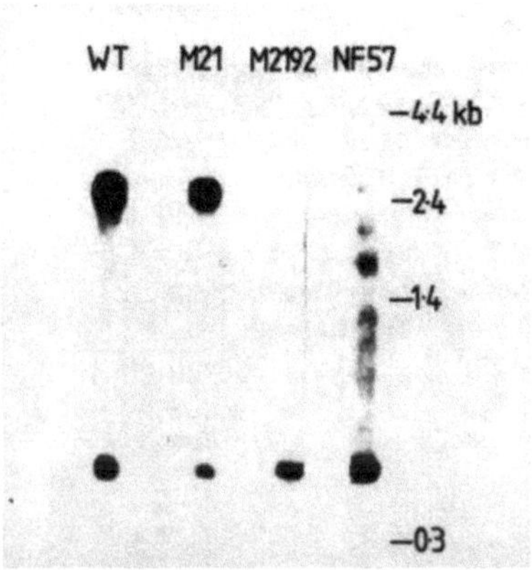

Figure 1. Northern blot of total RNA from wild type and mutant strains of Rb. sphaeroides. The blot was probed with pSRC2 (see figure 5).

photosynthetic growth. The 0.5 kb transcript, which encodes LH1 polypeptides, is detected in both these mutants however, although LH1 is present only in M2192. The absence of LH1 polypeptides in mutant NF57 must therefore arise from the instability of LH1 in the absence of a gene product encoded on the clone pSCN5H-1 which restores NF57 to photosynthetic growth (see Figures 5 and 6). Such a possiblity is consistent with results from the mutagenesis of this region of the Rb. capsulatus genome (11,12) and leads to the conclusion that the gene product of puh A or of a flanking region plays an important role in the assembly of the LH1-reaction centre complex. The 0.55 kb transcript from puc A,B genes encoding LH2 polypeptides (5,13,14) is absent from M21 and M2192 (not shown). The synthesis and assembly of LH1 does not appear to be dependent on the presence of reaction centre L or M subunits, despite the close genetic and functional relationship between these polypeptides, and the fact that such a dependence is observed in Rb. capsulatus (15).

Cells of wild type and mutant strains were disrupted in a French pressure cell, and the extracts subjected to rate-zone sedimentation in a sucrose gradient in order to compare the membrane fractions that could be obtained. The normal profile of a gradient contains two fractions which are thought to be derived from sites of initiation of membrane invagination (the upper fraction) and from mature, fully invaginated membranes (7). This analysis is presented in Figure 2

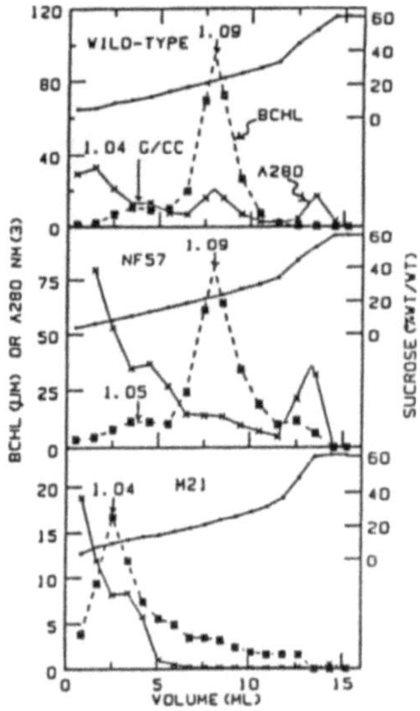

Figure 2. Profile of pigment distribution after rate-zone sedimentation of membranes from wild-type and mutant strains.
The gradient is 5-35% w/w on a cushion of 60% sucrose. The density of each pigmented peak is shown.

and it can be seen that there is a striking absence of the lower fraction in M21, whereas NF57 and the wild-type exhibit a normal profile. Further analysis of these mutants by electron microscopy (C.N. Hunter, J.D. Pennoyer, J.N. Sturgis, D. Farrelly and R.A. Niederman unpublished results) indicates that a tubular membrane system is present in M21 and that the absence of LH2 does not retard membrane invagination but instead appears to distort the process in some way. Similar distortions have been noted for the carotenoid-less mutant R26 which also lacks the LH2 complex (16). Restoration of LH2 to M21 by either reversion or gene transfer results in a normal membrane system (not shown). Overall, this demonstrates that the synthesis of normal intracytoplasmic membranes is obligately linked to LH2 formation and insertion.

SDS polyacrylamide gel electrophoresis of membranes from wild-type and mutant strains (Figure 3) demonstrates that reaction centre polypeptides H,M and L are not present in mutants M2192 and NF57. The region of the gel containing LH1 and LH2 polypeptides shows

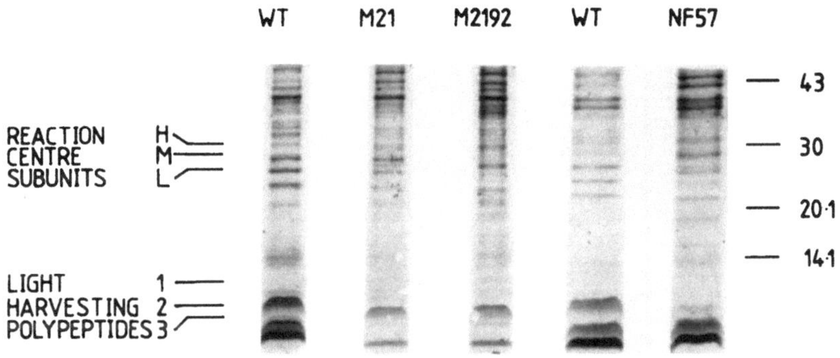

Figure 3. SDS-polyacrylamide gel electrophoresis of membranes from wild type and mutant strains.

marked differences in the mutants with respect to the wild-type. The profiles observed are consistent with the migration of LH1 and LH2 polypeptides described earlier (1) and indicate that LH1 α is absent from NF57, and that LH2 α is missing from M21 and M2192. Comigration of the β polypeptides (1) renders further interpretation difficult.

Absorption spectroscopy at 4K was performed in order to characterise these mutants further (Figure 4). The spectrum of membranes from the wild-type, included here for comparison, shows the bands of the LH2 complex near 800 and 850 nm, and of the LH1 complex at 884 nm. Mutants NF57 and M2192 exhibit the bands characteristic of isolated LH2 and LH1 complexes respectively, while M21 also possesses minor peaks at 806 and 756 nm which arise from the presence of reaction centres. Also included here for comparison are the spectra of the B800-850 (LH2) complex isolated using the detergents LDAO and Triton X-100 according to the method of Clayton and Clayton (2), and the B875 (LH1) complex isolated according to the method of Broglie et al., (1). Significant differences are observed between detergent solubilised complexes and the mutants. The bands of NF57 are broader than for the LDAO/Triton complex and the maximum near 850 nm is

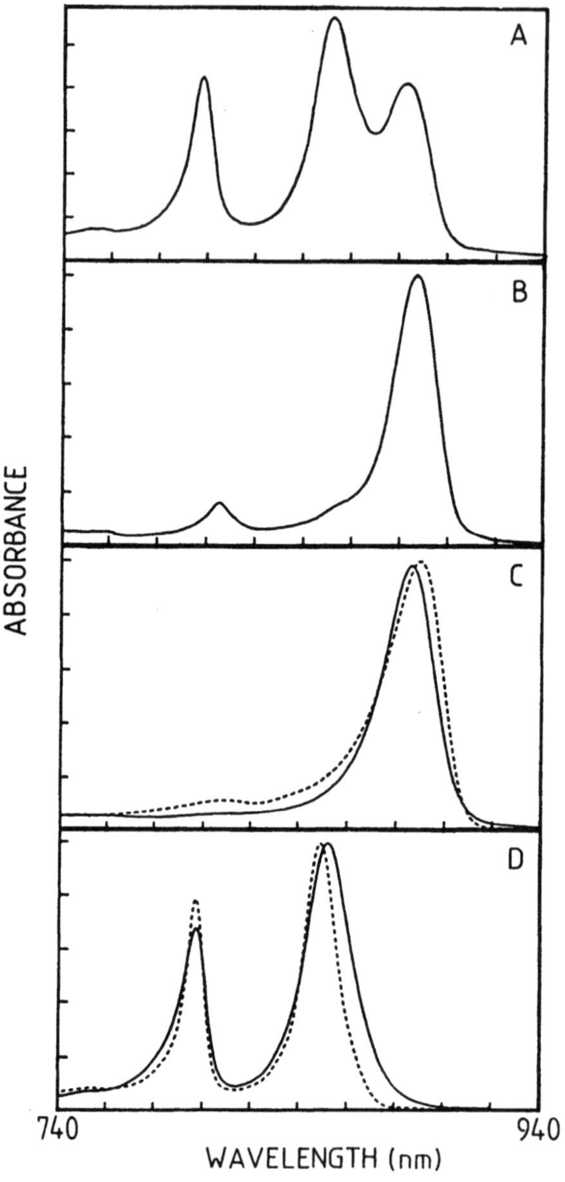

Figure 4. Absorption spectra of membranes at 4K (A) wild-type (B) M21 (C) M2192 (D) NF57. The dotted lines in (C) and (D) show absorption spectra of the complexes prepared according to 1 and 2 respectively.

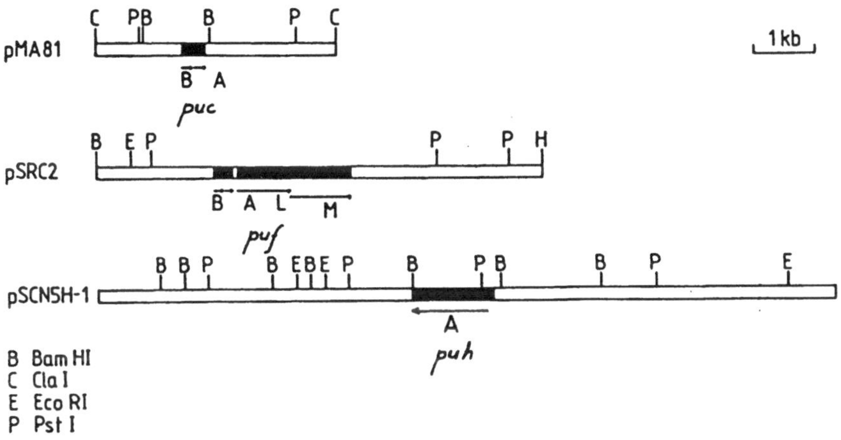

Figure 5. Restriction maps of Rb. sphaeroides DNA. Only the inserts of each plasmid are shown. pMA81 and pSCN5H-1 restore M21 and NF57, respectively to wild-type phenotype. pSRC2 restores reaction centre activity to M2192.

shifted by 3 nm to a longer wavelength. In contrast, the main absorption band is narrower in M21 than in the solubilised complex. The differences are emphasised by the results displayed in Figures 5 and 6; here the mutant M21 is restored to wild-type phenotype by transfer of genes puc A and B which code for LH2 polypeptides, borne on plasmid pMA81 (Figure 5; also 13). In Figure 6, the spectrum of transconjugant strain M21 (MA81) is compared with linear combinations of spectra of either M21 plus NF57 (top panel) or with the LH1 and LH2 complexes (lower panel) isolated by detergent treatment, as for Figure 4. A better agreement is clearly obtained using the spectra of the mutants, and it can be concluded that the spectral properties of antenna in the mutants give a better representation of the properties of bacteriochlorophyll in vivo than do the detergent solubilised preparations.

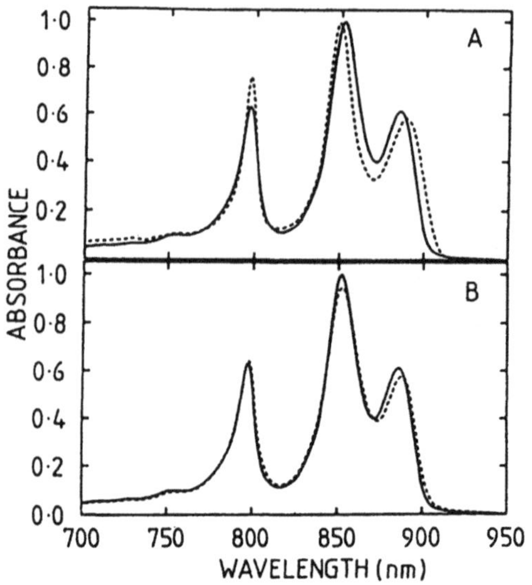

Figure 6. Spectral analysis of the absorption spectrum of transconjugant strain M21 (MA81). The dashed lines give a linear combination of the spectra of (A) isolated LH1 and LH2 complexes (1,2) and (B) mutants NF57 and M21.

Gaussian deconvolution of the 4K absorption spectra has been undertaken (17) in an attempt to account for the relative breadth of the absorption band of NF57 near 850 nm, and to investigate whether the species Bchl 896 (18,19) could be identified in such an analysis of M21 and M2192. It was found that in general the absorption bands of all three mutants could be deconvoluted into several components. For NF57, a component near 870 nm was identified which is also observed in recombinant strain NF57 [NH1491] and in the wild type. This species represents about 15% of the total absorption of the 850 nm band and appears to be absent from the LDAO-Triton LH2 complex. For mutants M21 and M2192, a long wavelength component at 899 nm or 897 nm respectively is observed, in addition to the major transitions at 883 and 891 nm. This component may be the species Bchl 896 which has been reported in membranes of the wild type and the LDS complex (19,20).

Table 1. Domain Sizes of LH1 and LH2 Complexes of Rb. sphaeroides

The numbers for LH2 refer to Bchl 850 pigments.

	LH1		LH2	
	298K	4K	298K	4K
Membranes				
Wild-type	≥1000	≥100		45
M21	≥330	100	-	
M2192	150	125	-	
NF57		-	≥365	30
LDS - LH1[a]	8			-
LDS - LH2[a]		-	30	
LDAO/Triton LH2[b]		-	600	

[a] Prepared according to the method in ref 1.
[b] Prepared according to the method in ref 2.

In view of the data available on the size of energy transfer domains in the wild type (21,22) and in detergent solubilised complexes (23), it is of interest to examine this parameter in the mutants. Singlet-singlet exciton annihilation was measured at room temperature and at 4K; at the lower temperature rates of energy transfer are reduced, and excitations tend to be confined to smaller domains.

The data in Table 1 emphasise how detergent treatment can markedly affect the aggregation properties of LH1 and LH2 complexes. Whereas the size of domains within the mutants bears some similarity to the wild type, the LDS-LH1 complex contains extremely small domains which can be attributed to a 'minimal unit' of 3 : 3 polypeptides : 6 bchl : 6 carotenoid (C.N. Hunter, R. van Grondelle and J.D. Olsen, in preparation). It is also remarkable how these detergents can influence the domain size of LH2; in the LDAO/Triton complex more than one thousand bacteriochlorophyll a molecules can form a domain, whereas the number is reduced to around 30-40 in LDS. A similar phenomenon has been reported recently by Miller et al., (24) who observe a 4 to 5 -fold decrease in the apparent molecular weight of the R. rubrum LH1 complex upon a change of detergent concentration.

Interestingly, this change was accompanied by a shift in absorbance maximum from 873 nm to 820 nm.

The data for wild type and mutant strains shows that the smallest LH1 domain is composed of approximately 100-150 molecules at 4K. This is observed in membranes from the wild type, M21 and M2192, and in membranes at an early stage of photosynthetic development (22). 'Dilution' of the membrane structure by addition of phospholipids *in vitro* also yields a similar result (25). With the exception of M2192, larger domains are observed at 298K; in the wild-type core structures are presumably connected by LH2 units as proposed by Vos *et al*., (36). LH2 units in NF57 exhibit a tendency to form relatively large domains consisting of at least 300 bacteriochlorophyll molelcules. The result for M21 shows that LH1-reaction centre cores also possess the capacity to form larger domains, although not of the same size as for the wild type. This may be due to limitations imposed by the tubular membrane structure rather than the incapacity of the cores. For example, there is a two-fold enrichment of the phospholipid : protein ratio for the M21 membrane in relation to wild-type intracytoplamic membranes. (C.N. Hunter, J.D. Pennoyer, J.N. Sturgis, D. Farrelly and R.A. Niederman, in preparation) However, no such limitation exists for LH1-reaction systems that exist in nature such as in *Rhodospirillum rubrum*, the photosynthetic apparatus of which consists of domains of several thousand bacteriochlorophyll molecules (21). M2192 cannot form domains extending beyond the minimal core structure, and it can be concluded that this inability arises from the absence of reaction centres.

Some of the data in Table 1 is represented in Figure 7; in particular, these models attempt to show how it is possible that the absence of reaction centres in M2192 perturbs the normal geometry of LH1 assemblies. In this way, arrangements favourable for aggregation may be affected. It is also noteworthy that some LH1 units (Bchl 896) normally attached to reaction centres in wild-type and M21

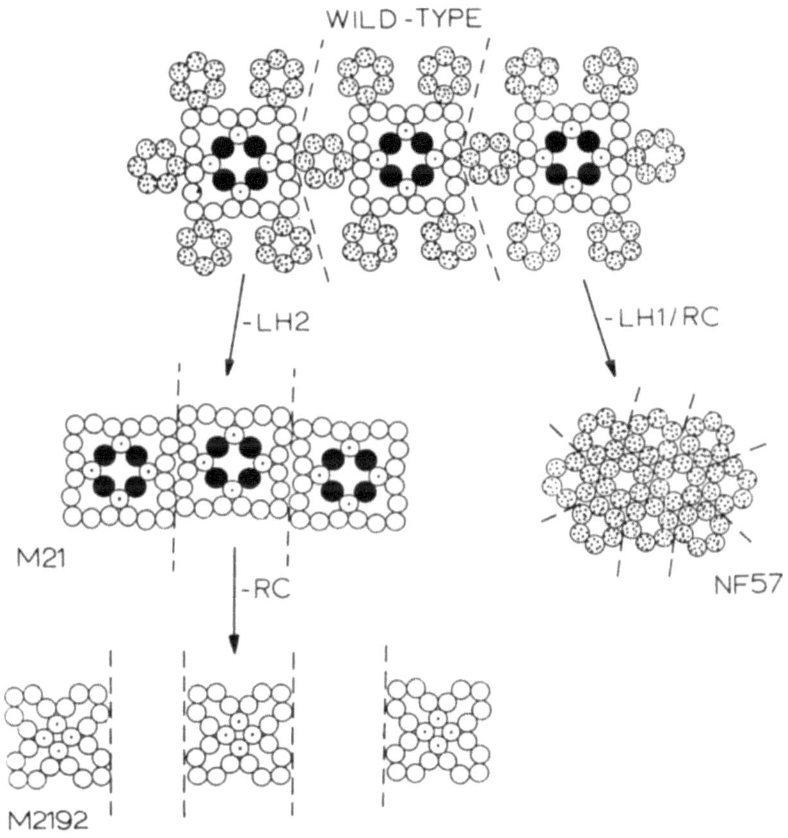

Figure 7. Models for the arrangement of antenna and reaction centre complexes in membranes from wild type and mutant strains. The dashed lines indicate domains at 4K.

- ◎ LH2 – 6 Bchl 850, 3 Bchl 800
- ○ LH1 – 6 Bchl 875
- ⊙ LH1 – 6 Bchl 896
- ● reaction centre – 4 Bchl

strains may be found in close proximity for energy transfer in M2192. This may explain the fluorescence polarisation properties of LH1 in M2192 (R. van Grondelle, personal communication). The polarisation values observed are significantly less than those for M21 (27); this difference may arise from energy transfer among Bchl 896 molecules which normally donate excitation energy solely to reaction centres in M21.

Conclusions

Several conclusions may be drawn from the study of these mutants.

First, the synthesis and assembly of LH1 depend on a product of the gene borne on clone pSCN5H-1, which may be puh A encoding the reaction centre H subunit or could be a gene flanking puh A.

Second, there is an absence from mutant M21 of the invaginated intracytoplasmic membrane that normally gives rise to a population of vesicles ('chromophores') from the wild-type. Thus, the latter stages of photosynthetic membrane assembly which are characterised by the addition of peripheral LH2 units to LH1-reaction centre core (28) are obligately linked to the maturation of the membrane initiation sites to form normal intracytoplasmic membranes.

Third, many spectroscopic properties of the antennae in the mutants differ from those described for detergent solubilised antennae. In LH2 and LH1 the species Bchl 870 and Bchl 896 respectively may serve to funnel excitations to the next complex in the sequence.

Finally, reaction centres appear to exert a strong influence on the capacity of LH1 to form large domains for energy transfer.

Acknowledgements

CNH gratefully acknowledges financial support from the Agricultural and Food Research Council of Great Britain, and a short-term fellowship from the European Molecular Biology Organisation. The author is grateful to R. van Dorssen, M. Vos and R.A. Niederman for their advice and to John Olsen for his technical expertise.

References

1. Broglie, R.M., C.N. Hunter, P. Delepelaire, R.K. Clayton, R.A. Niederman and N.-H. Chua, (1980) Proc. Natl. Acad. Sci. USA 77, 87-91.

2. Clayton, R.K. and B.J. Clayton (1972) Biochim. Biophys. Acta 283, 492-504.

3. Dawkins, D.J., L.A. Ferguson and R.J. Cogdell. (1987) This volume.

4. Hunter, C.N. (1987) manuscript in preparation.

5. Hunter, C.N., M.K. Ashby and S.A. Coomber (1987) Biochem. J. 247, 489-492.

6. Laemmli, U.K. (1970) Nature 227, 680-685.

7. Niederman, R.A., D.E. Mallon and L.C. Parks (1979) Biochim. Biophys. Acta 555, 210-220.

8. Kramer, H.J.M. and J. Amesz (1982) Biochim. Bophys. Acta 682, 201-207.

9. Vos, M. (1987) Thesis, University of Leiden, The Netherlands.

10. Zhu, Y.S., P.J. Kiley, T.J. Donohue and S. Kaplan (1986) J. Biol. Chem. 261, 10366-10374.

11. Youvan, D.C., J.E. Hearst and B.L. Marrs (1983) J. Bacteriol. 154, 748-755.

12. Zsebo, K.M. and J.E. Hearst (1984) Cell. 37, 937-947.

13. Ashby, M.K., S.A. Coomber and C.N. Hunter (1987) FEBS Lett. 213, 245-248.

14. Kiley, P.J. and S. Kaplan (1987) J. Bacteriol. 169, 3268-3275.

15. Jackson, W.J., P.J. Kiley, C.E. Haith, S. Kaplan and R.C. Prince (1987) FEBS Lett. 215, 171-174.

16. Lommen, M.A.J. and J. Takemoto (1978) Arch. Microbiol. 118, 305-308.

17. van Dorssen, R.J., C.N. Hunter, R. van Grondelle, A.H. Korenhof and J. Amesz (1987) submitted for publication.

18. Borisov, A.Yu, R.A. Gadonas, R.V. Danielus, A.S. Piskarskas and A.P. Razjivin (1982) FEBS Lett. 138, 25-28.

19. Sundstrom, V., R. van Grondelle, H. Bengstrom, E. Akesson and T. Gillbro (1986) Biochim. Biophys. Acta 851, 431-446.

20. Kramer, H.J.M., J.D. Pennoyer, R. van Grondelle, W.H.J. Westerhuis, R.A. Niederman and J. Amesz (1984) Biochim. Biophys. Acta. 767, 335-344.

21. Bakker, J.G.C., R. van Grondelle and W.T.F. Den Hollander (1983) Biochim. Biophys. Acta 725, 508-518.

22. Hunter, C.N., H.J.M. Kramer and R. van Grondelle (1985) Biochim. Biophys. Acta 807, 44-51.

23. van Grondelle, R., C.N. Hunter, J.G.C. Bakker and H.J.M. Kramer (1983) Biochim. Biophys. Acta. 723, 30-36.

24. Miller, J.F., S.B. Hinchigeri, P.S. Parkes-Loach, P.M. Callahan, J.R. Sprinkle, J.R. Riccobono and P.A. Loach (1987) Biochemistry 26, 5055-5062.

25. Westerhuis, W.H.J., M. Vos, R.J. van Dorssen, R. van Grondelle, J. Amesz and R.A. Niederman (1986) In: Progress in Photosynthesis Research (J. Biggins, ed.) Vol. I Martinus Nijhoff, Dordrecht, The Netherlands. pp 29-32.

26. Vos, M., R. van Grondelle, F.W. van der Kooij, D. van de Poll, J. Amesz and L.N.M. Duysens (1986) Biochim. Biophys. Acta. 850, 501-512.

27. Hunter, C.N., R. van Grondelle and R.J. van Dorssen (1987) manuscript in preparation.

28. Hunter, C.N., J.D. Pennoyer and R.A. Niederman (1982) In: Cell Function and Differentiation, Part B. Bioenergetics of Energy Transducing Membranes and Membrane and Protein Energetics (Akoyunoglou, G. et al., eds.) Alan R. Liss, Inc., New York. pp 257-266.

MECHANISMS OF PLASTID AND NUCLEAR GENE EXPRESSION DURING THYLAKOID MEMBRANE BIOGENESIS IN HIGHER PLANTS

P. Westhoff, H. Grüne, H. Schrubar, A. Oswald, M. Streubel, U. Ljungberg & R.G. Herrmann

Botanisches Institut der Universität München, Menzinger Str. 67, D-8000 München 19, Federal Republic of Germany

Introduction

Thylakoid membranes of higher plants are complex structures consisting of more than 50 different polypeptides. The majority of these are assembled into four major complexes: ATP synthase, cytochrome *b*/*f* complex, and photosystems I and II (1-3). Genetic and, more recently, molecular analysis have shown thylakoid membranes to be genetic mosaics (1). About one third of their constituent polypeptides are products of plastid genes (Tab. 1), the remainder is encoded by the nuclear genome. Because most of the catalytically active subunits of thylakoid membranes are among those plastome-encoded proteins, plastid DNA is a central genetic determinant of this membrane (1).

Since thylakoid membranes are of dual genetic origin, their biogenesis requires the coordinated expression of plastid and nuclear genes (4, 5). Incompatibility of nuclear and plastid genomes in certain interspecies hybrids within the genus *Oenothera* suggests a tight interplay between the two genetic systems (6). Its molecular basis is, however, almost unknown.

Photosynthetic Light-Harvesting Systems

Table 1: Plastid genes encoding thylakoid membrane proteins

Gene	Protein	Membrane Complex	Reference
psaA	P-700 chlorophyll a-apoprotein	photosytem I	7
psaB	P-700 chlorophyll a-apoprotein		8
psaC	9 kDa FeS polypeptide		9, 10
psbA	"D1" protein	photosystem II	11
psbB	51 kDa chlorophyll a-apoprotein (CP47)		12
psbC	44 kDa chlorophyll a-apoprotein (CP43)		12
psbD	"D2" protein		13
psbE/F	cytochrome b-559		14, 15
psbH	10 kDa phosphoprotein		16
petA	cytochrome f	cytochrome b/f complex	17
petB	cytochrome b6		17
petD	"subunit 4"		17
atpA	α subunit of CF1	ATP synthase	18
atpB	β subunit of CF1		18
atpE	ε subunit of CF1		18
atpF	CFo-I		19
atpH	CFo-III		20
atpI	CFo-IV		21, 22

In addition to complexity in structure and genetics, thylakoid membranes of higher plants are subject to developmental control and can adapt to various environmental constraints (23-25). Light is such an external factor influencing composition and thus function of this membrane (25-27). Light also triggers or modulates developmentally controlled processes in thylakoid membrane biogenesis with phytochrome, the blue/UV light receptor, and protochlorophyllide being involved as sensor systems (27-31).

In the present publication we summarize our investigations on the biogenesis of thylakoid membranes in higher plants. First, we describe the patterns of plastid and nuclear gene expression observed during light-induced greening of etiolated spinach and *Sorghum* seedlings. Then we focus on one aspect of plastid gene expression, i.e. the processing of polycistronic transcripts into mature mRNAs. Detailed information on mechanisms and function of RNA processing is of particular relevance in plastid gene expression, because the vast majority of plastome-encoded genes seems to be organized in operons and our data indicate that post-transcriptional regulation plays a major role in thylakoid membrane biogenesis (1, 38).

Results and Discussion

Light-induced biogenesis of the thylakoid membrane in etiolated seedlings

Spinach:

Etiolated spinach seedlings do not accumulate chlorophyll, and hence are not capable to carry out photosynthesis. Nevertheless, they already contain a part of the photosynthetic machinery as judged by immunoblot analysis of polypeptides from all four major thylakoid membrane complexes (1, 32). Two major classes of proteins can be distinguished with regard to their mode of accumulation. All subunits of photosystem I reaction center (2) - except subunit 2 which is present in trace amounts -, the four chlorophyll *a*-binding proteins of photosystem II reaction center core (ref. 33; 51 kDa, 44 kDa, D1 and D2 proteins), and a 10 kDa polypeptide probably associated with the water-splitting activity (34) are absent from etioplasts. Their accumulation requires light. A strictly light-dependent increase in steady state levels is also observed for subunit 2 of photosystem I reaction center and, to a lesser extent, for cytochrome *b*-559 (33) which is detectable in substantial amounts in spinach etioplasts. In contrast, subunits of the ATP synthase, the cytochrome b/f complex (1) and other photosystem II components, e.g. the 34, 23, 22 and 16 kDa polypeptides (35), do not depend on light for accumulation. Their abundance does not change more than 2- to 4-fold during greening. Fig. 1 illustrates the kinetics of accumulation for representative members of both classes of proteins.

While plastids in etiolated and greening spinach are clearly distinguishable with respect to their protein composition, they are less different, if both, levels and patterns of plastome-encoded transcripts for thylakoid membrane proteins are compared. All of these transcripts, including those for the chlorophyll a-apoproteins of both reaction centers, can be detected in the dark-grown seedlings. Their abundance does not change dramatically during greening. The absence of plastome-encoded photosystem I and II reaction center polypeptides in spinach etioplasts, therefore, cannot be explained by a lack of the corresponding mRNAs. Hence, the expression of these genes must be controlled post-transcriptionally and not primarily by transcription (1, 32).

In contrast, expression of nuclear genes coding for thylakoid membrane proteins seems to be regulated transcriptionally, since the abundance of the corresponding mRNAs increases drastically upon illumination of etiolated spinach (Fig. 2). However, the changes in steady state levels of some mRNAs, as for example those encoding the delta subunit of ATP synthase, the Rieske iron sulfur protein of the cytochrome b/f complex and the 34, 23 and 16 kDa polypeptides of the water-splitting activity, are not paralleled at the protein levels (Fig. 2). Expression of nuclear genes, therefore, may also involve post-transcriptional regulation.

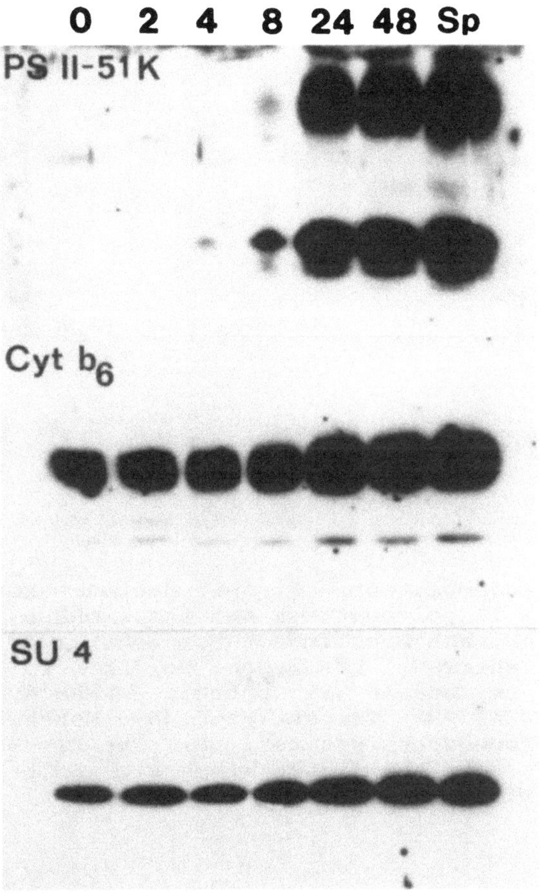

Fig. 1: Light-induced greening of etiolated spinach seedlings: accumulation of 51 kDa chlorophyll a-apoprotein of photosystem II reaction center, cytochrome b6 and subunit 4 of the cytochrome b/f complex. Spinach seedlings were grown for 6 days in the dark and then illuminated with continuous white light. Cotyledons were harvested during greening as indicated on top of the figure (in hrs) and a total membrane extract was prepared. Equal aliquots of these extracts (related to fresh weight) were electrophoresed on SDS/urea gels and the separated proteins were transferred electrophoretically onto nitrocellulose membranes. Filters were incubated with monospecific antisera and the immune complexes formed were detected by ^{125}I labelled protein A followed by fluorography.

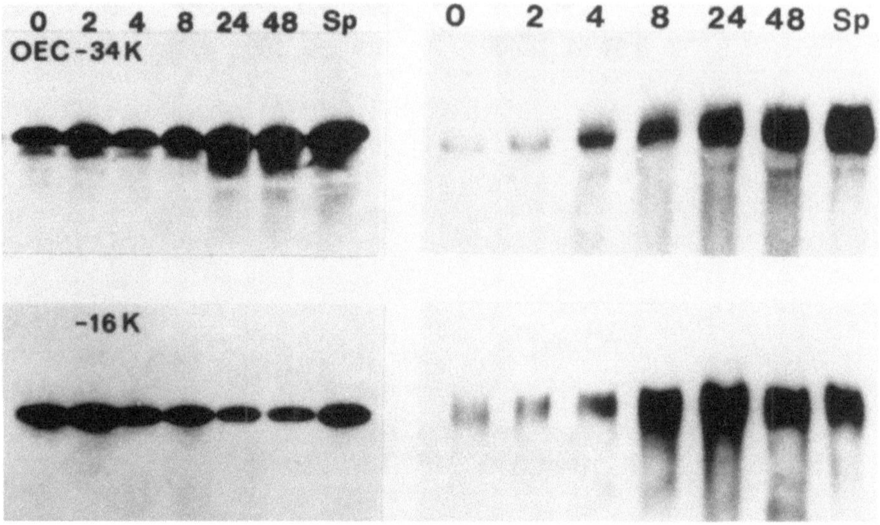

Fig. 2: Accumulation of the nuclear-encoded 34 and 16 kDa photosystem II polypeptides and their mRNAs during greening of etiolated spinach seedlings. For details of plant growth and immunoblot analysis see legend to Fig. 1. To determine the amounts of 34 and 16 kDa protein mRNAs during the greening, total cellular RNA was analyzed by Northern blotting using standard procedures (see ref. 38). The time-course of greening is given on top the figure (in hrs). Left: accumulation of proteins. Right: accumulation of RNAs.

Sorghum:

To analyze whether this mode of gene expression is representative for greening of etiolated seedlings, levels of plastid and cytosolic transcripts encoding thylakoid membrane proteins as well as the abundance of the corresponding proteins have been investigated in the monocotyledous C4 plant Sorghum bicolor. Figs. 3 and 4 demonstrate that this plant behaves quite differently from spinach (cf. Figs. 1 and 2). None of the subunits of the ATP synthase and the cytochrome b/f complex, nor cytochrome b-559, the 16, 22, 23 and 34 kDa proteins associated with the water-splitting activity accumulate in Sorghum seedlings grown in the dark for as long as 7 days. Only trace amounts of these proteins can be detected in the etiolated plants (Fig. 3).

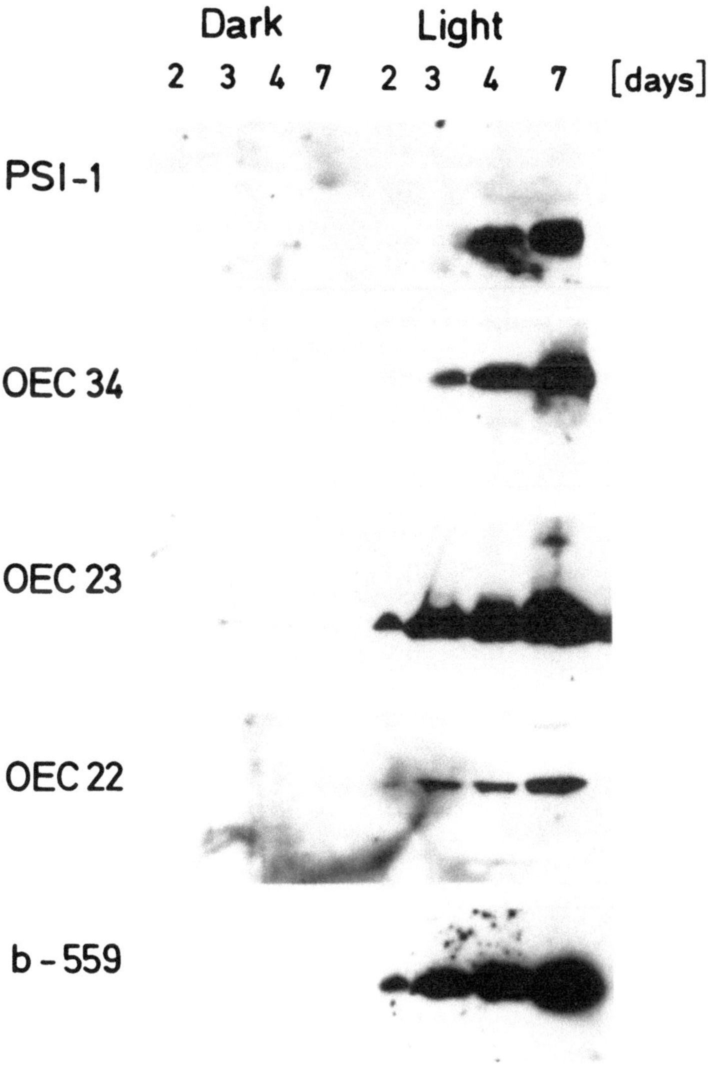

Fig. 3: Steady-state levels of chloroplast proteins in dark and light-grown Sorghum seedlings. Sorghum seedlings (hybrid Sorghum Sudan forage, variety "Golden Harvest") were grown on moistured filter paper in the dark or under continous white light at 22 °C. Shoots were harvested at days 2 - 7, and immunoblot analysis was performed as described in the legend to Fig. 1. PSI-1: P-700 chlorophyll a-apoprotein of photosystem I (plastome-encoded); b-559: cytochrome b-559 (plastome-encoded); OEC 34, 23 and 22: nuclear-encoded polypeptides of the water-splitting apparatus.

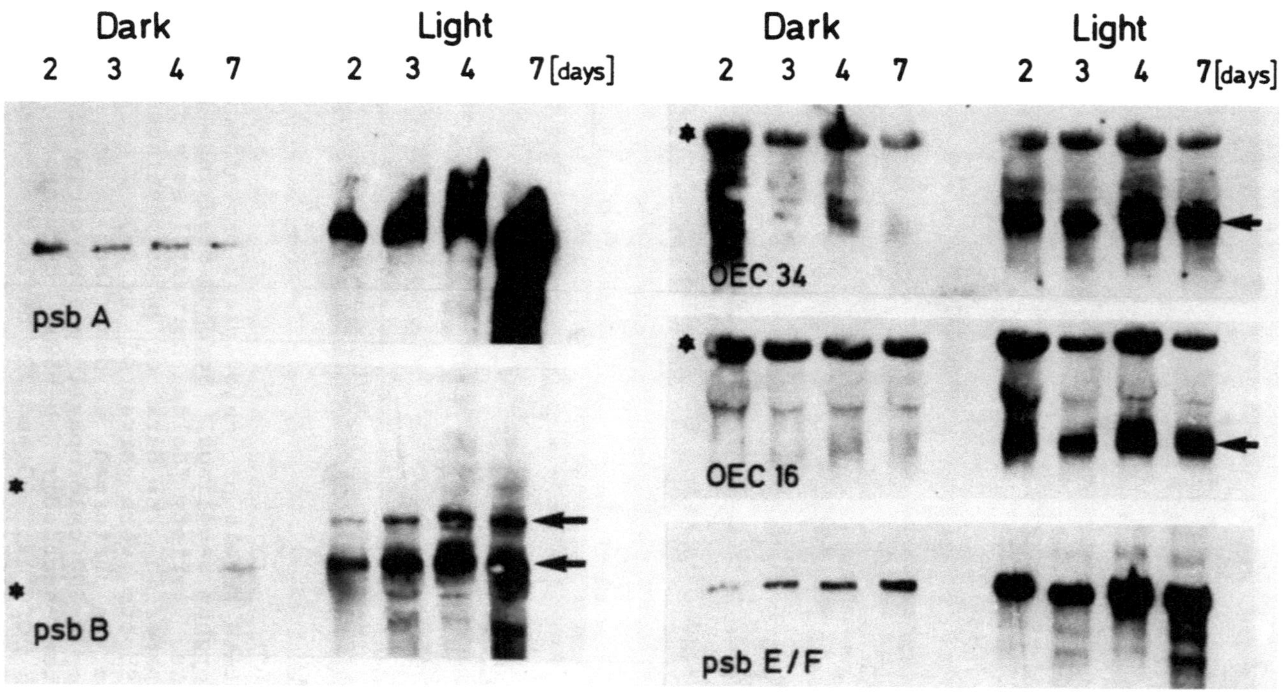

Fig. 4: Steady-state levels of plastid- and nuclear-encoded transcripts for thylakoid membrane proteins in dark- and light-grown *Sorghum* seedlings. For plant growth and details of the Northern blot analysis see legends to Figs. 2 and 3, respectively. For the nature of plastid transcripts being analyzed see Table 1. OEC 16 and 34 denote the nuclear-encoded transcripts for the respective photosystem II polypeptides. Asterisks label the ribosomal RNAs which hybridize unspecifically. Specific hybridization signals are marked by arrows. Note that transcripts of the psbB gene are heterogenous because of multiple processing events (see ref. 38).

Titration of plastid RNA levels (Fig. 4) supports this view: only small amounts of transcripts are present in the etiolated seedlings indicating that light is required to activate transcription of plastid genes in this plant. It should be mentioned that the expression patterns in Sorghum are also different from those described for two other monocots, namely barley (36) and maize (37) which more closely resemble spinach. We conclude from our comparative analysis that greening of etiolated plant seedlings may not necessarily involve the same regulatory controls of gene expression.

The differences in gene expression between Sorghum and the two other monocots barley and maize are also reflected at the morphological level. While the primary leaves of barley and maize expand in the dark, Sorghum leaf development is arrested. Even after 7 days of growth in the dark, the Sorghum leaf is still small and buried within the coleoptile. This correlation between plastid differentiation and leaf development suggests a strong coupling of the two processes (cf. 36).

Operon structure and maturation of polycistronic transcripts in plastids

The vast majority of plastid genes are contained within polycistronic transcription units as is characteristic for a procaryotic-type of gene organization (1). But the mode of organization is complex, since most, if not all, transcription units are non-homogenous, i.e. their component cistrons encode subunits of quite different protein complexes (Fig. 5).

Northern analysis revealed that the polycistronic primary transcripts are not the dominant RNA species found in chloroplasts. Most of these transcripts are not stable, they are generally subject to extensive processing events finally yielding small mono- or oligocistronic RNAs (1, 38).

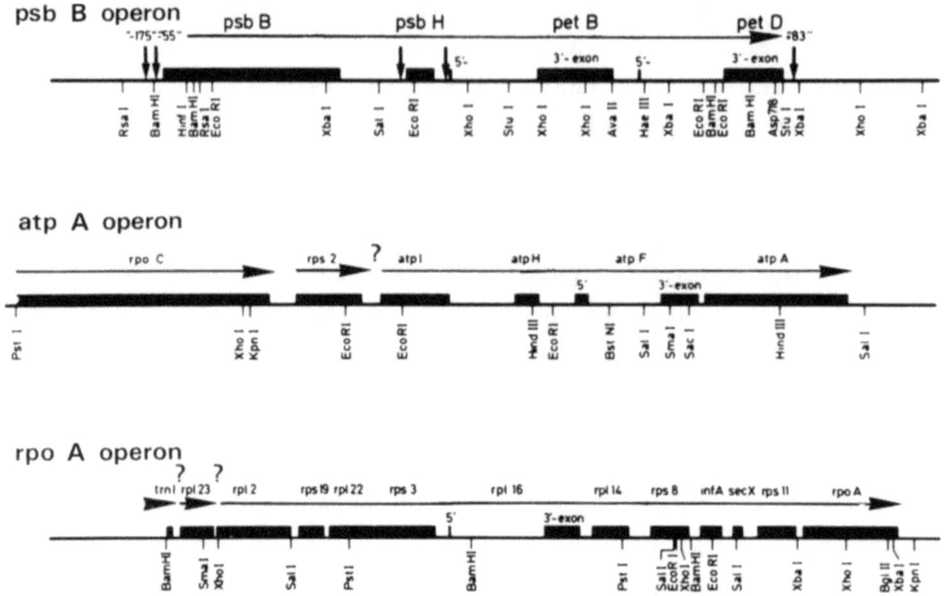

Fig. 5: Operon structure of three polycistronic transcription units in spinach plastid DNA. For the composition of the psbB operon see ref. 38. The atpA operon contains four ATP synthase genes and probably also the rps2 cistron encoding the ribosomal protein S2. Whether the genes are co-transcribed with the rpoB and rpoC genes (β and β' subunits of plastid RNA polymerase) has not yet been solved. Also not finally settled is the structure of the rpoA operon. It encodes the α subunit of RNA polymerase (rpoA), 8 ribosomal protein genes (rpl2 to rps11), a translational initiation factor (infA) and an open reading frame showing homology to the secX gene in E. coli (39, 40). Preliminary data indicate that rpl23 and the gene for tRNA-isoleucin (trnI) belong to that operon as well (Grüne and Westhoff, unpublished data). The arrangement of its cistrons is thus homologous to those in the E. coli S10, spc and α operons (39, 40). Transcription polarities are indicated by horizontal arrow. Vertical arrows (in the psbB operon) label transcription initiation ("-175") and processing sites ("-55"; two sites upstream and downstream of psbH), as well the putative termination site ("+83").

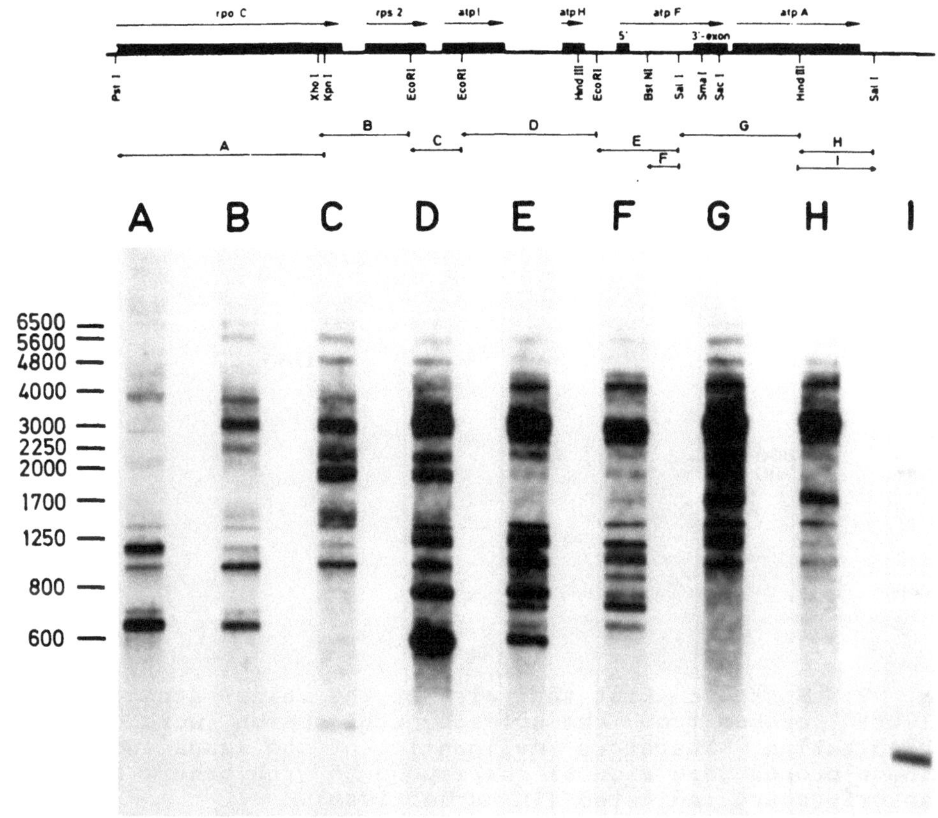

Fig. 6: Northern blot analysis of the spinach atpA operon. Spinach chloroplast RNA was sized on an 1.2 % agarose gel and transferred to Biodyne A membranes. Individual filter strips were hybridized with single-stranded RNA probes prepared by <u>in vitro</u> transcription of SP6/T7/T3 promoter plasmids. Location and polarity of the RNA probes are indicated on top of the figure below the physical map. Designation of probes and lanes corresponds to each other. Sizes of transcripts (in nucleotides) are indicated at the left. RNAs identified include:

Size of RNA	Nature of transcript	lanes
3000 a	rps2, atpI, atpH	B - E
3000 b	atpH, atpF, atpA	D - H
2000	atpI, atpH	C - D
1700	atpA	G - H
600	atpH	D - E

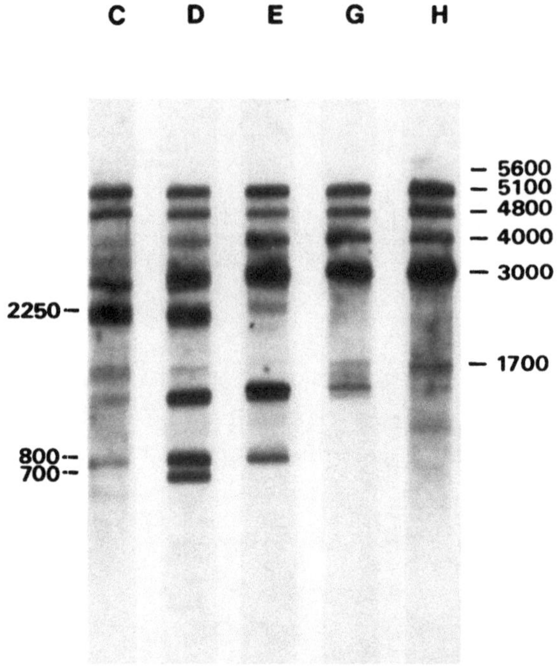

Fig. 7: Northern blot analysis of the maize atpA operon. Selected probes from the spinach atpA operon were used for hybridization. Therefore, designation of the lanes and of the spinach probes (see Fig. 6) correspond to each other. Sizes of transcripts are indicated (in nucleotides).

For example, the spinach atpA operon contains four cistrons encoding ATP synthase subunits (Fig. 6) and probably also the rps2 gene (cf. also 41). While the putative primary transcript has not been delineated with certainty, several of the processed oligocistronic RNAs have been identified (Fig. 2). Comparison of the RNA profiles obtained with spinach and maize reveals striking similarity (Figs. 6 and 7). Interestingly, the large precursors of 5 600 and 4 800 bases are much more prominent in the maize pattern, indicating different rates of processing in the two plants.

Detailed analysis of RNA maturation in the psbB operon (38)

has allowed to identify two new processing activities: endonucleolytic, intercistronic cleavage of polycistronic transcripts at discrete sequence motifs and exonucleolytic trimming of discrete RNA 3' ends up to inhibitory hairpin loops. In addition, processing in 5' untranslated leader sequences and splicing of class II introns has been observed. The occurence of these processing activities in other transcription units, i.e. the atpA, rpoA and petA operons, indicates that they represent an integral part of the RNA modification machinery of the organelle. They should, therefore, play an essential role in the expression of plastid DNA. This view is supported by two additional lines of evidence:

(a) An Oenothera plastome mutant which cannot process the polycistronic psbB-psbH-petB-petD transcript into the mature dicistronic petB-petD RNA does not contain any functional cytochrome b/f complex, while the other thylakoid membrane complexes are not affected (Grüne, Olesch-Kirchhof & Westhoff, unpublished data). At least in this instance, fully processed RNAs seem to be required for proper translation.

(b) The processing patterns of plastid operons encoding photosystem II proteins, e.g. the psbB operon, differ in mesophyll and bundle sheath cells of maize with respect to abundance of the processed RNAs. While the steady state levels for photosystem II transcripts are reduced in the bundle sheath chloroplasts, those for other thylakoid membrane complexes (e.g. the dicistronic petB-petD RNA) do not show any significant differences between the two cell types (Westhoff & Eskins, unpublished data). The data are compatible with the idea that once released from a polycistronic transcript, its component segments can exhibit different stabilities. Differential mRNA stability may be, therefore, of central importance in the regulation of plastid gene expression as it is in prokaryotic systems (cf. 40).

Acknowledgments

This work was supported by the Deutsche Forschungsgemeinschaft (SFB 184, We 1008). P. W. thanks K. Liedl for preparing the photographs and Dr. K. Eskins (USDA Northern Regional Research Laboratory, Peoria) for kindly providing the Sorghum seeds.

References

1. Herrmann, R.G., P. Westhoff, J. Alt, J. Tittgen, N. Nelson. 1985. In: Molecular Form and Function of the Plant Genome (L. van Vloten-Doting, G.S.P. Groot, T.C. Hall, eds.). Plenum Publ. Corp., New York. pp. 233.

2. Staehelin, L.A. 1986. In: Encyclopedia of Plant Physiology, Photosynthesis III. Photosynthetic Membranes and Light-Harvesting Systems, Vol 19 (L.A. Staehelin, C.J. Arntzen, eds.). Springer Verlag, Berlin. pp. 1.

3. Thornber, J.P. 1986. see ref. 2. pp. 98.

4. Parthier, B. 1982. Biochem. Physiol. Pflanzen 177, 283.

5. Sears, B.B. 1983. In: Stadler Symp., Vol. 15, University of Missouri, Columbia. pp. 119.

6. Stubbe, W. 1959. Z. Vererbungsl. 90, 288.

7. Westhoff, P., J. Alt, N. Nelson, W. Bottomley, H. Bünemann, R.G. Herrmann. 1983. Plant Mol. Biol. 2, 95.

8. Fish, L.E., U. Kück, L. Bogorad. 1985. J. Biol. Chem. 260, 1413.

9. Oh-oka, H., Y. Takahashi, K. Wada, H. Matsubara, K. Ohyama, H. Ozeki. 1987. FEBS Lett. 218, 52.

10. Hayashida, N., T. Matsubayashi, K. Shinozaki, M. Sugiura, K. Inoue, T. Hiyama. 1987. Curr. Genet. 12, 247.

11. Zurawski, G., H.J. Bohnert, P.R. Whitfeld, W. Bottomley. 1982. Proc. Natl. Acad. Sci. USA 79, 7699.

12. Westhoff, P., J. Alt, R.G. Herrmann. 1983. EMBO J. 2, 2229.

13. Rochaix, J.D., M. Dron, M. Rahire, P. Malnoe. 1984. Plant Mol. Biol. 3, 363.

14. Westhoff, P., J. Alt, W.R. Widger, W.A. Cramer, R.G. Herrmann. 1985. Plant Mol. Biol. 4, 103.

15. Herrmann, R.G., J. Alt, B. Schiller, W.R. Widger, W.R. Cramer. 1984. FEBS Lett. 176, 239.

16. Westhoff, P., J.W. Farchaus, R.G. Herrmann. 1986. Curr. Genet. 11, 165.

17. Alt, J., P. Westhoff, B.B. Sears, N. Nelson, E. Hurt, G. Hauska, R.G. Herrmann. 1983. EMBO J. 2, 979.

18. Westhoff, P., N. Nelson, H. Bünemann, R.G. Herrmann. 1981. Curr. Genet. 4, 109.

19. Westhoff, P., J. Alt, N. Nelson, R.G. Herrmann. 1985. Mol. Gen. Genet. 199, 290.

20. Howe, C.J., A.D. Auffret, A. Doherty, C.M. Bowman, T.A. Dyer, J.C. Gray. 1982. Proc. Natl. Acad. Sci. USA 79, 6903.

21. Hennig, J., R.G. Herrmann. 1986. Mol. Gen. Genet. 203, 117.

22. Cozens, A.L., J.E. Walker, A.L. Phillips, A.K. Huttly, J.C. Gray. 1986. EMBO J. 5, 217.

23. Leech, R.M. 1984. In: Chloroplast Biogenesis (N.R. Baker, J. Barber, eds.). Elsevier Publ., Amsterdam. pp. 1.

24. Wellburn, A.R., I. Gounaris, J.H. Owen, J.E.M. Laybourn-Parry, F.A.M. Wellburn. 1986. In: Regulation of Chloroplast Differentiation (G. Akoyunoglou, H. Senger, eds.). Alan R. Liss Inc., New York. pp. 371.

25. Anderson, J.M. 1986. Ann. Rev. Plant Physiol. 37, 93.

26. Eskins, K., M. Duysen, L. Dybas, S. Mc Carthy. 1985. Plant Physiol. 77, 29.

27. Eskins, K., S. Mc Carthy. 1987. Physiol. Plant. 71, 100.

28. Mohr, H. 1984. see ref. 23. pp. 305.

29. Harpster, M., K. Apel. 1985. Physiol. Plant. 64, 147.

30. Mohr, H. 1986. see ref. 24. pp. 623.

31. Fluhr, R., N.-H. Chua. 1986. Proc. Natl. Acad. Sci. USA 83, 2358.

32. Westhoff, P., H. Schrubar, C. Bisanz-Seyer & R.G. Herrmann, manuscript in preparation.

33. Arntzen, C.J., H.B. Pakrasi. 1986. see ref. 2. pp. 457.

34. Ljungberg, U., H.E. Akerlund, B. Andersson. 1986. Eur. J. Biochem. 158, 477.

35. Andersson, B. 1986. see ref. 2. pp. 447.

36. Klein, R.R., J. E. Mullet. 1987. J. Biol. Chem. 262, 4341.

37. Rodermel, S.R., L. Bogorad. 1985. J. Cell Biol. 100, 463.

38. Westhoff, P., R.G. Herrmann. Eur. J. Biochem., in press.

39. Sijben-Müller, G., R.B. Hallick, J. Alt, P. Westhoff, R.G. Herrmann. 1986. Nucleic Acids Res. 14, 1029.

40. Shinozaki, K., M. Ohme, M. Tanaka, T. Wakasugi, N. Hayashida, T. Matsubayashi, N. Zaita, J. Chunwongse, J. Obokata, K. Yamaguchi-Shinozaki, C. Ohto, K. Torazawa, B.Y. Meng, M. Sugita, H. Deno, T. Kamogashira, K. Yamada, J. Kusuda, F. Takaiwa, A. Kato, N. Tohdoh, H. Shimada, M. Sugiura. 1986. EMBO J. 5, 2043.

41. Hudson, G.S., J.G. Mason, T.A. Holton, B. Koller, G.B. Cox, P.R. Whitfeld, W. Bottomley. 1987. J. Mol. Biol. 196, 283.

42. Belasco, J.G., J.T. Beatty, C.W. Adams, A. von Gabain, S.N. Cohen. 1985. Cell 40, 171.

Section III

Organization: Special Spectroscopy Techniques and Models

ASSIGNMENT OF SPECTRAL FORMS IN THE PHOTOSYNTHETIC ANTENNAS TO CHEMICALLY DEFINED CHROMOPHORES

A.Scherz

The Biochemistry Department, Wiezmann Institute of Science, Rehovot, Israel, 76100

The photosynthetic antennas need to have a large cross section for absorption of the prevailing light, stability for prolonged illumination and the ability to transfer the collected energy with minimum dissipation and within the lifetime of the photoexcited singlet state into the site of electron allocation (1). These principles are implemented by special networks of tetrapyrroles, carotenoids (Cr) and polypeptide (PP) that are termed "Light-Harvesting Complexes" - (LHCs) (2).
Relative to the spectral properties of the *in vitro* monomers, the carotenoids retain their spectral properties (they may have strong optical activity that is induced by interactions with nearby chlorophyll (Chl) or bacteriochlorophyll (Bchl) molecules (4)) but the spectral properties of the *in vivo* tetrapyrroles can be significantly different from their spectral properties as isolated monomers *in vitro*. For example, the Qy transition of the *in vivo* Bchla can be shifted by 1200 cm^{-1} relative to the Qy transition of Bchla in dry ether or aceton (3,4); The Qy transtion of Chla can be shifted by 800 cm^{-1} (e.g. in the core LHC of P-700 (5)) and the lowest energ transitions of the open tetrapyrroles can be shifted by 1000-2000 $^{cm-1}$ (6).
The bands' shift regulates the primary interaction of the photosynthetic apparatus with light in two ways:

(a) It tunes the photosynthetic antennas to extraneous illumination (7);

(b) The antennas usually contain several LHCs with slightly different energies for the Qy transitions and as energy migrates from short-wavelength absorbing pigments to longer wavelength absorbing ones, a mechanism for unidirectional energy flow is established.

Unfortunately, the overlap transitions of different LHCs within a single antenna (a prerequisite for the efficient migration of the captive photons to the RCs) hampers a straightforward identification of the spectral forms that are involved in the energy transfer process with well defind chemical entities. Consequently, the *in vivo* pigments are phenomenologically rather than chemically defined. For example, B850,

Photosynthetic Light-Harvesting Systems

B870 and B890 are LHCs of purple photosynthetic bacteria termed by their lowest energy optical absorption.

The phenomenological definitions refer to spectral features that can be isolated in time or in frequency domains. In time-resolved spectroscopy (e.g. 8,9) one usually perturbe the studied LHCs by breif illumination and followes the resulted spectra at different time intervals. Transient absorptions with similar lifetimes that are monitored at the same time interval are assigned to the same spectral form. The uniqueness of this identification depends on the time duration of the perturbing event, the perturbation bandwidth, the resolution of the detection system and the probability for the occurence of biphotonic processes. The last point is of particular importance in the studies of Chls and Bchls since the optical absorption of their excited states is quite similar to the absorption of their ground states (Hoffmann and Leupold, this meeting and references therin). In general, one tries to shorten the pulse duration in order to follow-short lived transients but the accompanying increase of the perturbation bandwidth has to be considered.

In frequency-domain spectroscopy, one perturbes the system with highly monochromic light and probes the concomitant absorption or emission at different wavelengths. Narrow perturbation burns a hole in the sample absorption due to the bleaching of a single spectral form (Friedrich et al this volume). By shifting the perturbation light to different wavelengths in the absorption region one can resolve the spectrum into several components. Although the frequency-domain experiment utilizes CW methodology, it gives an insight into verious relaxation processes of the photoexcited singlet states. For example: coupling to the matrix phonons(10,11), coupling to charge transfer states (12) and energy barriers between vibronic states (Friedrich, this meeting).

The above mentioned techniqes are mostly applied to the ground states and the lowest excited singlet state. The triplet states of the *in vivo*chromophores probably do not participate in the energy funneling and the electron transfer processes but they are involve in protection of the *in vivo*Chls and the Bchls from destructive illumination (3) and they offer valuable information about the delocalization of the photoexcited electron (13,14) (the |D| and |E| values measure the Π electron delocalization in the three canonical coordinates of the photoexcited triplet state). There are several chromophores in the LHCs that can be excited into the triplet states with relatively similar |D| and |E| values and the traditional flash-photolysis-EPR detection experiment may be insufficient for their resolution. A powerful way to overcome these difficulties is offered by the Optical Detection Magnetic Resonance (ODMR) (Ulrich et al this volume) which has been widely used in studies of RCs.

Once we have isolated the spectral forms in the LHCs we may wish to correlate them with specific chromophores and their setting *in vivo*. Obviously, X-ray data will give us the best insight to the *in vivo* chromophores and their vicinity. Unfortunatly, the homogeneous crystallisation of LHCs is still limited. On the other hand, X-ray studies of the bacterial RCs confirmed most of the organizational principles suggested by optical and paramagnetic resonance techniques (15). One may therefore try to deduce the *in vivo*setting of the extracted chromophores from their spectral properties by three complementary methods:

(a) *In vitro*study of model compounds that have similar spectra to the resolved forms (Rosenbach-Belkin, Braun, Kovatch and Scherz, this volume)

(b) Theoretical simulation of the *in vivo* spectra considering the possible effects of amino-acid residues (Pearlstein, this volume, 16,17), exciton coupling between degenerate and non-degenerate states (18,19) and perturbation of intermolecular charge-transfer states (20).

(c) Comparative studies of chromophores' clusters in RCs that have known structures and in LHCs.

References

1. Zuber, H. (1985) Photochem. Photobiol. *42*, 821-844.
2. Zuber, H. (1986) TIBS *11*, 414-419.
3. Cogdell, R.J. and Thornber, J.P. (1980) FEBS Lett. *122*, 1-8.
4. Sauer, K. and Austin, L.A. (1978) Biochem. *17*, 2011-2019.
5. Cogdell, R.J, and Scheer, H., (1985) Photochem. Photobiol. 42, 669-678
6. Wittmershaus, B.P., (1987) in "Progress in Photosynthesis Research; Proceeding of the VIIth International Congress on Photosynthesis " (ed. J.biggins), vol I, pp 75-82

7. Scheer, H., (1982) in "Light Reaction path of photosynthesis" (ed. Fong, K.F.), 7-45 Springer-Verlag, Berlin.

8. Breton, J., Vermeglio, A., Garrigos, M. and Paillotin, G. (1981) Proc. 5th Intr. Cong. Photosyn. (ed. Akoyunoglou, G.) Vol. 3, pp. 445-459, Balaban Intr.

9. Holzwarth, A.R., Brock, H. and Schatz, G.H.

10. Beddard, G.S. and Cogdell, R. (1982) in "Light Reaction path of photosynthesis" (ed. Fong, K.F.), pp 46-79, Springer-Verlag, Berlin. pp 91-94.

11. Meech, S.R., Hoff, A.J. and Wiersma, D.A. (1985), Chem. Phys. Lett. *121*, 287-292

12. Meech, S.R., Hoff, A.J. and Wiersma, D.A. (1986), Proc. Nat.Acad.Sci. U.S.A. *83*, 9464-9468

13. Boxer, S.G., Middendorf, T.R., Lockhart, D.J., (1986) FEBS Lett.

14. Levanon, H. and Norris, J.R. (1982) *200*,237-241 in "Light Reaction path of photosynthesis" (ed. Fong, K.F.), pp 46-79, Springer-Verlag, Berlin.

15. Gonen, O., Regev, H., Levanon, H. Thurnauer, J.R., Norris, J.R. and Closs, G. (1985) Chem. Phys. Lett. *179*, 325-328.

16. Deisenhofer, J., Epp, O., Miki, K., Huber, R. and Michel, M. (1985) Nature *318*, 618-624.

17. Fajer, J., Barkigia, K.M., Fujita, E., Davis, M.S., Goff, D.A., Hanson, L.K. Head, J.D., Horning, T., Smith, K.M. and Zerner, M.C., (1985) in "Antennas and Reaction Centers of Photosynthetic Bacteria." (e.d. Beyerle, M.E.), pp. 324-338, Springer-Verlag,Berlin.

18. Eccles, J. and Honig, B. (1983) Proc. Natl. Acad. Sci. USA *80*, 4959-4962.

19. Scherz, A. and Parson, W.W. (1984) Biochim.Biophys. Acta *766*,

20. Scherz, A. and Parson, W.W. (1986) Photosynthetic Res. *9*, 21-32. 653-665.

21. Parson, W.W. and Warshel, A. (1987) J. Am. Chem. Soc. *109*, 6154

LINEAR DICHROISM AND ORIENTATION OF PIGMENTS IN PHYCOBILISOMES AND THEIR SUBUNITS.

Laura Juszczak, Nicholas E. Geacintov

Chemistry Department, New York University
New York, NY 10003, USA

Barbara A. Zilinskas

Department of Biochemistry and Microbiology
Cook College, Rutgers University
New Brunswick, NJ 08903, USA

Jacques Breton

Service de Biophysique, Departement de Biologie
Centre d'Etudes Nucleaires de Saclay
91191 Gif sur Yvette Cedex, France

Introduction

In the cyanobacterium Nostoc sp., the phycobilisome is hemi-discoidal in shape and contains the three phycobiliproteins allophycocyanin (APC), phycocyanin (PC), and phycoerythrin (PE), when grown in white light. The phycobilisomes consist of an APC-containing core composed of three stacked cylinders arranged so that their longitudinal axes are parallel to one another and their ends form a triangle. Each of these core cylinders consists of four disk-like aggregates of trimeric pigment-protein complexes. This stack of cylinders is attached to the thylakoid membrane by a linker polypeptide. Furthermore, six rod-like structures are attached to this core so that their axes are perpendicular to the long axes of the core cylinders. Each of these rods contains PC proximal to the core and PE distal to the core. Linker proteins are required to attach the rods to the core and rod elements to each other. Within each trimeric disk, each of the monomers consists of two different polypeptide chains labeled α and

Photosynthetic Light-Harvesting Systems

β, to which are attached varying numbers of bilin chromophores. A knowledge of the relative orientation of the transition moments of the different chromophores in phycobilisomes is important for an understanding of the high efficiencies of energy transfer in these organisms. Recently, the three-dimensional structure of PC trimers (without linker polypeptides) derived from two different cyanobacteria have been determined by X-ray analysis at 3 Å resolution (1,2). From this work, the relative orientations of the transition dipole moments of each chromophore within the pigment-protein complexes can be derived. This information has been utilized to interpret the fluorescence polarization (3) and energy transfer characteristics (4) of these complexes.

Polarized light optical spectroscopy methods utilizing oriented samples can also provide information on the orientations of chromophores, although not at near atomic resolution as X-ray crystallography. However, linear dichroism techniques are simpler, and can be applied to samples of any size or state of aggregation, as long as some degree of macroscopic orientation of the samples can be achieved.

There are relatively few linear dichroism studies on phycobilisome-containing systems. Gagliano et al. (5) utilized short, 5 ms duration electric field pulses, as well as squeezed polyacrylamide gels, to orient various submembrane fractions. In order to avoid dissociation of phycobilisomes and sub-phycobilisome fractions in the polyacrylamide gels, it was necessary to fix these preparations with glutaraldehyde. Bruce and Biggins (6) successfully oriented whole cells of Anacystis nidulans in gelatin/sucrose gels. Frackowiak and her co-workers (7,8) routinely embed phycobiliproteins in polyvinyl alcohol films which, when stretched, provide good alignment of the particles for linear dichroism studies.

We have investigated in detail the linear dichroism properties of phycobilisomes and phycobiliproteins isolated from Nostoc sp. The size of the particles varied from the small PE trimers to intact phycobilisomes, and thus suitable methods of orienting these systems were devised. The larger phycobilisome aggregates, which are suspended in an aqueous buffer medium, orient well in the hydrodynamic flow gradients of a Couette cell; on the other hand, thermal Brownian motion of the smaller, sub-phycobilisome particles does not allow for a satisfactory degree of alignment under the same conditions. However, all of the pigment-protein aggregates can be oriented in squeezed gels (10% gelatin), and LD signals of excellent quality can be obtained. Furthermore, as judged by comparisons of circular dichroism and fluorescence emission spectra in these gels and in aqueous suspensions, all of these preparations, even the rather fragile phycobilisomes, remained intact in 10% gelatin. However, as the polyvinyl alcohol and phycobiliprotein solution dries to a film, the fluorescence emission characteristics of the phycobiliproteins change. We have therefore not utilized the stretched polyvinyl alcohol technique in this work.

Materials and methods

Phycobilisomes and phycobiliproteins were isolated as described earlier from Nostoc sp. grown in cool-white fluorescent light (General Electric, F48PF17/CW Power Groove; light intensity 8 w/m^2) (9,10).

The linear dichroism apparatus was assembled from various components and is described in detail elsewhere (11). In the flow orientation experiments, the whole phycobilisomes suspended in buffer, were placed in the 0.5 mm annular gap between a stationary, transparent, 23 mm diameter outer cylinder and an inner, rotating (600 rpm), 22 mm diameter cylinder. In the gel orientation experiments, phycobiliproteins in buffer solution were mixed into gelatin-buffer solutions at 32°C. At this temperature, the gelatin solution is sufficiently liquefied to allow for thorough mixing of the phycobiliprotein, and yet this temperature is not high enough to cause denaturation of the protein. The 10% (w/v) gelatin (potassium phosphate buffer) solutions were prepared from porcine skin gelatin (300 Bloom strength, Sigma Chemical Company, St. Louis, MO); the potassium buffer solutions were either 0.1M, pH 5.0, or 0.5 M, pH 7.0, to accommodate the different specific phycobiliprotein requirements (9), i.e., larger phycobiliprotein aggregate conformations are maintained only at higher buffer strengths. After mixing, the phycobiliprotein-gel was cast into 1 cm x 1 cm x 2 cm blocks. The resulting gels were oriented by squeezing in a cell similar to the one described by Bruce and Biggins (6).

The dichroism was probed utilizing linearly polarized light propagating in a direction perpendicular to the axis of rotation of the Couette cell, or parallel to the direction of the force applied to the gel, i.e., perpendicular to the direction of elongation. The linear dichroism ΔA is defined as

$$\Delta A = A_{\parallel} - A_{\perp}$$

In the case of the Couette cell, $A_{\parallel}$ is the absorbance of the sample measured with the polarization of the light beam oriented parallel to the flow lines, while $A_{\perp}$ denotes the absorbance measured with the polarization perpendicular to the flow lines. In the case of the gels, $A_{\parallel}$ refers to an orientation of the polarization of the light beam along the direction of maximal elongation of the squeezed gel, while $A_{\perp}$ refers to the perpendicular orientation.

Results

The absorption and LD spectra for PE trimeric disks in 10% gelatin, 0.1 M potassium phosphate buffer, pH 5.0 are shown in Fig. 1. Both the absorption maximum at 550 nm, and the shape of the spectrum, are very similar in this gel and in 0.1 M phosphate buffer solution, pH 5.0 (12). The LD spectrum is S-shaped and displays a minimum at 544 nm and a maximum at 568 nm. Blank gels gave generally structure-less, but variable base lines; thus, all LD spectra are shown as measured, i.e. without corrections for varying base lines; this accounts for the relatively high, constant positive LD above 600-650 nm in fig. 1. Since the normal to the disk face is oriented perpendicular to the elongation direction (See inset, figure 1), and since the measuring beam propagation direction is coincident with the direction of the applied squeezing force, the sign of the LD signal suggests that the shorter wavelength transition moments tend to be oriented perpendicular, while the longer wavelength transitions are oriented closer to the plane of the disk. It has been shown that there are no spectroscopically significant changes in chromophore environment, nor any new chromophore-chromophore interactions induced upon association of α and β subunits into monomers, or monomers into trimers (11); therefore, there are no excitonic contributions to the spectrum which might have accounted for the S-shaped spectrum. The negative and positive LD signals in Fig. 1 can be attributed to the individual shorter-wavelength sensitizing pigments, and the longer wavelength energy-accepting pigments, respectively. The absorption and LD spectra of a PE hexamer are similar to those shown for the trimer in Fig. 1, except that the minimum in the LD spectrum occurs at 545 nm, while the positive LD maximum is observed at 571 nm; in addition, a weak absorption and a positive LD maximum are observed near 630 nm which is due to some PC contamination when the phycobiliproteins are separated.

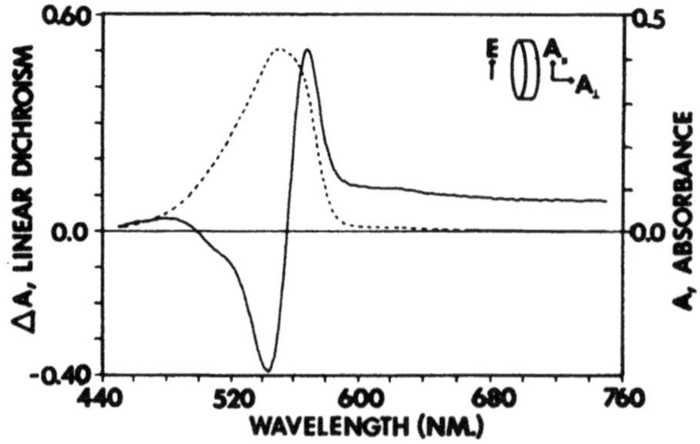

Figure 1: PE trimer in 10% gelatin (w/v), 0.1M potassium phosphate buffer, pH 5.0. (-------)... absorption spectrum, OD=0.46 at 550 nm, pathlength=1 cm. (———).. LD spectrum, pathlength = 0.5 cm; E...direction of gel elongation.

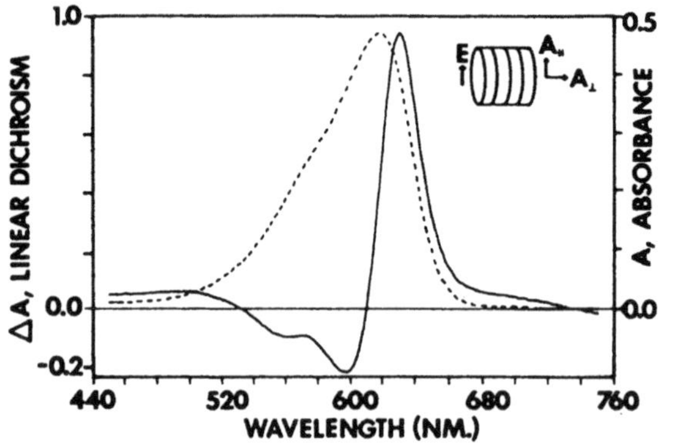

Figure 2: PC complex in 10% gelatin (w/v), 0.1M potassium phosphate buffer, pH 5.0. (-------)... absorption spectrum, OD=0.44 at 620 nm, pathlength=1 cm. (———).. LD spectrum, pathlength = 0.5 cm; E...direction of gel elongation.

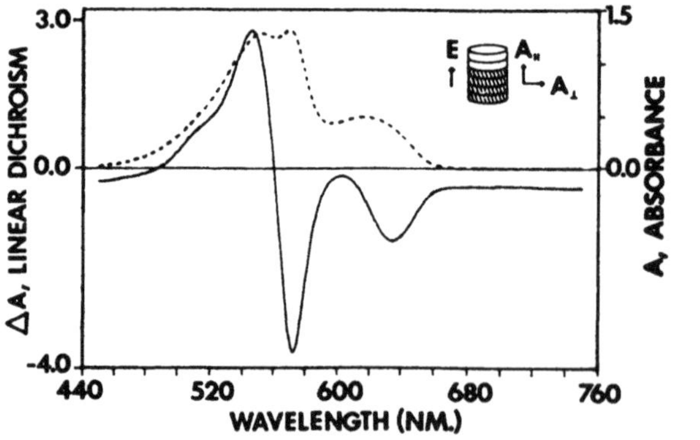

Figure 3: PE/PC complex in 10% gelatin (w/v), 0.5M potassium phosphate buffer, pH 7.0 (-------). absorption spectrum, OD=1.38 at 570 nm, pathlength=1 cm. (———).. LD spectrum, pathlength =0.5 cm; E...direction of gel elongation.

C-PC in cyanobacteria contains one phycocyanobilin chromophore on its α subunit and two on its β subunit (13). In 0.1M phosphate buffer, pH 5.0, PC trimers aggregate to dodecamers, i.e., four stacked disks (9), and are characterized by an absorption maximum at 620 nm (13). The absorption and LD spectra of these PC complexes in 10% gelatin gels are shown in Fig. 3; the absorption maximum is at 617 nm, rather than at 620 nm, and a small shoulder is also observed at 572 nm. The LD signal in the squeezed gel exhibits a pronounced minimum at 598 nm, a weaker minimum at 560 nm, and a strong positive peak at 631 nm. The LD maximum at 631 nm nearly coincides with the absorption maximum at 632 nm obtained by Mimuro et al. (14) when the PC monomer spectrum is subtracted from the PC trimer spectrum in M. laminosus; this maximum is attributed to the β_f chromophores in the trimer. A 598 nm maximum in the PC trimer-monomer absorption spectrum is attributed to β_s (14), or β_{155} (4) chromophores, and the corresponding LD signal is negative in sign. X-ray analysis has shown that the β_{155} chromophore in M. laminosus is almost parallel to the axis of the PC trimer (1). If the same orientation can be assumed for the PC aggregates extracted from Nostoc sp. as for the PE trimers (See inset, fig. 2), then the negative LD signal at 598 nm is consistent with an orientation of the PC dodecamer disk normals perpendicular to the elongation direction of the gel.

The phycobiliprotein aggregate called the PE/PC complex is a rod consisting of two hexamers of PE attached to one hexamer of PC (9). The absorption and LD spectra for this complex in a 10% gelatin, 0.5 M potassium phosphate buffer, pH 7.0 gel are shown in Fig. 3. The absorption maxima seen in the gel at 550, 570 and 615 nm are also observed when these same PE/PC preparations are suspended in aqueous buffer solution (0.75M phosphate buffer, pH 7.0) (9), and the fluorescence emission spectrum for the gel shows no change in the energy transfer characteristics. A positive LD maximum occurs at 546 nm, with a shoulder at 512 nm; negative LD peaks are observed at 573 and 635 nm. Comparing the LD spectrum of the PE/PC complex to those of the PE trimer and PC dodecamer, it is apparent that the major PE and PC LD signals near 546, 573, and 630 nm are inverted in sign in the PE/PC complex. This inversion suggests that the rod-like PE/PC complex orients with its axis parallel to the elongation direction of the gel, so that the planes of the individual disks are now rotated by 90° with respect to their orientations in the PE and PC aggregates (See inset, fig. 3). The LD spectrum of the PE/PC complex also shows that the long wavelength transition moments of PE and PC tend to be parallel to one another. Finally, the absorption maxima of the phycobiliprotein components of PE/PC are red shifted as compared to the same transitions in the PE trimer and the PC complex.

Several APC core-containing fractions can be isolated from dissociated phycobilisomes; each of these fractions has a different polypeptide composition and some of their spectroscopic properties are also different (15). In general, the APC disks are trimers in which each α and β subunit bears one phycocyanobilin chromophore (16). We have studied the absorption and LD spectra of these APC core-containing fractions in 10% gels. The absorption spectrum is very similar to the LD spectrum for each of these three preparations, and a typical example, for a fraction which is a mixture of the so-called trimeric APC II and trimeric APC III, is shown in Fig. 4. The absorption and positive LD maxima coincide at 651 - 652 nm, and a shoulder with a positive LD signal is also observed at about 620 nm. Since the LD signal is uniformly positive in the entire spectral range (560-690 nm), the transition dipole moments of all six phycocyanobilin chromophores within the APC trimers appear to lie entirely in a plane parallel to the disk face.

The absorption and LD spectra for intact phycobilisomes in a 10% gelatin, 0.5 M potassium phosphate buffer, pH 7.0 gel are

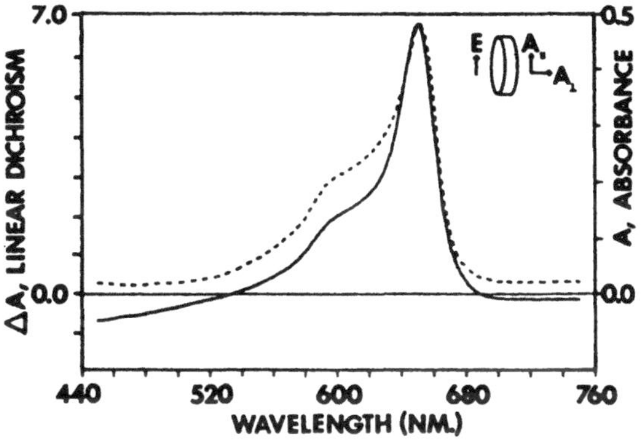

Figure 4: APC II and APC III in 10% gelatin (w/v), 0.5M potassium phosphate buffer, pH 7.0. (--------)...absorption spectrum, OD=0.50 at 650 nm, pathlength=1.0 cm. (————)...LD spectrum, pathlength=0.5 cm; E... direction of gel elongation.

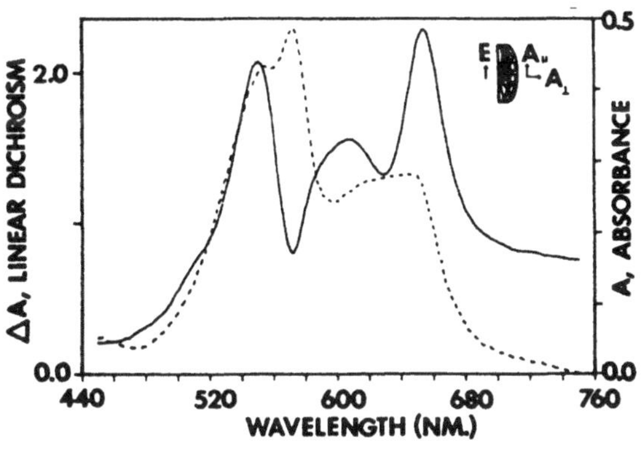

Figure 5: Phycobilisomes in 10% gelatin (w/v), 0.5M potassium phosphate buffer, pH 7.0. (-------)...absorption spectrum, OD=0.42 at 570 nm, pathlength= 1.0 cm. (————)...LD spectrum, pathlength= 0.5 cm; E...direction of gel elongation.

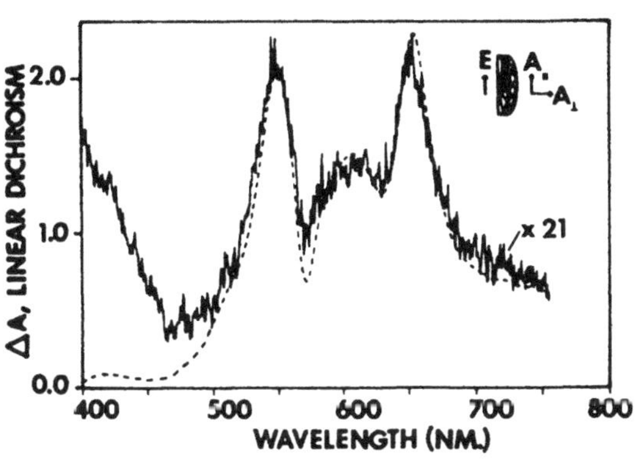

Figure 6: LD spectra of phycobilisomes: (-------)...in 10% gelatin (w/v), 0.5M potassium phosphate buffer, pH 7.0, same conditions as in fig. 5. (————).. in 0.75M potassium phosphate buffer, pH 7.0 OD=0.44 at 570 nm, pathlength=0.1 cm; normalized to gel LD at 550 nm. E...direction of gel elongation; direction of rotational flow for suspension.

shown in Fig. 5. The absorption spectrum in the gel is similar to the absorption spectrum for whole phycobilisomes suspended in 0.75 M phosphate buffer, pH 7.0. The LD spectra for phycobilisomes in squeezed 10% gel and in 0.75M phosphate buffer oriented hydrodynamically in the Couette cell are compared in Fig. 6. Both LD spectra exhibit maxima at 550 nm and 654 nm, and minima at 572 nm and 630 nm. The excellent coincidence of the two spectra suggest that there is no measurable dissociation of these phycobilisomes in gelatin gels. Furthermore, the near-coincidence of the two LD spectra, suggests that the hemidiscoidal phycobilisomes must align themselves in a similar manner in the squeezed gel and in the hydrodynamic flow gradient of the Couette cell (See inset, fig. 6). Comparing figs. 3 and 5, it is evident that in the phycobilisomes the signs of the LD contributions of PE and PC chromophores are the same in the PE/PC rods; thus, the PE/PC rods tend to be oriented in a similar manner in both cases, i.e. with their long axes parallel to the elongation direction. The LD of the APC disks is the same in Figs. 4 and 5; consequently, in the intact phycobilisomes the APC disks of the core cylinders tend to be oriented with their faces parallel to the elongation direction. It can be thus concluded that the hemidiscoidal cross-section of the phycobilisomes aligns itself parallel to the elongation direction in the gel, while in the hydrodynamic flow gradient it aligns itself parallel to the flow lines. In this mode of orientation, the APC core chromophores are aligned as they were in the individual oriented APC disks, whereas the six radiating PE/PC rods all lie in a plane parallel to the elongation direction. In essence, only some of the rods or parts of them, contribute to the phycobilisome LD signal. Since the ratio of PE:PC:APC trimeric disks in this phycobilisome is 2:1:1 and the rods are not optimally oriented, it is not surprising that the PC LD signal at 630 nm and the PE LD signals at 550 nm and 572 nm seem somewhat weak as compared to the APC LD signal at 654 nm.

Discussion

In their investigation of the kinetics of energy flow within the tricylindrical phycobilisome core of the cyanobacterial mutant Synechocystis 6701, strain UV 16, Glazer and coworkers (17) found that the organizational difference between rod and core substructures leads to more rapid inter-disk energy transfer within the core. In the core, the trimeric disks in adjacent cylinders are in edge-to-edge contact, whereas in the rods, trimeric disks are in face-to-face contact. Our spectral results allow us to further clarify the relationship between the chromophore and apoprotein conformation within the APC disk. Given that the APC chromophores all lie in the planes of the disks, it is feasible that the chromophores or some portion of them roughly form a concentric ring. The 620 nm shoulder seen in the absorption spectra of APC might be assigned to one set of subunit chromophores and designated as "sensitizing", while the 651-652 nm absorption maximum might be assigned to another set of subunit chromophores designated as "fluorescing" (18). Such an arrangement would facilitate energy transfer from core cylinder to core cylinder since transition moments of the chromophores would lie in parallel planes and possibly have some parallel vector component. The 635 nm absorption band of PC tends to lie parallel to the disk face (See fig. 3); however, it is not clear how this particular orientation, and the orientation of the APC transition moments, maximize rod to core energy transfer.

Acknowledgements

This work was supported in part by the National Science Foundation, Grant PCM 083190 and in part by USDA Competitive Grant 87-CRCR-1-2318.

References

1. Schirmer, T., W. Bode, R. Huber, W. Sidler, H. Zuber. 1985. J. Mol. Biol. 184, 257-277.

2. Schirmer, T., R. Huber, M. Schneider, W. Bode, M. Miller, M.L. Hackert. 1986. J. Mol. Biol. 188, 651-676.

3. Mimuro, M., P. Fuglistaller, R. Rumbeli, H. Zuber. 1986. Biochim. Biophys. Acta 848, 155-166.

4. Sauer, K., H. Scheer, P. Sauer. 1987. Photochem. Photobiol. 46, 427-440.

5. Gagliano, A.G., J. Hoarau, J. Breton, N.E. Geacintov. 1985. Biochim. Biophys. Acta, 808, 455-463.

6. Bruce, D., J. Biggins. 1985. Biochim. Biophys. Acta,810, 295-301.

7. Frackowiak, D., K. Fikinski, H. Pienkowska. 1977. Stud. Biophys. 63, 183-287.

8. Frackowiak, D., L.G. Erokhina, G. Picard, R.M. Leblanc. 1987. Photochem. Photobiol. 46, 277-285.

9. Zilinskas, B., D.A. Howell. 1983. Plant Physiol. 71, 379-387.

10. Troxler, R.F., L.S. Greenwald, B.A. Zilinskas. 1980. J. Biol. Chem. 255, 9380-9387.

11. Geacintov, N.E., V. Ibanez, M. Rougee, R. V. Bensasson. 1987. Biochemistry 26, 3087-3092.

12. Zilinskas, B.A., J. Grabowski, S. Campbell. 1984. In: Advances in Photosynthetic Research (C. Sybesma, ed.). Martinus Nijhoff/Dr. W. Junk, The Hague, Vol. II, 687-690.

13. Zilinskas, B.A., L.S. Greenwald. 1986. Photosynth. Res. 10, 7-35.

14. Mimuro, M., R. Rumbeli, P. Fuglistaller, H. Zuber. 1986. Biochim. Biophys. Acta 851, 447-456.

15. Zilinskas, B.A. 1982. Plant Physiol. 70, 1060-1065.

16. Zilinskas, B.A., L.S. Greenwald, C.L. Bailey, P.C. Kahn. 1980. Biochim. Biophys. Acta 592, 267-276.

17. Glazer, A.N., C. Chan, R.C. Williams, S.W. Yeh, J.H. Clark. 1985. Science 230, 1051-1053.

18. Glazer, A.N., S. Fang, D.M. Brown, 1973, J. Biol. Chem. 248, 5679-5685.

LOW TEMPERATURE SPECTROSCOPY OF CYANOBACTERIAL ANTENNA PIGMENTS

W. Köhler, J. Friedrich

Physikalisches Institut und Bayreuther Institut für Makromolekülforschung (BIMF), Universität Bayreuth, D-8580 Bayreuth, F.R.G.

R. Fischer, H. Scheer

Botanisches Institut, Universität München
D-8000 München 19, F.R.G.

Abstract

Disorder on a microscopic scale leads to inhomogeneous line broadening of the optical spectra of chromophores, which prevents high resolution spectroscopy in a straightforward manner. In this paper emphasis is put on hole burning experiments on C-phycocyanin (PC) and phycobilisomes (PBS) of *Masticogladus laminosus*. This technique is capable of resolving the zero phonon fine structure in spite of disorder. From the measured hole profiles energy transfer times within a broad frequency range of the phycobilisome absorption could be estimated. From fluorescence line narrowing experiments combined with hole burning, details about the level structure and the loss of correlation in energy transfer processes could be elucidated. From temperature dependent hole filling experiments the distribution of conformational barriers of the chromophore attached proteins could be measured.

Introduction

In the case of antenna pigments the knowledge of the details of the electronic-vibrational level structure is a prerequisite for understanding the

Abbreviations: PE phycoerythrin, PC phycocyanin, APC allophycocyanin, PBS phycobilisome

Photosynthetic Light-Harvesting Systems

energy transfer mechanisms and, hence, the functioning of these pigments in the photosynthetic process (1,2). Spectroscopy at liquid Helium temperatures can usually provide this information in organic molecular systems. This results from two facts: First, the linewidths may narrow dramatically as the temperature is lowered, and second, most of the intensity may be confined to the so-called 'zero phonon transitions'. Hence, low temperatures principally allow for highly resolved spectra with very good signal to noise ratio (3).

Unfortunately, disorder on a microscopic scale prevents straightforward high resolution spectroscopy in the pigments studied, e.g. PC and PBS from M. laminosus, because of the concommitant inhomogeneous line broadening. Inhomogeneous line broadening not only obscures all the information contained in the homogeneous line shape function but also wipes out the vibrational pattern to a high degree. Here we wish to report how some of these drawbacks can be overcome by spectral hole burning and fluorescence line narrowing techniques.

How disorder on a microscopic scale obscures spectroscopic information

To understand the kind of information one can obtain from a homogeneous line shape function of guest molecules in a host matrix we assume that the host-guest system is perfectly ordered and the concentration is low enough so that each guest molecule has the same microscopic environment. In this case the absorption line shape has the typical form shown in Fig. 1. It consists of a narrow so-called zero phonon line and a broad so-called phonon side band. The relative intensity in the narrow zero phonon line determines the Debye-Waller-factor, a characteristic parameter of the considered host-guest system, which is, for many systems on the order of 0.4-0.8 for temperatures around 4K. Above 40 K, or so, it becomes extremely small so that the zero phonon structure tends to vanish. The phonon side band is an outcome of the Franck-Condon-principle applied to the vibrations of the host material: The excited guest molecule has a different charge distribution. Hence, the matrix is polarized and tends to assume a new equilibrium configuration while the guest molecule is in the

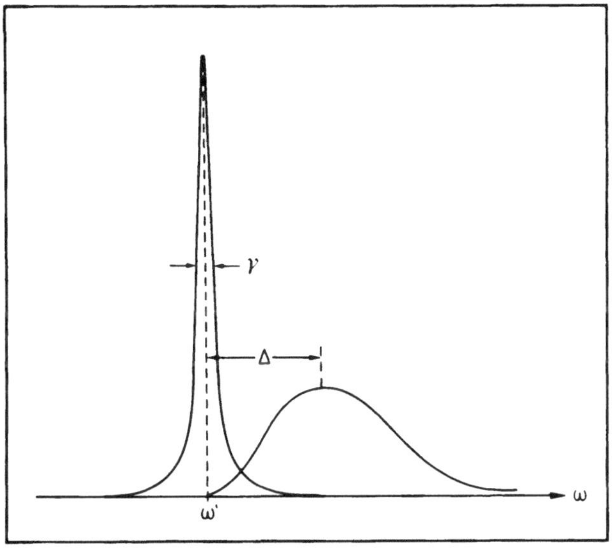

Fig. 1. Zero phonon line with homogeneous width γ and phonon sideband. Δ is half the Stokes-shift

excited state. This is equivalent to say that the matrix vibrates as a consequence of the guest excitation. Matrix vibrations are usually called phonons. If the matrix or, strictly speaking, the immediate environment of the excited chromophore is very rigid, the coupling of the electronic excitation to the matrix environment (to the lattice) is small and, hence, the phonon side band carries little oscillator strength and the Debye-Waller-factor will be close to 1. The spectrum is then dominated by the narrow and intense zero phonon lines.

The zero phonon line is a purely electronic transition, i.e. the vibrational state of the lattice does not change during this kind of excitation. Since the lifetime of an electronic level is orders of magnitude longer than that of a vibrational level, its width γ is, according to the uncertainty relation, much narrower, and, hence, its peak intensity may be orders of magnitude higher than the phonon side band. Hence, we have a typical absorption line shape as depicted in Fig. 1. Along these lines of reasoning it seems to be clear that one can determine the lifetime of an electronic level in case one succeeds in measuring its homogeneous zero phonon line shape. This, however, is not an easy task because of several reasons: First, only at extremely low temperatures on the order of 1 K is the width determined by the true lifetime of the electronic level excited. With increasing temperatures the dynamics of the host matrix broaden the line without changing its lifetime. Second, many dye molecules have lifetimes on the order of 1 nsec, hence the width will be on the order of 10 to 100 MHz (0.0003 - 0.003 cm^{-1}) and the required resolution of the spectrometer has to be on the order of 10^7. Third, and most important, the presence of microscopic disorder on a molecular level obscures the homogeneous line shape function. Hence, a straightforward spectroscopy is not possible (for a review, see (1)). How disorder changes the spectral properties of a chromophore in a host matrix is discussed in the following paragraph.

Fig. 2 symbolizes a disordered matrix. The solvent cage of molecule 1 is different from that of molecule 2 and 3. Hence, as schematically shown

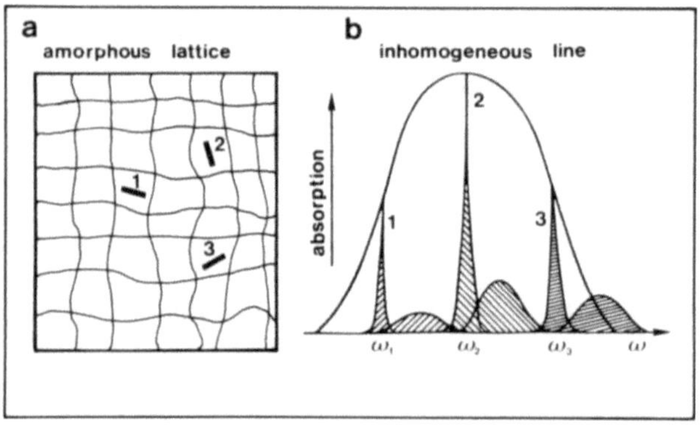

Fig. 2. Schematic representation of inhomogeneous line broadening (b) as a result of microscopic disorder (a)

in Fig. 2b, these molecules absorb at different frequencies. As a consequence, spatial disorder leads to a spread of absorption frequencies, which is called inhomogeneous line broadening. At low temperatures, the inhomogeneous width may be larger than the homogeneous width by more than 4 orders of magnitudes. Hence, all details of the homogeneous molecular line shape are wiped out due to disorder.

Fluorescence line narrowing (FLN) and hole burning (HB) spectroscopy

Both FLN and HB are high resolution techniques designed to overcome inhomogeneous line broadening. The basic idea of FLN is sketched in Fig. 3a. A narrow bandwidth laser with frequency ω_L excites only those centers in the inhomogeneous band which are accidentally tuned with their absorption freuqency to ω_L. Then, only molecules absorbing in a frequency range on the order of the homogeneous width around ω_L can emit fluorescence. If the fluorescence is detected with a high resolution spectrometer, the lines will show a structure as shown in Fig. 3a. A drawback of FLN is the fact that it is very difficult to observe resonance fluorescence and, hence, the lines are broadened by vibrational relaxation and other processes.

Hole burning is shown in Fig. 3b. It always works in case the guest molecules are photoreactive, and the zero phonon line carries enough oscillator strength. Then, those molecules excited within a frequency range

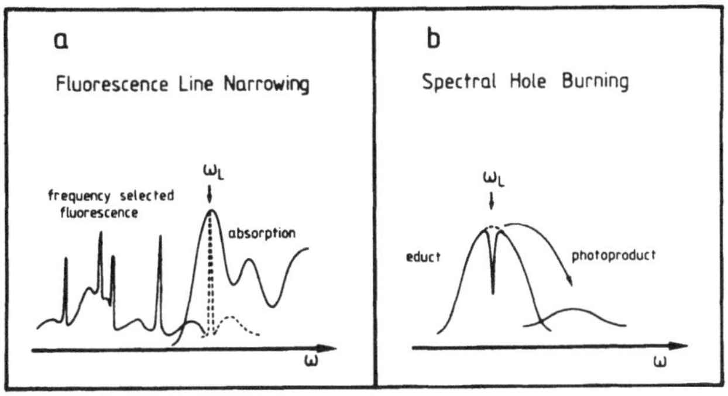

Fig. 3. Schematic representation of fluorescence line narrowing (a) and spectral hole burning (b).

of the homogeneous linewidth γ around ω_L are transformed to a product, hence, the number of molecular absorbers at ω_L is diminished and a hole appears in the spectrum. Contrary to FLN, HB allows for a resonant detection at the laser frequency; therefore, this technique has the capability of measuring the homogeneous linewidth. One has, however, to be careful to rule out all slow relaxation processes such as spectral diffusion, because the time scale of HB-experiments is slow and a lot of processes may occur which broaden the hole.

Hole burning experiments on phycobilisomes and spectrally resolved energy transfer times

Fig. 4 shows an absorption spectrum of phycobilisomes of M. laminosus at a temperature of 4 K in a saccharose/phosphate buffer solution. The

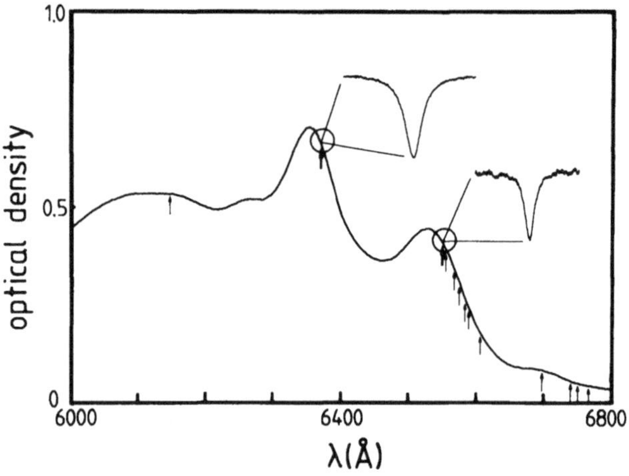

Fig. 4. Hole burning in PBS of M. laminosus. Burning positions are indicated by arrows. Typical hole shapes obtained in the PC and APC-peaks are shown on an enlarged scale. PBS preparation, modified from Nies and Wehrmeyer (4), was made in 0.9 M phosphate buffer (pH 6). Centrifugation was repeated twice. Isolated PBS solution was saturated with saccharose, to ensure coupling in a low temperature glass.

spectrum is characterized by a clearly resolved structure which originates from the various pigments which build the highly organized phycobilisome assembly APC, PC, PE. In spite of the highly organized structure, there is a remarkable amount of disorder as documented by the inhomogeneous linewidths. It is a question of great interest whether the observed disorder is an intrinsic property of the chromoprotein or of the proteinaceous environment. In case of the PBS, most of the chromophores are known to interact strongly with the protein. Consequently, we believe that disorder is intrinsic and reflects the different conformational states of the latter (5). The arrows in Fig. 4 indicate the positions where hole burning was performed. Typical holes are shown in the same figure as insets. Their width is on the order of 0.4 cm^{-1}. It is interesting to compare this width with that measured for isolated phycocyanine (6). This is done in Fig. 5 for phycobilisome and PC in saccharose/buffer solution. The hole width in

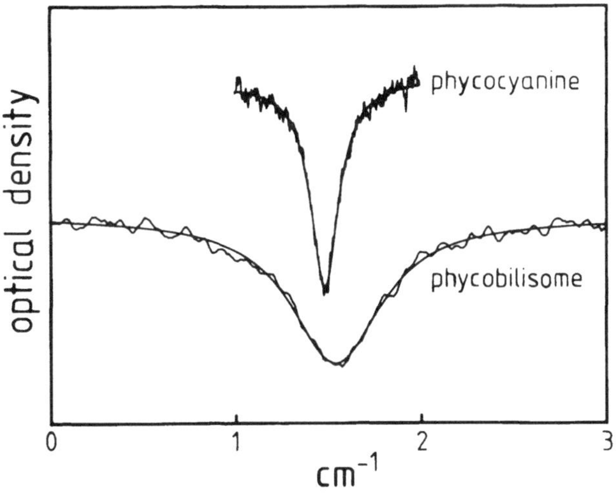

Fig. 5. Comparison of spectral holes in isolated PC and phycobilisomes (PC-peak).

the phycobilisome system is much larger than in the isolated PC. We interprete this finding in the following way: Hole burning on the PC-trimer occurs effectively only on the red edge of the visible band. Hence, it seems that it is only the fluorescing chromophore(s) (or chromophore states) which are effectively burnt. In these chromophores, the fluorescence lifetime is mainly governed by intramolecular decay processes, which occur on the order of nanoseconds. At sufficiently low temperatures, where dephasing is small, these decay processes determine the width of the hole. In the phycobilisomes there is, apart from these decay processes, a fast energy transfer to acceptor chromoproteins, like APC. In case these energy transfer processes dominate the dynamics of the excited level, the transfer times can be determined from the measured hole width. From our results we would estimate transfer times on the order of 30 ps. It is interesting to note that these results do not depend very much on the position of excitation (see Fig. 4).

We conclude this section with a few remarks on the advantages and disadvantages of the hole burning method in determining lifetimes. Unlike fluorescence detection methods, which suffer from overlapping contributions from the various pigments, hole burning is a resonant method which directly yields information on the burnt state. No kinetic model is necessary in evaluating this information. However, the width of the hole is not only determined by the lifetime of the state considered, but depends also on the pure dephasing time. It is difficult to determine the latter one separately and, hence, the hole burning results have to be considered as lower bounds for the transfer times.

Spectral distribution of the photoproduct

The nature of the hole burning reaction is, as of yet, not clear. Basically one discerns between photochemical - and photophysical reactions. In the first case it is assumed that photochemical changes of the dye molecule itself (i.e. of the chromophore) occur (e.g. bond breaking, proton transfer reactions, etc.). In the second case it is assumed that the dye molecule itself is unchanged whereas the surrounding changes (e.g. a change in the conformational state of the protein). Photochemical reactions usually lead to large spectral shifts, while photophysical reactions lead to small spectral shifts (7). In many cases (including chromoproteins) it may be difficult to distinguish both types of reactions in a clear cut way. Fig. 6 shows (together with the absorption spectrum) the difference in optical

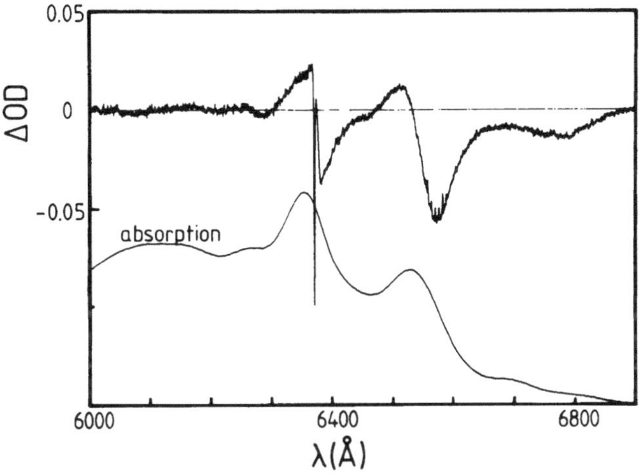

Fig. 6. Difference in optical density (ΔOD) before and after burning the phycobilisome spectrum around 6370 A.

density before and after burning. Note, that most of the product occurs in a rather narrow range around the educt which favours a photophysical hole burning reaction. An interesting observation is the phototransformation which occurs in the APC-range around 6560 A. Note, that laser irradiation is performed at 6370 A in the PC-range. As obvious, the spectral range burnt in the APC-spectrum is almost as broad as the inhomogeneous band. Hence, frequency selection is lost in this case. Interestingly, only the long wavelength edge of the APC- band is burnt. It seems, that only this long wavelength range of the band is photoreactive.

Conformational barriers

In case the reaction product is determined by the interaction between probe molecule (i.e. chromophore) and environment, we expect that, due to the disorder of the system, the reaction barriers are not well defined, but are rather characterized by a broad distribution. Hole burning offers a possibility for directly measuring this distribution of barriers (5,7). To this end we make use of temperature cycling experiments: A hole is burnt at a low temperature T_b ('burn'). Then, the temperature of the sample is raised to some value T (the 'excursion temperature') and cycled back to T_b, where the hole is measured again.

In our case burning was performed at the APC-peak, near 6550 A (see Fig. 4). T_b was 4 K. The changes of the hole area during such a cycle is a measure for the number of product molecules having returned to their educt state. The experiment is performed as a function of the excursion temperature T. For a fixed excursion temperature T, all those centers return to the educt state characterized by barriers

$$V_T = kT \ln R_o \tau \qquad (1)$$

R_o is the attempt frequency and τ the experimental time needed to drive the system through one cycle. $\ln R_o\tau$ is on the order of 30. Centers with barriers higher than V_T stay unaffected. The variation of the excursion temperature enables a sampling of the distribution of barrier heights. Fig. 7a shows the result for PBS. For comparison, the results obtained for a series of organic glasses are also shown.

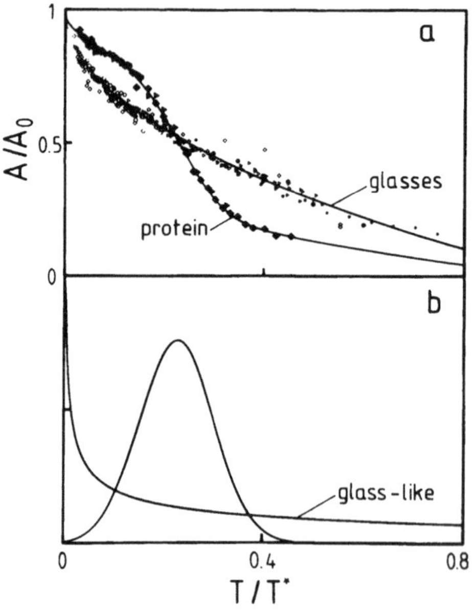

Fig. 7. Annealing of a spectral hole burnt into the PBS absorption (a). For comparison the same behavior for organic glasses is also shown. The distribution of reaction barriers can be well described by a superposition of glass-like states and a discrete feature distributed in a Gaussian fashion (b).

The experiments on the protein sample can be perfectly fitted by assuming that the distribution of barrier heights consists of a superposition of a glass-like distribution and a Gaussian distribution which is centered around 650 cm^{-1} and which has a width of 210 cm^{-1}. We interprete these findings in the following way: We assume that hole burning is due to a conformational change of the protein environment around the irradiated chromophore (photophysical HB- mechanism). On one hand, a protein is very much like a glass, i.e. it is disordered on a microscopic scale and can exist in many conformational states with different barriers. We have recently shown that in a glass the probability of finding a barrier with height V is proportional to $1/\sqrt{V}$ (5,7). On the other hand a protein has also very specific conformational states, i.e. 'functionally important states', which are related to well defined motions, say of a large part of the peptid chain (8). It is plausible that the barrier height of such a specific conformational state depends on the actual substate of the glass-like states which the protein occupies, and, hence the barrier is spread around some mean value, which in our case is 650 cm^{-1}. The decomposition of the experimental results in a glass-like distribution and a symmetrically broadened discrete feature is shown in Fig. 7b.

Fluorescence line narrowing experiments and level structure in C-phycocyanin

The hole burning experiments document well the zero phonon nature of the optical transitions in chromoproteins (9-11) in contrast to the results obtained for reaction centers of Rhodopseudomonas viridis and Rhodopseudomonas sphaeroides (12-14). However, the fluorescence emission is broad irrespective of the mode of excitation and detection. A simple explanation of this phenomenon is the assumption of a loss of energy correlation due to transfer processes, that is, a specific level with a sharp frequency in a sensitizing state populates the whole inhomogeneous band of the fluorescing state. The idea was to narrow the fluorescence by resonant excitation of the fluorescing state. Fig. 8 shows that this idea did not work out. There is no sharply structured fluorescence even in case excitation is carried out far in the red edge of the PC-absorption. This phenomenon can be understood on the basis of the special level structure involved (15) and the loss of energy correlation. As to the special level structure we have to assume that the fluorescing state consists in reality of (at least) two states with strongly overlapping inhomogeneous bands. The loss of energy corre-

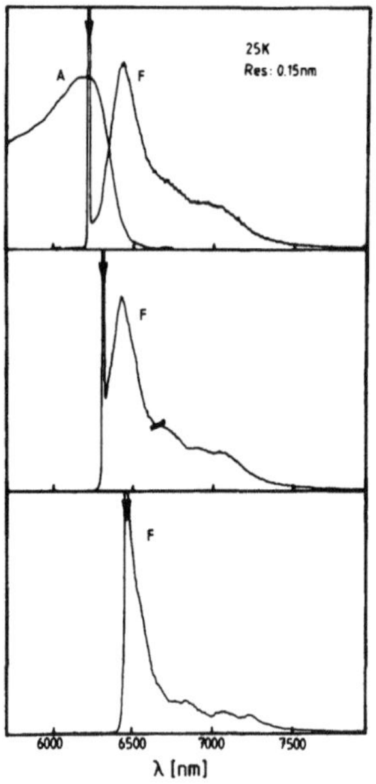

Fig. 8. Selectively excited fluorescence of PC-trimer as a function of excitation wavelength.

lation means that, though the inhomogeneous bands are strongly overlapping, the energy spacing between the two states in an individual molecule can have any value between zero and (roughly) the inhomogeneous width. Since these states are coupled via energy transfer processes the maximum of the fluorescence does not shift with excitation frequency and, as the frequency is scanned into the red, the fluorescence shows a characteristic cut off at the laser frequency. Another consequence of this special level structure is a wavelength dependent hole burning efficiency: As long as the excitation frequency is far enough to the blue of the states involved, there is a high probability that the second level is lower in energy. Hence, the probability for energy transfer is high and, correspondingly, the probability for hole burning or resonance fluorescence is low. As the laser frequency is turned to the red, the probability of finding an acceptor state with lower energy decreases, hence, the efficiency for hole burning and resonance emission increases. Both phenomena can indeed be observed.

Acknowledgement

The authors acknowledge grants from the Deutsche Forschungsgemeinschaft, SFB 213 (Bayreuth) and SFB 143-A1 (Munich).

References

1. Scheer, H. 1981. Ang.Chemie 93, 230; Int.Ed.Engl. 20, 241
2. Cogdell, R.J. and H. Scheer. 1985. Photochem.Photobiol. 42, 669
3. Friedrich, J. and D. Haarer. 1984. Ang.Chemie 96, 96; Int. Ed. Engl. 23, 113
4. Nies, M. and W. Wehrmeyer. 1980. Planta 150, 330
5. Köhler, W., J. Friedrich and H. Scheer. 1987. Phys.Rev.A, submitted
6. Füglistaller, P., H. Widmer, G. Frank and H. Zuber. 1981. Arch. Microbiol. 129, 268
7. Köhler, W. and J. Friedrich. 1987. Ber. Bunsenges. Phys. Chemie 91, 858
8. Fraunfelder, H. 1984. Helvet.Phys.Acta 57, 165
9. Friedrich, J., H. Scheer, B Zickendraht-Wendelstadt and D. Haarer. 1981. J.Am.Chem.Soc. 103, 1030
10. Friedrich, J., H. Scheer, B. Zickendraht-Wendelstadt and D. Haarer. 1980. J.Chem.Phys. 74, 2260
11. Köhler, W., J. Friedrich and H. Scheer. 1987. Chem.Phys. Lett., submitted
12. Meech, S.R., A.J. Hoff and D.A. Wiersma. 1985. Chem.Phys. Lett. 21, 287

13. Boxer, S.G., D.J. Lockart and T.R. Middeldorf. 1986. Chem.Phys. Lett. 123, 476

14. Gillie, J.K., B.L. Fearey, J.M. Hayes, G.J. Small and J.H. Golbeck. 1986. Chem.Phys.Lett. 134, 316

15. Zickendraht-Wendelstadt, B., J. Friedrich and W. Rüdiger. 1979. Photochem.Photobiol. 31, 367

CHROMOPHORE CONFORMATIONS IN PHYCOCYANIN AND ALLOPHYCOCYANIN AS STUDIED BY RESONANCE RAMAN SPECTROSCOPY

B. Szalontai, V. Csizmadia, Z. Gombos, K. Csatorday

Biological Research Center, Hungarian Academy of Sciences
H-6701 Szeged, P.O.B. 521, Hungary

M. Lutz

Service de Biophysique, Departement de Biologie C.E.N. Saclay
91191 Gif-sur-Yvette, Cedex, France

Introduction

Phycocyanin and allophycocyanin are light-harvesting biliproteins of blue-green and red algae (1). They are organized into supramolecular complexes called phycobilisomes which assure the efficient energy transfer toward the photosynthetic reaction centers.

It is interesting that there is only one type of chromophore, the covalently bound phycocyanobilin (Fig. 1) in both proteins (2,3) The absorption range of the two proteins in the visible region covers the 580-650 nm part of the spectrum. This wide absorption range is the result of chromophore-chromophore and chromophore-protein interactions which alter the chromophore absorption properties.

These proteins are in aggregated states in the phycobilisome, phycocyanin forms trimers/hexamers, allophycocyanin trimers. There are evidences in both proteins for chromophore-chromophore interactions (for phycocyanin see (4), for allophycocyanin (5,6)). In the case of phycocyanin these interactions are not obvious from the absorption spectra, while in the allophycocyanin absorption spectrum there is a strong band around 650 nm which was attributed to strongly exciton coupled interactions between two of the six chromophores of the trimer molecule (6).

Less is known about the chromophore-protein interactions. Recently the three dimensional structure of phycocyanin from X-ray crystallographic

Photosynthetic Light-Harvesting Systems

studies became available (7,8) which may help in interpreting in vivo studies of this problem as well.

We have studied the chromophore-protein, chromophore-chromophore interactions in phycocyanin and in allophycocyanin and the conformational states of the chromophores in these chromoproteins by resonance Raman spectroscopy (9,10).

The UV and the VIS-excited resonance Raman spectra gave somewhat complementary informations. In the UV-excited spectra there are only minor changes which can be attributed to alterations in the chromophore-protein interactions. We could find a resonance Raman band (at 1640-1650 cm^{-1} in the native protein) which is sensitive in its position to the degree of folding of the chromophores. The origin of this band is rather complex. It is a sum of different normal modes of the molecule which are differently active in the different chromophore conformations.

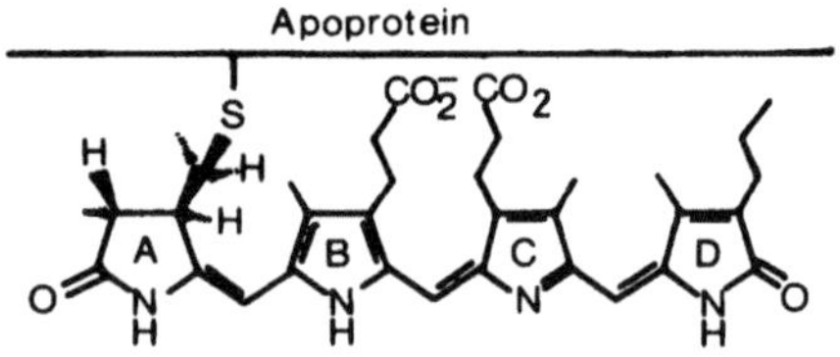

Fig. 1. Phycocyanobilin

On the other hand, the visible excited resonance Raman spectrum in this region shows a band (at 1595 cm^{-1}) which is absent from the UV-excited spectra and this band is more sensitive to alterations in the chromophore-protein interactions than the ones found in the UV-excited spectra.

The study of the nitrogen isotope sensitivity gave additional informations about the origin of certain bands in the resonance Raman spectrum.

Resonance Raman studies

UV-excited (363.8 nm) native phycocyanin (Fig. 2a) and allophycocyanin (Fig. 3a) resonance Raman spectra are similar. Nevertheless there are distinct differences between the two spectra which point to different conditions for the chromophores in the two proteins. The 1248 cm^{-1} and the

1400 cm^{-1} bands in allophycocyanin are stronger. The shoulder around 1605-1610 cm^{-1} is also more pronounced in allophycocyanin. These differences are important because the 1248 cm^{-1} band and the 1605 cm^{-1} shoulder are the most sensitive features for changes in the chromophore-protein interactions. Both completely disappear upon denaturation of the proteins and, in the case of allophycocyanin they change already during monomerisation (compare Fig. 2b to Fig. 3b). Regarding the higher contributions from these bands in the trimer allophycocyanin resonance Raman spectrum one may say that the chromophore conformation in allophycocyanin is in a sense 'beyond' of phycocyanin.

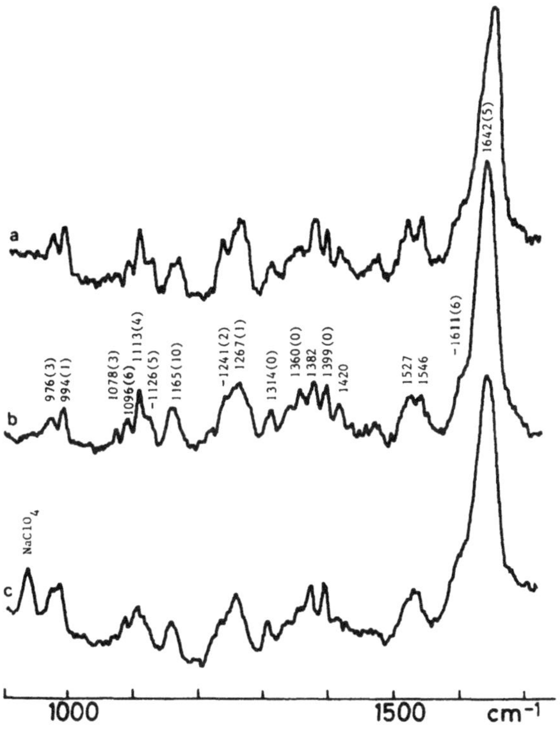

Fig. 2. UV-excited resonance Raman spectra of phycocyanin. a - trimer at pH 7.5, b - monomer at pH 3.5, c - monomerized by addition of 1M $NaclO_4$. Numbers in brackets indicate shifts upon N-isotope substitution.

We have shown (10) that in phycocyanin the chromophore conformation is not effected considerably by the monomerisation of the protein, when this monomerisation was achieveed by lowering the pH of the medium from 7.5 to 3.5 (Fig. 2a,b). It is known, however, that chaotropic salts can also monomerize biliproteins (11). Such a resonance Raman spectrum is shown in the presence of 1 M $NaClO_4$ for phycocyanin and allophycocyanin on Fig. 2c and 3c, respectively. The resonance Raman spectra of monomers produced either by decreasing the pH or by chaotropic salt are very similar in both proteins. This means that the two methods yield the same monomers.

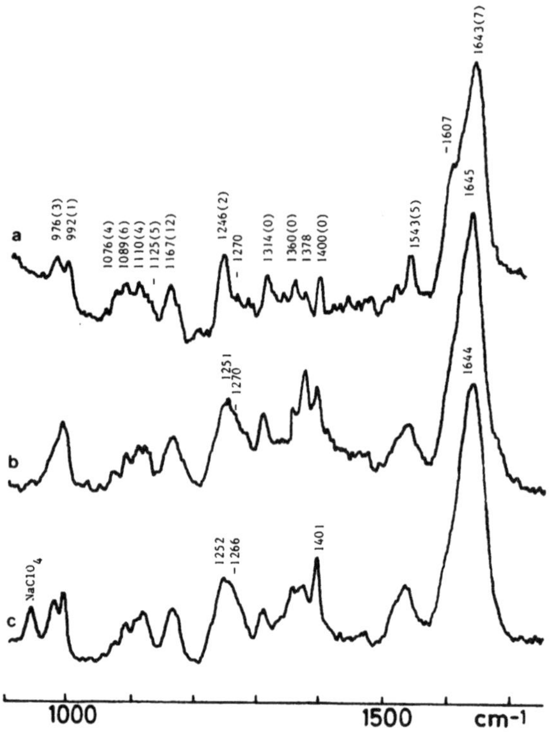

Fig. 3. Low temperature (40 K) UV-excited (363.8 nm) resonance Raman spectra of allophycocyanin. a - trimer allophycocyanin at pH 7.5, b - monomer allophyacocyanin at pH 3.5, c - allophycocyanin monomerized by addition of 1M $NaClO_4$.

(Certain, minor, differences nevertheless exist in the case of allophycocyanin in the 1350-1400 cm^{-1} region but we can not connect those to conformational differences so far.)

In a series of studies where the authors tried to reproduce in phycocyanin and allophycocyanin the light induced absorption changes resembling those found in phytochromes and phycochromes some deterioration of the native protein structure was necessary (12-15).

For a better understanding of chromophore conformational changes in these proteins we studied the process of denaturation by decreasing the pH. We have chosen this line of investigation because resonance Raman data were available only for similarly treated model compounds (16) which have cyclic conformation similar to the one postulated for denatured biliproteins (17).

Indeed, more interesting is the behaviour of the resonance Raman spectra during denaturing these proteins. The changes are the same in both proteins, but show different pH dependence.

In the 1050-1130 cm^{-1} region the complex four band structure is transforming into a three band structure. During this transformation the 1124 cm^{-1} band gains intensity as compared to the other two bands which can be found at 1076 and 1095 cm^{-1} in both denatured protein (Fig. 4).

The band around 1248 cm^{-1} disappears, a new band emerges around 1265-1270 cm^{-1}. This process is more obvious in allophycocyanin where the 1248 cm^{-1} band is stronger and the 1265 cm^{-1} is practically absent in the native protein (Fig. 3a), but in its denatured state allophycocyanin has the same resonance Raman spectrum as phycocyanin (Fig. 4).

The band around 1400 cm^{-1} shifts to 1405 cm^{-1}. This shift is continuous as the pH is decreased. An increase of its relative intensity can be observed at the same time.

The greatest changes are in the 1590-1650 cm^{-1} region of the spectra. The shoulder around 1600-1610 cm^{-1} disappears in both proteins. More interesting is the behaviour of the 1645 cm^{-1} band which shifts downwards upon decreasing the pH. The shape of this band indicates that it is the sum of several narrower bands. The downshift can be interpreted as the result of changes in the relative intensities of these components which are differently active in the different chromophore conformations. This interpretation is supported by our finding that the nitrogen isotope sensitivity of the band at positions around 1645-50 cm^{-1} is high, while in denatured proteins this band can be found around 1625 cm^{-1}, with negligible isotope sensitivity (see Figs. 2,3 and 4.).

Investigating the shift of this band during denaturing the proteins in details, it turned out that it can be used as a marker of the state of chromophore folding. Lower the position of the originally 1645 cm^{-1} band, more folded the chromophore is. The lower limit we found in biliproteins was around 1625 cm^{-1}. This is in good correlation with the value found in free phycocyanobilin (1627 cm^{-1}) and in biliverdin dimethyl ester (1620 cm^{-1}) (10).

Theoretical studies on the absorption properties of open-chain tetrapyrroles (4,18,19) showed that the ratio of the UV and VIS oscillator strengths is a good measure of the overall molecule conformation. Since we

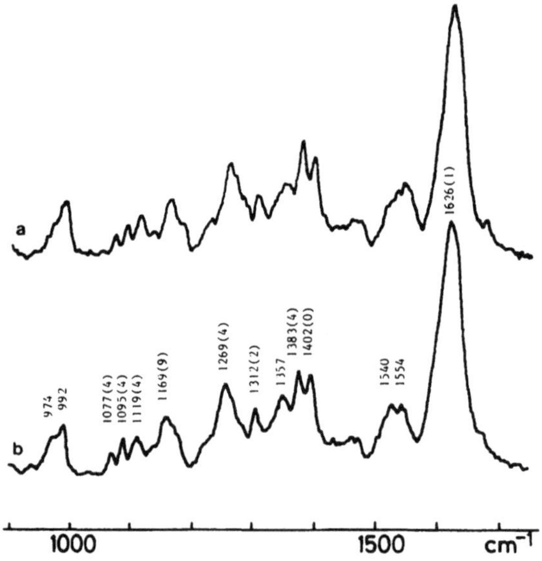

Fig. 4. Denatured phycocyanin (a) and allophycocyanin resonance Raman spectra. Excitation with 363.3 nm at 40 K. The pH was 2.2 and 2.5 for phycocyanin and allophycocyanin, respectively.

found that there is no change in the position of the 1645 cm^{-1} band during monomerisation of either proteins but at lower pH values a downshift starts and this shift appears 0.3-0.5 pH unit higher pH values in allophycocyanin than in phycocyanin we plotted the UV/VIS absorption ratio as a function of the pH for both proteins. It can be seen on Fig. 5 that indeed the UV/VIS absorption ratio starts to increase earlier in allophycocyanin than in phycocyanin in agreement with the resonance Raman data.

This plot on Fig. 5 offered an other possibility to check the similarity of the pH and chaotropic salt monomerized proteins. We plotted the UV/VIS ratio of the salt treated proteins as well at pH 6 which was the value of the solutions after adding the 1 M $NaClO_4$ to the pH 7.5 20 mM phosphate buffered samples. The ratios correspond perfectly to undisturbed chromophores.

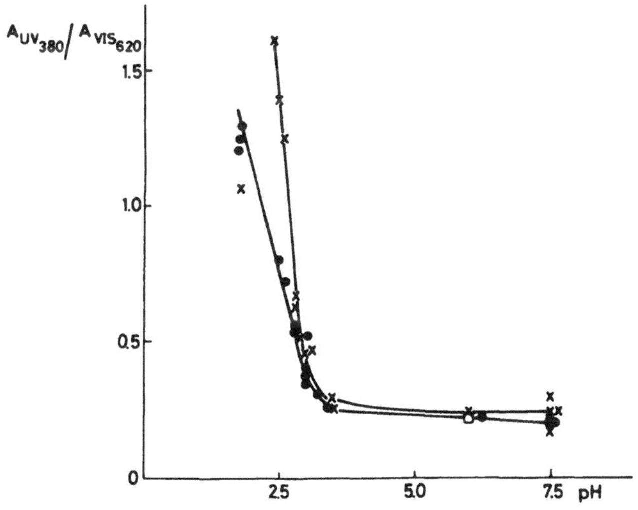

Fig. 5. The UV/VIS absorption ratio in the case of phycocyanin (●) and allophycocyanin (X) as the function of the pH. Open square indicates the value got for chaotropic salt treated allophycocyanin.

In our first report (9) we have presented VIS excited resonance Raman spectra of phycocyanin and allophycocyanin. They are slightly different from the UV-excited ones. The differences can be explained by the fact that distinct electronic transitions are excited in the two cases and certain normal modes can be selectively coupled to one of these transitions. This phenomenon is most obvious in the 1590-1650 cm^{-1} region, where the VIS-excited spectrum shows a strong band at 1595 cm^{-1} which is absent from the UV-excited spectra (compare Fig. 6 and 2.) The example is given for phycocyanin but the same difference exists for allophycocyanin as well.

Upon decreasing the pH a continuous change can be observed in the VIS-excited resonance Raman spectrum. This change is the disappearance of the 1595 cm^{-1} band from the spectrum. In contrast to the UV-excited spectra where no change was found between the native and the monomer spectra i.e. between pH 7.5 and 3.5, VIS-excited spectra showed a disappearance of the 1595 cm^{-1} band which went parallel with the decrease of the pH. This change should reflect more alterations in the chromophore-protein interactions than in that of the overall chromophore conformation since in this pH range there is no change in the UV/VIS absorption ratio (Fig. 5).

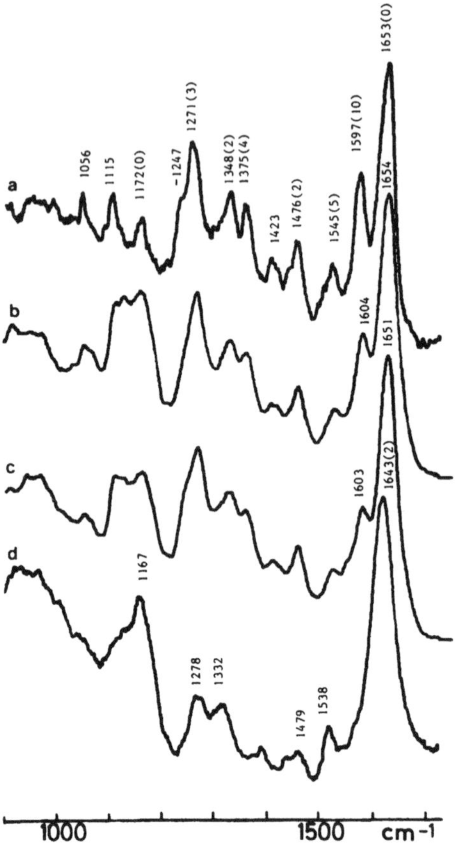

Fig. 6. Visible excited (488 nm), low temperature (40 K) resonance Raman spectra of phycocyanin. a - trimer phycocyanin at pH 7.5, b - phycocyanin at pH 4.0, c - monomer phycocyanin at pH 3.0, d - denatured phycocyanin at pH 1.8.

The VIS-excited resonance Raman spectra show new aspects of the complexity of the 1590-1650 cm^{-1} region since the isotope sensitivity of the bands is different from that found in the UV-excited case. The main band of the spectrum, around 1651-1653 cm^{-1} shows very limited sensitivity, but the 1595 cm^{-1} band has a large, 10 cm^{-1} downshift upon ^{14}N-^{15}N substitution. Chromophore nitrogens can form H-bonds with the protein environment. Changes in the chromophore-protein interactions probably would effect these H-bonds. Via these type of interactions with the protein moiety a mode involving large nitrogen displacements can be, as it was observed, sensitive to environmental changes.

Conclusions

The chromophore conformations in the native phycocyanin and allophycocyanin are different. During monomerisation there is no change in the chromophore conformation of phycocyanin. In allophycocyanin a change occurs in a way that in the monomer allophycocyanin the chromophore conformation becomes very similar to that of native/monomer phycocyanin.

In the denatured biliproteins the chromophore conformation is the same, since phycocyanin and allophycocyanin have identical resonance Raman spectra in this state.

The main band of the spectrum, which is around 1645 cm^{-1} in the native proteins can be used as a marker of the overall chromophore conformation. Its behaviour is in a good agreement with the UV/VIS absorption ratio.

Allophycocyanin is more sensitive to the decrease of the pH than phycocyanin.

In the VIS-excited resonance Raman spectra a band can be found which is closely related to vibrations involving nitrogen displacement and seems to be sensitive to chromophore-protein interactions. According to its behaviour, changes in the protein environment around the chromophore start immediately with decreasing pH.

Acknowledgement

This work was supported by a grant of the Hungarian Academy of Sciences (OTKA 559). B. Sz. is thankful for a three months EMBO fellowship at the Service de Biophysique, Departement de Biologie, C.E.N. de Saclay.

References

1. Gantt E. 1981 Ann. Rev. Plant Physiol. 32, 327-34

2. Brown A.S., J.A. Foster, P.V. Voynow, C.F. Franzblau and R.F. Troxler. 1975 Biochemistry 14, 3581-3588

3. Scheer H. 1982 In Light Reaction Path of Photosynthesis (F.K. Fong ed.) ch. 2.

4. Scheer H., H. Formanek and S. Schneider. 1982 Photochem. Photobiol. 36, 259-262

5. MacColl R., K. Csatorday, D.S. Berns and E. Traeger. 1980 Biochemistry 19, 2817-2820

6. Csatorday K., R. MacColl, V. Csizmadia, J. Grabowski and Cs. Bagyinka. 1984 Biochemistry 23, 6466-6470

7. Schirmer, T., W. Bode, R. Huber, W. Sidler and H. Zuber. 1985 J. Mol. Biol. 184, 257-277

8. Schirmer, T., R. Huber, M. Schneider, W. Bode, M. Miller and M.L. Hackert. 1986 J. Mol. Biol. 188, 651-676

9. Szalontai B., Z. Gombos and V. Csizmadia. 1985 Biochem. Biophys. Res. Commun. 130, 358-363

10. Szalontai B., Z. Gombos, V. Csizmadia and M. Lutz. 1987 Biochim. Biophys. Acta 893, 296-304

11. MacColl R., K. Csatorday, D.S. Berns and E. Traeger. 1981 Arch. Biochem. Biophys. 208, 42-48

12. Ohad I., H.-J.A.W. Schneider, S. Gendel and L. Bogorad. 1980 Plant Physiol. 65, 6-12

13. Ohki K. and Y. Fujita. 1979 Plant and Cell Physiol. 20, 483-490

14. Scharnagl C., E. Kost-Reyes, S. Schneider, H.P. Kost and H. Scheer. 1983 Z. Naturforsch. 38c, 951-959

15. de Kok J. 1982 Photochem. Photobiol. 35, 849-851

16. Margulies L. and M. Toporowitcz. 1984 J. Am. Chem. Soc. 106, 7331-7336

17. Scheer H. and W. Kufer. 1977 Z. Naturforsch. 32c, 513-519

18. Burke M.J., D.C. Pratt and A. Moscowitz. 1972 Biochemistry 11, 4025-4031

19. Falk H., and N. Muller. 1983 Tetrahedron 39, 1875-1885

COHERENT ANTI-STOKES RAMAN SPECTROSCOPY OF PHYCOBILISOMES, PHYCOCYANIN AND ALLOPHYCOCYANIN FROM MASTIGOCLADUS LAMINOSUS

S. Schneider, F. Baumann, W. Steiner
Institut für Physikalische und Theoretische Chemie der TU-München, Lichtenbergstr. 4, D-8046 Garching

R. Fischer, S. Siebzehnrübl, H. Scheer
Botanisches Institut der Ludwig-Maximilians-Universität Menzingerstr. 67, D-8000 München 19

Introduction

In order to provide a large geometrical and spectral cross-section for the absorption of light, red and blue-green algae produce the so-called phycobilisomes (PBS), which contain the chromoproteins phycoerythrin (or phycoerythrocyanin), phycocyanin (PC) and allophycocyanin (APC). Although PC and APC contain the same chromophore, phycocyanobilin, the spectral properties of these biliproteins (tetrapyrrol chromophore bound to apoprotein) are quite different. Their excitation energies are modified according to their special function in the light harvesting system (see e.g. 1,2) by chromophore-protein interactions (14).
Resonance-enhanced Coherent Anti-Stokes Raman Spectroscopy (CARS) has proved to be a very suitable technique to produce vibrational spectra of highly fluorescing chromophores as e.g. light harvesting pigments (3,4,5). In this contribution CARS-spectra of room-temperature solutions of phycobilisomes and of phycocyanin and allophycocyanin trimers from Mastigocladus laminosus are presented and the implications to chromophore structure are discussed.

Photosynthetic Light-Harvesting Systems

Experimental

Resonance-enhanced CARS spectra are obtained by focussing two laser beams with different wavelengths λ_{pump} (fixed) and λ_{Stokes} (variable) in the sample, where by a non-linear effect the CARS beam is generated, which holds the equality

$$\tilde{\nu}_{CARS} = \tilde{\nu}_{pump} + (\tilde{\nu}_{pump} - \tilde{\nu}_{Stokes})$$

$$\tilde{\nu}_{CARS} > \tilde{\nu}_{pump} > \tilde{\nu}_{Stokes} \text{ , with } \tilde{\nu} = 1/\lambda \quad [cm^{-1}]$$

A plot of the CARS intensity as function of $\tilde{\nu}_{pump} - \tilde{\nu}_{Stokes}$ reveals a vibrational spectrum of the chromophore. As the CARS frequency is higher than either pump and Stokes frequency, no flourescence problems as in spontaneous Raman spectroscopy can arise. (For more experimental details see ref. 3).

Preparation

PBS were isolated according to the method published by Nies and Wehrmeyer (6), except that the pH of the phosphate buffer was 6 and the purification on the sucrose gradient was done twice. APC was prepared very similar to a procedure described by Füglistaller et. al. (7).
The aggregation state of PC and APC was determined by sedimentation runs according to Martin and Ames (8). Both biliproteins were found to be trimeric. Linker peptides were not present as proved by absorption spectra, sedimentation coefficients and SDS-gel electrophoresis.

Results

As both the fingerprint (1100 - 1300 cm^{-1}) and the double bond stretching region (1500 - 1750 cm^{-1}) have proved to be most suitable for the investigation of chromophore geometry by CARS (3,4,5,9) or Resonance Raman Spectroscopy (10,11), both regions are considered in the following.

The phycobilisome spectra (PBS) are dominated by the contribution of the constituent phycocyanin trimers (PC) and are therefore rather similar to the spectra of the latter. Most of the vibrational frequencies are reproduced within experimental accuracy, though with different intensities (especially between 1200 and 1300 cm^{-1})(fig 1). At 1623 cm^{-1} an additional weak peak is found in PBS, which is due to allophycocyanin (APC). A second interesting feature is that for the pronounced bands a smaller width is found in PBS than in PC. It possibly indicates a lower structural inhomogeneity in PBS due to higher aggregation and the presence of linker peptides.

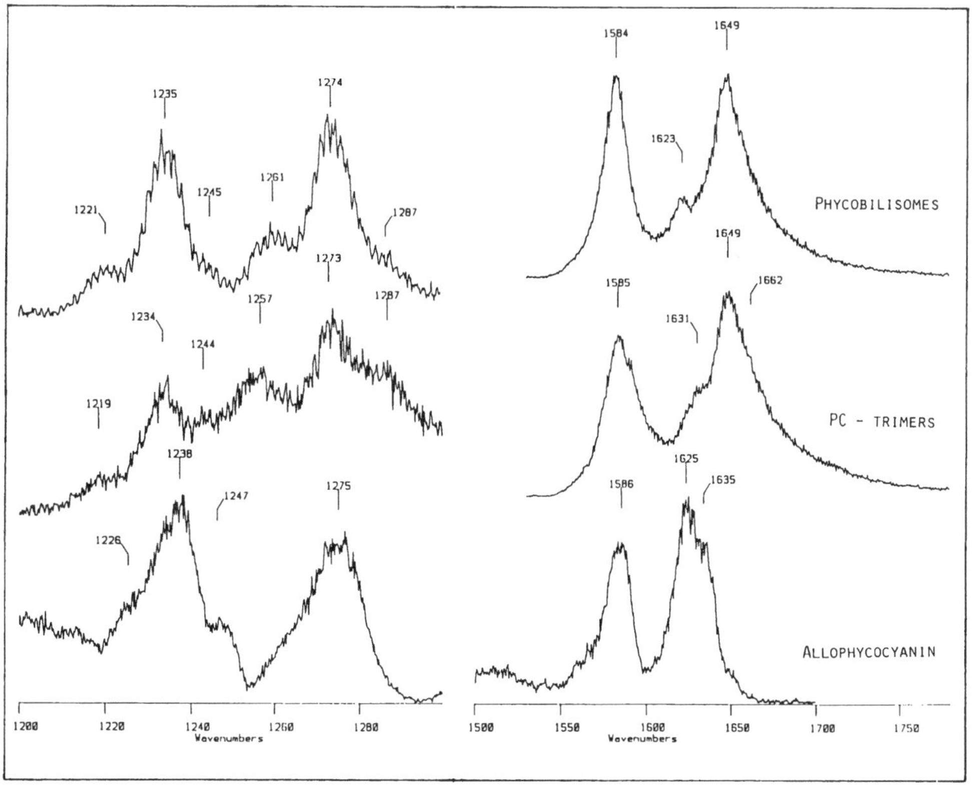

Fig. 1: Resonance-enhanced CARS spectra of phycobilisomes, phycocyanin trimers and allophycocyanin trimers from Mastigocladus laminosus. Pump wavelength 640 nm.

APC spectra are very different from PC spectra (see also ref. 10). The 1257 cm^{-1} and 1287 cm^{-1} bands observed in PC are absent in APC. In the double bond stretching region the dominant peak in PC is found at 1649 cm^{-1} with shoulders at ≈1631 and ≈1662 cm^{-1}. In APC spectra, however, the dominant peak is found at 1625 cm^{-1} with a strong shoulder at ≈1635 cm^{-1} and a weak one at ≈1650 cm^{-1}.

Discussion

According to preliminary normal mode calculations using the coordinates provided by X-ray analysis (12,13), the bands found between 1200 and 1300 cm^{-1} represent extensively mixed C-C stretch / C-H in plane bending modes and can presently not be interpreted in a straight-forward manner. The band around 1585 cm^{-1} is related to C=C stretching vibrations in the ring B/C fragment. That this band exhibits nearly no change in frequency, most likely indicates a conservation of the geometry of this part of the chromophore.
The C=C double bonds of the methine bridges between ring A-B and ring C-D, on the other hand, should give rise to localized modes with frequencies between 1650 and 1700 cm^{-1}. The C=O stretch mixes with the C=C stretch in ring D to produce vibrational frequencies around 1630 cm^{-1}. The actual value of the calculated frequencies is found to be sensitive to structural changes.
From these theoretical results and the comparison of CARS spectra of model compounds (9) with the presented CARS spectra a significant difference in chromophore geometry and/or Π-electron distribution between PC and APC must be postulated. Interpretation of e.g. fluorescence decay data of APC should therefore not be based on structural information gained from PC.

Additional measurements on antenna pigments and a more elaborate normal mode analysis are in preparation.

Acknowledgement

Financial support by Deutsche Forschungsgemeinschaft (SFB 143) and Fonds der Chemie is gratefully acknowledged.

References

(1) Gantt E. 1981. Phycobilisomes . Ann. Rev. Plant Physiol. 32, 327 - 347.

(2) Scheer H. 1982. In: Light Reaction Path of Photosynthesis (F.K. Fong, ed.). Springer, Berlin. pp. 7 - 45.

(3) Schneider S.,F. Baumann, U. Klüter, 1987. Z. Naturforsch. 42c(11/12).

(4) Schneider S., F. Baumann, P. Geiselhart, S. Siebzehnrübl, R. Fischer, H. Scheer, 1987. Proc. of the Sec. Europ. Conf. on the Spectr. of Biol. Mol., J.Wiley, Chichester.

(5) Schneider S., F. Baumann, P. Geiselhart, H. Kayser, H. Scheer, 1987. Photochem. Photobiol., submitted.

(6) Nies M. and W. Wehrmeyer, 1980. Planta 150, 330 - 337.

(7) Füglistaller P., H. Widmer, G. Frank and H. Zuber, 1981. Arch. Microbiol. 129, 268 - 274.

(8) Martin R.G. and B.N. Ames, 1961. J. Biol. Chem. 236, 1372 - 1379

(9) Schneider S., P. Geiselhart, F. Baumann, H. Falk and W. Medinger, 1987. J. Biochem. Biophys., submitted.

(10) Szalontai B., Z. Gombos, V. Csizmadia, 1985. Biochem. Biophys. Res. Com. 130, 358-363.

(11) Szalontai B., Z. Gombos, V. Csizmadia and M. Lutz, 1987. Biochim. Biophys. Acta 893, 296-304.

(12) Schirmer T., W. Bode and R. Huber, 1985. J. Mol. Biol. 184, 257-277.

(13) Schirmer T., W. Bode and R. Huber, 1987. J. Mol. Biol., 196, 677-695.

(14) Schneider S., C. Scharnagl, M. Duerring, T. Schirmer and W. Bode, 1987. Contribution in this volume.

OPTICAL ABSORPTION AND CIRCULAR DICHROISM OF BACTERIOCHLOROPHYLL OLIGOMERS IN TRITON X-100 AND IN THE LIGHT- HARVESTING-COMPLEX B850; A COMPARATIVE STUDY.

V. Rosenbach-Belkin, P. Braun, P. Kovatch, A.Scherz *
The Biochemistry Department, Weizmann Institute of Science, Rehovot, Israel

Abstract

The Bacteriochlorophyll a (Bchla) monomeric units in the light harvesting complexes (LHCs) of photosynthetic bacteria have local Qy transitions at 780-800 nm. The Bchla monomers self-organize into dimers with a geometry that is similar to the geometry of P-860 and a Bchla dimer that is formed in Formamide:Water - 3:1 (Bchl-853). Δ G for the self-organization of the *in vitro* dimer and probably the *in vivo* forms is about -3.6 Kcal/mole. Two *in vivo* Bchla dimers are weakly coupled in a polypeptide (PP) tetramer. Three tetramers form a complete LHC. This model is shown to account for the optical absorption and circular dichroism of B850 that is treated with verious detergents.

Introduction

The *in vivo* Bchls show a bathochromic shift of the Qy transition (1,2,3) accompanied by induced optical activity (4) and increased oscillator strength (4-6), relative to their spectral properties as isolated monomers *in vitro*. Variations in the Qy shift of the Bchls are usually accompanied by some modifications in the composition of their liganding polypeptides (PP) and characterize different LHCs (1,7).

The variable Qy shift has been used to probe the *in vivo* setting of Bchls and Chlorophylls (Chls) in one of the following three approaches:

Photosynthetic Light-Harvesting Systems

(a) The first approach suggests that the shift of the Qy transition is induced by interactions among ground states in large aggregates of Bchls *in vivo* accompanied by weak (230-300 cm^{-1}) exciton coupling of the Qy transitions (8-9).

(b) Another approach proposes that specific interactions between the occlluded Bchls and the protein side chains (e.g. interactions with charged amino acid residues (10a), Schiff base formation (10b), or interactions with aromatic amino residues (Pearlstein, Bronisholtz and Zuber, this volume)), result in modification of the chromophores and thereby a large shift of their Qy transition (3,10c,11).

Starting with Bchls that have locally shifted Qy transitions, Kramer *et al* (12) suggested a PP tetramer for the LHC B800-850 where 4 Bchls that absorb at 850 nm are positioned symmetrically 15 Å apart (center to center separation) near the cytoplasmic side of the photosynthetic membrane and two Bchls, with local Qy transitions at 800 nm, are positioned symmetrically close tothe periplasmic side of the PPs. A modification of this model is presented in the current volume by van Grondelle (13).

Similar assumptions on the nature of the locally excited states have lead Zuber and Pearlstein to propose a hexamer of α-β PP dimers with 12 Bchls that absorbe at 850 nm, positioned 12-15Å apart (center to center) near the cytoplasmic side of the photosynthetic membrane and 6 Bchls that absorb at 800 nm near the periplasmic side of the photosynthetic membrane (1,2). Both groups considered weak exciton coupling among the Bchl monomers that absorb at 850 nm to account for the apparent CD (14 a).

(c) Updated sequencing (14 a-c) and hydropathy plots (16) of PPs from several LHCs, also presented in this meeting (16), show no charged groups that can induce a large bathochromic shift of the Qy transition in the Bchls' neighbourhood. The aromatic residues that are proximal to the Bchls are similar to those that are positioned near the Bchls of the "special pair" in reaction centers (RCs) (16). However, it is clear now that the bathochromic shift in the Qy transition of P-860 is mostly induced by interactions among the Bchls that make the "special pair" (17-19) and have local Qy transitions at 780-800 nm. Therefore, realizing the similarity in spectra of the "special pair" Bchls and the Bchls of the LHC, Scherz and Parson proposed that individual Bchls *in vivo* with local Qy transitions at 780-800 nm are tightly packed into dimers that are attached to the α - β PP pair of the LHCs. The locally excited states of the Bchls are so strongly coupled that the Qy wavelength shifts to 852 nm (20). Further oligomerization of the Bchl-PP pairs into complete LHCs (e.g. B850) should result in a weak exciton coupling of the shifted Qy transitions that introduces a strong and conservative CD around 852 nm and a small shift of the Bchl dimer exciton (the exciton shift can be bathochromic or hypsochromic). Under these assumptions Scherz and Parson found a unique geometry of for a Bchla dimer (with Bchl monomers separated by 7.5 Å) for which the calculated OD and CD in the UV-VIS-NIR fit well the spectra of partly bleached B850 (20). The spectra (optical absorption and CD) of the complete LHC (e.g. the SDS-B850) were predicted reasonably well with two such dimers that are separated by 18 Å (center to center).

In looking for further support for the theory of Scherz and Parson (20),*it seemed useful to compare the spectral variations upon the dissociation of* LHCs *into single* Bchl-PP*units with the variations in spectra when in vitro Bchls gather into tightly packed dimers and the dimers form loose aggregates.*

Materials and Methods

Formamide (FA) (fluka) and *Pyridine* (Merck) were used with no further purification. Bchla was extracted from lyophilized cells of *R. rubrum* and purified as previously described (6). The sample purity was checked by TLC and reversed phase HPLC (as in ref. 21). No impurities (e.g. 10-OH Bchl) could be traced.

Large oligomers of Bchla with optical absorption at 860 nm were prepared under dim light conditions as previously described (22). To dissociate the large oligomers into smaller units (dimers), Triton X-100 (Tx-100) (Sigma) was added to this Bchla solution (22).

The LHC B850 and whole chromatophores were extracted from *Rb. sphaeroides* strain R-26.1 following the method of Austin and Sauer with some modifications (4,23).

Molecular weight of the LHCs in Tris buffered solutions (0.01 M pH 7.5) that contained SDS or LDS were determined by non-denaturing transverse polyacrylamide gel electrophoresis (TPAGE) (24) using Ferguson plots (24), by non denaturing vertical PAGE (24) and by gel filtration chromatograpy on preswollen Sephacryl SF 300 and Sephadex G-50 (fine) (4,24).

Circular dichroism (CD) spectra were measured on a computerized home-built spectropolarimeter (8). Ellipticity was calibrated by an aqueous solution of 1 mg/ml d-camphor sulphonic acid ($\Theta_{292\ nm} = 0.308^{\circ}$).

Absorbance (OD) was measured with computerized Bausch and Lomb (Milton-Roy) spectronics 1201 and Cary 219 (Varian) spectrophotometers.
Fluorescence was measured with a Perkin-Elmer MPF-44A spectrofluorimeter.
Atomic absorption was recorded by a Perkin-Elmer 306 spectrometer.

Results

<u>In vitro preparations</u>

Bchla makes two spectral forms in Formamide:Water-3:1 (Fw) (17): one, termed Bchl-780, has its NIR maximum absorption at 780 nm and the other one, termed Bchl-860 has its NIR maximum absorption at 860 nm. The notations: [Bchl-780], [Bchl-860] and $[\text{Bchla}]_T$ stand for the concentrations of single molecules of Bchla in the Bchl-780 form, Bchl-860 form and in all forms together respectively. The absorption spectra of 1 M [Bchl-780] and 1 M [Bchl-860] are illustrated in Fig.1.

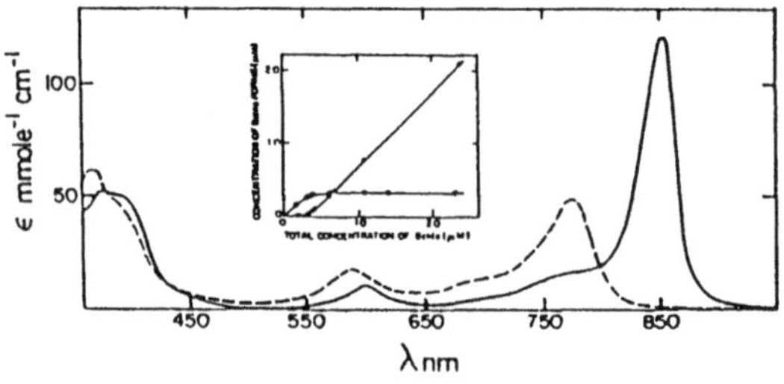

FIGURE 1. *The optical absorption of Bchl-780 (- - -) and Bchl-860 (-----) in Fw that contains* $5x10^{-2}$ *M Tx-100 and no Tx-100 respectively. Inset - [Bchl-780] ( ) and [Bchl-860] ( ) respectively.*

When the $[\text{Bchla}]_T$ is above $3x10^{-7}$M, the [Bchl-780] becomes fixed at an asymptotic value of $3x10^{-7}$ and the [Bchl-860] increases linearly (fig.1 inset) (17,22). The introduction of increased concentrations of Tx-100 to Fw solutions that contain about $8x10^{-6}$M $[\text{Bchla}]_T$ either before or after the formation of Bchl-860, has the following consequences:
(a) The [Bchl-780] increased.

(b) There was no change in the lineshapes of the Bchl-780 absorption and circular dichroism as the value of [Bchl-780] increased.

(c) The quantum yield of the Bchl-780 fluorescence (at 793 nm) increased to 0.10, close to the yield of Bchla monomers in Methanol (0.12) (25)

(d) The absorption maximum in the NIR region shifted to 853 nm and except for some differences in the bandwidth and the slight blue shift of the Qy transition, the optical absorption spectra of Bchl-860 and Bchl-853 looked the same. We shall use the notation [Bchl-853] for the concentration of Bchla single molecules in the new spectral form.

(e) The triple-banded CD of Bchl-860 at 860 nm (17,22) was replaced by a doublet with negative and positive extrema at 877 and 850 nm respectively (Fig. 2). The rotational strength was net positive in the NIR region and net negative in the UV and VIS regions.

When we held [Tx-100] at 5.77×10^{-3}M and changed $[Bchla]_T$ the following were observed:

(a) The CD in the NIR region changed from being almost single-banded and positive at low $[Bchla]_T$ values into an almost conservative double-banded CD at high values of $[Bchla]_T$ (Fig. 2).

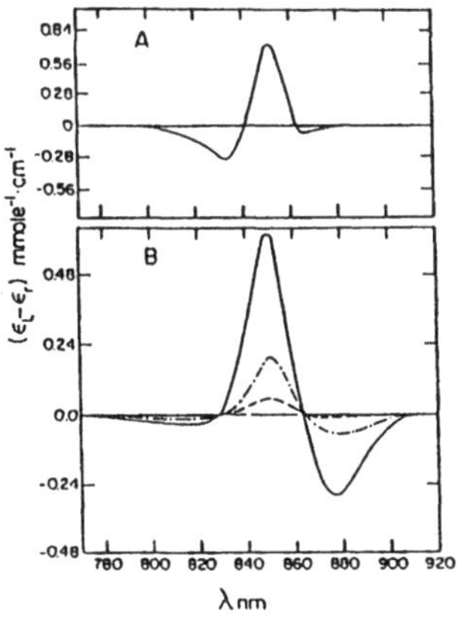

FIGURE 2. *(A) - CD spectrum of [Bchl-860] in Fw; (B) - CD spectra of [Bchl-853] in Fw that contains* 5.77×10^{-3} *M [Tx-100] and (a)* 0.86×10^{-6}*M (- - -); (b)* 1.50×10^{-6}*M (- · -); (c)* 4.15×10^{-6}*M ( ); [Bchl 853].*

(b) For $[Bchla]_T < 10^{-5}M$, the value of [Bchl-780] increased (Fig. 3A) and could be related to the $[Bchla]_T$ by the equation:

$$[Bchla]_T = 4.90x10^6x[Bchl\text{-}780]^2+[Bchl\text{-}780] \quad (II)$$

At higher values the [Bchl-780] approached an asymptotic value - $[Bchl\text{-}780]=1.5x10^{-6}$ M and the 853 nm absorption broadened and shifted towards 860 nm.

(c) The [Bchl-853] was given by:

$$[Bchl\text{-}853] = 4.90x10^6x[Bchl\text{-}780]^2 \quad (III)$$

and was qudratically related to $[Bchla]_T$.

(d) The quantum yield for the fluorescence of [Bchl-780] was close to that of Bchla in methanol up to $[Bchla]_T = 0.8x10^{-6}M$ and dropped down at higher values. No fluorescence could be resolved for Bchl-853.

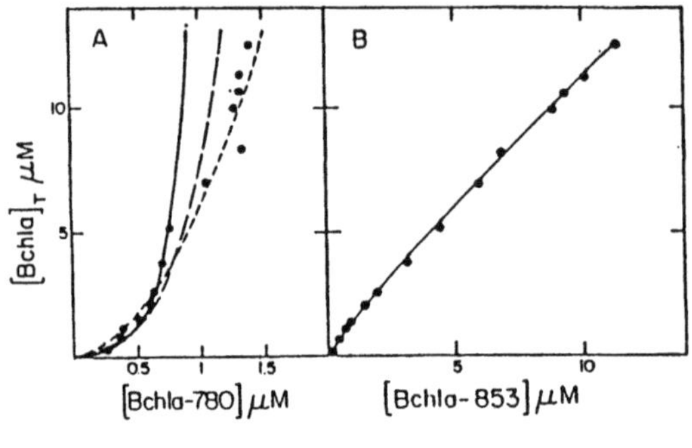

FIGURE 3. *(A) - The $[Bchla]_T$ as a function of [Bchl-780] in Fw that contains $5.77x10^{-3}M$ [Tx-100]. ( ) - experimental values; (- - -) - Values calculated for [Bchl-853]= K_2[Bchl-780]^2; (-- --) - Values calculated for [Bchl-853]= K_3[Bchl-780]^3; (-----) - Values calculated for [Bchl-853]= K_4[Bchl-780]^4. (B) - The $[Bchla]_T$ as a function of the [Bchl-853]; ( ) - experimental values; (----) - values calculated for $[Bchla]_T = 4.9x[Bchl\text{-}853]^2 +[Bchl\text{-}780]$.*

In vivo preparations

Treatment of chromatophores with the same concentration of TX-100 (up to 2.5%), but with different incubation times at different temperatures resulted in variations of the B850 optical absorption and CD (i.e. "B850" in table 1.) The net rotational strength however, of all samples, is positive and is equal to about 0.5-1.0 DBM. The deconvoluted contributions of the positive and negative bands are much higher than in whole chromatophores (table 1.)

TABLE 1. *Optical absorption and CD of B850 from R-26.1 under different conditions. The resulted fragments are signified by the maximum absorption of their Qy transition. The last three entries were recorded directly from SDS gels .*

	O.D. λ(max)	Relative Anisotropies and Maximum(nm) of pos.CD and neg.CD band		Rotational strength [DBM] R+ and R-	
"B864" in chromatophores	864	2.3(858)	-1.9(890)	2.3	1.2
"B855" in Tx-100	855	4.0(851)	-3.5(888)	5.7	4.3
"B850" in Tx-100	850	7.1(844)	-6.2(875)	10.7	9.3
"B850" in 1% SDS	850	10.9(849)	-9.3(878)	16.0	13.4
"B850" in 1% LDS	850	7.6(849)	-6.5(870)	19.0	14.3
"B850" in 1%SDS-PAGE	850	5.8(848)	-6.3(888)	15.0	14.4
"B846" in 1%SDS-PAGE	846	7.9(844)	-6.9(875)	16.8	16.4

SDS and LDS dissociate the B850 that was treated with Tx-100 (M.W. = 320 Kd (23)) into four fragments termed according to their optical transitions in the NIR region: Two heavy units (M.W. 125 and 105 Kd both termed "B850" in table 1.) two medium units (72 Kd and 60 Kd termed "B856" and "B846" in table. 1 respectively); and one light unit (6 Kd). Despite the relatively high detergents' concentrations, the complexes were not dissociated into Bchls and PPs but into PP-Bchl units of different sizes.

The differences in molecular weights are accompanied by different wavelengths and rotational strengths for the Qy transition (table 1). In general the Cotton effect in the Qy region of the B850 treated with SDS and LDS, is larger than the Cotton effect of the B850 treated with TX-100 and whole chromatophores by factors of 1.5 and 3 respectively. The 6 Kd unit has the optical absorption and CD of *in vitro* Bchla.

Dialysis of the detergent-treated chromatophores against Tris buffer solution in Amicon, resulted in heavy complexes (M.W $>$300 Kd) with Qy transition at 862 nm and weak CD (as in whole chromatophores).

Discussion

Sauer and Austin dissociated B850 (4,23) that was solubilized in Tx-100, from R-26, (M.W. = 320 Kd) into protein-chromophore complexes of approximately 22 Kd (4,23) by SDS treatment. The B850 fragments in SDS had a bathochromically shifted Qy transition with an almost conservative CD at 852 nm (4,23) and (1 Bchla)/(1 PP) (M.W. 7-10 Kd). Sauer and Austin suggested that the SDS B850 complex is made up of two PP and two Bchls. Therefore, to account for the conservative CD around 852 nm they had to assume that the local Qy transitions of each Bchla are at 852 nm and that the two Bchls are separated by about 12-15 Å. Rafferty *et al* supported this hypothesis by showing an asymptotic shift of the Qy transition to 852 nm accompanied by a conversion of the conservative CD at 852 nm to a single band CD by destructive illumination of whole chromatophores (10 b.).

The PPs subunits in the LHCs B850 of R-26.1 should be similar to those of R-26 but the molecular weights of the B850 from R-26.1 that were treated with SDS, are multiples of 6 Kd units. The smallest unit (M.W.= 6 Kd) has the spectral proprties of Bchl monomers *in vitro*, in particular, Qy transition at 780 nm and practically no optical activity. We therefore suggest that the major contribution to the bathochromic shift of the Qy transition in B850 comes from interactions among Bchls with locally excited Qy transitions at 780-800 nm, provided that there is no major change in the amino acid configuration near the Bchls site upon dissociation into the 6 Kd PP-Bchl units.

Our conclusion seems to be in contrast with the conclusions of Sauer and Austin (4,23) and Rafferty *et al* (10 b.) but in good agreement with their observations and the nature of the single PP in the LHC. The updated molecular weights of the α and β subunits are 5.5 and 6.5 Kd units respectively (17,19); therefore, the 22 Kd protein isolated by SDS treatment of B850 from R-26 is probably composed of 4 PP-bchl units (each with a M.W. of approximately 6 Kd), rather than 2 PP-Bchl units (each with a M.W. of 10 Kd) as Sauer and Austin originaly proposed (4,23).

The CD curves for the different preparations of B850 from R-26.1 (table 1. and ref 23) can always be deconvoluted into negative and positive (slightly stronger) bands that are centered at 860 - 870 nm (table 1). Therefore ,the CD should originate from the weak exciton coupling of two degenerate transitions at 860-870 nm. However, the spectra of the 6 Kd PP-Bchla units show that the *in vivo* Bchl are locally excited at 780 nm hence, we have to assume that the pool of Bchls in B850 is grouped into equal oligomers that have Qy transitions at 860 nm with a single, positive, Cotton effect (0.5 DBM). Since the smallest number of Bchls that produce the double band CD is 4 (in the 22 Kd SDS-B850) they should be grouped into two dimers as proposed by Scherz and Parson (29). The CD and optical absorption of the different B850 fragments could be simulated reasonably well (using the extended exciton approach with dimers of Bchla that are oriented relative to each other as in table. 2

The spectral properties of the *in vivo* Bchls can be better understood by a comparison with the spectral properties of the Bchls in the Fw system, but for doing so we first have to find the molecular composition of the *in vitro* forms.

The Bchl-780 is probably a Bchla monomer because: (a) The lineshapes of its absorption and circular dichroism do not depend upon its concentration and are typical to Bchla monomers in protic solvents. (b) When no other forms (e.g. Bchl-853 or Bchl-860) are present it has the normal quantum yield expected for monomers in protic solvents (26).

TABLE 2. *Configuration of the lower energy Qy excitons that is needed to simulate the CD spectra of the different B850 preparations. One Bchla dimer is at the origin and R gives the displacement of the other dimer center. β, α and γ are the angles between the the Q excitons of the two dimers (21).*

Anntenna complex	Distance $R=(r_x, r_y, r_z)$ [A]	Orientation of Q_x-and Q_y β_y γ_y α_x γ_x	Rotational strength [DBM]	Wavelength λ [nm]	Dipole strength [D^2]
"B864" in Chromatophores	1.9, 10. -14	69 23 98 70	2.5 -1.2	863 889	16.2 8.4
"B850" in TX-100	-2, 10, -14	69 23 99 70	11.0 -10.3	854 877	16.2 8.5
"B850"in 1% SDS	-2, 10, -14	69 26 99 70	14.1 -13.6	854 877	16.2 8.4
"B850" in 1% SDS-PAGE	-2, 10, -14	69 30 97 70	17.2 -16.9	855 876	16.2 8.4

(c) HPLC and TLC of the Bchla that was extracted from the Fw solutions show no chemical modification (e.g. 10-OH Bchl formation). The concentration increase of Bchl-780 in the presence of Tx-100 above its threshold value in pure Fw (fig. 3), results from its partial solubilization by micelles of Tx-100 (20) The solubilization results in having two forms of Bchla monomers in equilibrium:

$$[\text{Bchla}] + [\text{MT}] \overset{K_1}{\rightleftarrows} [\text{Bchla-MT}] \qquad \text{(IV)}$$

Both forms have the spectra of Bchl-780, therefore

$$[\text{Bchl-780}] = [\text{Bchla}] + [\text{Bchla-MT}] \qquad \text{(V)}$$

With [Bchla] being the concentration of monomers in the free solution, [Bchlaa-MT]- the concentration of solubilized monomers and [MT] being the micelle concentration - ([Tx-100]-$3\text{x}10^{-3}$)/1000 (27). Since [Bchla]$\leq 3\text{x}10^{-7}$M, the [Bchl-780] should have an asymptotic value given by:

$$[\text{Bchl-780}]_{\infty} = 3\text{x}10^{-7}(1+K_1[\text{MT}].) \qquad \text{(VI)}$$

We found that for $[Tx\text{-}100]=5.77x10^{-3}M$, and $[Bchla]_T=12x10^{-6}M$, $[Bchl\text{-}780]_{\infty}$ approaches $1.5x10^{-6}M$, hence $K_1[MT]=4$.

Obviously, as long as $[Bchl\text{-}780]<1.5x10^{-6}M$, the [Bchla] value is smaller than $3x10^{-7}M$, which is the onset for the Bchl-860 formation (Fig. 1 inset), there can be no oligomers in the free Fw solution and the Bchl-853 must be solubilized by the micelles together with monomers. This conclusion is further supported by the quenching of fluorescence from Bchl-780 that accompanied the increased concentration of Bchl-853. Within the micelles the Bchl-853 maintains quadratic equilibrium with the Bchl-780 (eq. II,III) and since the Bchl-780 is a Bchla monomer, the Bchl-853 should be a Bchla dimer.

At $2x10^{-7}M$[Bchl-853], most micelles should contain one dimer or monomers since [MT] is approximately 2.8 x $10^{-6}M$ or more. Consequently the CD pattern at that concentration should mainly reflect isolated dimers that apparently have the predicted (20,22) positive Cotton effect at about 852 nm (Qy transition) (fig. 2).

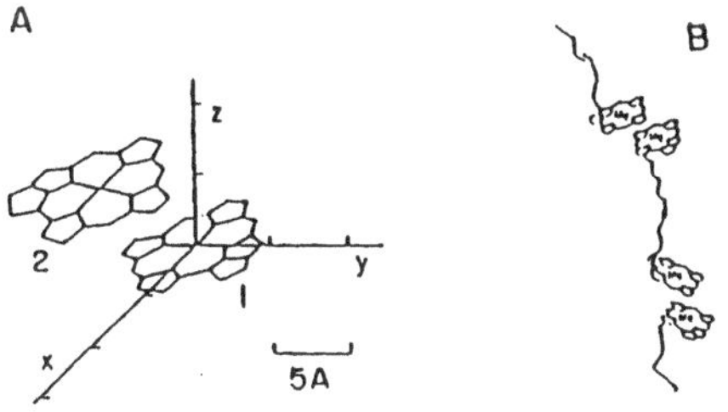

Figure 4. *Possible structure for the Bchl-853 dimer (left) and tetramer (right). Bchla 1 is at the origin; Bchla 2 is at R = (-0.5X -5.7Y +3.5Z). The angles between the transition dipoles (in degrees) are :* $Qy_2Qy_1(\beta_y) = 165$; $Qx_2Qy_1 = 105$; $Qy_2Qx_1 = 102$; $Qx_2Qx_1(\beta_x) = 29$; *The angles between the transition dipoles of molecule 2 and the normal to the Bchla 1 are:* $Qy_2Z_1(\gamma_y) = 80$; $Qx_2Z_1(\gamma_x) = 115$. *(B) - Possible organization of two dimers in vitro.*

As the [Bchl-853] increases some micelles are populated with several dimers and dipolar interaction among the shifted Qy transitions of the trapped dimers adds the predicted (20,22) conservative contribution to the CD spectrum (Fig. 2). The optical absorption and circular dichroism of the isolated dimers ($[Bchl\text{-}853]\leq 2x10^{-7}M$) and the dimer pairs within the micelles, could be best fitted by the geometry shown in Fig. 4 which is very close to the geometry proposed for the dimer and the pairs of dimers in B850 (20). The similar properties of the *in vivo* and *in vitro* clusters of Bchla (fig. 5) do not prove the hypothesis of Scherz and Parson (20) but strongly support it. We may therefore go one step further trying to correlate the chromophore arrangment with the PP stoichiometry as deduced from the molecular weight determination.

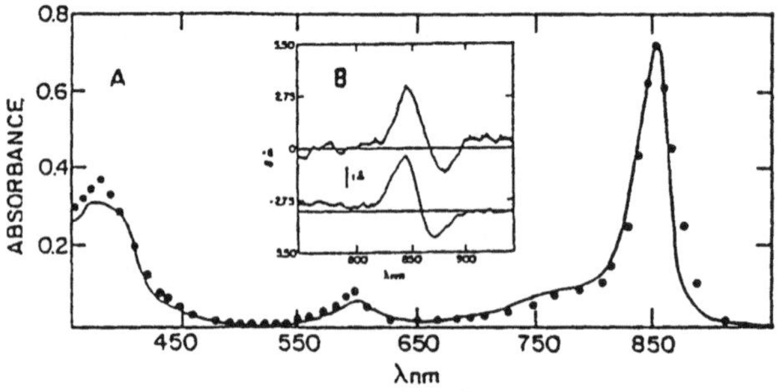

FIGURE 5. *The optical absorption of [Bchl-858] (----) and SDS-B850 (0 0). Inset - the CD of Bchl-858 (upper) and SDS-B850 (lower).*

To orient the Bchl dimers relative to the membrane plane we assumed the following: (a) The Bchls are liganded to the hystidins as suggested by Zuber (1,2); (b) The PP are similarily positioned in the photosynthetic membrane (2) ; (c) The 72 Kd LHC is an α-β hexamer that contains 12 Bchls. The Bchls-self organize into dimers as in fig. 4 A with their Qy and Qx transitions almost parallel and perpendicular respectively to the membrane surface. The self-organization of the Bchls ($\Delta G =$ -3.6 Kcal/mole) resembles the dimerization of aromatic side chains within proteins ($\Delta G =$ -2 Kcal/mole (26-29)) and probably results from qudropole-monopole interactions among the ground states of the Bchl monomers (28,29). Two Bchl dimers are positioned at a center to center separation of 16 Å with their Qy excitons at an angle of 120 ^{o} forming a PP-Bchla tetramer which go through further oligomerization into PP-Bchl dodecamers. (d) The 60 Kd LHC is an α-β pentamer that contains 10 Bchls. The Bchls are organized into dimers as in the 72 Kd B850 but the angle between the Qy excitons is 110 ^{o} ; A model that fits these assumptions and our observation is shown in fig. 6. The proposed structure can be further expanded to large arrays (fig. 6C)

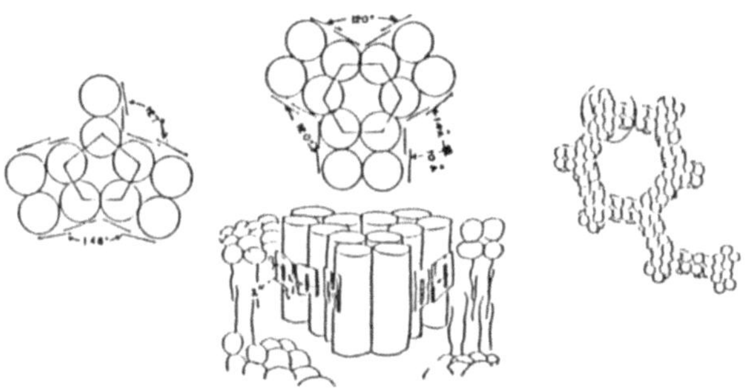

A B C

Figure 6. *A suggested model for B850* (A) - 60 Kd form; (B) - 72 Kd form; *(C) - large arrayes.*

Acknowledgment.

This grant was financially supported by the U.S.-Israel Binational Science Foundation (Grant No. 84-00144). Helpful discussions with Prof. H. Scheer are greatly appreciated. P. Braun is aM.Sc. student of Prof. H. Scheer and A. Scherz; A. Scherz has a Recanetii Career Development Chair.

References

1. Zuber, H. (1985) Photochem. Photobiol. 42, 821-844.

2. Zuber, H. (1986) TIBS 11, 414-419.

3. Cogdell, R.J. and Thornber, J.P. (1980) FEBS Lett. 122, 1-8.

4. Sauer, K. and Austin, L.A. (1978) Biochem. 17, 2011-2019.

5. Bolt, J. (1980) Thesis, Univ. California, Berkeley.

6. Scherz, A. and Parson, W.W. (1984) Biochim.Biophys. Acta 766, 653-665.

7. Thornber, J.P., Cogdell, R.J., Pierson, B.K. and Seftor, R.E.B. (1983) J. Cell. Biochem. 23, 159-169.

8. Ballschmiter, K., Truesdell, K. and Katz, J.J. (1969) Biochim. Biophys. Acta 184, 604-613.

9. Worcester, D.L., Michalsky, T.J. and Katz, J.J. (1986) Proc. Natl. Acad. Sci. USA 83, 3791-3795.

10. (a) Eccles, J. and Honig, B. (1983) Proc. Natl. Acad. Sci. USA 80, 4959-4962.
(b) Davis, R.C., Ditson, S.L., Fentiman, A.F. and Pearlstein, R.M. (1981) J. Am. Chem. Soc. 103, 6823-6826.
(c) Rafferty, C.N., Bolt, J., Sauer, K. and Clayton, R.K. (1979) Proc. Natl. Acad. Sci. USA 76, 4429-4432.

11. Pearlstein, R.M. and Zuber, H. (1985) in "Antennas and Reaction Centers of Photosynthetic Bacteria" ed. Michel-Beyerle, M.E. Springer Series in Chemical Physics 42, pp. 53-61, Springer-Verlag.

12. Kramer, J.M., Van Grondelle, R., Hunter, C.N., Westerhuis, W.H.J. and Amesz, J. (1984) Biochim. Biophys. Acta 765, 156-165.

13. van Grondelle, R., Sundstrom, V., van Dorssen, R.J., vos, M. and Hunter, C.N. (1988), This volume.

14. (a) Zuber, H. (1985) in "Antennas and Reaction Centers of the Photosynthetic Bacteria", ed. Michel-Beyerle, M.E. Springer Series in Chemical Physics, Vol. 42, pp. 2-14, Springer-Verlag.
(b) Cogdell, R.J. and Valentine, J. (1983) Photochem. Photobiol. 38, 769-772
(c) Theiler, R., Suter, F., Wiemken, V. and Zuber, H. (1984) Hoppe Seyler]s Z. Physiol. Chem. 365, 703-719.

15. Theiler, R. and Zuber, H. (1984) Hoppe-Seyler]s Z. Physiol. Chem. 365, 721-729.

16. Brunisholz, R. and Zuber, H. (1988), This volume.

17. Scherz, A. Rosenbach, V. and Malkin, S. (1985) in "Antennas and Reaction Centers of The Photosynthetic Bacteria", ed. Michel-Beyerle, M.E., Springer Series in Chemical Physics, Vol. 42, pp. 314-323, Springer-Verlag.

18. Parson, W.W. Warshel, A. and Scherz, A. ibid pp. 122-130.

19. Parson, W.W. and Warshel, A. (1987) To be published in J. Am. Chem. Soc. 109, 6154

20. Scherz, A. and Parson, W.W. (1986) Photosynthetic Res. 9, 21-32.

21. Haidl, H. Knodlmayr, K., Rudiger, W., Scheer, H. and Schoch, S. (1985) Z. Naturforsch. 40c, 685-692.

22. Scherz, A. and Rosenbach-Belkin, V. (1987), submitted.

23. Austin, L.A., (1979), Thesis, Univ. California, Berkeley

24. "Gel electrophoresis of protein; Practical approach." eds. Hames, B.D. and Rickwood, D. (1985), IRL press, Oxford.

25. Scherz, A. and Parson, W.W. (1984) Biochim. Biophys. Acta 766, 665-678.

26. Connoly, J.S., Samuel, E.B. and Janzen, A.F. Photochem. Photobiol. 36, 565-574.

27. The reported values for cmc of Tx-100 in H_2O range from $3x10^{-4}M$ to $5x10^{-3}M$. We calculated similar values in the Fw solution (ref. 20). A spectral form of Bchla that has maximum NIR absorption at 865 nm was found in H^2O in equilibrium with Bchl-780 depending on the Tx-100 content (Gottstein, J. and Scheer, H. (1983) Proc. Natl. Acad. Sci. USA 80, 2231-2234). Paradies, H.H. ((1980) J. Phys. Chem. 84, 599-607)), calculated that about 40 molecules of Tx-100 make one micelle in H_{2O} The number of TX-100 molecules that make one micelle in Fw (1000), was therefore determined by a comparison of the amount of TX-100 needed to dissociate the same amount of [Bchl-780] in the two systems.

28. Allen, J.P., Feher, G., Yeates, T.O., Rees, D.C., Deisenhofer, J., Michel, H. and Huber, R. (1986) Proc. Natl. Acad. Sci. USA 83, 8589-8593.

29. Burley, S.K. and Petsko, G.A. (1985) Science 229, 23-28.

30. Singh, J. and Thornton, J.M. (1985) FEBS Lett. 191, 1-6.

31. Burley, S.K. and Petsko, G.A. (1986) J. Am. Chem. Soc. 86, 7995-8001.

32. Thomas, K.A., Smith, G.M., Thomas, T.B. and Feldman, R.J. (1982) Proc. NAtl. Acad. Sci. USA 79, 4843-4847.

Absorption Detected Magnetic Resonance in Zero Magnetic Field on Antenna Complexes from Rps. Acidophila 7050 - The Temperature Dependence of the Carotinoid Triplet State Properties

J. Ullrich, J. U. v. Schütz and H. C. Wolf

3. Physikalisches Institut, Universität Stuttgart
Pfaffenwaldring 57, D 7000 Stuttgart 80

Using ADMR, the triplet state of carotinoid in antenna complexes of Rps. acidophila has been detected up to 200 K. We observe a small temperature dependence of the fine structure constant E and an unexplained temperature dependence of the ADMR signal intensity between 80 K and 200 K.

Introduction

Carotinoids play two known roles in photosynthesis. On one side they serve as additional light harvesting pigments, transfering their excitation energy to the BChl molecules of the antenna complexes [1]. On the other hand they play an important protective role. By efficently quenching the BChl triplet state, the carotinoids prevent the formation of very destructive singlet oxygen [2]. Due to the fast triplet - triplet transfer from BChl to carotinoid no BChl triplet state can be observed in carotinoid containing antennas.

The formation of the carotinoid triplet state in antennas has been investigated by flash photolysis experiments. The transient triplet - triplet absorption spectra and the triplet yield of the carotinoids have been determined [1]. Very recently high field ESR measurements on antenna complexes and model compounds have been published [3]. By comparison to the spectra of the model compounds, the ESR spectra that have been measured on the antenna complexes have been ascribed to the carotinoid triplet state.

Absorption detected magnetic resonance (ADMR) correlates the optical and magnetic properties of the triplet state [4]. Here we present new ADMR

Photosynthetic Light-Harvesting Systems

experiments which allow us to found the assignment of the triplet signals in the B800 - 850 antenna complex onto the known absorption spectra of the carotinoids.

Principles of ADMR

Due to the magnetic dipolar interaction of the two unpaired electrons, the triplet state splits into three sublevels even in the case of no external magnetic field. This zero field splitting (zfs) depends on the spatial distribution of the two electrons on the molecule of interest. The determination of the zfs parameters D and E can help to identify the molecule which carries the triplet state. On the other hand, changes of the zfs parameters under certain conditions indicate, that the geometry of the electronic distribution has changed.

Optically detected magnetic resonance in zero magnetic field is a powerful method to determine the zfs values D and E. The optical detection of magnetic resonance is based on the fact that the resonant microwave field changes the total triplet concentration. Concomitantly, the concentration of the molecules in the singlet ground state is changed as well. The changes of the populations can be monitored by the fluorescence, the singlet absorption or the triplet - triplet absorption.

The principle of the optical detection is illustrated in figure 1:

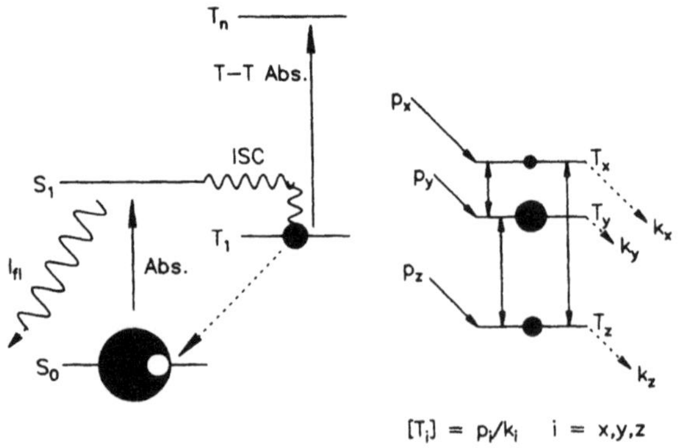

Fig. 1

Under continuous illumination the triplet state is populated. The relative population of the three triplet sublevels depends on the population - (p_i) and decay - (k_i) rates of each individual sublevel, with $n_i = p_i / k_i$ the population of the sublevel i. If spin lattice relaxation (SLR) can be neglected, the steady state distribution among the sublevels is far from thermal equilibrium and the triplet state appears spin polarized. When two sublevels are connected by resonant microwaves, population is transfered from one level to the other. If these sublevels differ in their decay rate constants, the total triplet concentration is changed. As an example: if transitions are induced from a slowly decaying level to a faster decaying level, the total triplet concentration will be lowered, the singlet concentration will increase and this increase will be reflected by an enhanced absorption of the sample.

If the SLR rates are comparable to or faster than the triplet lifetime, the spin polarization on the triplet state vanishes and no ODMR signals can be detected any more at high temperatures. Since the triplet states of BChl and carotinoids examined in photosynthesis research have rather short lifetimes, especially the carotinoid triplet state [1], spin polarization can be observed up to temperatures above 77 K [5,6]. On application of ADMR to the reaction center triplet it has been demonstrated very recently that ADMR is the method of choice to determine the temperature dependence of the zfs parameters far above nitrogen temperature [7].

The correlation between magnetic resonance and the optical absorption is achieved via the microwave induced absorption (MIA): the microwaves are fixed at the resonance frequency and amplitude modulated while the detection wavelength is scanned. By lock-in detection only the absorption of those molecules is monitored which contribute to the ADMR signal at the fixed microwave frequency.

Materials

The antenna preparation was a kind gift from Dr. R. J. Cogdell. It was prepared as described in [8]. For the ADMR measurements 0.1 ml of the preparation was diluted with 0.1 ml Tris (20 mM; pH 8) and 0.4 ml glycerol. The samples were stored at -20° C until they were used.

Results

Fig. 2 shows the absorption spectrum of the B800 - 850 antenna complex from Rps. acidophila 7050. The absorptions at 591 nm, 806 nm and 865 nm can be ascribed to the BChl molecules in this complex, whereas the absorptions at 467 nm, 495 nm and 531 nm stem from the carotinoid molecules.

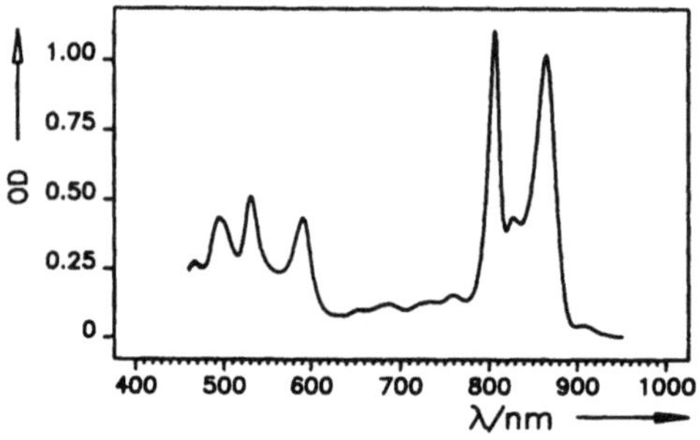

Fig. 2
Absorption spectrum of the B800 - 850 antenna complex from Rps. acidophila 7050 at T = 10 K.

Fig. 3 shows ADMR spectra, recorded at 11 K and 140 K. The detection wavelength was 558 nm. Two resonances could be detected. Both transitions are positive, corresponding to an increase in the transmission of the sample. The resonance frequencies are 185 MHz and 749 MHz at 11 K, they shift to 172 MHz and 757 MHz at 140 K, respectively.

The microwave induced absorption (MIA) spectra of the low frequency transition taken at 10 K and 140 K (fig. 4) have negative maxima at 466 nm, 494 nm and 531 nm. Positive bands appear at 519 nm and 556 nm. The position of the bands as well as their relative intensities do not change going from 10 K to 140 K.

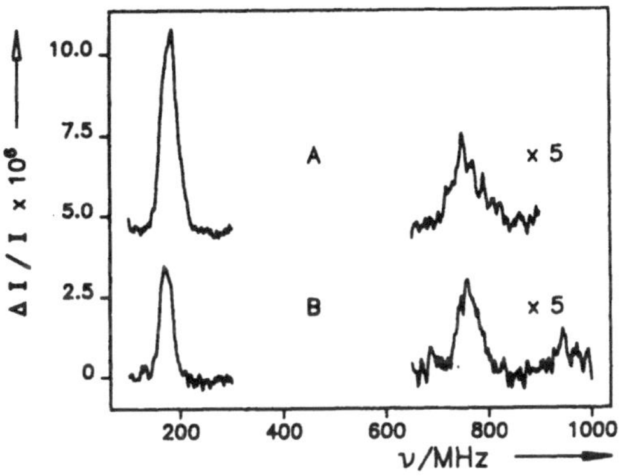

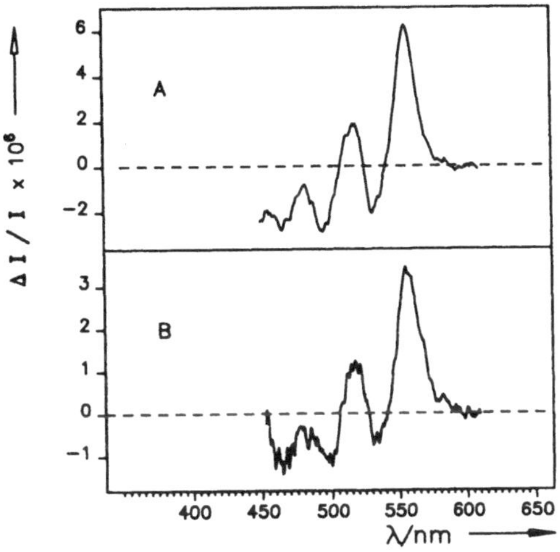

Fig. 3 (left)
ADMR spectrum of the B800 - 850 antenna complex from Rps. acidophila 7050. (a) T = 11 K; excitation by the white light from a 150 W tungsten-iodine bulb, filtered by 5 cm water; detection at 558 nm; RF power .8 W; sweeprate 15 MHz/sec; number of sweeps 74 (low frequency spectrum) and 628 (high frequency spectrum). (b) same as (a) except: T = 140 K; number of sweeps 62 and 576 respectivly. The high frequency spectra are magnified by a factor of 5. The scale at the left refers to the low frequency spectra and represents the relative change of the transmitted light intensity.

Fig. 4 (right)
Microwave induced absorption spectrum of the B800 - 850 antenna complex from Rps. acidophila 7050. (a) temperature 10 K; microwave frequency 184 MHz. (b) 140 K and 172 MHz respectivly . A positive band corresponds to an increase of the transmittance of the sample.

The temperature dependence of the low frequency ADMR signal exhibits a remarkable feature (fig. 5). From 8 K up to 25 K there is a strong decrease by a factor of 10. Between 30 K and 80 K the intensity remains unchanged at this low level. Above 80 K the intensity increases again. Between 110 K and 130 K the intensity reaches a maximum, which is 4 times higher than the level between 30 K and 80 K and which is as high as that at 12 K. At temperatures higher than 130 K the intensity falls continuously. Even above the melting point of the glass at about 200 K the ADMR signal is still detectable.

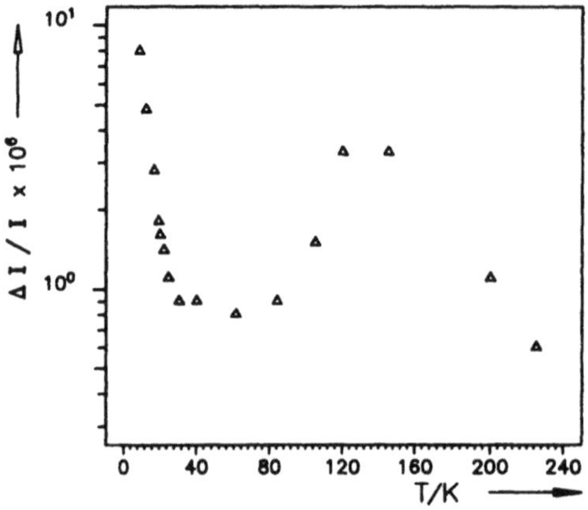

Fig. 5
Temperature dependence of the low frequency (2E) ADMR transition of the B800 - 850 antenna complex from Rps. acidophila 7050. Detection at 558 nm.

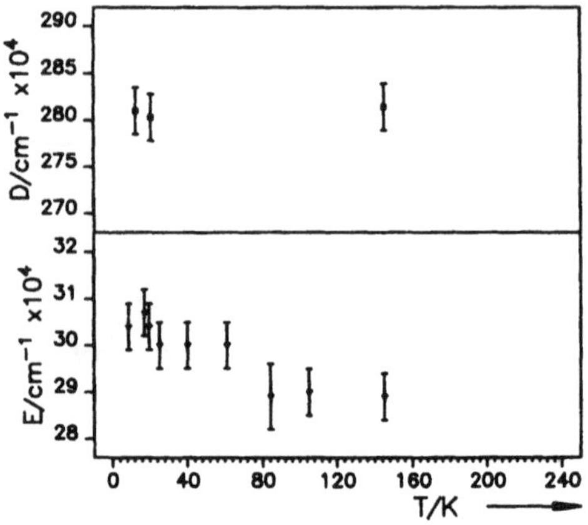

Fig. 6
Temperature dependence of the zero field splitting parameters from ADMR spectra of Rps. acidophila B800 - 850 antenna complexes. Detection wavelength 558 nm.

From the resonance frequencies the zfs constants D and E are calculated. D is $(281 \pm 2.5) \times 10^{-4}$ cm^{-1}. It remains unchanged within the limits of error in the whole temperature range (fig. 6). The zfs E value is $(30.2 \pm .5) \times 10^{-4}$ cm^{-1} in the temperature range from 8 K to 60 K. Above 80 K up to 150 K the E value is $(29.0 \pm .5) \times 10^{-4}$ cm^{-1}.

Discussion

For an identification of the triplet state, observed in the ADMR spectrum, we compare the absorption spectrum (fig. 1) of the sample with the MIA spectrum (fig. 4) of the low frequency (2E) transition. The negative bands, observed in the MIA spectrum at 466 nm, 495 nm and 531 nm are in excellent agreement with the carotinoid absorption bands visible in the absorption spectrum, the respective band of the antenna BChl would be at 591 nm. Thus it is unambigiously the carotinoid triplet state, what we see in the ADMR spectra. Consequently the positive bands in the MIA spectrum can be interpreted as T-T absorptions of the carotinoid. The signs of the ADMR signals indicate, that for both transitions the resonant microwave field reduces the triplet state concentration and by this enlarges the singlet absorption.

The zero field splitting of the triplet state shows a slight temperature dependence. The E value decreases going from 60 K to 100 K (fig. 6). This change in E might be caused by conformational changes in the antenna protein on raising the temperature. As E is a measure of the asymmetry of the electronic distribution, it is much more sensitive to changes in the surrounding than the D value, which reflects the distance between the two unpaired electron spins.

This change was not found in high field ESR measurements on the B800 - 850 antenna complex from Rps. acidophila 7750, which Frank et al. [3] have published very recently. The zfs values, computed from our ADMR resonance frequencies, correspond within the limits of error to those Frank et al. evaluated by a computer fit of their ESR 'powder' spectra. Concerning the slight change of the zfs E value we observed above 60 K, the difference between our results and those of Frank et al. may be explained by the higher accuracy of zero field ADMR in determining the zfs values.

The more striking temperature dependence is that of the ADMR intensity (fig. 5). According to the generally accepted model the following assumptions are made about the population and relaxation of the carotinoid triplet state:

- the carotinoid triplet decays via ISC directly to the carotinoid singlet ground state, i.e. no intermediate state is involved,

- the triplet sublevel decay rates are not temperature dependent and

- the triplet - triplet transfer from the BChl to the carotinoid is independent of temperature in the temperature range of interest.

The latter assumption is based on the finding that the carotinoid triplet yield in antennas is independent of temperature [9].

With this model spin lattice relaxation (SLR) is the only remaining temperature dependent process that influences the ADMR intensity. SLR destroys the spin polarization in the triplet state by diminishing the population difference between the triplet sublevels. As a result the ADMR intensity decreases. Applied to the temperature dependence of the ADMR intensity of the carotinoid triplet state, we interpret the decrease of the ADMR intensity above 130 K as controlled by increasing SLR rates. But the temperature dependence of the ADMR intensity below 130 K can not be explaind by SLR. To explain the strong decrease between 8 K and 25 K on one side and the increase of the ADMR intensity from 80 K to 110 K on the other side, further temperature dependent processes have to be introduced. Therefore the kinetic model of the carotinoid triplet state has to be extended and even processes have to be taken into account, which involve intermediate states in the population or the decay path of the carotinoid triplet state.

Acknowledgements

We would like to thank Dr. R. J. Cogdell for the kind gift of antenna preparations and valuable contributions to the discussion of the results. The assistance of R. Speer in taking the ADMR spectra is gratefully acknowledged. This work was supported by grants from the Deutsche Forschungsgemeinschaft (Wo 41/37-2).

References

[1] Cogdell, R.J., M.F. Hipkins, W. MacDonald, T.G. Truscott. 1981. Biochim. Biophys. Acta 634, 191-202.

[2] Krinsky, N.I.. 1971. In: Carotinoids (O. Isler ed.). Birkhauser Basel. pp. 670-716.

[3] Frank, H.A., B.W. Chadwick, J.J. Oh, D. Gust, T.A. Moore, P.A. Liddell, A.L. Moore, L.R. Makings, R.J. Cogdell. 1987. Biochim. Biophys. Acta 892, 253-263.

[4] Den Blanken, H.J., A.J. Hoff. 1982. Biochim. Biophys. Acta 681, 365-375.

[5] Frank, H.A., J. Machnicki, M. Felber. 1982. Photochem. Photobiol. 35, 713-718.

[6] Hoff, A.J., I.I. Proskuryakov. 1985. Chem. Phys. Lett. 115, 303-310.

[7] Ullrich, J., A. Angerhofer, J.U. von Schütz, H.C. Wolf. 1987. Chem. Phys. Lett., in press.

[8] Angerhofer, A., R.J. Cogdell, H.F. Hipkins. 1986. Biochim. Biophys. Acta 848, 333-341.

[9] Speer, R.. University Stuttgart. to be published

EFFECT OF LITHIUM DODECYL SULFATE ON B 800-850 ANTENNA COMPLEXES FROM RHODOPSEUDOMONAS ACIDOPHILA : A RESONANCE RAMAN STUDY

Bruno ROBERT * and Harry FRANK **

* : Service de Biophysique, Département de Biologie, C.E.N. Saclay, 91191 Gif sur Yvette CEDEX FRANCE

** : Department of Chemistry, 215 Glenbrook Road, University of Connecticut, U-60, Storrs, Connecticut 06268 USA

Introduction

In 1981, it was reported by Clayton and Clayton (1) that the 800 nm absorbance of the B 850-800 complexes from Rhb sphaeroides could be attenuated using Lithium Dodecyl Sulfate (LiDS), and that the 800 nm absorption peak was reinstated when LiDS was dialysed against LDAO. Kramer et al (2) extended this work and constructed a structural model for the B 850-800 complexes : they proposed that the B 850-800 minimal unit consists of regular aggregates of two BChl dimers, two BChl monomers and three carotenoids molecules noncovalently bound to two pairs of apoproteins. They also distinguished two spectrally distinct pools of carotenoids, one pool being associated with the BChl dimers. Chadwick et al (3) have argued recently that the LiDS-induced absorption changes in Rps acidophila may be brought about by a change in protein conformation. We present here a RR study of the B 850-800 complexes from Rps acidophila which supports this argument : moreover, examination of RR spectra of the carotenoids in the native and LiDS complexes allows the conclusion to be made that LiDS treatment induces a change in the configuration of the carotenoids bound to B 850-800 complexes.

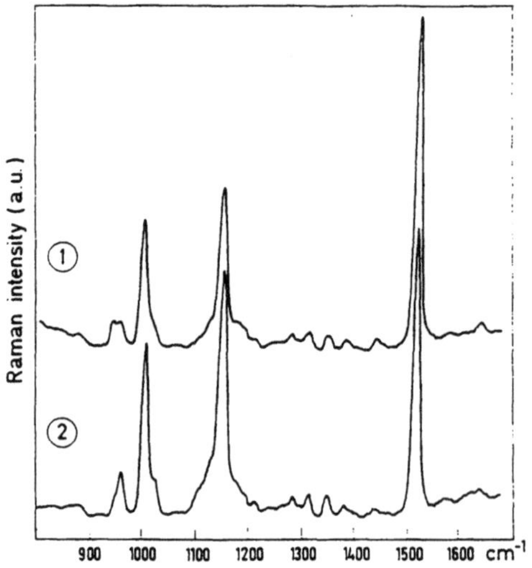

Fig 1 : RR spectra (488 nm excitation) of B 850-800 from Rps acidophila
1)after LDS treatment
2)before LDS treatment

Fig 2 : RR spectra (363.8 nm excitation) of B 850-800 from Rps acidophila
a) before LDS treatment (upper trace)
b) after LDS treatment (lower trace)

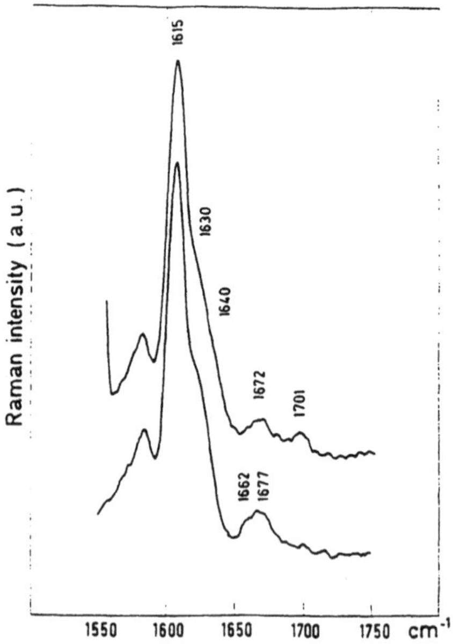

Effect of LDS on the Carotenoid

Fig 1 displays RR spectra of carotenoids bound to B 850-800 complexes from Rps acidophila before and after LiDS treatment (for experimental procedures see ref 3). Table 1 summarizes significant changes observed when treating these complexes : 2 cm^{-1} upshift of the 1517 cm^{-1} band reveals a change in the carotenoid configuration, affecting the end of the polyene chain ; upon association with the 3 cm^{-1} downshift of the 1004 cm^{-1} mode, it may be interpreted, considering the influence of the different isomerization of the beta carotene on its RR spectra (4), as a cis-trans isomerization around the 6-7 C=C double bond. On the other hand, the presence of a supplementary mode at 950 cm^{-1} demonstrates an out-of-plane distortion of the molecule. It has been calculated (5) that the 950 cm^{-1} mode mainly arises from out-of-plane vibrations of C-H bonds from the C_7-C_8 carbons. We thus conclude that LiDS also induces a twisting of the molecule at the end of the conjugated polyene chain. These changes of both the configuration and the conformation of the molecule may likely be responsible for the 4 nm upshift of the absorption peaks of the carotenoid observed during LiDS treatment. It is worth noting that these changes are observed at any excitation wavelength between 514.5 and 441.8 nm, i.e. that the carotenoids bound to B 850-800 complexes from Rps acidophila behave as a single pool.

TABLE 1

Positions of resonance Raman bands (cm^{-1} units).

Carotenoid			Bacteriochlorophyll		
untreated	*treated*	*Assigment*	*untreated*	*treated*	*Assigment*
-	950	ν_4	1614	1614	
960	960	ν_4	1630	1630	Acetyl C=O bound
1005	1002	ν_3	1640	-	
1150	1150	ν_2	-	1662	Acetyl C=O free
1517	1519	ν_1	1672	1669	bound keto
			-	1677	
			1701	-	free keto

Effect of LiDS on BChls

Fig 2 displays RR spectra of B 850-800 complexes extracted from Rps acidophila (type I) before and after LiDS exposure, excited near the top of the Soret transition of the BChl. Drastic changes occur in the 1620-1720 cm^{-1} C=O stretching region (see ROBERT et al, these proceedings). The 1640 and 1701 cm^{-1} disappears during LiDS treatment, and two new bands appear at 1662 and 1677 cm^{-1} . According to the frequency range in which keto and acetyl C=O respectively vibrate (6), it is possible to conclude that one keto group, free from intermolecular interactions before LiDS treatment (vibrating at 1701 cm^{-1}) is engaged in a H-bond during LiDS exposure, while there is a breakage of the intermolecular bond assumed by an acetyl C=O (that vibrates at 1640 and 1662 cm^{-1} respectively before and after LiDS treatment). LiDS thus affects the local environment of one BChl (two C=O vibrators). The fact that no difference appears in the frequency and in the width of the 1615 cm^{-1} band clearly indicates that the central Mg of this molecule is coordinated in a similar manner before and after LiDS treatment, which explains why this treatment is reversible. Fact that both C=O groups are changing their interactions is consistent with, and indicative of, a dramatic perturbation of the protein environment.

References

1) CLAYTON, R.K. and CLAYTON, B.J. (1981) Proc. Natl. Acad. Sci USA 78, 5583-5587

2) KRAMER, H.J.M., VAN GRONDELLE, R., HUNTER, C.N., WESTERHUIS, W. H.J. and AMESZ, J. (1984) Biochim. Biophys. Acta 765, 156-165

3) CHADWICK, B.W., FRANK, H.A., ZHANG, C., TAREMI, S.S. and COGDELL, R.J. (1987) in : Progress in Photosynthesis Research (BIGGINS, J. ed) Martinus Nijhoff, Dordrecht, The Netherlands

4) KOYAMA, Y., TAKII, T., SAIKI, K. and TSUKIDA, K. (1983) Photobiochem. Photobiophys. 5, 1939

5) SAITO, S. and TASUMI, M. (1983) J. Ram Spectrosc. 14-5, 310-318

6) LUTZ, M. (1983) in : Adv. in I.R. and Raman Spectrosc. (CLARK, R.J.H and HESTER, R.E. eds) Wiley Heyden, Chichester, U.K., Vol 11, pp 211-300

BChl a/b IN ANTENNA COMPLEXES OF PURPLE BACTERIA

Bruno ROBERT[*], André VERMEGLIO[**], Robert STEINER[+], Hugo SCHEER[+] and Marc LUTZ[*]

[*] : Service de Biophysique, Département de Biologie, C.E.N. Saclay, 91191 Gif sur Yvette CEDEX FRANCE

[**] : A.R.B.S., CEN Cadarache, B.P.1, St Paul Lez Durance,

[+] : Botanisches Institut der Universitat Munchen, D 8000 Munchen 19, GFR

Introduction

During the last five years, decisive progresses has been made in the knowledge of the structures of the protein-pigment complexes involved in the first steps of bacterial photosynthesis. However, this progress has largely concerned reaction centers (1), and has been slower for light harvesting complexes. As far as Rhodospirillales (purple bacteria) are concerned, the two subunits of many B 870 and B 850-800-type antenna complexes have now been sequenced (2). However, up until now, none of the several crystallization attempts yet allowed a highly diffracting crystals to be obtained. Resonance Raman spectroscopy is thus currently the only method able to provide information about the structure of these complexes. This technique indeed permits selective observation of the 'active site' of the antenna, i.e., the regions of the proteins which bind the BChl molecules and are responsible for their electronic absorption properties : studying the Raman-active vibration modes of BChl within the native light harvesting pigment-protein complexes indeed permits to characterize several of the interactions occurring between these pigments and their micro-environment, and thus gives some indications of the structures of BChl host sites within the protein. In this article, we will briefly summarize conclusions recently drawn from

Photosynthetic Light-Harvesting Systems

resonance spectroscopy of BChl a and b-containing complexes of Rhodospirillales.

Resonance Raman Spectroscopy of Antenna complexes from Rhodospirillales

Excitation of both BChl a and BChl b near the tops of their Soret transitions provides resonance Raman spectra which contain much structural information (3,4,5). The 1620-1720 cm^{-1} regions of the Raman spectra obtained in these conditions of resonance contain bands arising from the stretching modes of the 2-acetyl and 9-keto carbonyl groups, which are conjugated with the dihydrophorbin cycles of the molecules. The frequencies of these modes are sensitive to existence and strength of intermolecular bonds in which these chemical groups may be engaged (6) : typically, 2-acetyl groups of BChl a and b vibrates at 1665 cm^{-1} and 1670 cm^{-1} respectively when free from intermolecular bonding. These frequencies may shift down to ca 1630 when the groups assumes hydrogen bonding (6). The 9-keto groups of both BCHl a and b vibrates near to 1700 cm^{-1} when free from intermolecular interactions, and their stretching frequency may shift down to 1660 cm^{-1} when they are hydrogen bonded (6). Moreover, in the same spectral region, the frequency of a band located near to 1600 cm^{-1} , and which mainly arises from methine bridge stretching, is sensitive to the number of axial ligands that bind the central Mg of the molecule for both BChl a and b (6). This band indeed occurs at 1600 cm^{-1} when two axial ligands are bound to the Mg atom, and around 1615 cm^{-1} when it is five-coordinated. Thus, spectral analysis of the 1550-1750 cm^{-1} region brings information about the interaction states of the central Mg and of both the conjugated C=O of BChl a or b. These groups are known to be predominantly involved in intermolecular interactions of chlorophylls both in vivo and in vitro (6).

BChl a-containing Complexes

Core antenna (B 890,870 complexes)

The very close similarity of resonance Raman spectra of core antenna complexes from any species of Rhodospirillales that wa have studied so far indicates that, in all these complexes, the proteic host sites of BChl are extremely similar, most probably providing the same binding groups to the conjugated carbonyls of BChl a molecules (3). For more than ten species and strains (Rhb sphaeroides, capsulatus, Rps palustris, acidophila, Rsp rubrum, Chromatium vinosum Thiocapsa roseopersicina) the RR spectra of the core antenna obtained at Soret resonance are extremely similar to that presented in fig 1a. The strong 1645 cm^{-1} band most probably arises from both of the acetyl carbonyl groups of both of the BChl a molecules known to be present in B 870-type complexes. These groups thus assume similar if not identical intermolecular interactions. The frequency of this band is species-dependent, being up or downshifted by a few wavenumbers from the 1645 cm^{-1} mean value. These variations likely arise from slight variations in the environment or geometry of the intermolecular C=O --- X bond rather than from a variability in the nature of the site X. Two bands are observed in the keto stretching region of RR spectra : interactions assumed by keto groups of both BChl in B 870-type are thus not the same. However both of them are most probably H-bonded, vibrating respectively at 1667 and 1678 cm^{-1} . The 1615 cm^{-1} frequency of the methine bridge stretching mode clearly indicates that, in B 880-type complexes as well as in any BChl-containing antenna complexes studied to date (3,4,5), the central magnesiums of the BChls are five-coordinated.

Peripheric antenna (B 850-800 or 820-800-type complexes)

From the BChl host-site structures revealed by RR spectroscopy at least two families must be distinguished among the B 850-800-type complexes : intermolecular interactions assumed by the three BChl in B 850-800 complexes extracted from Rhodospirillaceae indeed differ from those extracted from Chromatiaceae (fig 2) (3). In RR spectra of B 850-800 complexes from Chromatiaceae, the C=O stretching modes consitute as a complex cluster located around 1656 cm^{-1}, with a shoulder at 1671 cm^{-1}, and involving very weakly bound acetyl and H-bonded keto groups. RR spectra of the Rhodospirillaceae complexes, exhibit resolved bands at 1633, 1641, 1666-1670 and occasionally 1700 cm^{-1}.

Moreover, intergeneric variability has been observed among the RR spectra of B 850-800 complexes from Rhodospirillaceae thus revealing intergeneric structural variability (4): in RR spectra of the B 850-800 complexes extracted from bacteria of the Rhodobacter genus (sphaeroides, capsulatus), the keto C=O group of the 800-nm absorbing BChl is free from intermolecular bonding, vibrating at 1700 cm^{-1} whereas it is intermolecularly bound in B 850-800 complexes of bacteria from the Rhodopseudomonas genus (palustris, acidophila, type II complexes). This intergeneric variability is also observed for the acetyl groups of the 800 nm-absorbing and one of one of the 850 nm-absorbing BChls. In this case only changes in the strengths of the intermolecular interactions of these groups are observed. Two types of B 850-800 complexes have been extracted from Rps acidophila (type I and type II (7)). Host sites of BChls within type I complexes are very similar to those of B 850-800 complexes extracted from the Rhodobacter genus (B. ROBERT and R.J. COGDELL, unpublished results).

B 830-800 complexes (or 'high 800nm-absorbing complexes') have been purified from bacteria of the Rhodopseudomonas genus grown in low light. These complexes exhibit absorption spectra very similar to those of B 850-800 complexes extracted from Chromatiaceae. However, on the basis of the structures of the BChl host sites, revealed by RR spectroscopy, these complexes appear to be actually related to 'regular' B 850-800 complexes of Rhodospirillaceae, genus Rhodopseudomonas, from which they appear to only differ by the presence of an additional BChl molecule (4).

Fig 1 : RR spectra (363.8nm excitation) of
a)(lower trace) B 880 from Rps acidophila
b)(middle trace)B 1015 from Rps viridis
c)(upper trace) B 910 from Chromatium tepidum

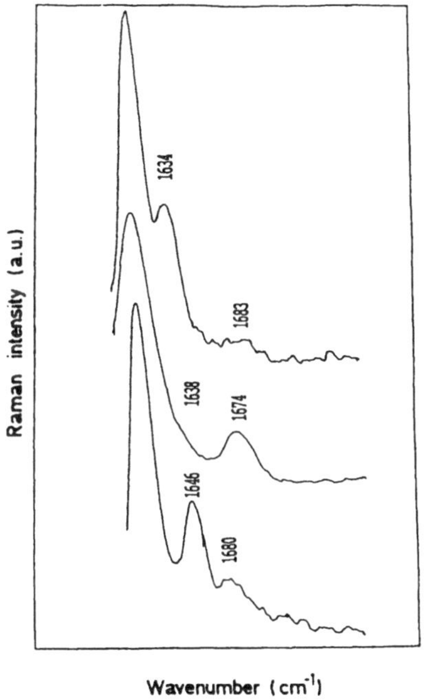

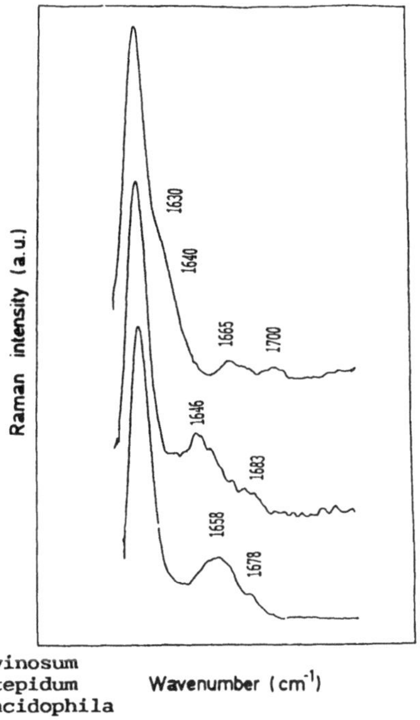

Fig 2 : RR spectra (363.8 nm excitation) of
a)(lower trace) B 850-800 from Chromatium vinosum
b)(middle trace) B 850-800 from Chromatium tepidum
c)(upper trace) B 850-800 type I from Rps acidophila

The thermophilic bacterium Chromatium tepidum

We have studied antenna complexes extracted from the thermophilic bacterium Chromatium tepidum : this bacterium synthesizes both a core antenna (B 920) and a peripheric one (B 855-800) which are thermostable up to 70 C in the intact membrane (8). In both these complexes, the central magnesiums of all of the BChl molecules are five-coordinated, as in the mesophilic strains. On the other hand, in both of these complexes, a ca 13 cm^{-1} downshift of the whole set of frequencies of the acetyl carbonyl vibrators is observed, indicating that these groups are all engaged in stronger intermolecular interactions than in the mesophilic Chromatium vinosum strain. At such wavenumbers, a 13 cm^{-1} downshift of a C=O vibrator approximately corresponds to an increase in H-bond strength of 2 kcal/M. In B 920 complexes, the keto carbonyl stretching frequencies are also downshifted (16 cm^{-1}) relative to the mesophilic strain. Hence, in antenna complexes extracted from this thermophilic strain, proteic host sites appear to provide BChl molecules with noticeably stronger interactions on most of their conjugated carbonyls.
Following suggestion by BABCOCK and CALLAGHAN (9) it would be tempting to relate the 880->920 nm absorption difference to the difference in ligation strength of the carbonyl groups of the BChls in the core complexes of the mesophilic and thermophilic species. Nevertheless, the liganding of these groups (at least of the acetyl ones) appears to have only a limited influence on the electronic absorption of the pigments since the peripheric antennae of both tepidum and vinosum species exhibit the same absorption spectra, whereas the ligation strength of the acetyl carbonyl of their BChl a also differ by hte same amounts.

BChl b-Containing Antenna Complexes

Core antenna of Rps viridis (B 1015)
RR spectra of Bchl b-containing antenna indicate that, as in the case

of BChl a-containing complexes, the central Mg of the pigments within these complexes generally are singly liganded (methine bridge frequency at 1615 cm^{-1}). However, the ligation of BChl b in B 1015 from Rps viridis appears to be different from that of BChl a in B 880-type complexes : indeed, in the carbonyl stretching region of the former complex, a weak 1639 cm^{-1} arises from a H-bonded acetyl C=O, a strong 1670 cm^{-1} most probably arises from a free acetyl group and from a H-bonded keto grouping, and a 1682 cm^{-1} , band arises from a weakly interacting keto carbonyl (fig 1c).

B 800-1020 complex from Ectothiorhodospira halochloris

B 800-1020 complex from Ectothiorhodospira halochloris is a BChl b-containing core antenna. Biochemical analysis (10) indicated that it contains five BChl b, two of them being responsible for the 1015 nm absorption peak. RR data largely confirm these hypothesis : at least four unequivalently interacting BChl b are present in these complex, as far as 8 different frequencies may be observed in the C=O stretching region. Moreover, HCl-treatment, which results in shifting the 1015 nm transition down to 960 nm (10), affects the stretching vibrations of no more than two acetyl and two keto carbonyl groups.

Overview

RR spectroscopy shows that, although the sequences of antenna complexes of purple bacteria appear to be highly conservative (2), interactions assumed by BChls in many of these complexes are largely species- (or genus-) dependent. Despite this interspecific variability, antenna complexes of purple bacteria share common structural features at their active sites :

- all BChl molecules have singly liganded Mg. From both sequence data and RR spectra (3), these ligands are most probably imidazole side chains of histidines in most cases. The 800 nm-absorbing BChl of Rhb sphaeroides B 850-800 complexes may constitute an exception in this respect.

- In all of these complexes, BChl appears not to be directly exposed to the lipidic phase or aqueous detergent : indeed, RR spectra are insensitive to the isolation procedures, to the nature of the detergent as well as to the size of the complex multimers. Moreover, when direct comparison are possible (Rsp rubrum) RR spectra of the membrane-bound and extracted complexes are identical.

Now that several sequences are available for (2) antenna complexes of purple bacteria hypotheses can be drawn from RR spectroscopy about the nature of the aminoacids interacting with the pigments For example, B 1015 complex from Rps viridis is the only core antenna for which no 1645 cm^{-1} RR band is observed but for which 1639 cm^{-1} and 1670 bands are observed. This specificity may well be related to the fact that its polypeptides each have specific amino acid changes relative to all of the other polypeptides of core antenna : phenylalanine -8 (taking the conserved His as 0 position) in the alpha polypeptide is replaced in viridis by a tyrosine which could provide a H-bond to the C=O acetyl group so that its frequency could be downshifted to 1639 cm^{-1} ; in the beta one the +6 Trp is also replaced by an Ala. These aminoacids are thus likely candidates as binding sites of acetyl C=O groups. Similarly, when more antenna complex polypeptides are sequenced, it will most probably be possible to locate the BChl molecules in the polypeptide matrix taking into account both orientations of electronic transition moments and RR data.

References

1) MICHEL, H., EPP, O. and DEISENHOFER, J. (1986) EMBO J. 5-10 2445-2451

2) ZUBER, H., SIDLER, W., FUGLISTALLER, P., BRUNISHOLZ, R. and THEILER, R. (1987) Trends in Biochem. Sc., 11, 414-419

3) ROBERT, B. and .LUTZ, M. (1985) Biochim. Biophys. Acta 807, 10-23

4) ROBERT, B. ANDRIANAMBININTSOA, S. and .LUTZ, M. (1985) Biophys J. (Tokyo), 100, 6-11

5) ROBERT, B., STEINER, R., ZHOU, Q., SCHEER, H. and LUTZ, M. (1987) in : Progess in Photosynthesis Research (BIGGINS, J. ed) Martinus Nijhoff, Dordrecht, The Netherlands

6) LUTZ, M. and ROBERT, B. (1987) in Biological applications of Raman spectroscopy (SPIRO, T.G. ed) Vol III Chap 9

7) COGDELL, R.J., DURANT, I., VALENTINE, J., LINDSAY, J.G. and SCHMIDT, K. (1983) Biochim. Biophys. Acta 722, 427-455

8) MADIGAN, M.T. (1984) Science 225, 313-315

9) CALLAGHAN, P.M. and BABCOCK, G.T. (1983) Biochemistry 22, 452-461

10)STEINER, R (1985) Thesis, Ludwig Maximilian University (Munchen)

BACTERIOCHLOROPHYLL C AGGREGATES IN CARBON TETRACHLORIDE AS MODELS FOR CHLOROPHYLL ORGANIZATION IN GREEN PHOTOSYNTHETIC BACTERIA

J. M. Olson and J. P. Pedersen

Institute of Biochemistry, Odense University
Campusvej 55, DK-5230 Odense M, Denmark

Introduction

Energy capture in green photosynthetic sulfur bacteria begins with the absorption of light by bacteriochlorophyll (BChl) c, which is localized in oblong bodies (chlorosomes) appressed to the inside surface of the cytoplasmic membrane [1,2]. Excitation energy is transferred from BChl c to BChl a in the chlorosome [3], then to BChl a in the baseplate between the chlorosome and the cytoplasmic membrane, and finally to the reaction centers embedded in the membrane [1].

Our working hypothesis based partly on absorbance and circular dichroism (CD) spectra is that the BChl c in the chlorosome is organized in aggregates of 10-12 strongly interacting molecules which are anchored to specific sites on specific proteins [4]. The major protein in the chlorosome is a 7.5-kDa polypeptide containing 74 amino acid residues [5]. *In vitro* BChl c aggregates without protein have also been investigated, because the absorption spectrum maximum of BChl c in chlorosomes (740-750 nm) is so close to that of BChl c aggregates (747 nm) in CCl_4, CH_2Cl_2 and CH_2Cl_2-hexane [6,7]. A fundamental relationship between the two systems is inferred.

BChl c from the green sulfur bacterium *Chlorobium limicola* f. *thiosulfatophilum* is a mixture of 4 homologs of farnesyl BChl c

Photosynthetic Light-Harvesting Systems

(F-BChl c): 4-ethyl-5-methyl F-BChl c (EMF-BChl c), 4-ethyl-5-ethyl F-BChl c (EEF-BChl c), 4-n-propyl-5-ethyl F-BChl c (PEF-BChl c), and 4-isobutyl-5-ethyl F-BChl c(iBEF-BChl c). The exact proportion of the four components can vary with the age of the bacterial culture, but the two main components are usually EEF-BChl c and PEF-BChl c [8].

Materials and Methods

F-BChl c was extracted from wet cell of Cb. limicola f. thiosulfatophilum 6230 (Tassajara) with methanol/diethyl ether/petroleum ether (5:2:1). The four homologs were isolated by isocratic HPLC using 75% acetone with an RP18 column (9). Each homolog was dried and then washed 3 times with CCl_4 before use.

Absorption spectra were recorded with a Perkin-Elmer 330 recording spectrophotometer; fluorescence emission and excitation spectra were recorded with a Spex Fluorolog 111A in the front-face-emission mode; and CD spectra were recorded with a Jobin Yvon dichrograph. Cuvettes with the following light paths were used: 0.02, 0.10, 0.50, 1.00, 2.00, 5.0, and 10.0 cm.

Results

Preliminary experiments on concentration effects were carried out with mixtures of all 4 homologs. At very low concentrations (ca. 1 µM) in CCl_4 only one F-BChl c species could be detected. This species (I), thought to be monomeric F-BChl c, was characterized by absorption and fluorescence peaks both at 666 nm. At higher F-BChl c concentrations (7-90 µM) additional species could be detected: Species II, characterized by absorption and fluorescence maxima at 680 and ca. 692 nm respectively, had a positive CD band at 682 nm (10). Species III was characterized by broad absorption bands at 633 and 705 nm, a shoulder at 720 nm, and fluorescence emission at 730 nm. The corresponding CD bands were positive at 632 nm and negative at 706 nm (10). As the F-BChl c concentration was raised from 7 to 40 μM the mole fraction of III was about doubled, while the other mole fractions changed relatively little.

In the experiments described above it was not possible to determine which homologs participated in the formation of the three spectroscopic species (I-III). Therefore the absorption spectrum of each homolog in CCl_4 was examined. Each of the first 3 homologs showed a monomer peak at 668-673 nm (I) and an aggregate peak at ca. 705 nm (III) as shown in Fig. 1. The fourth homolog (iBEF-BChl c) showed an additional peak at 743-747 nm shown in Fig. 2 due to the formation of a new aggregate (IV).

Absorption and fluorescence spectra of the third homolog (PEF-BChl c) in CCl_4 were recorded for BChl c concentrations between 0.1 and 770 μM, and the results are shown in Table 1. CD spectra (not shown) were also recorded over a similar concentration range. Our interpretation of the data concerning PEF-BChl c follows. At very low BChl c concentrations (ca. 0.1 μM) the solution contains only one species, the monomer (I) characterized by absorption peaks at 434 and 666 nm, and a fluorescence peak at 668 nm. At somewhat higher BChl c concentrations (ca. 1 μM) a second species (II) appears with absorption and fluorescence peaks at ~685 and ~690 nm respectively. Species II has a conservative CD signal at ~661 nm (−) and 682 nm (+) and may be a BChl c dimer. As the BChl c concentration is raised to 5 μM, a third species (III) appears with broad absorption bands at ~450, ~630 and 705-710 nm and a

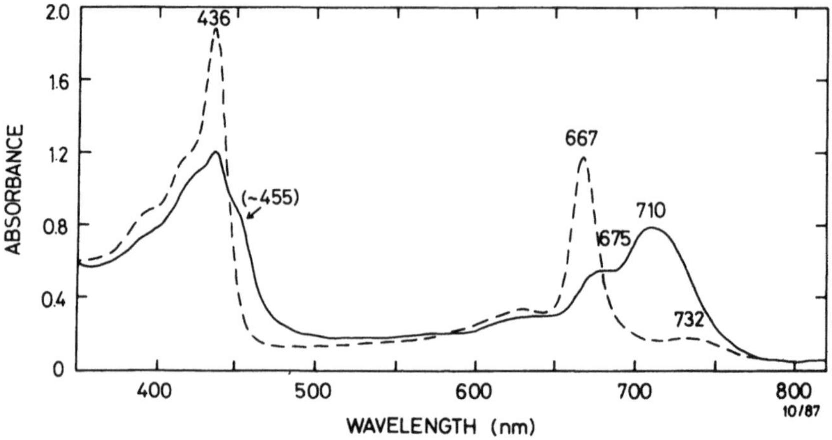

Fig. 1. Absorption spectrum of PEF-BChl c (770 μM) in CCl_4 before (solid line) and after (dashed line) addition of methanol (1:200). Light path = 0.02 cm.

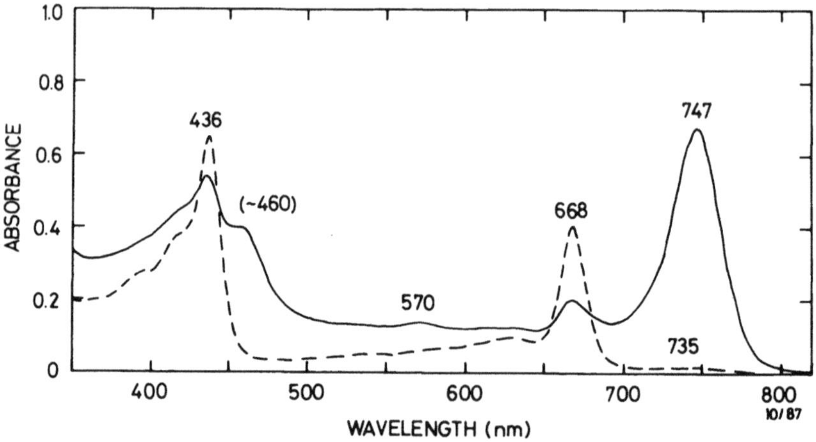

Fig. 2. Absorption spectrum of iBEF-BChl c (106 μM) in CCl_4 before (solid line) and after (dashed line) addition of methanol (1:200). Light path = 0.10 cm.

Table 1. Absorption and Fluorescence Characteristics of PEF-BChl c in CCl_4. Parentheses denote shoulders. All wavelengths are given in nm. nd = not determined.

Spectrum	BChl c concentration (µM)				
	0.1	1	12	100	770
Absorption	434	434	434	435	436
	-	-	(~455)	(~455)	(~455)
	666	666	675	(675)	(~675)
	-	(~685)	-	-	-
	-	-	705	708	710
$\frac{A710-A800}{A675-A645}$	0	0	1.4	2.2	2.4
Emission					
λ_{ex} = 436	nd	668	668	(~666)	(~666)
	nd	(~687)	~690	(~684)	-
	nd	(~723)	~738	740	741
λ_{ex} = 460	nd	689	(~688)	-	-
	nd	(~734)	738	740	740
Excitation					
λ_{em} = 660	nd	433	434	433	434
λ_{em} = 690	nd	434	434	~436	nd
λ_{em} = 740	nd	nd	~450	~450	~450
	nd	nd	nd	~631	~631
	nd	nd	nd	(~676)	(~675)

Table 2. Absorption and Fluorescence Characteristics of iBEF-BChl c in CCl_4. Parentheses denote shoulders. Wavelengths in nm. nd = not determined. *precipitating.

Spectrum	BChl c concentration (µM)				
	2	14	55	106	360*
Absorption	436	437	435	436	436
	-	(~460)	(~460)	(~460)	(~460)
	~620	625	~620	~630	-
	667	667	667	668	668
	(~685)	(~685)	(~685)	(~680)	nd
	-	742	743	747	747
(A747-A850) / (A668-A645)	nd	1.2	2.2	8.1	8.6
Emission					
λ_{ex} = 436	668	668	668	668	668
	(724)	(725)	(725)	(724)	(725)
λ_{ex} = 460	~678	(680)	~676	nd	~680
	~686	(690)	(~690)	nd	-
	~734	732	~730	nd	-
Excitation					
λ_{em} = 667	nd	434	434	434	435
		629	nd	nd	nd
λ_{em} = 690	nd	434	435	436	nd
λ_{em} = 730[a]	nd	[b]434	[a]434	nd	[c]434(weak)
750[b]		(~450)	(~450)	nd	-
760[c]		~630	~630	nd	-
		~670	~666	nd	-
		~705	nd	nd	-

conservative CD signal at ~630 nm (+) and 706 nm (-). Species III is highly fluorescent with peak emission at ~740 nm. As the BChl c concentration is raised above ca. 10 μM up to ca. 800 μM species III becomes the major species in solution (Fig. 1). We suppose that species III is a higher aggregate (n > 2) of PEF-BChl c.

Absorption and fluorescence spectra of iBEF-BChl c in CCl_4 were recorded for BChl c concentrations between 2 and 360 μM, and the results are shown in Table 2. Our interpretation of this data follows. At low BChl c levels (ca. 2 μM) the solution already contains two spectral species (I and II). Species I is characterized by absorption peaks at 434 and 667 nm, and a fluorescence emission peak at 668 nm. Species II has absorption and emission bands at 685 and 690 nm respectivley. As the BChl c concentration is raised to 14 μM, two new species (III and IV) appear. Species III has absorption bands at 450 and 705 nm, and an emission band at 730 nm. Species IV has absorption bands at 460 and 742-747 nm, but no detectable fluorescence emission. Species IV has a conservative CD signal at 737 nm (+) and 755 nm (-), similar to that for the mixed BChl c aggregate in CH_2Cl_2-hexane (4). As the BChl c concentration increases to 55 μM and beyond, species IV becomes the dominant species in solution (Fig. 2). Species III essentially "disappears" at very high BChl c levels (360 μM). We suppose that species II and III of iBEF-BChl c are similar to the corresponding species II and III of PEF-BChl c. Species IV appears to be a higher aggregate (n > 10) of iBEF-BChl c, which resembles in its absorption characteristics the F-BChl c antenna in the chlorosome of *Chlorobium*.

Some questions which remain to be answered are the following:

1. Is species II a dimer?

2. Are the aggregate species III and IV aggregates of monomers or dimers?

3. Are the aggregates formed in CCl_4 related to the BChl c-protein complexes in chlorosomes?

Acknowledgement

We thank Steen G. Melberg (Novo Research Institute) for generously making available a dichrograph for recording CD spectra.

References

1. Olson, J.M. 1980. Biochim. Biophys. Acta 594, 33.

2. Gerola, P.D., J.M. Olson. 1986. Biochim. Biophys. Acta 848, 69.

3. van Dorssen, R.J., P.D. Gerola, J.M. Olson, J. Amesz. 1986. Biochim. Biophys. Acta 848, 77.

4. Olson, J.M., P.D. Gerola, G.H. van Brakel, R.F.Meiburg, H. Vasmel. 1985. In: Antennas and Reaction Centers of Photosynthetic Bacteria (M.E. Michel-Beyerle, ed.) Springer-Verlag, Berlin, p. 67.

5. Gerola, P.D., P. Højrup, J. Knudsen, P., P. Roepstorff, J. M. Olson. 1988. In: Green Photosynthetic Bacteria (J. M. Olson, ed.) Plenum Press, New York, in press.

6. Krasnovsky, A.A., M. I. Bystrova. 1980. BioSystems 12, 181.

7. Smith, K.M., L. A. Kehres, J. Fajer. 1983. J. Am. Chem. Soc. 105, 1387.

8. Caple, M.B., H. Chow, C.E. Strouse. 1978. J. Biol. Chem. 253, 6730.

9. Brune, D.C., T. Nozawa, R.E. Blankenship. 1987. Biochemistry, (in press).

10. Olson, J.M., G.H. van Brakel, P.D. Gerola, J.P. Pedersen. 1986. In: Progress in Photosynthesis Research (J. Biggins, ed.) Martinus Nijhoff, Dordrecht, p I.3.341.

ORIENTATION OF THE PIGMENTS IN THE REACTION CENTER AND THE CORE ANTENNA OF PHOTOSYSTEM II

J. BRETON and J. DURANTON

Service de Biophysique, Département de Biologie, CEN Saclay, 91191 Gif-sur-Yvette cedex, France

K. SATOH

Department of Biology, Faculty of Science, Okayama University, Okayama, Japan

The photosynthetic membrane of green plants, cyanobacteria and algae is made up of a mosaic of chlorophyll-protein complexes differing in their composition and serving various functions. Some of these complexes serve only as light-harvesting antenna while others contain the Photosystem I or Photosystem II (PS II) reaction centers together with a core of associated antenna pigments. The specific arrangement of the pigment molecules inside the complexes determines the high photosynthetic efficiency ; thus the description of these macromolecular structures is an important step towards the understanding of the primary reactions.

The primary photochemistry occuring in the PS II is currently thought to involve a charge separation between the primary donor chlorophyll(s) of P 680 and a pheophytin acceptor. This event is followed by a stabilization step where the electron moves from the pheophytin to the first quinone acceptor (Q_A). With respect to these primary reactions, PS II bears striking analogies with purple photosynthetic bacteria (1). In purple bacteria, the isolation and purification of a functional reaction center complex lacking its associated antenna has greatly facilitated our understanding of the structural and energetic factors governing these primary reactions. While the spatial organization of the bacterial reaction center chromophores has been inferred from various optical and EPR spectroscopic measurements, the recent crystallographic determination of

Photosynthetic Light-Harvesting Systems

the structure of the reaction center from Rhodopseudomonas viridis at 3A resolution has provided an invaluable picture of the organization of the constituent pigments and proteins (2).

Up to very recently the purification of a minimal PS II reaction center comparable to that of photosynthetic purple bacteria had not been achieved and the photochemical activity was associated with a specific chlorophyll- protein complex bearing typically 40 chlorophyll a per P 680. Such PS II particles, which contain the functional components involved in the charge separation and stabilization steps, consist of five protein subunits with apparent molecular weights of about 47, 40, 30-34 and 9-10kDa (3-11). The two large polypeptides of 47 and 40 kDa carry antenna chlorophyll a, whereas the 9-10 kDa polypeptide constitutes the apoprotein of cytochrome b_{559}. The two polypeptides in the 30-34 kDa range are named D_1 and D_2 and, in view of the analogies in the primary sequences of these polypeptides with those of the L and M subunits which constitute the scaffold of transmembrane alpha-helices holding together the essential cofactors in the reaction center of purple bacteria, it has been suggested that D_1 and D_2 play a similar role in forming the reaction center of PS II (12,13).

We have investigated by polarized light absorption and emission spectroscopies at cryogenic temperatures the organization of the pigments in small unfractionated PS II particles from Chlamydomonas (14) and Synechococcus (15). Such studies can reveal useful relationships between the presence of some spectral components, the polypeptide composition and the biological function associated with each of these particles. Amongst the various observations reported in these studies, a correlation was established between the magnitude of a small dip around 691nm in the 5 K linear dichroism spectra and the amplitude of the characteristic PS II emission band at 695 nm (F 695). Because the orientation of the F 695 emission, rather perpendicular to the plane of the thylakoid membrane (16-18), is both different from that observed for the Q_Y absorption and emission dipoles of most of the chlorophylls in vivo and identical to the orientation of the Q_Y transition of the pheophytin acceptor (19), it has been hypothesized that F 695 was emitted directly from this last molecule (20). The negative linear dichroism signal around 691nm (14,15,21) corresponding also to dipoles oriented rather perpendicular to the membrane

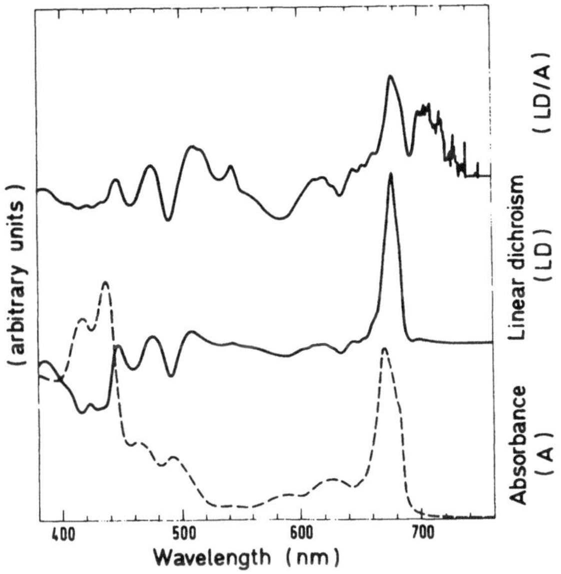

Fig. 1 : Absorption (A), linear dichroism (LD) and LD/A spectra of the PS II core complex isolated from spinach according to (6). The particles were oriented in squeezed polyacrylamide gel and the spectra were recorded at 10K.

plane, it has been further proposed that this spectral feature could also be assigned to this pheophytin molecule (21). This negative LD 691 signal, although somewhat variable in wavelength and amplitude, is present in the spectra at 10K of all the PSII core particles investigated so far, including the first isolated particle prepared from spinach (6) as shown in Fig. 1.

Several laboratories have reported the fractionation of PS II core particles from various organisms into two smaller chlorophyll-protein complexes containing one or the other of the two large polypeptides (4-11). In general the PS II activity, although significantly reduced compared to that of the native particles, was found in association with the complex carrying the largest polypeptide (5,7,10). Using a thermophilic strain of the cyanobacterium Synechococcus, Yamagishi and Katoh have been able to isolate and characterize two such subparticles, named CP2-b and CP2-c, which exhibit a markedly increased stability compared to that of their counterparts isolated from green plants and algae (7-9). CP2-c contains only the 40 kDa chlorophyll a-binding polypeptide, is lacking pheophytin and exhibits no PS II activity, while CP2-b which contains the 47,30-34 and

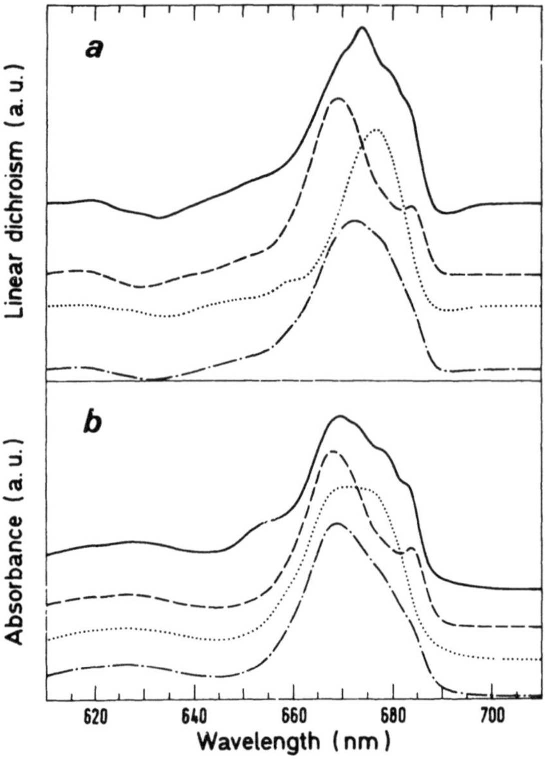

Fig. 2 : Absorption (a) and linear dichroism (b) spectra at 10 K of the core complex (——), the photochemically active CP2-b (...) and the photochemically inactive CP2-c (---) subparticles isolated from the thermophilic cyanobacterium Synechococcus. (-.-.-) linear combination of the spectra of the CP2-b (50 %) and CP2-c (50 %) subparticles. The spectra are taken from ref. 15.

9-10 kDa polypeptides but not the 40 kDa subunit, is able to photoreduce Q_A and to photoaccumulate reduced pheophytin (9). Using these two subparticles, the absorption and linear dichroism at 10K of the core PS II particles could thus be compared to those of its two main constituent chlorophyll-protein complexes (15).

In general, the linear dichroism and absorption spectra of each of the two subparticles (Fig. 2) appear significantly simpler (less spectral components) than the spectra of the core particle. Furthermore, the various features seen in the spectra of the core PS II particle appears in the spectra of either one of the two subparticles. This is also evident when comparing the spectra of the intact particle to the corresponding sum of the spectra of the two smaller complexes (Fig. 2). Apart from the cytochrome b_{559}, the 655-nm component assigned to the allophycocyanin present in the intact particle but absent in the two subparticles and the amplitude of LD 691 relative to the amplitude at the red maximum, the calculated sum spectra resemble much more closely the spectra of the intact particle than those of the two smaller subparticles. In fact the calculated spectra with their rather smooth shoulders very closely resemble the spectra of a core PS II particle measured at 100 K (J.Breton, unpublished results) and it appears as if the SDS treatment used to split the two subparticles had slightly loosened or altered the binding sites of the pigments. The observation that the general shape of the linear dichroism and absorption spectra of the intact PS II particle can be closely mimicked by the algebraic sum of the corresponding spectra of CP2-b and CP2-c clearly implies that the orientation axis of these complexes in the gel is the same whether they are attached to or separated from each other. In previous studies (14,21,22), we have demonstrated that the isolated intact PS II particles behave in the gel like flattened ellipsoids, the normal of which (oriented along the gel squeezing direction) coincided in vivo with the normal to the thylakoid plane. This is further verifed in (15) by using the characteristic S-shape of the cytochrome b_{559} linear dichroism signal with its negative lobe at 558 nm and positive lobe at 554 nm, which is identical for the native particle (15) and for the intact thylakoids (23). We thus conclude that in the gel the isolated CP2-b and CP2-c complexes adopt the same orientation with respect to the squeezing direction as they do when they are still embedded in the photosynthetic membrane. Furthermore, while the use of a one-to-one contribution of CP2-b and CP2-c to calculate the absorption spectrum is justified by the similar chlorophyll a content of the two subparticles (9), the identical coefficients used for the two normalized linear dichroism spectra to obtain a satisfactory fit of the linear dichroism spectrum of the intact complex most probably indicate a rather homogenous orientation of the chlorophylls in the two subparticles.

In agreement with our previous report (14) that two populations of carotenoids with dichroism of opposite sign and with slightly different absorption maxima were present in the intact PS II particles isolated from Chlamydomonas, we have confirmed this observation in the case of core particles isolated from spinach (Fig. 1) or from Synechococcus (15) and furthermore demonstrated that CP2-c contains the carotenoid pool absorbing

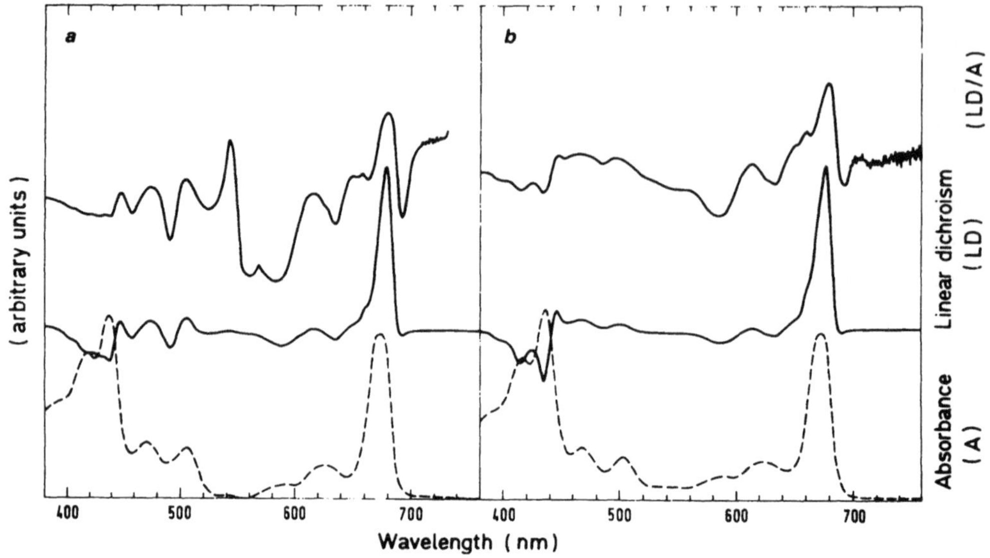

Fig. 3 : Absorption, linear dichroism and LD/A spectra at 10K of the photochemically active CP2-b complex from Synechococcus (a) and of the photochemically inactive CP 47 particle isolated from spinach (b). The spectra in 3a and 3b are taken from refs. 15 and 29, respectively.

at 495 and 465 nm and oriented rather perpendicular to the membrane plane, while CP2-b (Fig. 3a) contains a major carotenoid population absorbing around 505 and 470 nm and oriented close to the membrane plane and a minor one absorbing at 490nm and 455nm and oriented very much perpendicular to the membrane (15). The different absorption maxima of the various carotenoid populations cannot be ascribed to a difference in the type of carotenoid molecules because highly purified PS II complexes contain only beta-carotene and no xanthophylls (24).

In a previous study (14) we had noticed the identical wavelength position of the well-resolved shoulder at 682nm in the spectra of the intact PS II particle and of the peak of the absorption detected magnetic resonance (ADMR) spectrum recorded at 1.2 K on the same material (25). This ADMR signal has been assigned to the ground state of P 680 which disappears upon triplet formation. In ref. 14 we have thus used this previous assignment, together with the known property of the in-plane orientation of the Q_Y transition of P 680 (26), to propose that the 682-nm shoulder (which is also characterized by a positive linear dichroism) in our spectra could be assigned to P 680 either alone or together with a small number of core chlorophyll a molecules surrounding it. The observation (Fig. 2) that the 682 nm component of the intact particle is specifically segregated in the CP2-c photosynthetically inactive complex, while no evidence for such a component can be found in the spectra of CP2-b is thus no longer consistent with our previous assignment (14). As shown in Figs. 2a and 3a, the negative LD 691 component present in the intact PS II particle (14,15,21) is segregated in CP2-b together with the photoactive pheophytin molecule responsible for the 544 nm positive linear dichroism (LD 544) signal (14,19,21,22). It has been shown previously that the Q_Y transition of this molecule is oriented rather perpendicular to the plane of the membrane (19), in the same way as is oriented the Q_Y transitions of the two bacteriopheophytin molecules in the reaction center of photosynthetic purple bacteria (27). It has been previously demonstrated that F 695, a characteristic emission of PS II at low temperature, was specifically associated with CP2-b (7). The observation that LD 691 is preferentially segregated in CP2-b together with F 695 and LD 544 thus does not invalidate our hypothesis on the origin of F 695 (20). Furthermore we note that the amplitude of LD 691 relative to the dichroism at the red maximum is smaller (by about a factor of 2) in CP2-b than it is in the intact particle. However, the ratio of the amplitudes of LD 691 to LD 544 is comparable in CP2-b and in the intact particle (15). Assuming LD 691 is due to the pheophytin acceptor as proposed in (14,21), this observation can be taken as an indication that a large fraction of this acceptor has been either lost or disoriented during the splitting of the intact particle into its CP2-b and CP2-c constituents. This observation goes along the report that about half of the primary charge stabilization on Q_A is lost during the isolation of CP2-b (9). It has also been previously noticed that a large fraction of both the photoreducible pheophytin and the PS II activity was lost during the splitting of an intact particle from mesophilic organisms (5,10,11). Furthermore it has been reported that the isolation of CP2-b was accompanied by a significant decrease in the content of D_1 and D_2 compared to that of the core complex (9). Our observation that only a fraction of the LD 691 signal initially present in the spectrum of the core particle (relative to the amplitude of the oriented chlorophyll a signal in the Q_Y region) is recovered in CP2-b can be interpreted as an indication that LD 691 is specifically associated with D_1 and D_2. Alternatively our data cannot exclude that LD 691 is associated with the 47 kDa polypeptide of CP2-b but that the SDS treatment utilized to isolate CP2-b from the intact particle leads to the observed partial loss of LD 691 and of PS II activity. Thus it could not be concluded from the work on the Synechococcus subparticles (15) whether LD 691, and thus the 695-nm emission (F695) to which it is correlated, is specifically associated with the 47kDa polypeptide or with D_1-D_2.

The recent isolation of a photoactive D_1-D_2 PS II reaction center (28) which does not contain the 47kDa or 40kDa polypeptides, together with the purification of a chlorophyll-protein complex (CP 47) devoided of D_1-D_2 (Fujitani, M. and Satoh, K., unpublished results), led us to assess this question of the localization of the pigment responsible for F 695 in vivo (29). The A, LD and LD/A spectra at 4K of CP 47 are shown in Fig. 3b while those of a D_1-D_2 complex prepared by K. Satoh in Leyden are depicted in Fig. 4a. The orientation of the D_1-D_2 particles in the squeezed gel must be compared to the geometry of this complex when it is still embedded in the photosynthetic membrane. In this respect the 544nm transition assigned to pheophytin a and the 490nm carotenoid signal, which exhibit linear dichroism of opposite signs, can be used as specific markers for this particle. This 544nm transition gives rise to a positive linear dichroism in thylakoids (22), grana membranes (30), the core PS II particle (14,15), the CP2-b complex (Fig. 3b and (15)) and D_1-D_2 (Fig. 4a). The 490nm negative linear dichroism is clearly present in the core particle ((14,15) and Fig. 1), the CP2-b complex and D_1-D_2. It has been previously demonstrated that the core and the CP2-b particles exhibit the same orientation with respect to the squeezing direction in the gel as they have with respect to the normal to the photosynthetic membrane in vivo (14,15,22,30). The sign of the linear dichroism for the 544nm and 490nm transitions observed for the D_1-D_2 particle (Fig. 4a) thus unambiguously demonstrates the same behavior for the D_1-D_2 particle, i.e., the directions of the complex which are in vivo parallel to the membrane plane, orient in the gel perpendicular to the squeezing direction. The close resemblance

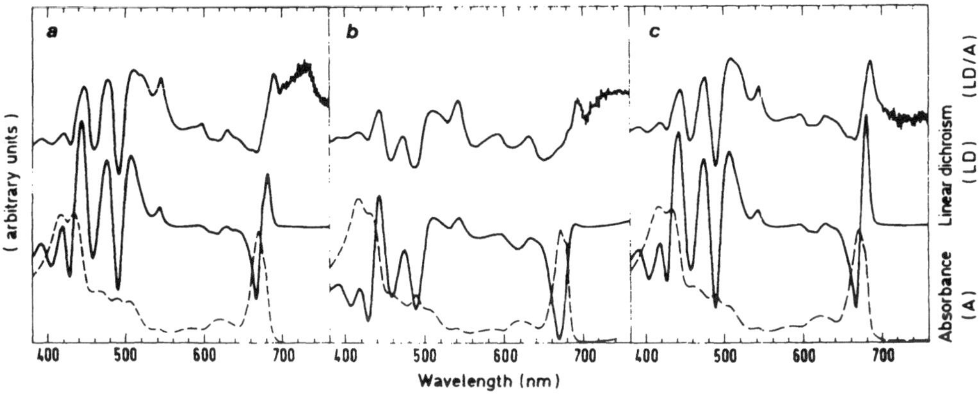

Fig. 4 : Absorption, linear dichroism and LD/A spectra of various preparations of the PS II reaction center (D_1-D_2-cyt b_{559}). The spectra in 4a are taken from ref. 29.

between the LD and LD/A spectra of CP 47 (Fig. 3b) and CP2-b (Fig. 3a), except for the contribution of D_1-D_2 pigments in the latter, demonstrates a similar orientation property also for the CP 47 particle.

It is clear from the spectra of CP 47 (Fig. 3b) and D_1-D_2 (Fig. 4a) that LD 544, which is assigned to the photoactive pheophytin, is segregated in D_1-D_2, while LD 691 belongs to the photochemically inactive CP 47 complex. In addition D_1-D_2 is observed to emit F 685 polarized rather parallel to the direction of the membrane plane, while F 695 is emitted by CP 47 with a polarization perpendicular to the same plane (29). Thus, the hypothesis on the pheophytin origin of F695, which has been instrumental in developing the investigation of the structural and functional organization of the PS II pigments (14,15,22,29-33), has been essentially disproved. In view of the absence of a pheophytin a band in the absorption spectrum of CP 47, it is reasonable to assume that the species responsible for both the LD signal at 691nm and F695 is a chlorophyll a molecule. Its function, however, remains unknown.

The linear dichroism spectrum of the D_1-D_2 complex clearly indicates the presence of two spectrally distinct carotenoid molecules with different orientations. The three bands of the first carotenoid population are located at 506, 474 and 443nm and the positions of the maxima of the second one are blue shifted by 12-14nm to 489 and 546nm. The short-wave band of this second carotenoid cannot be resolved, due to the large overlap with other pigments below 440nm. The band at 435nm in Fig. 4a is accounted for by the Soret transitions of chlorophyll a, and two minor components around 520 and 540nm represent bands of pheophytin a, respectively. Chemical analysis of the D_1-D_2 complex reveals the presence of 1.0 $\pm$ 0.3 beta-carotene molecules per reaction center (28). The observation that two spectrally different carotenoid species with different orientation can be distinguished suggests that two carotenoid molecules are associated with the PS II reaction center, in contrast to the bacterial reaction center which contains only one. Part of the carotenoids, especially of the short-wave species, then would be solubilized and lost during the isolation of the D_1-D_2 complex.

The spectra of the D_1-D_2 complex, especially in the Q_Y region, are much less resolved than those of bacterial reaction center preparations where the absorption and linear dichroism measurements reveal well-separated bands of the bacteriopheophytin molecules, the accessory bacteriochlorophylls and the primary electron donor (27). In contrast, the low temperature absorption spectrum of various D_1-D_2 complexes shows only one band at 670nm and a shoulder at 680nm as shown in Fig. 4a-c. The particles used for the spectra of Fig. 4b were prepared by K. Satoh in Japan and the ones used for Fig. 4c were prepared by J. Duranton in Saclay according to (28). While the absorption spectra of these three D_1-D_2 preparations are quite similar, the linear dichroism spectra in the Q_Y region are strickingly different, especially in the ratio of the amplitude of the negative linear dichroism signal around 666nm to the positive linear dichroism signal at longer wavelength. This negative linear dichroism band at 666nm cannot be ascribed to the photoactive pheophytin a, the Q_Y transition of which makes a large angle with the membrane (19), because the available evidence indicates that this transition, at least at room temperature, is at considerably higher wavelength, around 680nm (19,34). Furthermore, there is no evidence for a component with negative dichroism around 666nm in the spectra at 10K of the core PS II (14,15) and of CP2-b (15). It is thus possible that this spectral component observed in D_1-D_2 represents chlorophyll molecules, the orientation of which has been disturbed during the isolation procedure, which involves high Triton X-100 concentrations (28). In this respect the negative linear dichroism component seen around 610nm in the spectra of the D_1-D_2 particles, which has no counterpart in the spectra of the core and CP2-b complexes (15), can be assigned to the vibrational $Q_{Y\ 0-1}$ component of the main $Q_{Y\ 0-0}$ transition at 666nm. Such short wavelength negative linear dichroism components are commonly observed for a variety of PS I fractions which have received rather harsh treatments with detergents (35,36 and J. Breton, unpublished results). With the exception of the chlorophyll a band mentioned above, the linear dichroism and LD/A spectra of CP2-b (15) show the combined features of the corresponding spectra of D_1-D_2 and CP47. This extends the earlier observation that the spectra of the core particle of Synechococcus could be constructed from those of CP2-b and CP2-c (15). It thus appears that the orientation of the chromophore pigments in PS II can be efficiently used to probe the integrity of the pigments during fractionation processes and may help in the development of more gentle isolation schemes.

ACKNOWLEDGEMENTS

J.B. acknowledges stimulating exchanges of information with A.W. Rutherford and N.E. Geacintov. Parts of the work described in this article have been done in collaboration with B. Diner, F.-A. Wollman, P. Tapie and Y. Choquet for the preparations from Chlamydomonas, with S. Katoh for the Synechococcus fractions and with R. van Dorssen, J.J. Plijter, H. van Gorkom and J. Amesz for the particles from spinach.

REFERENCES

1. Rutherford, A.W. (1986) Biochem. Soc. Trans. 14, 15-17
2. Deisenhofer, J., Epp, O., Miki, K., Huber, R. and Michel, H. (1984) J. Mol. Biol. 180, 385-398
3. Diner, B.A. and Wollman, F.A. (1980) Eur. J. Biochem. 110, 521-526
4. Delepelaire, P. and Chua, N.H. (1979) Proc. Nat. Acad. Sci. USA 76, 111-115
5. Camm, E.L. and Green, B.R. (1983) J. Cell. Biochem. 23, 171-179
6. Satoh, K. and Butler, W.L. (1978) Plant Physiol. 61, 373-379
7. Yamagishi, A. and Katoh, S. (1983) Arch. Biochem. Biophys. 225, 836-846
8. Yamagishi, A. and Katoh, S. (1984) Biochim. Biophys. Acta 765, 118-124
9. Yamagishi, A. and Katoh, S. (1985) Biochim. Biophys. Acta 807, 74-80
10. Nakatani, H.Y., Ke, B., Dolan, E. and Arntzen, C.J. (1984) Biochim. Biophys. Acta 765, 347-352

11. De Vitry, C., Wollman, F.A. and Delepelaire, P. (1985) Biochim. Biophys. Acta 767, 415-422
12. Michel, H. and Deisenhofer, J. (1986) in Photosynthesis III : Photosynthetic Membranes. Encyclopedia of Plant Physiology, Vol. 19 (Arntzen, C.J. and Staehelin, L.A., eds.), pp 371-381, Springer-Verlag, Berlin
13. Trebst, A. (1986) Z. Naturforsch. 41c, 240-245
14. Tapie, P., Choquet, Y., Wollman, F.A., Diner, B. and Breton, J. (1986) Biochim. Biophys. Acta 850, 156-161
15. Breton, J. and Katoh, S. (1987) Biochim. Biophys. Acta 892, 99-107
16. Garab, G.I. and Breton, J. (1976) Biochem. Biophys. Res. Commum. 71, 1095-1102
17. Vasin, Y.A. and Verkhoturov, U.N. (1979) Biophysik 24, 269-273
18. Kramer, H.J.M. and Amesz, J. (1982) Biochim. Biophys. Acta 682, 201-207
19. Ganago, I.B., Klimov, V.V., Ganago, A.O., Shuvalov, V.A. and Erochim, Y.E. (1982) FEBS Lett. 140, 127-130
20. Breton, J. (1982) FEBS Lett. 147, 16-20
21. Breton, J. (1986) in Photosynthesis III : Photosynthetic Membranes. Encyclopedia of Plant Physiology, Vol. 19 (Arntzen, C.J. and Staehelin, L.A., eds.), pp. 319-326, Springer-Verlag, Berlin
22. Tapie, P., Haworth, P., Hervo, G. and Breton, J. (1982) Biochim. Biophys. Acta 682, 339-344
23. Vermeglio, A., Breton, J., Barouch, Y. and Clayton, R.K. (1980) Biochim. Biophys. Acta 593, 299-311
24. Ohno, T., Satoh, K. and Katoh, S. (1986) Biochim. Biophys. Acta 852, 1-8
25. Den Blanken, H.J., Hoff, A.J., Jongenelis, A.P.J.M. and Diner, B. (1983) FEBS Lett. 157, 21-27
26. Mathis, P., Breton, J., Vermeglio, A. and Yates, M. (1976) FEBS Lett. 63, 171-173
27. Breton, J. and Vermeglio, A. (1982) in Photosynthesis : Energy Conversion by Plants and Bacteria (Govindjee, ed.), vol. I pp 153-194, Academic Press, New York
28. Namba, O. and Satoh, K. (1987) Proc. Natl. acad. Sci. USA 84, 109-112
29. Van Dorssen, R.J., Breton, J., Plijter, J.J., Satoh, K., Van Gorkom, H.J. and Amesz, J. (1987) Biochim. Biophys. Acta 893, 267-274
30. Van Dorssen, R.J., Plijter, J.J., Dekker, J.P., Den Ouden, A., Amesz, J. and Van Gorkom, H.J. (1987) Biochim. Biophys. Acta 890, 267-274
31. Tapie, P., Acker, S., Arntzen, C.J., Choquet, Y., Delepelaire, P., Diner, B., Wollman, F.A. and Breton, J. (1984) in Advances in Photosynthesis Research (Sybesma, C., ed.), Vol. I pp. 693-696, Martinus Nijhoff/Dr. W. Junk Publishers, Dordrecht, The Netherlands
32. Renger, G., Koike, H., Yuasa, M. and Inoue, Y. (1983) FEBS Lett. 163, 89-93
33. Kyle, D.J., Arntzen, C.J., Franck, F. and Ynoue, Y. (1983) Photochem. and Photobiol. 38, 609-614
34. Nuijs, A.M., Van Gorkom, H.J., Plijter J.J. and Duysens, L.N.M. (1986) Biochim. Biophys. Acta 848, 167-175
35. Gagliano, A.G., Geacintov, N.E., Breton, J., Acker, S. and Rémy, R. (1979) Photochem. Photobiol. 29, 415-418
36. Tapie, P., Choquet, Y., Breton, J., Delepelaire, P. and Wollman, F.A. (1984) Biochim. Biophys. Acta 767, 57-69.

NON-LINEAR ABSORPTION SPECTROSCOPY OF ANTENNA CHLOROPHYLL A IN HIGHER PLANTS

D. Leupold, H. Stiel

Central Institute of Optics and Spectroscopy, Academy of Sciences of the GDR, Berlin 1199, GDR

P. Hoffmann

Department of General Botany, Section of Biology, Humboldt-University, Berlin 1040, GDR

Introduction

Whereas the information contained in non-linear emission of higher plants or green algae is subject of continuous interest since the first application of laser spectroscopy to photosynthesizing systems (1 - 4), much less attention has been paid to non-linear absorption (5 - 7).

In this paper experiments and results of non-linear absorption and related excite- and probe-beam measurements with chloroplast suspension of Triticum aestivum L. in the high intensity range will be described. Besides evaluation of rate constants and cross section of excited state processes of chlorophyll a (chl a) antenna molecules it will be shown that these molecules behave like a spectral uniform species. This is in conflict with the various concepts of "Chlorophyll forms" (8 - 12) and therefore will be discussed more in detail.

Photosynthetic Light-Harvesting Systems

Methodical Aspects of Non-linear Absorption

If the transmission of a sample (e. g. chl a in solution) is measured at a fixed wavelength λ within an absorption band with (laser-)light pulses from the low intensity range, a constant ("low signal") value T_0 is obtained irrespective of variations in the pulse intensity. This linear absorption takes place as long as the change of intensity is proportional to this intensity I_λ itself when the pulse passes along x through the sample:

$$\frac{\partial I_\lambda (x,t)}{\partial x} = - \alpha I_\lambda (x,t) \qquad (\underline{1})$$

where the constant α is the absorption coefficient of the ground state ($\alpha = \sigma \cdot N$) at the wavelength of the pulse. For example when measuring the transmission of a solution of chl a in dioxane within the Soret or red absorption band with nanosecond (ns) pulses, this region of linear absorption (constant transmission) is extended up to pulse intensities of about 10^4 Wcm^{-2} (corresponding to photonflux-densities of some 10^{22} cm^{-2} s^{-1}, or photon densities per pulse of some 10^{13} cm^{-2}).

Further increase of intensity of the measuring pulse effects an non-vanishing population of excited state(s), this means the factor α in ($\underline{1}$) becomes a function of intensity, i. e. the absorption becomes non-linear, or the transmission becomes a function of pulse intensity. In the mentioned chl a-example non-linear absorption is manifested in a (transient) bleaching (13) (cp. also Fig. 3).

The course of transmission vs. intensity in the non-linear region is very informative because it reflects implicitly all excited state absorption-, relaxation-, and interaction processes which are involved in the light-matter interaction process under investigation. To transform this implicit information into explicit results at first one has to replace

Eq. (1) by the more general photon transport equation (2)

$$\left(\frac{1}{c}\frac{\partial}{\partial t}+\frac{\partial}{\partial x}\right) I_{\lambda}(x,t) = -NI_{\lambda}(x,t)\sum_{i<j}\sigma_{ij}(s_i - s_j) \quad (2)$$

where

N - particle density

σ_{ij} - cross section of the absorptive transition $i \rightarrow j$ of wavelength λ ; with $\sigma_{ij} = \sigma_{ji}$

s_i - normalized population density of the level i, i. e. $\sum^{n} s_i = 1$

n - number of energy levels

Eq. (2) combines with the system of rate equations (3),

$$\frac{\partial s_i(x,t)}{\partial t} = -I_{\lambda}(x,t)\cdot\sum_{j\neq i}\sigma_{ij}(s_i-s_j)+\sum_{j>i}^{n} k_{ji}s_j-\sum_{j<i}^{n} k_{ij}s_i+ + q\varsigma \quad (3)$$

to form the coupled system of differential equations for the description of non-linear absorption (and of non-linear emission).
In (3), k_{ij} means the relaxation constant for the transition $i \rightsquigarrow j$, and the last term describes the annihilation process in an excited state i_2, with the final products i_1 and i_3, including the inverse process according to

$$\varsigma(s(x,t)) = \gamma s_{i_2}^2(x,t)-\tilde{\gamma}\, s_{i_1}(x,t)\cdot s_{i_3}(x,t) \quad (4)$$

and $q = \{q_j\}$, where

$$q_j = \begin{cases} 1 \text{ for } j = i_1,\ i_3, \\ -2 \text{ for } j = i_2, \\ 0 \text{ otherwise.} \end{cases}$$

The system (2), (3) is the basis of a "Physico-Mathematical Methods Package of Nonlinear Absorption and Emission", which has been developed to extract the full content of information from the non-linear functions (e. g. energy level scheme of the sample, excitation and de-activation processes, excited state absorption cross sections and rate constants, population densities of excited states ...). This package consists of 3 parts

- Measurement of nonlinear functions with suitable devices
- Expertsystem (dialog with computer) based on analytical investigations of (2), (3)
- Numerical software package

The Package is described in detail elsewhere (14) (cp. this reference also for explicit results concerning chl a in solution).

Survey of Investigations

In the following this Methods Package (esp. the first and third part) is used for the investigation of non-linear absorption of chl a in a Chloroplast suspension. At first the transmission of this sample is investigated at λ = 670 nm (within the longest-wave absorption band) with pulses of 2 ns and 0.5 ns duration in a high intensity range corresponding to the non-linear region of chl a in solution ($I > 10^4$ W.cm^{-2}). These experimental results are analyzed on the basis of a 3-level scheme of chl a antenna molecules including excited state absorption and exciton annihilation. In a second type of investigations excite and probe beam measurements are used to decide on the nature of broadening of the $s_o - s_1$ absorption band of chl a antenna molecules (problem of "Chlorophyll forms"). The principle of this experiment, which is well-known from hole-burning investigations (15) and which was adapted to the problem of chl forms already earlier (16), is illustrated in Fig. 1:

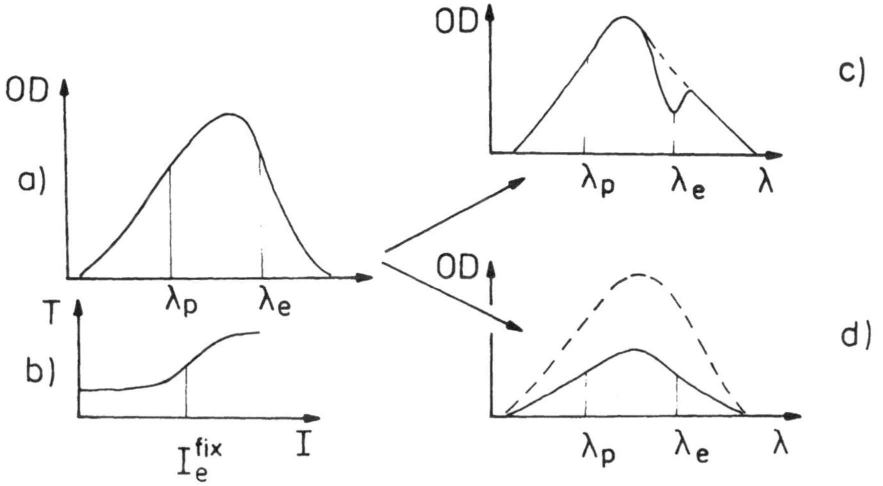

Fig. 1: Principle of laserspectroscopic sub-band analysis

a) Shape of the absorption band of interest, spectral location of excite (λ_e)- and probe (λ_p)-pulse
b) Intensity-dependent transmission (non-linear absorption) at λ_e; fixation of a suitable excite pulse intensity (I_e^{fix}) for the sub-band analysis
c) Result of measurement in case of inhomogeneous broadening (e. g. sub-bands)
d) Result of measurement in case of homogeneous broadening (spectral uniform sample or rapid spectral cross relaxation)

Within the absorption band of interest a non-linear absorption is induced by an intense pulse of wavelength λ_e. With a synchronous λ-tunable probe pulse the spectral profil of this band is analyzed. In case of inhomogeneous broadening (e. g. by several sub-bands corresponding to different "forms") the nonlinearity is restricted to a small range

around λ_e, outside the profil is unchanged (cp. Fig. 1c). In case of homogeneous broadening of the total band (e. g. with a spectral uniform sample or fast spectral cross relaxation) a change in the whole band shape is induced (cp. Fig. 1d). It is noteworthy that Fig. 1 is a simplified version, neglecting e. g. excited state absorption.
In the corresponding experiment described below λ_e is fixed at the longwave side of the band at 695 nm and λ_p is tuned between 650 and 680 nm.

Characterization of the Sample

On the basis of earlier experiences we used chloroplast suspensions which are more proper than leaves. The organells were isolated according to Reevers and Hall (17) in 50 mM tricin buffer, p_H 7,5 from primary leaves of 8 d old Triticum aestivum se dlings, cultivated on earth in continous white light (40 W m^{-2}). The functional state of the isolated chloroplasts was checked in parallel measurements: chlorophyll a content in 1 ml suspension amounted to ca. 1 mg, in order to ensure the nessessary density (chl b $\simeq$ 250 μg; Car. $\simeq$ 230 μg; chl a/chl b $\simeq$ 4,3; Chl/Car $\simeq$ 6,0).
The electron transport rate testet with 1 mM DCPIP amounted to $\simeq$ 4mM DCPIP x g chl^{-1} x Min^{-1}. Encoupling the electron transport by 250 mM methylamin increased the rate up to 6 (coupling degree $\simeq$ 1,7 fold), indicating the sufficient photochemical intactness of the chloroplasts used.

Experimental Set-up

The non-linear absorption measurements with the 670 nm/2 ns-pulses were made with a microcomputerized device allowing automatical registration and comprehensive data handling (18).
The set-up for measurements of non-linear absorption with 0.5 ns-pulses and for the excite- and probe-pulse investi-

gations is shown in Fig. 2:

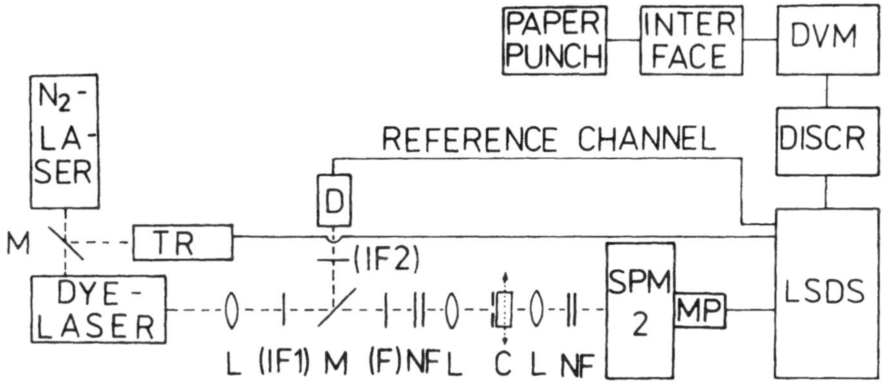

Fig. 2: Experimental set-up for measurement of non-linear absorption and excite- and probe-measurements with single dye laser pulses of $\Delta t = 0.5$ ns.

Abbreviations: L - lens, M - mirror, (IF 1), (IF 2) - narrow band interference filters, (F) - RG 5 filter (Schott, Jena), NF - neutral filters, C- sample cell, movable by step motor, D - diode, SPM-2 - double monochromator (Zeiß, Jena), MP - multiplier, TR - trigger, DISCR -discriminator, LSDS - Laser spectroscopy detector system (Molectron)

It is based on a nitrogen laser (type IGT-101, ZWG, Acad. Sci. GDR) pumped grazing incidence dye laser (home made), which allows on the one hand to work with powerful , narrowband 670 or 695 nm pulses for the non-linear absorption measurements (additional suppression of background radiation by home made interference filters), on the other hand to use the background emission as a probe pulse in the excite- and probe-measurements. Due to the restricted cavity round trips with the short excitation pulse, the change from the regime excite- plus probe-pulse to that of probe pulse alone can simply be realized by covering the grating without affecting the background intensity.

It has to be noted that in Fig. 2 the filters in parenthesis are changed according to the working regime in the following way:

non-linear absorption at 695 nm:	use of IF 695 in the indicated position, absence of IF 670 and F
non-linear absorption at 670 nm:	use of IF 670 instead of IF 695
excite- and probe-pulse :	absence of IF 695, filter F is used for attenuation of the background radiation to a real low-level intensity, whereas there is only minimum change of the intensity at 695 nm (cut off-filter), IF 670 is used to adjust the reference channel to the probe beam radiation.

The excite- and probe-experiments were made with the highest available intensity of the 695 nm - excite pulse of $8 \cdot 10^7$ $W.cm^{-2}$. The intensity variations in nonlinear absorption measurements were realized by suitable neutral filters and careful calibration procedures.

A discriminator is used to select pulses of nearly equal intensity. With the chloroplast sample, each measurement was made on a formerly non-illuminated part of the cell, realized in a single shot laser regime and by a pin-hole diaphragm and a step motor controlled motion of the cell perpendicular to the direction of the light beam. Each point in the measured curves shown below is a meam value of at least ten measurements.

Results and Discussion of Non-linear Absorption at λ = 670 nm

The results of measurement of intensity-dependent transmission (non-linear absorption) with the chloroplast suspension at λ = 670 nm with 0.5 ns-pulses and with 2 ns-pulses are shown in Fig. 3a, b. In both cases with increasing intensity a very small transmission increase was found. For comparison the strong non-linear response of a chl a-solution in dioxane in the same intensity range is shown in the box in Fig. 3.

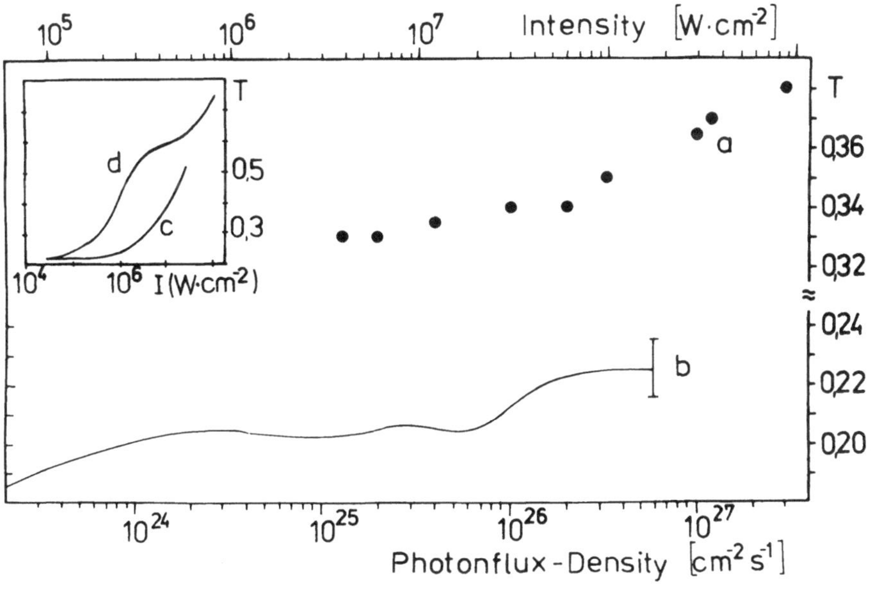

Fig. 3: a) Intensity-dependence of transmission of the chloroplast suspension at λ = 670 nm, measured with 0.5 ns-pulses. Cell length 10^{-2} cm, low-level transmission T_o = 0.30 (measured at Beckmann DK-2A).

b) the same as a) but measured with 2 ns-pulses, T_o = 0.18.

Box: Intensity-dependent transmission of chl a in dioxane (4 . 10^{-3}cm), measured with 0.2 ns-pulses (c) and 2 ns-pulses (d) (13).

Increase of concentration of the solution to in-vivo-conditions results in a diminution of the bleaching effect (14), but the transmission change is still higher than with the chloroplasts.

For a quantitative interpretation of the results shown in Fig. 3a, b numerical investigations of the differential equation system (2), (3) for a 3-level-model of chl a antenna molecules including exciton-annihilation (cp. Fig. 4) were made.

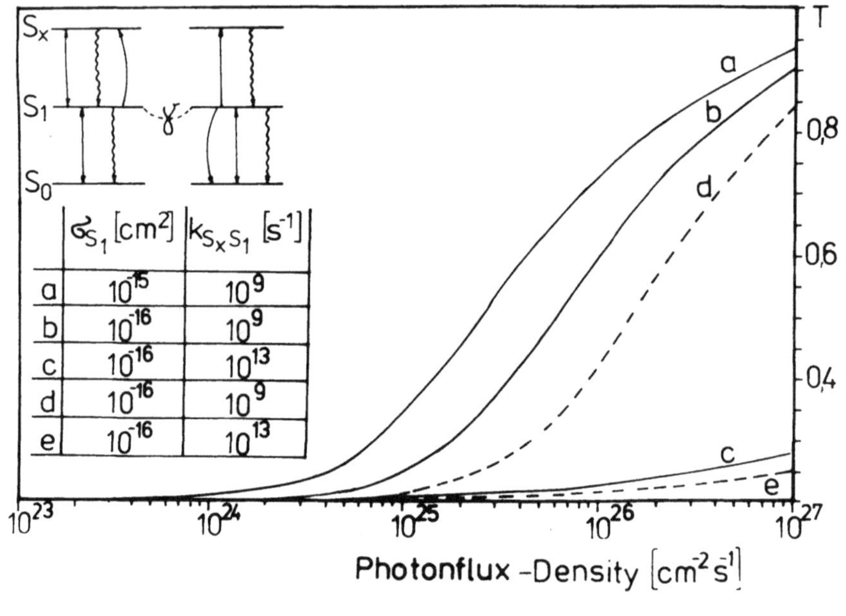

Fig. 4: Numerical results on intensity-dependent transmission for the model shown in the box. The underlying differential equation system is given in (2), (3). Parameters assumed as fixed are

$\sigma_{s_0 s_1}$ (670 nm) = 10^{-16} cm^2, $k_{s_1 s_0} = 2 \cdot 10^9$ s^{-1} and

$\gamma = 1{,}9 \cdot 10^{11}$ s^{-1}.

The other parameters are shown in the figure. The full line represents the case of 2 ns-pulses, the broken line represents the case of 0.5 ns-pulses.

By using the annihilation constant of chl in vivo from earlier investigations (19) as $\gamma = 1,9 \cdot 10^{11}\ s^{-1}$ ($\hat{=} 3 \cdot 10^{-9}\ cm^3\ s^{-1}$), the s_1-absorption cross section $\sigma_{s_1}(670)$ and the s_x-relaxation constant $k_{s_x s_1}$ are varied. The calculations show (as can be seen in part in Fig. 4) that the course of the intensity-dependent transmission is strongly influenced by $k_{s_x s_1}$ whereas $\sigma_{s_1}(670)$ is not sufficiently fixed by this type of measurements. When using $\sigma_{s_1} = \sigma_{s_0}$ (as suggested by results of chl a in solution (20)), best simulation of both experimental curves (Fig. 3a, b) can be obtained with $k_{s_x s_1} = 10^{13}\ s^{-1}$. With chl a in dioxane the corresponding relaxation constant is much smaller ($7,6 \cdot 10^{10}\ s^{-1}$ (20)), but is very sensitive to the environment (21).

Whereas in the non-linear absorption measurements described in this chapter $\lambda = 670$ nm was selected for the purpose of comparison with the known behaviour of chl a in solution, for the excite- and probe-pulse experiments described below it was necessary to shift the excite-pulse wavelength λ_e to the longwave side of the absorption band in order to avoid simple energy transfer from λ_e to λ_p.

With the 695 nm-excite-pulse a non-linear behaviour of the chloroplast suspension similar to that described here with $\lambda = 670$ nm was observed.

Results and Discussion of Excite- and Probe-Pulse-Experiments

The experimental result of change of the red absorption band profile of the chl a antenna molecules with powerful 0.5 ns/695 nm-pulse illumination ($8 \cdot 10^7\ W.cm^{-2}$) is shown in Fig. 5. It is obvious that the change takes place over the entire range of probe wavelengths (650 - 680 nm). Taking into account the explanations given above (cp. also Fig. 1), this result is a clear indication of homogeneous broadening of the absorption band at least between 650 and

695 nm for time intervals ≥ 0.5 ns.

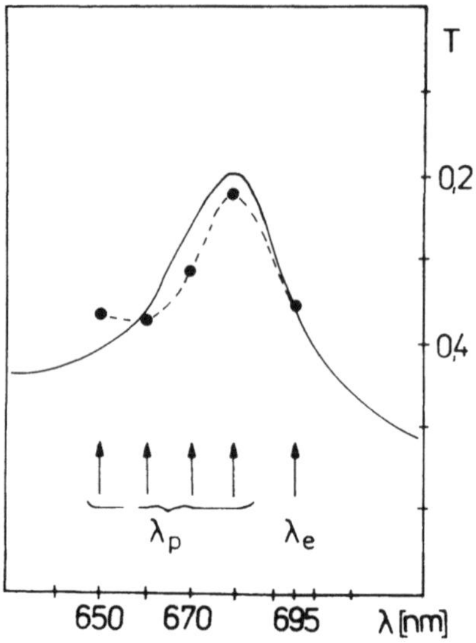

Fig. 5: Full line: wavelength-dependent transmission profile of the longest-wave absorption band of the chloroplast suspension used in the excite- and probe-experiments, measured at the Beckman DK-2A spectrophotometer, cell length 10^{-2} cm.
Full circles: probe pulse (0.5 ns)-transmission at simultaneous irradiation with an intense pulse at 695 nm (0.5 ns, 8 . 10^7 $W.cm^{-2}$).
Broken line: Free-hand connection of measuring points (for sake of illustration only).

Similar results were obtained already earlier for time intervals ≥ 2 ns (22), but for the first time transmission increase (λ = 660 - 680 nm) as well as transmission decrease (λ = 650 nm) was observed. Especially the latter finding excluds the explanation, that the homogeneous behaviour is caused by ultrafast spectral cross relaxation between subbands. This would result only in transmission increase. Therefore there is no "hope" to find an inhomogeneity of the red absorption band with still shorter pulses.
This means with respect to red light absorption most if not all[x] chl a antenna molecules behave like a single, uniform

species. This is in contrast to the results of conventional sub-band analysis. In that field there was in the beginning a large spread in the results (9 - 12) (up to 16 sub-bands), later a certain unification to "four universal forms" (23) but than again the many sub-bands concept was preferred (8) (cp. in this respect the careful investigations of Meister (24) concerning possible over-interpretations of band profils).

From the present results it follows that the different chl a vibrational sub-species found by Raman methods (25, 26) are not reflected in the longest-wave electron transition.

The present results do'nt support any model of a special spectral-spatial organization of the chl a antenna (based on "chl-forms") to direct the energy transfer to the reaction centers.

x The quantitative interpretation based on calculation of excited state populations is in preparation.

References

1. Knox, R.S. 1977. In: Primary Prozesses of Photosynthesis (J. Barber ed.) Elsevier/North-Holland Biomedical Press, pp. 56 - 97

2. Paillotin, G., C. E. Swenberg, J. Breton and N. E. Geacintov. 1979. Biophys. J. 25, 513

3. Geacintov, N. E., and J. Breton. 1986. In: Encyclopedia of Plant Physiology, New Series, Vol. 19 Photosynthesis III (L. A. Staehelin and C. J. Arntzen eds.). Springer-Verlag, Berlin/Heidelberg, pp. 310 - 318

4. Leupold, D., E. Brecht and P. Hoffmann. 1987. Photosynthetica (Prague) 2 (in press)

5. Leupold, D., B. Voigt, S. Mory, P. Hoffmann and B. Hieke. 1978. Biophys. J. 21, 177

6. Breton, J. and N. E. Geacintov. 1980. Biochim. et Biophys. Acta 594, 1

7. Seifert, B., B. Wiesner, P. Hoffmann and D. Leupold.1983. studia biophys. 97, 149

8. Brown, J. S. 1983. Photosyn. Res. 4, 375

9. Cederstrand, C. N., E. Rabinovitch and Govindjee. 1966. Biochim. et Biophys. Acta 126, 1

10. French, C. S. 1971. Proc. Nat. Acad. Sci USA 68, 2893

11. Litvin, F., B. A. Gulyaev and V. A. Sineshchenkov. 1971. Dokl. Akad. Nauk SSSR 199, 1428

12. Katz, J. J. 1976. Phil. Trans. Roy. Soc. London 80, 877

13. Leupold, D., M. Scholz and J. Ehlert. 1985. Chem. Phys. Letters 115, 434

14. Leupold, D. 1987. Dissertation B.' Humboldt-University of Berlin, Section of Physics

15. Röss, D. 1966. Laser-Lichtverstärker und Oszillatoren. Akad. Verlagsgesellschaft, Frankfurt/Main

16. Voigt, B., D. Leupold, P. Hoffmann and B. Hieke. 1979. Proc. 3. Conference on Luminescence, (L. Kozma ed.), Szeged. Vol. 1, 189

17. Reevers, S. G. and D. O. Hall. 1980. In: Methods in Enzymology 69, Photosynthesis and Nitrogen Fixation (A. San Pietro ed.), Part C, Academic Press New York

18. Dähne, S., D. Leupold and H. Stiel. 1987. Acta Phys. Polonica A71, 777

19. Jahnke, R. and D. Leupold. 1981. studia biophys. 86, 227

20. Leupold, D., J. Ehlert, S. Oberländer and B. Wiesner. 1983. Chem. Phys. Letters 100, 345

21. Leupold, D., J. Ehlert, S. Oberländer and H. Stiel. 1984. In: Teubner Texte zur Physik. Vol. 4 (D. Fassler, K. H. Feller and B. Wilhelmi eds.), Teubner Verlagsgesellschaft, Leipzig, 111

22. Hoffmann, P. and D. Leupold. 1984. In: Advances in Photosynthesis Research. Proc. VI. International Congress on Photosynthesis (C. Sybesma ed.). Martinus Nijhoff/W. Junk Publishers, 1.6.709

23. Brown, C. S., J. S. Brown and M. Lawrence. 1972. Plant Physiol. 49, 421

24. Meister, A. 1986. Dissertation B. Acad. Sci. GDR

25. Lutz, M. 1977. Biochim. et Biophys. Acta 460, 408

26. Höxtermann, E., W. Werncke, I. N. Stadnichuk, A. Lau and P. Hoffmann. 1982. studia biophys. 92, 147

Section IV

Function: Electronic Excitation and Energy Transfer

EXCITATION ENERGY TRANSFER IN PHOTOSYNTHESIS

R. van Grondelle[a] and V. Sundström[b]

[a]Department of Biophysics, Physics Laboratory of the Free University, Amsterdam, The Netherlands
[b]Department of Physical Chemistry, University of Umeå, Umeå, Sweden

1. Introduction

Photosynthetic organisms have succeeded in reducing significantly the number of complex protein assemblies capable of converting solar energy into chemical free energy (reaction centers, cytochrome b-c_1-oxido-reductase) by using an efficient light-harvesting antenna. Excitations absorbed by the pigment molecules that are part of this antenna are transferred to the reaction center (RC) within at most a few hundred picoseconds, but often less, where they can be used for the charge separation. Because the turnover rate of the RC exceeds 10^3 s^{-1} and because even in bright sunlight the number of photons absorbed per pigment molecule is not more than 1 s^{-1}, the light-harvesting antenna can contain over a hundred pigment molecules per RC without saturating the energy conversion process.
Photosynthetic organisms show a high variability in their pigment content and the spectroscopic properties of the pigments are usually such that they suit the light conditions under which the organism grows. In addition, many photosynthetic organisms have the ability to adapt their pigment composition to the circumstances (oxygen, light intensity etc.). Either a different pigment appears or the same pigment absorbs at a different wavelength. The major pigments involved are a variety of derivatives of chlorophyll (Chl) and bacteriochlorophyll (Bchl). In addition, there are numerous other pigments with absorption spectra

Photosynthetic Light-Harvesting Systems

overlapping the solar spectrum (phycobilins, β-carotene, carotenoids etc.) that have important light-harvesting functions.

Since the early work of Duysens [1-3] it is established that the antenna pigments transfer their excitation energy to the RC with an efficiency that in most cases is high. Because the excitation transfer is intimately associated to losses, such as fluorescence and radiationless processes, the measurement of the fluorescence yield provides detailed information on the yield of the energy transfer process. Duysens used mainly fluorescence excitation spectra in combination with so-called 'action spectra' to deduce the preferred energy transfer pathways in photosynthetic organisms. His conclusions are widely accepted and form a basis for the interpretation of energy transfer measurements.

(i) The excitation of pigments absorbing at relatively high energy results in the transfer of energy to pigments absorbing at relatively low energy. Duysens specifically studied the multiple Bchl forms in *Chromatium (C.) vinosum*, the energy transfer in phycobilisomes, the β-cartene, Chl *b* and Chl *a* in higher plants and algae and the carotenoids in photosynthetic purple bacteria.

(ii) Excitation of the pigment with the lowest excitation energy, e.g. Chl *a* that absorbs around 680 nm in green algae and higher plants and Bchl *a* that absorbs around 870 nm in purple bacteria, results in the transfer of excitation energy among more or less identical pigment molecules.

(iii) A special pigment P, was postulated that occurs only in a relatively low ratio to the other pigments, that traps the excitations and produces a charge separation.

After the successful isolation of photosynthetic RCs from purple bacteria, the properties of P were extensively studied. Only the recent crystallization of the RC of *Rhodopseudomonas (Rps.) viridis* [4] and *Rhodobacter (Rb.) sphaeroides* [5] revealed all the details of Duysens' original proposal.

(iv) On the basis of the Förster mechanism [6] efficient energy transfer was calculated from high energy absorbing chromophores to those absorbing at lower energy. In addition, using the Förster mechanism, Duysens showed that transfer among at least 100 identical chromophores surrounding the RC could easily result in efficient trapping, assuming that the RC has the correct optical properties.

Since that time our knowledge of the energy transfer process has increased tremendously. Especially the application of modern biochemistry and molecular biology has revealed that in virtually all cases the photosynthetic pigments are associated to specific proteins [see for reviews 7-12], which tune their spectral properties and determine their assembly into larger units. Many pigment-protein complexes have been isolated; for a significant fraction the primary polypeptide structure is available and concrete ideas exist about their secondary and tertiary structure. In three cases the three-dimensional structure of antenna pigment-proteins was determined by X-ray diffraction on crystals: the Bchl *a* protein from a green sulfur bacterium [13] and the C-phycocyanin trimer from two cyanobacteria [14,15].

Although only a few exact structures are known, many of these pigment-protein complexes have been studied extensively with biochemical and biophysical techniques and detailed ideas exist about their structure and function in photosynthesis. One of the major advances has been the application of picosecond laser spectroscopy to these systems, allowing the accurate determination of excited state decays, detected either via polarized fluorescence or polarized absorption. In addition, high intensity picosecond or nanosecond laser flashes are used to probe the topography of the photosynthetic antenna and to estimate the rate of excitation diffusion in such systems.

We shall review here the latest developments in energy transfer research in a variety of photosynthetic systems. Because energy transfer in Photosystem (PS) I and II was extensively discussed in recent reviews [16-19,28] we shall limit ourselves to a short overview in these cases. Phycobilisomes, green photosynthetic bacteria and purple bacteria are considered in some more detail, with an emphasis on recent time-resolved measurements. Comprehensive reviews on these and related subjects can be found in refs. 16-28.

The subsequent discussion shall be based on the following fundamental concepts.

(i) Energy transfer within a pigment-protein complex that contains chemically or spectroscopically different chromophores is generally fast. As a consequence the excitation density is equilibrated among the different pigments within at most a few picoseconds. Energy transfer between identical pigments within a pigment-protein occurs beyond the available time-resolution ($\leq 10^{-12}$s).

(ii) Energy transfer among different pigment-proteins is generally slower ($5\text{-}50.10^{-12}$s), probably due to the larger distances involved. Examples are the antenna of purple bacteria or the rod-to-core transfer in phycobilisomes. Of course in special cases excited state decay rates may increase upon aggregation into tight complexes. This is due to the opening of new decay channels as for instance occurs in the different phycobilin trimers or hexamers upon assembly in a phycobilisome rod (see below).

(iii) Trapping of the excitations absorbed by the antenna pigments by the RC takes place on a time scale of 50-200 ps. In some cases slower components are observed that may be due to the fact the the charge separated state is not far lower in energy than the antenna excited state. Specific examples are purple bacteria and PS II.

(iv) The transfer of excitations into the RC is not a priori fast. Therefore, a focussing of the excitation density close to the RC may be required to obtain a high trapping efficiency. This has resulted in minor, low energy, antenna species closely associated to the RC. Specific examples may be the pigments responsible for the long-wavelength emission in PS I (F720) and PS II (F695) and $Bchl_{896}$ in purple bacteria.

2. Energy transfer in PS I

The pigment system associated with photosystem I of oxygen evolving organisms is most conveniently divided into two main components: (a) a core complex termed CC I which contains the PS I RC and a central antenna of approximately 65 Chl *a* molecules; and (b) a light-harvesting Chl *a/b*-protein complex termed LHC I (for a review see (10)). At room temperature the fluorescence of intact thylakoids is dominated by PS II emission, and the PS I fluorescence is therefore not easily detected. Upon lowering the temperature to 77K there is a strong increase of the fluorescence intensity above 700 nm. A fluorescence band around 720 nm in green plants and algae was attributed to the core comples CC I whereas the 735 nm band in green plants was assigned to LHC I [10,29]. In some algae the latter emission is replaced by a band centered around 710 nm [29], and similarly assigned to LHC I.

As a result of the weak PS I fluorescence at room temperature, and partly overlapping much stronger PS II emission, fluorescence from PS I in intact thylakoids at room temperature was only recently resolved [30-32]. Picosecond time-resolved fluorescence measurements revealed the presence of a short-lived decay component (τ=50-100 ps) having a broad fluorescence spectrum peaking at about 700 nm [30,32]. Since this decay component was unaffected by the state of the PS II RC (closed or open) and also was observed in a photosynthetic mutant of *Chlamydomonas reinhardtii* lacking PS II [31], it was assigned to excitation migration and trapping in PS I. Owens *et al.* [33] also investigated the PS I antenna size dependence of this decay component. By measuring the fluorescence decay time of PS I core complexes isolated from barley, having varying Chl *a*/RC ratios, a linear relationship between the lifetime and antenna size was found. Applying a random-walk model for the exciton dynamics [34] values of the microscopic rate constants of migration and trapping were obtained. Thus, the single-step transfer between antenna molecules was estimated to be very fast: $\tau_h = 0.2$ ps, and the time constant for the primary charge separation process in the RC was calculated to be 2.8-3.4 ps, depending on whether P_{700} is a monomeric or dimeric species. The estimated rate of charge separation is similar to that measured in isolated RCs of purple bacteria [35,36], and the hopping rate in the antenna is of the same order of magnitude as that estimated for Bchl antennas of purple bacteria [37-39]. From the comparison between measured PS I fluorescence lifetimes and the predictions of the random-walk model it was also concluded that an excitation originating in the PS I core antenna on the average visits the RC 2.4 times before it is photoconverted [33]. This suggests that the overall exciton dynamics in PS I is largely determined by the diffusion towards the RC. However, it should be noticed that in order to reach the above cited conclusions all effects on the PS I exciton dynamics due to a possible heterogeneity of the PS I antenna were neglected.

The excitation transfer and trapping in isolated PS I particles (Chl *a*/RC ≈ 70) was also studied by Nuijs *et al.* [40] by measuring time resolved absorbance difference spectra. They obtained a lifetime of 40 ps for the excitations within the PS I antenna, which is in reasonable agreement with

Owens *et al.* [33]. This lifetime was not affected by the oxidation state of P_{700} [40].

The low-temperature 735 nm fluorescence assigned to PS I was extensively studied in the past (for reviews see [16-19] and from its non-exponential decay it was concluded to be inhomogeneous in nature [41]. At 77 K the decay was found to be approximately described by a 3 ns lifetime [42]. The 735 nm fluorescence was also observed to have a finite risetime; values in the range 20-150 ps where obtained from several different experiments [43-45]. Recent experiments on chloroplasts of spinach revealed that this risetime could be resolved into two components with time constants of about 20 and 80 ps [45]. Similar experiments at 77 K performed on PS I particles (Chl *a*/RC ≈ 100) [46] isolated from pea chloroplast and the blue-green alga *Phormidium luridum* displayed very fast (≈ 15 ps) fluorescence decay at 685 nm and a correspondingly fast risetime of the 735 fluorescence. These observations where attributed to efficient energy transfer from the LHC I and/or core antenna of PS I to a low lying antenna component, characterized by the F735 fluorescence [45,46]. By relating electrophoretically separated chlorophyll-protein complexes of several different mutants of *Chlamydomonas reinhardtii* to low temperature fluorescence bands, Garnier *et al.* [47] proposed a model for energy transfer in PS I. A sequential energy transfer involving the following steps was suggested: LHC II → LHC I → CC I → RC. Using the terminology related to fluorescence band maxima this would correspond to F680 → F710 → F720 → RC. The F710 feature observed in this case would correspond to F735 in higher plants [10,29]. Thus, the low temperature data strongly suggest the existence of a low energy minor antenna surrounding the PS I RC, possibly needed to focus the excitations close to the RC. As with purple bacteria, this will have an effect on the trapping and transfer characterics extracted from room temperature experiments and random walk considerations [48].

3. Energy transfer in PS II

The pigment system associated to PS II of oxygen evolving organisms consists of a PS II core and a variable amount of LHC II [10]. The total number of Chl *a* plus Chl *b* molecules may easily exceed 200; most of them occur in the LHC II. The dominant form of LHC II is LHC II *β*, the classical LHC Chl *a/b* complex. LHC II *β* exists as a 72 kDa polypetide trimer, in total binding 40-45 Chl *a*/ Chl *b* molecules [10,49-52]. The core of PS II contains the light-harvesting polypeptide CP 29, two other light-harvesting polypeptides CP 47 and CP 43, each with about 20 Chl *a* molecules and the PS II RC, which consists of the D_1 and D2 polypeptides binding 4 Chl *a*, 2 pheophytin *a* and 1 *β*-carotene [10,53-60]. In addition the polypeptide containing cytochrome b_{559} is associated to the PS II RC.

In intact thylakoids the PS II fluorescence dominates over that of PS I at room temperature. Numerous time-resolved picosecond fluorescence and absorption measurements have been performed on a variety of PS II preparations, with variable antenna size, and over a wide range of experimental conditions. These measurements were extensively reviewed by Karukstis and Sauer [16], Van Grondelle [17], Holzwarth [18,28], Moya *et al.* [25] and Geacintov and Breton [19].

Most of the experimental fluorescence decays measured with intact thylakoids or purified PS II particles (mutants, grana fractions) suggest that the red-shifted fast decay phase (50-100 ps) must largely be ascribed to PS I [18,28,30,61,62]. With open PS II RCs (Q oxidized) a second decay phase (200-300 ps) is observed, which by most of the authors is ascribed to trapping by the PS II RCs. For the case of fast energy transfer in an almost homogeneous antenna, the measured time constant would correspond approximately to N k_t^{-1}, where N is the number of Chl *a* molecules per RC and k_t the rate of formation of the state P^+_{680} $Pheo^-$ (P_{680} is the primary donor, Pheo is the primary acceptor of the PS II RC). For N=200, k_t is about $(1\text{-}2).10^{12}$ s^{-1}; a number not dissimilar to that calculated for PS I [33] and photosynthetic purple bacteria [37] and rather close to the measured rate of charge separation in isolated bacterial RCs [63,64]. The idea that the 200-300 ps component represents the formation of the state P^+_{680} $Pheo^-$ and not the equilibration between the excited antenna and P^*_{680} is essentially identical to the hypothesis that Schatz and Holzwarth [65,66] used in their calculation of the PS II fluorescence kinetics. It explains nicely why in PS II particles with only about 80 Chl *a* per PS II RC the trapping time decreases to 80-100 ps [66].

Recently, Trissl *et al.* [67] measured directly the formation of the PS II charge separated state by time-resolved photovoltaic experiments. They observed that the primary radical pair is formed in about 200 ps and concluded that the intrinsic time constant of charge separation (k_t^{-1}) must be less than 35 ps. In addition, a second phase in the charge separation process of about 0.4-0.5 ns was found that contributes 33-50% to the total potential change. In our opinion, this time-constant is surprisingly similar to a third phase in the time-resolved fluorescence decay observed in intact thylakoids and purified PS II preparations with open RCs. In isolated PS II particles with about 80 Chl*a*/RC Schatz *et al.* [66] also report biphasic trapping kinetics. In their view the fast 100 ps phase represents the formation of P_{680}^{+} $Pheo^{-}$ (as discussed above) and the slow phase (0.5 ns) the reduction of Q_A by $Pheo^{-}$. This slow phase appears in the fluorescence decay because the radical pair is energetically not far below the excited antenna.

Finally, in their kinetic modelling of the PS II fluorescence decay kinetics Schatz and Holzwarth [65] state that a choice of 3.5-7.0 ns^{-1} for the rate constant of the recombination reaction (Chl *a* P_{680}^{+} $Pheo^{-}$ $\rightarrow$ Chl *a* P_{680} Pheo), in combination with a forward rate constant $k_tN^{-1} = 5$ ns^{-1} yields a good fit to the experimental fluorescence decays with open and closed PS II RCs. This choice of the rate constants leads for open PS II RS's to a significant concentration of Chl**a* after the first equilibration step. With their value for the rate of Q_A reduction ($k_2 = 5$ ns^{-1} [65]) a marked 0.5 ns decay in the time-resolved fluorescence of intact thylakoids should be present, carrying a significant fraction of the total amplitude.

Sofar the slow (0.5 ns) fluorescence decay phase with all the PS II RCs active was not interpreted in this manner, although this possibility was extensively discussed in ref. 17. Maybe those authors who ascribe it to "excitations that have visited the PS II RC at least once" [19,68] have a similar view. Originally, this slow decay phase was ascribed to loosely coupled LHC II [16,69], but this does not seem realistic, especially because LHC II-less mutants also exhibit the slow decay phase [31,70]. In addition, strong evidence exists that LHC II is well coupled to the PS II core [71,72]. Maybe a fraction of the very fast decay represents the equilibration between LHC II and PS II core, similar to what is observed in purple bateria [38], but in view of the small energy difference, this is not expected to be a large component.

Holzwarth and coworkers [18,28,30] have interpreted the multi-exponential decay in terms of heterogeneity of the PS II RCs (see for instance ref. 73 for a review on α- and β- PS II). Their conclusion is mainly based on the fact that Chl *b* does not seem to contribute to the excitation spectrum of the 0.5 ns component. This view is shared by various other research groups [31,74,75]. In addition Wendler and Holzwarth [62] were able to relate reversible changes in the amplitudes of the 0.2-0.3 ns and 0.6 ns components to State I - State II transitions, coupled to changes in the relative antenna sizes coupled to PS II α and PS II β. Surprisingly, the antenna directly coupled to PS I remained unaffected.

The assignment of the multi-exponential fluorescence decay kinetics to PS II heterogeneity must be taken with some caution. First of all, it is not clear why Holzwarth cs. ignore the fact that even PS II particles with only one type of RC do show multiphasic fluorescence kinetics [66,76]. Secondly, it is not clear why PS II β would give a much slower trapping than PS II α, while the antenna size is presumably smaller [73]. Thirdly one wonders why changes in antenna size only affect the amplitudes of the fluorescence decay components but not the lifetimes.

With closed PS II RCs (P_{680} Pheo Q_A^-) the fluorescence decay is characterized by a fast (100 ps) and two slow components (1 ns and 2.5 ns). The fast decay is due to PS I. The 2.5 ns decay has been ascribed to recombination luminescence arising from the radical pair P_{680}^+ $Pheo^-$ Q_A^- in analogy to purple bacteria [16,77,78]. However, for small PS II particles rather convincing results were presented that argue in favour of a strongly decreased rate of trapping in the state P_{680} Pheo Q_A^- [66].

In systems with a large PS II antenna, the yield of the radical pair becomes negligible after the reduction of Q_A [67]. In the view of Holzwarth *et al.* [18,28,65,66] the rate constant of trapping decreases about ten fold, resulting in an effective trapping rate of about 0.5 ns^{-1}. Thus, the 2.5 ns lifetime represents 'normal' fluorescence from PS II, or from PS II α (according to Holzwarth cs). In addition, they assign the 1.2 ns phase to closed PS II β centers. However, Hodges and Moya [61,76] claim that this 1.2 ns phase is also present in PS II preparations from grana stacks with Q_A reduced. Moreover, they conclude that the lifetime of this component increases gradually from about 0.15 ns to 1 ns upon closing the PS II RCs, thus representing well connected PS II antenna. It is clear that no agreement exists among the different research groups concerning the interpretation of these fluorescence decays. In that respect we consider the combination of absorption and fluorescence measurements as a major step forward towards the final solution of this problem.

From several higher plants, green algae and cyanobacteria a PS II particle can be isolated that is photochemically active, contains about 40-50 Chl *a* per PS II RC and probably represents the PS II core [10,51,54-60]. This pigment protein is enriched in long-wavelength absorbing Chl *a* (676 nm, 683 nm) that in the intact system is oriented with the Q_y close to the plane of the membrane [79-81]. The emission of the PS II core shows a peak at 685 nm at 293 K, but at lower temperatures an additional peak at 695 nm appears [80]. Disagreement exists about the origin of the 695 fluorescence: it may either be due to a component absorbing at about 691-693 nm [79-81] or to one of the pigments contributing to the absorbance at 683 nm [80]. The polarization of the 695 nm emission is perpendicular to the plane of the membrane, contrary to the overall orientation of most of the Chl *a* Q_y transition moments [82]. The core of PS II can be further split into three pigmented particles; CP 47, CP 43 and the D1 D2 complex [10,51,54-60]. The first two contain about 20 Chl *a* molecules each and are photochemically inactive. The D1 D2 probably represents the PS II RC. It turns out that the 695 nm emission and 691 nm LD component are associated to CP 47 [83]. Possibly, the 691 nm component has a role in PS II similar to B896 in photosynthetic purple bacteria [48].

4. Energy transfer in phycobilisomes and phycobiliproteins

Blue-green and red algae contain phycobilisomes as major light-harvesting antenna [23,85-87]. A phycobilisome is composed of disc-shaped aggregates (trimers, hexamers) of different phycobiliproteins which form rod-shaped elements that are attached to a central core. This composition typically renders the phycobilisome an overall hemidiscoidal shape, with an approximate diameter of 40 nm. The phycobilisomes are attached to the outer side of the thylakoid membrane and transfer energy preferentially to PS II. The phycobiliprotein building blocks of the phycobilisomes may be classified into three major groups on the basis of their spectral properties [88-90]. Phycoerythrin (PE) (in some species replaced by phycoerythrocyanin) and phycocyanin (PC) are found in the rods, whereas allophycocyanin (APC) only occurs in the core of the phycobilisome. When assembled into phycobilisomes the phycobiliproteins are organized with the high energy pigments (PE) at the periphery and the low energy pigments (PC) closer to the core. This constitutes a very efficient structure for directed energy transfer towards the core of the phycobilisome and eventually to the RC. In addition to the pigmented polypeptides small amounts of other polypeptides lacking chromophores are found in the rods and the core. These so-called linker polypeptides play an essential role in the assembly of rods and core [9].

Extensive experimental work during the last decade, concerning spectral, biochemical and structural properties has resulted in a detailed picture of phycobilisome architecture and energy transfer dynamics. Several recent reviews on these aspects are available (for reviews on structure see [9,23,84-86,91,92]; and for summaries of energy transfer see [16-18,93,94].

4.1. Isolated phycobilisomes

As a result of numerous picosecond time-resolved studies (see [16-18,93-94] and references therein) the following general picture of energy transfer dynamics in phycobilisomes has evolved. Energy transfer from the rods to the core occurs on the time-scale of several tens of picoseconds (typically 50-100 ps) and depends on the length of the rods). The fastest processes observed in isolated phycobilisomes are in the order of 10 ps [95-97] and tentatively attributed to s → f transfer between neighbouring chromophores within a hexameric unit, or possibly between two adjacent units [97]. Similar fast decays of induced absorption anisotropy supported this assignment [97]. Two different models have been suggested for the energy transfer along the phycobilisome rods. Glazer *et al.* [94] described the energy transport as an unidirectional transfer along a linear array, where the disc-to-disc transfer represents the rate limiting step. Holzwarth *et al.* [18,95,96,117] on the other hand, suggested that the energy is equilibrated rapidly within the rod due to fast interdisc transfer and that the transfer step from the rod to the core constitutes the rate limiting process. Assuming a random walk model [21,112,117], it should be possible to distinguish between the two cases on the basis of the size-dependence (number of discs) of the rod → core transfer rate. However, the small number of transfer steps in these systems makes this difficult in practice. Time-resolved experiments performed on a mutant of *Synechoccus* 6301 (AN112), having only one hexameric disc in the rod [97] and experiments on a sub-particle of the phycobilisome, the 18-S particle [98], composed of part of the core and only one trimeric disc of a rod offers additional information on this point. It was concluded that rod → core transfer in the AN112 phycobilisome occurs with a time constant of 40 ps [97], whereas rod → core transfer as measured in the 18-S particle occurs within 20 ps [98]. If the 10 ps depolarization time observed in whole phycobilisomes [97,99] is assumed to represent the disc-to-disc transfer time it follows that disc-to disc transfer and the final rod → core transfer step are of approximately equal magnitude and therefore will both contribute to the overall rod → core transfer time.

The energy transfer within the core of the phycobilisome has been sudied to a much lesser extent. However, the existing data indicates that allophycocyanin itself does not function as a terminal emitter in the isolated phycobilisome [95,97]. A special form of allophycocyanin, allophycocyanin B, and the coloured 75 kD linker polypeptide may act as parallel terminal emitters [95,96,100,101]. Excitation energy is transferred from allophycocyanin to these two emitters with a time-constant of about 50 ps [95,97].

4.2. Isolated phycobiliproteins

The chromophores of the phycobiliprotein building blocks of the phycobilisomes differ widely in their absorption and fluorescence spectra, as a consequence of differing chemical nature and conformational properties [102,103]. Each phycobiliprotein generally holds more than one type of chromophore. On the basis of steady-state fluorescence measurements the concept of sensitizing (s) and fluorescing (f) chromophores was introduced [104,105] to account for the observation that energy was efficiently transferred from high energy absorbing chromophores (s) to low energy ones (f). This simple picture was essentially confirmed later by many time-resolved studies on isolated phycobiliproteins [106-114]. Isolated phycobiliproteins may occur in various degrees of aggregation: monomer, trimer and hexamer. A monomer of C-PC consists of two different polypeptide subunits α and β, together holding three chromophores [14,15], one on the α-subunit (α-84) and two on the β-subunit (β-84, β-155). The chromophores are covalently bound to amino acid residues of the protein [102,103] and interact strongly with several neighbouring amino acids [118]. The three chromophores of a monomer unit are all chemically identical, but have slightly different absorption spectra due to variation in the interactions with the surroundings. Consequently, already the simplest particle represents a complicated problem from the viewpoint of energy transfer. With increasing degree of aggration ($(\alpha\beta)_3$-trimers and $(\alpha\beta)_6$-hexamers) the number of interactions and possible energy transfer pathways increases dramatically. Therefore, it must be realized that when characterizing the energy transfer dynamics in such complex systems using time-resolved methods, the kinetic parameters extracted from the usual exponential analysis of the measured decay curves generally represent a combination of several elementary processes. Time-resolved measurements with high wavelength resolution (both excitation and analyzing beams) combined with measurements of the equilibrated excitation density can considerably improve this situation.

A major step forward towards a detailed understanding of the energy transfer processes in phycobiliproteins and phycobilisomes was taken when the three-dimensional structure of $(\alpha\beta)_3$- trimeric C-PC [14] and $(\alpha\beta)_6$-hexameric C-PC [15] became available. This, for the first time, provides unique possibilities to obtain a detailed picture of the energy transfer pathways in a photosynthetic antenna system, as well as a test of the assumed mechanism of energy transfer.

Sauer *et al.* [115] performed energy transfer calculations based on these structures, and concluded that energy transfer in C-PC occurs solely by the Förster weak interaction mechanism. At least qualitative agreement between calculated rates and results obtained from several time-resolved measurements was observed (see below for some details). However, a refined structure of C-PC was recently published [116] and this new structure differed in one important respect from the original one. The orientation of one of the chromophores (β-84) was much better defined in the refined structure and its orientation relative to other chromophores had been changed dramatically. This has an important consequence for the energy transfer in trimeric and hexameric aggregates of C-PC, where the α/β-84 chromophores of two adjacent monomers are now in a very favourable orientation, at a center-to-center distance of 21 Å. Using Förster theory to calculate the rate of energy transfer between these two chromophores one obtains a value of about $10^{12}s^{-1}$, which is much faster than any experimentally observed rate in these systems. With the new crystallographic coordinates, the assessment of all transfer processes in C-PC as Förster energy transfer may perhaps be questioned.

Experimentally the ($\alpha\beta$)-monomeric C-PC shows two lifetimes: 50 and 200-300 ps, associated with energy transfer [112-114]. These are in reasonable agreement with the lifetimes calculated by Sauer *et al.* [115] for β-155 → β-84 and α-84 → β-84 transfer, respectively.

In $(\alpha\beta)_3$- trimers, the measured kinetic components shorten to about 25 and 120 ps [106,112-114]. This illustrates the feature predicted by the crystal structure and from the calculations [115], that more channels of energy transfer become available with increasing degree of aggregation. The faster of the two measured time constants was tentatively assigned to β-155 → β/α-84 transfer [112,113] or α-84 → β-84 transfer [114]. The longer lifetime was suggested to represent the equilibration of energy among the α/β-84 chromophores of different monomeric units [112]. However, considering the changes in predicted rates of energy transfer following the refinement of the C-PC structure, these assignments should now be reanalyzed. Thus, the α-84 ↔ β-84 transfer between two adjacent monomeric units may be too fast to be resolved with presently used time-resolved methods, having a time resolution of about 10 ps. Further, judging from the absorption spectra of the individual chromophores of C-PC [37], α-84 and β-84 should be excited in roughly equal amounts at most wavelengths, and the equilibrated energy distribution for the two chromophores should be about 1:2 in favour of β-84. These facts indicate that it probably will be difficult to resolve the single step α-84 ↔ β-84 in a kinetic measurement. The most likely assignment of the observed kinetic components appears, at this time, to be the following: The fast 25 ps component represents all transfer processes from a β-155 chromphore in one subunit to the α- and β-84 chromophores of all three subunits. This would account for the increased rate of this process (from 50 to 25 ps) in going from monomeric to trimeric C-PC. The slower 120 ps component tentatively represents the equilibration among all α/β-84 chromophores of the three monomeric units. This assignment of the 120 ps component, however, requires that the close-lying α/β-84 pairs of adjacent monomers are not all spectrally equivalent. This is seen from the following argument. Excitation initially produces a non-equilibrated distribution among the α/β-84 chromophores. This distribution equilibrates as energy is redistributed over all the α/β-84 chromophores of the trimer. This process will proceed in two steps with widely differing rates. A very fast (~1 ps) initial equilibration over the two close lying α/β-84 chromophores of two adjacent monomers, will be followed by a much slower process when energy is redistributed over the three α/β-84 pairs. It seems likely that this process occurs on the 100 ps time-scale. However, there is a problem with this suggestion. Assuming that all three monomers of the trimer are equivalent this process would not be observed in an isotropic measurement, since equilibrium among the two close-lying α/β-84 chromophores is maintained at all times due to the very fast ~1 ps process. If the energy equilibration over all the α/β-84 chromophores of the trimer is to be associated with the 120 ps lifetime we have to assume that there exists a spectral inhomogeneity among the different α/β-84 chromophores within the trimer. It is interesting to note that the 120 ps lifetime is also observed in the anisotropy decay, with an amplitude (≈25%), which is twice the amplitude of the corresponding isotropic decay (≈14%) [113,114]. This could indeed suggest that the isotropic 120 ps decay is due to a small spectral inhomogeneity, and that the anisotropic decay with the same lifetime represents depolarization when energy is equilibrated over all the α/β-84 pairs of the trimer. Experiments on C-PC trimers with sub-picosecond time resolution were performed in order to obtain more conclusive information about the α/β-84 process [113]. Sofar, the experiments did not provide a definite answer on this point; a very fast feature of low intensity was suggested, but it is unclear whether this can be attributed to energy transfer.

From the crystallographic structure of $(\alpha/\beta)_6$- hexameric C-PC [15] it is seen that additional efficient transfer channels between trimeric discs are present. The calculations by Sauer *et al.* [115] suggest the presence of two efficient transfer channels between adjacent trimers (along the β-155 and α-84 chromophores) and, consequently energy is rapidly transferred between the trimeric discs. Combining hexameric discs into phycobilisome rods then produces highly efficient energy flow along the long axis of the phycobilisome rod. These features are reflected in the observed increase of energy transfer rate in $(\alpha\beta)_6$-hexamers of C-PC and phycobilisome rods [112].

5. Energy transfer in green photosynthetic bacteria

In contrast to the purple photosynthetic bacteria, green photosynthetic bacteria typically contain a large number, 1000-2000, antenna pigments per RC [119]. The defining characteristic of these green bacteria is the presence in the cells of a chlorosome [120]. The chlorosome has an important function as a light-harvesting complex, it contains over 10.000 pigment molecules, is surrounded by a single lipid layer and attached to the cytoplasmic membrane [see for reviews references 121-123].
Green photosynthetic bacteria for which the energy transfer process was extensively studied during the past few years are the sulfur bacteria *Prostecochloris (P.) aestuarrii* and *chlorobium (C.) limicola* and the thermopholic, non-sulfur filamentous bacterium *Chloroflexus (C.) aurantiacus*. For these species the chlorosome contains mainly Bchl *c* as light-harvesting pigment. The protein structure of the chlorosome of *C. aurantiacus* shows an array of long rods, each about 5-6 nm in diameter, regularly spaced at 6 nm intervals, that run along the long axis of the chlorosome [124]. The basic protein unit of these rods is a dimer of a 5.6 kDa polypeptide [125], that has been sequenced and contains per monomer 7 possible binding sites for the Bchl *c* (glutamins and asparagins) [126]. The Bchl *c* can be arranged along the α-helical part of the polypeptide with their Q_y's almost parallel to the α-helical axis. Such a model for pigment organization within the chlorosome stands in sharp contrast to proposals where Bchl *c* aggregates form the basic structure [127,128] and

the protein functions as a matrix. Some of the *in vitro* aggregates fit the spectral features of the chlorosomes very well [129,130].
The chlorosome of *C. limicola* contains similar rod-like elements, but with a larger diameter (~10 nm) [131]. The protein sequence shows only little homology to that of *C. aurantiacus* [132], is not easily folded into an α-helix and suggests a different organization of the Bchl *c* pigments. Chlorosomes from these three species can be isolated under relatively mild conditions. The absorption spectra are dominated by Bchl *c* (740 nm in *C. aurantiacus*, 749 nm in *C. limicola*) but, in addition show the presence of a small amount of intrinsic Bchl *a*: ± 4% for *C. aurantiacus* [133,134] and ± 1% for *P. aestuarrii* and *C. limicola* [135,136]. This chlorosomal Bchl *a* is proposed to act as an intermediate for energy transfer from the chlorosome to the membrane associated pigments. The cytoplasmic membrane of the green sulfur bacteria contains several (≥10) RCs per chlorosome. Each RC occurs in a core complex with about 20 tightly coupled Bchl *a* per RC and at least three trimers of the membrane associated, water soluble Bchl *a* containing protein [137]. This Bchl *a* protein constitutes the 5-6 nm thick, quasi crystalline base plate, which links the chlorosome to the core. The base plate Bchl *a* complex of *P. aestuarrii* was crystallized and its three dimensional structure was solved with a high resolution [138]. The transfer of excitation energy from the water soluble Bchl *a* to the core seems to be inhibited in the isolated membrane fractions. Trapping of excitation energy in the core of green sulfur bacteria is fast (≤70 ps [158]).
The pigment organization in the membrane of *C. aurantiacus* is strongly reminiscent of that in the membrane of purple bacteria. A type I light-harvesting complex occurs, B808-866 [139] and transfer from the chlorosomal Bchl *a* is thought to proceed via $Bchl_{808}$ and $Bchl_{866}$ to the RC. For each RC there are about 25 $Bchl_{866}$ and, surprisingly, the photosynthetic units appear to be isolated. Trapping of excitation energy by the open RCs is efficient [140].

5.1. *Chlorobium limicola* and *Prostecochloris aestuarrii*

The organization of the Bchl *c* in chlorosomes of *C. auranticus* shows strong similarities. In both species the Q_y transition moments show a preferential orientation long the long axis of the chlorosome. For chlorosomes of *C. limicola* almost perfect alignment was claimed [141,142]. The Q_y of chlorosomal Bchl *a* is observed to make a relatively large angle to the long axis [141].

The transfer of excitation energy in chlorosomes of *C. limicola* from Bchl *c* to chlorosomal Bchl *a* is very efficient (~100% [141]), in contrast to *P. aestuarrii* (~40% in the presence of sodium dithionite [136]). In *C. limicola* transfer from the chlorosome to the membrane is about 70% efficient [143].

The measurement of fluorescence quenching due to singlet-singlet annihilation in chlorosomes of *C. limicola* has shown that at room temperature excitations can move more or less freely among all the Bchl *c* of the chlorosome [144]. The Bchl *c* → Bchl *c* transfer rate must be extremely high ($\geq 10^{13} s^{-1}$) to account for the annihilation behaviour. Even if it assumed that for instance clusters of 14 Bchl *c* exist over which the excitation is instantaneously delocalized, transfer between different clusters must still occur on a picosecond timescale.

Excited state lifetime measurements, either measuring the fluorescence or the time-resolved absorption in chlorosomes and chlorosome-membrane preparations of *C. limicola*, show that the Bchl *c* lifetime is very short ($\lesssim$ 20-40 ps [145-147,159]). This time probably reflects energy transfer to the chlorosomal Bchl *a*, and is again a manifestation of the extremely fast excitation diffusion in the chlorosome. Time-resolved absorption anisotropy measurements show some initial depolarization within a few ps after excitation, followed by a relatively constant anisotropy [$r \simeq 0.2$] for the remaining time [146]. This indicates that very rapid energy transfer occurs among a few strongly coupled, well-ordered Bchl *c*, followed by excitation diffusion among similar Bchl *c* clusters to the ultimate energy acceptor, Bchl *a*, not associated to further depolarization.

In chlorosomes of *P. aestuarrii* and *C. limicola* the Bchl *a* emission is strongly quenched [136,141]. This is also reflected by the short Bchl *a* excited state lifetime (~30 ps [146]). Some of this quenching can be undone by adding strong reductants to the chlorosomes yielding a Bchl *a* lifetime of about 100 ps, with additional slower components [144,146,159]. Bchl *a* oxidation products, capable of quenching excited states, probably

appear during chlorosome preparation and are hard to remove. Sofar, the spectroscopic evidence suggests that the chlorosomal Bchl *a* occurs in relatively small clusters of interacting pigment molecules [135,144]. The steady state fluorescence polarization of the Bchl *a* emission is low at room temperature [141,146] indicating a low degree of local order.
Not so much is known about the excited state lifetime of chlorosomal Bchl *a* in the intact cell. Energy transfer to the membrane is relatively efficient and so is transfer in the membrane. The excitation trapping time in the membrane is probably of the order of 50 ps [140,148]. A variety of pigment-protein complexes can be prepared from green sulfur bacteria that retain photochemical activity [137]. In general these complexes do contain some of the water soluble Bchl *a* protein, but excitation energy transfer from this Bchl *a* to the RC-core complex is inefficient, may be due to some change in the structure [149].
Finally, time-resolved absorption anisotropy measurements on the isolated water soluble Bchl *a* protein trimer indicate that at room temperature excitation energy transfer within the complex, probably even within the monomer, is fast ($\leq$10 ps) leading to a complete depolarization of the excited state within the available time resolution [146]. Excitation annihilation experiments with the same complex show efficient coupling of the 21 Bchl *a* molecules within the trimer [160,161].

5.2. *Chloroflexus aurantiacus*

The organization of Bchl *c* in chlorosomes of *C. aurantiacus* is probably not unlike *C. limicola* and *P. aestuarrii*: Bchl *c* Q_y preferentially along the long axis of the chlorosome [134,150], maybe even within less than 20° [151]. The Bchl *a* Q_y is less well organized and on the average makes a relatively large angle with the chlorosome long axis. Fluorescence polarization measurements support the high degree of local order for Bchl *c* [150]. Energy transfer from Bchl *c* to chlorosomal Bchl *a* has an efficiency of 50-60% in isolated chlorosomes. In cells of *C. aurantiacus* this efficiency can approach 100%, although strong variations are observed dependent on the specific conditions [129,152-154]. Cooling cells from 50°C, their normal growing temperature, to 10°C, shows a decrease in Bchl *c* → chlorosomal Bchl *a* transfer efficiency from 100% to 70%, possibly associated to a membrane phase transition [152]. Other disturbances, such as sonication of the cells or the preparation of membrane associated chlorosomes generally lead to losses in transfer efficiency from the chlorosome to the membrane, manifested by dramatic increases in Bchl *c*, but especially in Bchl *a* emission [154].

At 4 K in *C. aurantiacus* cells the overall transfer efficiency from Bchl *c* to the membrane bound chromophores is low, i.e. about 15%. Because the efficiency for excitation transfer from Bchl *c* to chlorosomal Bchl *a* (~50%) and from chlorosomal Bchl *a* to the membrane (~30%) could be separately determined, it is concluded that at least at 4 K the preferred excitation transfer pathway is from Bchl *c* via chlorosomal Bchl *a* to B808-866 in the membrane [152].

Although excitation annihilation measurements in chlorosomes of *C. aurantiacus* could not be interpreted unequivocally, they showed that energy transfer takes place over many Bchl *c* molecules, or Bchl *c* clusters, with a rate that is comparable to that observed in *C. limicola* [144]. From fluorescence lifetime measurements it follows that both in isolated chlorosomes and in the membrane associated chlorosomes the major part of the Bchl *c* emission lives less than 30 ps, while in intact *C. aurantiacus* cells this is even shorter ($\leq$15 ps [154,156]). The Bchl *a* excited state lifetime is about 100 ps in chlorosome membrane-fractions and increases to 160 ps in the isolated chlorosome [154-156]. A crude estimate indicates that in whole cells the Bchl *a* lifetime may be significantly shorter (~10 ps). An additional complication in the interpretation of these fluorescence data is that in chlorosomes the emission yields of Bchl *c* and Bchl *a* are extremely low and about equal [144]. The emission yields do not correspond to the observed lifetimes, suggesting a very heterogeneous and strongly quenched emission. Sofar, only anaerobically prepared chlorosomes show a reasonable Bchl *a* emission at room temperature [153].

The orientation and spectral properties of the membrane bound pigments is such that energy transfer from the chlorosome to the membrane is facilitated. The $Bchl_{808}$ Q_y makes an angle of about 45° with the membrane plane, and may be well oriented to accept excitation energy from the chlorosomal Bchl *a*. The Q_x of Bchl *a* is oriented at about 35° with the membrane plane [157]. $Bchl_{808}$ to $Bchl_{866}$ transfer takes only a few ps

[140,157]. Trapping in the $Bchl_{866}$ antenna occurs in ± 50 ps, while the excited state lifetime increases to 200 ps upon closing the RCs [140,154]. The ordering of $Bchl_{866}$ is very similar to that of LH1 in purple bacteria: Q_y in the plane of the membrane, Q_x makes a small angle (20-30°) with the membrane normal [157]. From the strong increase in fluorescence polarization upon excitation in the red wing of the $Bchl_{866}$ absorption band it is concluded that C. aurantiacus contains a special red shifted Bchl species that mediates energy transfer to the RC [157].

6. Energy transfer in purple bacteria

The primary photosynthetic processes and associated pigment-protein structures of purple bacteria are considerably better characterized than for green plants and algae. The main reason for this is the presence of only one photosystem and the fact that it has been possible to isolate functional parts of the photosynthetic apparatus. Purified antenna and RC preparations have been available since a long time. The crystallization and structure determination of the RC of *Rps. viridis* [4] and, recently, *Rb. spaeroides* [5] are major contributions to our understanding of the primary photosynthetic reactions. Although for no antenna complex from purple bacteria the three dimensional structure has been determined sofar, there exists a wealth of information on structure and energy transfer dynamics for the intact Bchl antenna, and for numerous pigment-protein complexes derived from it. This information has been obtained mainly through a multitude of spectroscopic measurements and recently several reviews on this subject have appeared. For more background information we refer to references [17,18,21,26,162] concerning energy transfer, and [9,10,163] on structure.

For the structure in relation to the process of energy transfer the following general picture has evolved during the last decade. In Bchl *a* containing purple bacteria a long-wavelength core antenna exists that surrounds and interconnects the RCs. In *R. rubrum* there are approximately 30-50 $Bchl_{880}$ molecules per RC and the B880 antenna seems to be organized as aggregates of basic units containing 6-8 $Bchl_{880}$ molecules, and half this number of carotenoid molecules [17]. Upon excitation of the antenna, energy is transferred to the RC with an overall efficiency of about 95%. The lifetime of the excitation in the B880 antenna depends on the oxidation state of the primary donor P_{870} of the RC [164,165]. Thus, for active RC (reduced P) the overall trapping time is about 60 ps, and when P is oxidized, either chemically or by continuous illumination, this time increases to about 200 ps.

In *Rb. sphaeroides* the antenna contains an additional component B800-850, which is assembled of pigment-protein complexes holding probably 9 Bchl molecules (6 Bchl 850 and 3 Bchl 800) and ≥ 5 carotenoid molecules [17,166]. Steady-state fluorescence measurements [167,168], suggested an equilibration of energy between the B800-850 and B875 antennas. However, phase fluorimetry measurements [169] gave no evidence for a decay component due to B800-850 → B875 transfer. This transfer step was consequently assumed to be very fast [169]. Transfer form B875 to the RC in *Rb. sphaeroides* is characterized by kinetics very similar to that observed for *R. rubrum*.

The polarization of fluorescence emission originating from the antenna is generally low ($p \approx 0.1$) [166,170-172], when measured on chromatophores and isolated pigment-protein complexes. Excitation in the long-wavelength wing of B875 constitutes an exception: at low temperatures (4 K) a maximum polarization of $p=0.4$ is observed [166]. The low polarization observed through most of the spectrum led to the suggestion that the Q_y transition dipole moments of the Bchl molecules of the repeating units of B875 and B800-850 lie in one plane or in several parallel planes [166,170,172].

The exciton dynamics in the purple bacteria antennas has recently been extensivey studied, using both picosecond fluorescence [164] and absorption techniques [38,39]. In both kinds of measurements sufficiently low excitation intensity was used to avoid excitation annihilation, and measure the undistorted kinetics of exciton dynamics. Upon excitation of the main antenna B880 of *R. rubrum* a very rapid initial depolarization ($r(0) \approx 0.1$) within less than 1 ps occurs. This is due to very fast energy transfer between identical $Bchl_{880}$ molecules, probably within a minimum unit of 6-8 $Bchl_{880}$ molecules. A 30-40 ps decay component observed in the isotropic absorption recovery at about 890 nm is tentatively assigned to equilibration of energy between B880 and a long-wavelength absorbing antenna component, B896. From the equilibration rate an average jumping time between different B880 pigment-protein complexes of about 10 ps can be estimated. For open reaction centers the overall trapping time is about 60 ps, in good agreement with fluorescence streak camera [164] and other absorption measurements [165]. The fact that the overall trapping time is not too much different from the equilibration time between B880 and B896 led to the suggestion that the overall trapping of excitation energy is largely limited by the diffusion of energy through the antenna towards the RC.

Results for *Rb. sphaeroides* [38] are explained with a very similar model, including fast (≤1 ps) energy transfer from Bchl 800 to Bchl 850. In the picosecond absorption experiment the equilibration between B800-850 and B875 is clearly observed (~40 ps).

This picture of the energy transfer dynamics in *R. rubrum* and *Rb. sphaeroides* is confirmed by picosecond time-resolved absorption [173] and fluorescence [174] measurements on chromatophores at 77 K. The absorption measurements offering high time and spectral resolution, yield detailed information on the forward energy transfer rates between the various pigment pools [173]. The existence of a low lying antenna component B896 is put on a firm basis by these picosecond measurements; both absorption [173] and fluorescence [174] measurements yield a time constant of about 15 ps for the energy transfer from the core antenna to the low lying B896. In agreement with earlier steady-state fluorescence depolarization measurements [166], B896 is characterized by a high absorption anisotropy, when directly excited. This points towards a highly ordered organization of B896, or strongly interacting chromophores, possibly resulting in a delocalized exciton state.

Several possibilities were discussed concerning the origin of B896 [173]. The function can be to concentrate the energy in the vicinity of the RC, and thereby increase the probability of trapping even in the case of a relatively low rate of excitation transfer from the surrounding antenna to the RC. At room temperature the energy difference between P_{870} and B896 is small enough so that this slightly uphill energy transfer is more than compensated by the advantages of the concentration effect [48].

It is interesting to notice that such a low-lying antenna component may be a more general feature shared by other photosynthetic systems. For PS I

and PS II the antenna components related to the fluorescence bands F720 and F695, respectively, may have a similar origin.
Steady-state fluorescence measurements [175] and room temperature picosecond absorption measurements [38,39] indicate a very fast and efficient energy transfer from $Bchl_{800}$ to $Bchl_{850}$ in the B800-850 pigment-protein complex. A slight temperature dependence of this process made it possible to resolve it at 77 K and a time constant of about 2 ps [173] or 5 ps [176] is estimated. This point towards a rate of about 10^{12} s^{-1} at room temperature. When the same process was studied with much shorter pulses (~150 fs) of high energy [177], it was concluded that $Bchl_{800} \rightarrow Bchl_{850}$ transfer occurs in less than 150 fs. However, these experiments were performed using pulse intensities high enough to cause extensive excitation annihilation. Therefore it seems likely that this extremely short time constant for the $Bchl_{800} \rightarrow Bchl_{850}$ transfer is substantially shortened by the annihilation process. This conclusion is strongly supported by the fact that low intensity absorption [38,39,173] and fluorescence [176] measurements yield essentially the same result, and that the process is associated with a measurable fluorescence quantum yield [175].
Carotenoids are integer parts of the Bchl-protein complexes [178], and energy transfer from carotenoids to Bchl proceeds with an efficiency between 30 and 100 percent [179-181]. Since carotenoids have intrinsically very short excited state lifetimes [182], energy transfer from carotenoids to Bchl must be fast in order to account for the measured yields. Picosecond pulse excitation of the carotenoids in a B800-850 complex, and monitoring the $Bchl_{850}$ bleaching suggested a transfer time of about 10 ps [183]. Monitoring the recovery of carotenoid absorption in an isolated B800-820 complex, after excitation with a 4 ps long pulse of low intensity results in an estimated transfer time of 3 ps [184]. These rates agree favourably with the measured yields of energy transfer.

The efficiency of excitation annihilation in a photosynthetic system is a measure of the number of connected pigment molecules that form a domain [17,185,186]. This phenomenon was extensively studied in several purple photosynthetic bacteria under a variety of conditions [37,160,187-191]. At room temperature the long-wavelength antenna, B880,

of *R. rubrum* forms a domain that comprises over 500 $Bchl_{880}$ molecules and more than 10 RCs. In
Rb. sphaeroides similar large domains were observed for the total B800-850 and B875 antenna. Recently, mutants of *Rb. sphaeroides* were investigated with this technique [189]. Mutants which contain only B800-850 or only B875 plus RCs show that the 'isolated' components also tend to form large aggregates, at least at room temperature. A surprising exception is a mutant of *Rb. sphaeroides* that only contains B875 and shows small aggregates of about 100-150 Bchl a molecules [189,190].
In all species investigated the domain sizes decrease to about 100 $Bchl_{875}$ / $Bchl_{880}$ molecules upon cooling to 4 K. Similarly, the B800-850 domain size becomes about 40 Bchls [191]. Results with lipid-diluted membranes of *Rb. sphaeroides* support the idea of a core structure with 3-5 RCs and 100 $Bchl_{875}$ molecules. Lipids modulate the aggregation of these cores into larger units and in addition are involved in connecting or disconnecting B800-850 [191]. Experiments with membranes of *Rb. sphaeroides* from a very early stage of development indicate that photosynthetic unit assembly starts from these core structures [188]. For all these reasons we believe that the B875-RC cores (100 Bchls, 3-5 RCs) and the B800-850 units (30 $Bchl_{850}$, 15 $Bchl_{800}$) represent basic elements that are assembled into a functional light-harvesting antenna [190,192].

References

1. Duysens, L.N.M. 1952 Doctoral Thesis, State University of Utrecht, The Netherlands

2. Duysens, L:N.M. 1964 Progr. Biophys. 14, 1-104

3. Duysens, L.N.M. 1986 in: Light Emission by Plants and Bacteria (Govindjee, J. Amesz and D.C. Fork, eds.) chapter 1, pp. 3-26, Academic Press, Orlando, Florida

4. Deisenhofer, J., Epp. O., Miki, K., Huber, R. and Michel, H. 1984 J. Mol. Biol. 180, 385-398

5. Allen, J.P., Feher, G., Yeates, T.O., Komiya, H. and Rees, D.C. 1987 Proc. Natl. Acad. Sci: USA 84, 5730-5734

6. Förster, T. 1948, Ann. Phys. Leipzig, 2, 55-75

7. Thornber J.P., Markwell, J.P. and Reinman, S. 1979 Photochem. Photobiol. 29, 1205-1216

8. Drews, G. 1985 Microbiological Reviews 49, 59-70

9. Zuber, H. 1985 Photochem. Photobiol. 42, 821-844

10. Thornber, J.P. 1986 in: Encyclopedia of Plant Physiology New Series, Volume 19, Photosynthesis III (L.A. Staehelin and C.J. Arntzen, eds.) chapter 3, pp. 98-142, Springer-Verlag, Berlin, Heidelberg

11. Scheer, H. 1981 Angew. Chem. 93, 230-250

12. Zuber, H., Brunisholz, R. and Sidler, W. 1987 in: Photosynthesis (Amesz, J. ed.) pp. 233-271, Elsevier, Amsterdam

13. Matthews, B.W., Fenna, R.E., Bolognesi, R.E., Schmid, M.F. and Olson, J.M. 1979 J. Mol. Biol. 131, 259-285

14. Schirmer, T., Bode, W., Huber, R., Sidler, W. and Zuber, H. 1985 J. Mol. Biol. 184, 257-277

15. Schirmer, T., Huber, R., Schneider, M., Bode, W., Miller, M. and Hackert, M.L. 1986 J. Mol. Biol. 188, 651-676

16. Karukstis, K.K. and Sauer, K. 1983 J. Cell. Biochem. 23, 131-158

17. Van Grondelle, R. 1985, Biochim. Biophys. Acta 811, 147-195

18. Holzwarth, A.R. 1986 Photochem. Photobiol. 43, 707-725

19. Geacintov, N.E. and Breton, J. 1987 CRC Critical Reviews in Plant Science, Vol. 5, No. 1, 1-44

20. Knox, R.S. 1977 in: The Intact Chloroplast, Vol. 2, (J. Barber, ed.) pp. 55-97, Elsevier, Amsterdam

21. Pearlstein, R.M. 1982 in: Photosynthesis, Energy Conversion by Plants and Bacteria (Govindjee, ed.) Vol. 1, pp. 293-329, Academic Press, New York

22. Petrich, J.W. and Fleming, G.R. 1984 Photochem. Photobiol. 40, 775-780

23. Glazer, A.N. 1984, Biochim. Biophys. Acta 768, 29-51

24. Sebban, P. 1985 Physiol. Vég. 23, 449-462

25. Moya, I., Sebban, P. and Haehnel, W. 1986 in: Light Emission by Plants and Bacteria (Govindjee, J. Amesz and D.C. Fork, eds.) ch. 7, pp. 161-186, Academic Press Inc., Orlando, Florida

26. Van Grondelle, R. and Amesz, J. 1986 ibidem, ch. 8, pp. 191-223

27. Amesz, J. and Vasmel, H. 1986 ibidem, ch. 15, 423-446

28. Holzwarth, A.R. 1987 in: Topics in Photosynthesis, Vol. 8, The Light Reactions (J. Barber, ed.) ch. 3, pp. 95-157, Elsevier Science Publishers b.v., Amsterdam, The Netherlands

29. Nechushtai, R., Nourizadeh, S.D. and Thomber, J.P. 1986 Biochim. Biophys. Acta 848, 193-200

30. Holzwarth, A.R., Weadler, J. and Haehnel, W. 1985 Biochim. Biophys. Acta 807, 155-167

31. Gulotty, R.J., Mets, L., Alberte, R.S. and Fleming, G.R. 1985 Photochem. Photobiol. 41, 487-496

32. Yamazaki, I., Mimuro, M., Tamai, N., Yamazaki, T. and Fujita, Y. 1985 FEBS Letters 179, 65-68

33. Owens, T.G., Webb, S.P., Mets, L., Alberts, R.S. and Fleming, G.R. 1987 Proc. Natl. Acad. Sci USA 84, 1532-1536

34. Pearlstein, R.M. 1982 Photochem. Photobiol. 35, 835-844

35. Breton, J., Martin, J.-L., Migus, A., Antonetti, A. and Orszag, A. 1986 Proc. Natl. Acad. Sci. USA 83, 5121-5125

36. Zinth, W., Nuss, M.C., Franz, M.A. and Kaiser, W. 1985 in: Antennas and Reaction Centers of Photosynthetic Bacteria (M.E. Michel-Beyerle, ed.) pp. 286-291, Springer Verlag, Berlin, Heidelberg

37. Bakker, J.G.C., van Grondelle, R., den Hollander, W.T.F. 1983 Biochim. Biophys. Acta 725, 508-518

38. Sundström, V., van Grondelle, R., Bergström, H., Åkesson, E. and Gillbro, T. 1986 Biochim. Biophys. Acta 851, 431-446

39. Bergström, H., Sundström, V., Van Grondelle, R., Åkesson, E. and Gillbro, T. 1986 Biochim. Biophys. Acta 852, 279-287

40. Nuijs, A.M., Shuvalov, V.A., van Gorkom, H.J., Plijter, J.J. and Duysens, L.N.M. 1986 Biochim. Biophys. Acta 850, 310-318

41. Reisberg, P., Nairn, J.A. and Sauer, K. 1982 Photochem. Photobiol. 36, 657-661

42. Butler, W.L., Tredwell, C.J., Malkin, R. and Barber, J. 1979 Biochim. Biophys. Acta 545, 309-315

43. Campillo, A.J., Shapiro, S.L., Geacintov, N.E. and Swenberg, C.E. 1977 FEBS Letters 83, 316-320

44. Pellegrino, F., Dagen, A., Sekulen, P. and Alfano, R.R. 1983 Photobiochem. Photobiophys. 6, 15-23

45. Wittmershaus, B., Nordlund, T.M., Knox, W.H., Knox, R.S., Geacintov, N.E. and Breton, J. 1985 Biochim. Biophys. Acta 806, 93-106

46. Wittmershaus, B. 1987 Progr. Photosynthesis Res. (J. Biggins, ed.) vol. I, pp. 75-82, Martinus Nijhoff, Dordrecht, The Netherlands

47. Garnier, J., Maroc, J. and Guyon, D. 1986 Biochim. Biophys. Acta 851, 395-406

48. Van Grondelle, R., Bergström, H., Sundström, V., Vos, M., Van Dassen, R.J. and Hunter, C.N. in: Organisation and Structure of Photosynthetic Antennas (H. Scheer and J. Schneider, eds.), Walter de Gruyter and Co., Berlin, in press

49. Kühlbrandt, W. 1984 Nature 307, 478-480

50. Mullet, J.E. 1983 J. Biol. Chem. 258, 9941-9948

51. Thornber, J.P., Peter, G.F., Peterson, C.C. and Nechushtal, R. in: Organization and Function of Photosynthetic Antennas (H. Scheer and S. Schneider. eds.) Walter de Gruyter and Co., Berlin, in press

52. Bürgi, R., Suter, F. and Zuber, H., Biochim. Biophys. Acta 890 1987 346-351

53. Delepelaire, P. and Chua, N.-H. 1979 Proc. Natl. Acad. Sci. USA 76, 111-115

54. Satoh, K. 1985 Photochem. Photobiol. 42, 845-853

55. Green, B.R., Camm, E.L. and Van Houten, J. 1982 Biochim. Biophys. Acta 681, 248-255

56. Trebst, A. 1986 Z. Naturforsch. 41C, 240-245

57. Satoh, K. and Nanba, O. 1987 in: Progress in Photosynthesis Research (J. Biggins, ed.), Vol. II, pp. 69-72, Martinus Nijhoff, Dordrecht, The Netherlands

58. Nanba, O. and Satoh, K. 1987 Proc. Natl. Acad. Sci. USA 84, 109-112

59. Barber, J., Chapman, D.J. and Telfer, A. 1987 FEBS Letters 220, 67-73

60. Murata, N. and Satoh, K. 1986 in: Light Emission by Plants and Bacteria (Govindjee, J. Amesz and D.C. Fork, eds.) ch. 6, pp. 137-159, Academic Press Inc., Orlando, Florida

61. Hodges, M. and Moya, I. 1986 Biochim. Biophys. Acta 849, 193-202

62. Wendler, J. and Holzwarth, A.R. 1987 Biophys. J., in press

63. Woodbury, N.W., Becker, M., Middendorf, D. and Parson, W.W. 1985 Biochemistry 24, 7516-7521

64. Martin, J.-L., Breton, J., Hoff, A.J., Migus, A. and Antonetti, A. 1986 Proc. Natl. Acad. Sci. USA 83, 957-961

65. Schatz, G.H. and Holzwarth, A.R. 1986 Photosynthesis Research 10, 309-318

66. Schatz, G.H., Brock, H. and Holzwarth, A.R. 1987 Proc. Natl. Acad. Sci. USA, in press

67. Trissl, H.-W., Breton, J., Deprez, J. and Leibl, W. 1987 Biochim. Biophys. Acta 893, 305-319

68. Dobek, A., Deprez, J., Geacintov, N.E., Paillotin, G. and Breton, J. 1985 Biochim. Biophys. Acta 806, 81-92

69. Karukstis, K.K. and Sauer, K. 1985 Biochim. Biophys. Acta 806, 374-388

70. Karukstis, K.K. and Sauer, K. 1983 Biochim. Biophys. Acta 725, 384-393

71. Rijgersberg, C.P., Amesz, J., Thielen, A.P.G.M. and Swager, J.A. 1979 Biochim. Biophys. Acta 545, 473-482

72. Knox, R.S. and Lin, S. in: Organization and Function of Photosynthetic Antennas (H. Scheer and S. Schneider, eds.) Walter de Gruyter and Co, Berlin, in press

73. Van Gorkom, H. 1986 in: Light Emission by Plants and Bacteria (Govindjee, J. Amesz and D.C. Fork, eds.) ch. 10, pp. 267-289, Academic Press Inc., Orlando, Florida

74. Berens, S.J., Scheele, J., Butler, W.L. and Magde, D. 1985 Photochem. Photobiol. 42, 51-57

75. Berens, S.J., Scheele, J., Butler, W.L. and Magde, D. 1985 Photochem. Photobiol. 42, 59-68

76. Hodges, M. and Moya, I. 1987 Biochim. Biophys. Acta 892, 42-47

77. Klimov, V.V. 1984 in: Advances in Photosynthesis Research (C. Sybesma, ed.) Vol. 1, pp. 131-138, Nijhoff/Junk, Dordrecht, The Netherlands

78. Van Bochove, A.C., Van Grondelle, R. and Duysens, L.N.M. 1981 in: Photosynthesis (G. Akoyunoglou, ed.), Vol. III, pp. 989-996, Balaban International Sciences Services, Philadelphia, PA

79. Tapie, P., Choquet, Y., Wollman, F.-A., Diner, B. and Breton, J. 1986 Biochim. Biophys. Acta 850, 156-161

80. Van Dorssen, R.J., Plijter, J.J., den Ouden, A., Amesz, J. and Van Gorkom, H.J. 1987 Biochim. Biophys. Acta 890, 134-143

81. Breton, J. amd Katoh, S. 1987 Biochim. Biophys. Acta 892, 99-107

82. Garab, G. and Breton, J. 1976 Biochim. Biophys. Res. Commun. 71, 1095-1102

83. Van Dorssen, R.J., Breton, J., Plijter, J.J., Satoh, K., Van Gorkom, H.J. and Amesz, J. 1987 Biochim. Biophys. Acta 893, 267-274

84. Gantt, E. 1980 Int. Rev. Cytol. 66, 43-80

85. Gantt, E. 1981 Ann. Rev. Plant. Physiol. 32, 327-347

86. Mörschel, E. and Riehl, E. 1987 in: Electron Microscopy of Proteins (Vol. 6): Membraneous Structures (J.R. Harris, and W. Home, eds.), pp. 209-254, Academic Press, London

87. Zuber, H. 1978 Ber. Deutsch. Bot. Ges. 91, 459-475

88. Glazer, A.N. 1982 Ann. Rev. Microbiol. 36, 173-198

89. Wehrmeyer, W. 1983 in: Photosynthetic Prokaryotes (G.C.Papageorgio, and L. Packer, eds.), pp. 1-22, Elsevier, Amsterdam

90. Wehrmeyer, W. 1983 in: Proteins and Nucleic Acids in Plant Systematics (U. Jensen and D.E. Fairbrothers, eds.), pp. 143-167, Springer Verlag, Berlin

91. Glazer, A.N. 1983 Ann. Rev. Biochem. 52, 125-157

92. Glazer, A.N., Lundell, D.J., Yamanaka, G. and Williams, R.C. 1983 Ann. Microbiol. 134B, 154-180

93. Glazer, A.M. 1985 Ann. Rev. Biochem. Biophys. 14, 47-77

94. Glazer, A.N., Yeh, S.W., Webb, S.P. and Clark, J.H. 1985 Science 227, 419-423

95. Suter, G.W., Mazzola, P., Wendler, J. and Holzwarth, A.R. 1984 Biochim. Biophys. Acta 766, 269-276

96. Wendler, J., Holzwarth, A.R. and Wehrmeyer, W. 1984 Biochim. Biophys. Acta 765, 58-67

97. Gillbro, T., Sandström, Å., Sundström, V., Wendler, J. and Holzwarth, A.R. 1985 Biochim. Biophys. Acta 808, 52-65

98. Sandström, Å., Gillbro, T., Sundström, V., Wendler, J. and Holzwarth, A.R. 1987 Biochem. Biophys. Acta, in press

99. Gillbro, T., Sandström, Å., Sundström, V. and Holzwarth, A.R. 1983 FEBS Letters 162, 64-68

100. Yamazaki, I., Mimuro, M., Murao, T., Yamazaki, T., Yoshihara, K. and Fujita, Y. 1984 Photochem. Photobiol. 39, 233-240

101. Mimuro, M., Yamazaki, I., Yamazaki, T. and Fujita, Y. 1985 Photochem. Photobiol. 41, 597-603

102. Glazer, A.N. 1981 in: The Biochemistry of Plants, Vol. 18, (M.D. Hatch and N.K. Boardman, eds.) pp. 51-96, Academic Press, New York

103. Scheer, H. 1982 in: Light Reaction Path of Photosynthesis (Fong, F.K., ed.) pp. 7-45, Springer Verlag, Berlin

104. Dale, R.E. and Teale, F.W.J. 1970 Photochem. Photobiol. 12, 99-117

105. Teale, F.W.J. and Dale, R.E. 1970 Biochem. J. 116, 161-169

106. Wendler, J., John, W., Scheer, H. and Holzwarth, A.R. 1986 Photochem. Photobiol. 44,

107. Wehrmeyer, W., Wendler, J. and Holzwarth, A.R. 1985 Eur. J. Cell. Biol. 36, 17-23

108. Holzwarth, A.R., Wendler, J. and Wehrmeyer, W. 1983 Biochim. Biophys. Acta 724, 388-395

109. Schneider, S., Geiselhart, P., Scharnagl, C., Schirmer, T., Bode, W., Sidler, W. and Zuber, H. 1985 in: Antennas and Reaction Centers of Photosynthetic Bacteria (M.E. Michel-Beyerle, ed.), pp. 36-44, Springer Verlag, Berlin

110. Hanzlik, C.A., Hancock, L.E., Knox, R.S., Guard-Friar, D. and MacColl, R. 1985 J. Lumin. 34, 99-106

111. Dagen, A.J., Alfano, R.R., Zilinskas, B.A. and Swenberg, C.E. 1986 Photochem. Photobiol. 43, 71-79

112. Holzwarth, A.R., Wendler, J. and Suter, G.W. 1987 Biophys. J. 51, 1-12

113. Gillbro, T., Sandström, Å., Sundström, V., Fischer, R. and Scheer, H. 1987 in: Organization and Function of Photosynthetic Antennas (H. Scheer and S. Schneider, eds.) Walter de Gruyter and Co., Berlin, in press

114. Sandström, Å., Gillbro, T., Sundström, V., Fischer, R. and Scheer, H. 1987 Biochim. Biophys. Acta, in press

115. Sauer, K., Scheer, H. and Sauer, P. 1987 Photochem. Photobiol. 46, 427-440

116. Schirmer, T., Bode, W. and Huber, R. 1987 J. Mol. Biol. 196, 677-695

117. Holzwarth, A.R. and Suter, G.W. 1987 Biophys. J., in press

118. Schneider, S., Scharnagl, C., Duerring, M., Schirmer, T. and Bode, W. in: Organization and Function of Photosynthetic Antennas (H. Scheer and S. Schneider, eds.) Walter de Gruyter and Co., Berlin, in press

119. Fowler, C.F., Nugent, N.A. and Fuller, R.C. 1971 Proc. Natl. Acad. Sci. USA 68, 2278-2282

120. Staehelin, L.A., Golecki, J.R., Fuller, R.C. and Drews, G. 1978 Arch. Microbiol. 119, 269-277

121. Pierson, B.K. and Castenholz, R.W. 1978 in: The Photosynthetic Bacteria (R.K. Clayton and W.R. Sistrom, eds.) pp. 179-197, Plenum Press, New York

122. Olson, J.M. 1980 Biochim. Biophys. Acta 594, 33-51

123. Blankenship, R.E. and Fuller, R.C. 1986 in: Encyclopedia of Plant Physiology, New Series, Vol. 19 (L.A. Staehelin and C.J. Arntzen, eds.) Springer Verlag, Berlin, New York, pp. 390-399

124. Feick, R.G., Fitzpatrick, M. and Fuller, R.C. 1982 J. Bact. 150, 905-915

125. Feick, R.G. and Fuller, R.C. 1984 Biochemistry 23, 3693-3700

126. Wechsler, Th., Suter, F., Fuller, R.C. and Zuber, H. 1985 FEBS Lett. 181, 173-178

127. Smith, K.M., Kehres, L.A. and Fajer, J. 1983 J. Am. Chem. Soc. 105, 1387-1389

128. Olson, J.M., Gerola, P.D., Van Brakel, G.H., Meiburg, R.F. and Vasmel, H. 1985 in: Antennas and Reaction Centers of Photosynthetic Bacteria - Structure, Interactions and Dynamics (M.E. Michel-Beyerle, ed.) Springer Verlag, Berlin 129. Brune, D.C., Nozawa, T. and Blankenship, R.E. 1987 Biochemistry, in press

130. Olson, J.M., van Brakel, G.H. and Pedersen, J.D. 1988 in: Proc. EMBO Workshop on Green Photosynthetic Bacteria (J.M. Olson, ed.) Plenum Press, New York, in press

131. Staehelin, L.A., Golecki, J.R. and Drews, G. 1980 Biochim. Biophys. Acta 305, 597-609

132. Gerola, P., Højrup, P., Knudsen, J. and Olson, J.M. 1988 in: Proc. EMBO Workshop on Green Photosynthetic Bacteria (J.M. Olson, ed.) Plenum Press, New York, in press

133. Schmidt, K. 1980 Arch. Microbiol. 136, 11-16

134. Betti, J.A., Blankenship, R.E., Natarajan, L.V., Dickinson, L.C. and Fuller, R.C. 1982 Biochim. Biophys. Acta 680, 194-201

135. Gerola, P.D. and Olson, J.M. 1986 Biochim. Biophys. Acta 848, 69-76

136. Van Dorssen, R.J., Vos, M. and Amesz, J. 1988 in: Organization and Function of the Photosynthetic Antennas (H. Scheer and S. Schneider, eds.) Walter de Gruyter and Co., in press

137. Swarthoff, T. and Amesz, J. 1979 Biochim. Biophys. Acta 548, 427-432

138. Matthews, B.W., Fenna, R.E., Bolognesi, M.C., Schmid, M.F. and Olson, J.M. 1979 J. Mol. Biol. 131, 259-285

139. Zuber, H. 1988 in: Proc. EMBO Workshop on Green Photosynthetic Bacteria (J.M. Olson, ed.) Plenum Press, New York, in press

140. Nuijs, A.M., Vasmel, H., Duysens, L.N.M. and Amesz, J. 1986 Biochim. Biophys. Acta 849, 316-324

141. Van Dorssen, R.J., Gerola, P.D., Olson, J.M. and Amesz, J. 1986 Biochim. Biophys. Acta 848, 77-82

142. Fetisova, Z.G., Khardenko, S.G. and Abdourakhmanov, I.A. 1986 FEBS Letters 199, 234-236

143. Olson, J.M. and Sybesma, C. 1963 in: Bacterial Photosynthesis (H. Gest, A. San Pietro and L.P. Vernon, eds.) pp. 413-422, Antioch. Press, Yellow Springs, Ohio

144. Vos, M., Nuijs, A.M., van Grondelle, R., van Dorssen, R.J., Gerola, P.D. and Amesz, J. 1987 Biochim. Biophys. Acta 891, 275-285

145. Fetisova, Z.G. and Borisov, A.Yu. 1980 FEBS Lett. 114, 323-326

146. Gillbro, T., Sandström, Å, Sundström, V. and Olson, J.M. 1988 in: Proc. EMBO Workshop on Green Photosynthetic Bacteria (J.M. Olsen, ed.) Plenum Press, New York, in press

147. Freiberg, A.M., Timpmann, K.E. and Fetisova, Z.G. 1988 in: Proc. EMBO Workshop on Green Photosynthetic Bacteria (J.M. Olson, ed.) Plenum Press, New York, in press

148. Borisov, A.Yu., Fetisova, Z.G. and Godik, V.I. 1977 Biochim. Biophys. Acta 461, 500-509

149. Kramer, H.J.M., Kingma, H., Swarthoff, T. and Amesz, J. 1982 Biochim. Biophys. Acta 681, 359-364

150. Van Dorssen, R.J., Vasmel, H. and Amesz, J. 1986 Photosynth. Res. 9, 33-46

151. Van Amerongen, H., Vasmel, H. and van Grondelle, R., submitted for publication to Biophys. J.

152. Van Dorssen, R.J. and Amesz, J. 1987 Photosynth. Res., in press

153. Wittmershaus, B.P., Brune, D.C. and Blankenship, R.E. 1988 in: Proc. of the Workshop on Organization and Function of the Photosynthetic Antennas (H. Scheer and S. Schneider, eds.) Walter de Gruyter and Co., Berlin, in press

154. Brune, D.C., King, G.H., Infosino, A., Steiner, T., Thewalt, M.L.W.

and Blankenship, R.E. 1987 Biochemistry, in press 155. Brune, D.C. and Blankenship, R.E. 1987 in: Progress in Photosynthesis Research, Vol. 1 (J. Biggins, ed.), pp. 419-422 Martinus Nijhoff, Dordrecht, The Netherlands

156. Blankenship, R.E., Brune, D.C., Freeman, J.M., King, G.M., McManus, J.H., Nozawa, T., Trost, J.T. and Wittmershaus, B.P. 1988 in: Proc. EMBO Workshop on Green Photosynthetic Bacteria (J.M. Olsen, ed.) Plenum Press, New York, in press

157. Vasmel, H., van Dorssen, R.J., de Vos, G.J. and Amesz, J. 1986 Photosynth. Res. 7, 281-294

158. Nuijs, A.M., Vasmel, H., Joppe, L.P., Duysens, L.N.M. and Amesz, J. 1985 Biochim. Biophys. Acta 807, 24-34

159. Hoover, S.M. 1985 M.Sc. Thesis, University of California, Berkeley, USA

160. Van Grondelle, R., Hunter, C.N., Bakker, J.G.C. and Kramer, H.J.M. 1983 Biochim. Biophys. Acta 723, 30-36

161. Vos, M. 1987 Doctoral Thesis, State University of Leiden, The Netherlands

162. Borisov, A.Yu. 1978 in: The Photosynthetic Bacteria (R.K. Clayton and W.R. Sistrom, eds.) pp. 323-331, Plenum Publishing Corporation, New York

163. Thornber, J.P., Cogdell, R.J., Pierson, B.K. and Seftor, R.E.B. 1983 J. Cell. Biochem. 23, 159-169

164. Borisov, A.Yu., Freiberg, A.M., Godik, V.I., Rebane, K. and Timpmann, K.E. 1985 Biochim. Biophys. Acta 807, 221-229

165. Nuijs, A.M., van Grondelle, R., Joppe, H.L.P., van Bochove, A.C. and Duysens, L.N.M. 1986 Biochim. Biophys. Acta 850, 286-293

166. Kramer, H.J.M., van Grondelle, R., Hunter, C.N., Westerhuis, W.H.J. and Amesz, J. 1984 Biochim. Biophys. Acta 765, 156-165 167. Rijgersberg, C.P., van Grondelle, R. and Amesz, J. 1980 Biochim. Biophys. Acta 592, 53-64

168. Zankel, K.L. and Clayton, R. 1969 Photochem. Photobiol. 9, 7-15

169. Sebban, P., Jolchine, G. and Moya, I. 1984 Photochem. Photobiol. 39, 247-253

170. Breton, J. and Vermeglio, A. 1982 in: Photosynthesis, Energy Conversion by Plants and Bacteria, Vol. 1 (Govindjee, ed.) pp. 153-194, Academic Press, New York

171. Bolt, J.D., Hunter, C.N., Niederman, R.A. and Sauer, K. 1981 Photochem. Photobiol. 34, 653-656

172. Kramer, H.J.M., Pennoyer, J.D., van Grondelle, R., Westerhuis, W.H.J., Niederman, R.A. and Amesz, J. 1984 Biochim. Biophys. Acta 767, 335-344

173. Van Grondelle, R., Bergström, H., Sundström, V. and Gillbro, T. 1987 Biochim. Biophys. Acta, in press

174. Freiberg, A., Godik, V.I. and Timpmann, K. 1987 in: Progress in Photosynthesic Research, Vol. 1 (J. Biggins, ed.) pp. 45-48, Martinus Nijhoff Publishers, Dordrecht, The Netherlands

175. Van Grondelle, R., Kramer, H.J.M. and Rijgersberg, C.P. 1982 Biochim. Biophys. Acta 682, 208-215

176. Godik, V.I., Freiberg, A., Timpmann, K., Borisov, A.Yu. and Rebane, K.K. 1987 in: Progress Photosynthesis Research, Vol. 1, (J. Biggins, ed.) pp. 41-44, Martinus Nijhoff Publishers, Dordrecht, The Netherlands

177. Petrich, W., Breton, J. and Martin, J.-L. 1987 in: Springer Series in Chemical Physics, in press

178. Siefermann-Harms, F.M. 1985 Biochim. Biophys. Acta 811, 325-355

179. Cogdell, R.J., Hipkins, F.M., MacDonald, W. and Truscott, T.G. 1981 Biochim. Biophys. Acta 634, 191-202 180. Cogdell, R.J. 1985 Pure and Appl. Chem. 57, 723-728

181. Angerhofer, A., Cogdell, R.J. and Hipkins, M.F. 1986 Biochim. Biophys. Acta 848, 333-341

182. Wasielewski, M.R. and Kispert, L.D. 1986 Chem. Phys. Lett. 128, 238-243

183. Wasielewski, M.R., Tiede, D.M. and Frank, H.A. 1986 in: Springer Series in Chemical Physics 46, Ultrafast Phenomena 5 (G.R. Fleming and A.E. Siegmann, eds.) pp. 388-392, Springer Verlag

184. Sundström, V., Bergström, H., Gillbro, T., van Grondelle, R., Westerhuis, W.H.J., Niederman, R.A. and Cogdell, R.J. 1987 in: Organization and Function of Photosynthetic Antennas (H. Scheer and S. Schneider, eds.) Walter de Gruyter and Co., Berlin, in press

185. Paillotin, G., Swanberg, C.E., Breton, J. and Geacintov, N.E. 1979 Biophys. J. 25, 513-533

186. Den Hollander, W.T.F., Bakker, J.G.C. and van Grondelle, R. 1983 Biochim. Biophys. Acta 725, 492-507

187. Vos, M., van Grondelle, R., van der Kooy, F.W., van de Poll, D., Amesz, J. and Duysens, L.N.M. 1986 Biochim. Biophys. Acta 850, 501-512

188. Hunter, C.N., Kramer, H.J.M. and van Grondelle, R. 1985 Biochim.

Biophys Acta 807, 44-51

189. Vos, M., van Dorssen, R.J., Amesz, J., van Grondelle, R. and Hunter, C.N., submitted for publication to Biochim. Biophys. Acta

190. Hunter, C.N. and van Grondelle, R. 1987 in: Organization and Function of Photosynthetic Antennas (H. Scheer and S. Schneider, eds.) Walter de Gruyter and Co., Berlin, in press 191. Westerhuis, W.H.J., Vos, M., van Dorssen, R.J., van Grondelle, R., Amesz, J. and Niederman, R.A. 1987 in: Progress in Photosynthesis Research (J. Biggins, ed.), Vol. 1, pp. 29-32, Martinus Nijhoff Publishers, Dordrecht, The Netherlands

192. Hunter, C.N., van Grondelle, R. and van Dorssen, R.J., submitted for publication in Biochim. Biophys. Acta

FLUORESCENCE SPECTROSCOPY OF ALLOPHYCOCYANIN COMPLEXES FROM *Synechococcus* 6301 STRAIN AN112

Patricia Maxson, Kenneth Sauer

Department of Chemistry and Chemical Biodynamics Division, Lawrence Berkeley Laboratory, University of California, Berkeley, CA, USA

Alexander N. Glazer

Department of Microbiology and Immunology, University of California, Berkeley, CA, USA

Abstract

Three trimeric core complexes (AP trimer, AP-$L_C^{8.9}$, and APB-$L_C^{8.9}$) and the monomer of AP of the phycobilisome of *Synechococcus* PCC 6301 strain AN112 have been isolated and characterized with respect to their steady state fluorescence polarization and time resolved fluorescence emission properties. The data for APB-$L_C^{8.9}$ support an "s" to "f" model of energy transfer. The data for the monomer and the AP trimers with and without linker show that new interactions are created upon trimer formation but that the linker has no detectable effect on the spectroscopy of the isolated complex. No evidence of "s" to "f" transfer of energy is seen for the AP and AP-$L_C^{8.9}$ trimers; however, exciton delocalization may play a significant role.

Introduction

Cyanobacteria and red algae possess a macromolecular protein complex that serves as a light-harvesting antenna to photosystem II. This complex, known as the phycobilisome, consists of phycobiliproteins, whose brilliant colors derive from covalently linked bilin prosthetic groups, and of linker polypeptides that are involved in assembly as well as light-harvesting functions. The location of the component polypeptides is known for phycobilisomes of several organisms, and is defined in the greatest detail for those of *Synechococcus* PCC 6301 [1]. In this phycobilisome, six rods made up of stacked hexameric complexes of phycocyanin radiate in a hemidiscoidal array from two contiguous core cylinders that consist of allophycocyanin-containing trimeric complexes [2]. The nitrosoguanidine-induced mutant, *Synechococcus* PCC 6301 strain AN112, is particularly useful for the study of the core

Photosynthetic Light-Harvesting Systems

complexes because it lacks the 30 and 33 kD polypeptides required for the assembly of the distal hexamers of the phycobilisome rods. Only the basal hexamer of each rod is present [3]. In consequence, the ratio of allophycocyanin to phycocyanin is much higher in this mutant than in the wild type, and purification of the core complexes is thereby facilitated.

The core complexes of strain AN112 are identical to those of the wild type. Two copies of four distinct complexes make up the two core cylinders: 1) $(\alpha^{AP}\beta^{AP})_3$ (AP trimer), 2) $(\alpha^{AP}\beta^{AP})_3$-$L_C^{8.9}$ (AP-$L_C^{8.9}$), 3) $(\alpha^{APB}\alpha_2^{AP}\beta_3^{AP})$-$L_C^{8.9}$ (APB-$L_C^{8.9}$), and 4) $(\alpha_2^{AP}\beta_2^{AP}\beta^{18.3}L_{CM}^{75})$ where the phycobiliprotein subunits and linkers are abbreviated as suggested by Glazer [4]. The spectroscopic properties of AP, AP-$L_C^{8.9}$ and APB-$L_C^{8.9}$ have been the subject of several recent studies [5-7] aimed at establishing the pathways by which energy is transferred within the core. In this report, we describe the steady state emission polarization and time resolved emission of these three complexes and of the AP monomer $(\alpha^{AP}\beta^{AP})$. These studies support the "s" to "f" model proposed by Dale and Teale [8,9], in which a sensitizing ("s") chromophore transfers its energy to a fluorescing ("f") chromophore, as applicable to energy transfer within APB-$L_C^{8.9}$ but not within the AP or AP-$L_C^{8.9}$ trimers. In addition, the presence of the $L_C^{8.9}$ linker polypeptide does not appear to influence the energy transfer properties of the isolated AP trimer.

Materials and Methods

Synechococcus PCC 6301 strain AN112 was grown on BG-11 medium under constant illumination and CO_2 flow. Phycobilisomes were prepared as described previously [10]. AP was purified directly from fractionation of phycobilisomes into the constituent phycobiliprotein complexes on a DEAE-cellulose (Whatman DE-52) column [11]. For the preparation of AP-$L_C^{8.9}$ and APB-$L_C^{8.9}$, partially dissociated phycobilisomes were separated by sucrose density gradient centrifugation into three components termed 6S, 11S, and 18S after their positions in the gradients [10]. Chromatography of the 6S component on hydroxylapatite (Bio-Rad, Richmond, CA) yielded pure AP-$L_C^{8.9}$ and a mixture of phycocyanin and APB-$L_C^{8.9}$ which were further separated on a DEAE-cellulose column as described in [12]. The complexes were found to be free of contaminating polypeptides, as determined by SDS-PAGE.

AP monomer was prepared by dissociation of the trimer with 0.6 M NH_4SCN in 0.65 M Na/K phosphate buffer pH 7.

For spectroscopic measurements the proteins were diluted in 0.65M Na/K phosphate pH 7 to a maximum absorbance between 0.05 and 0.1. Because dissociation occurs with time at such low concentrations, measurements were made as soon as possible after dilution.

Absorption measurements were made with a Varian 2300 UV-Visible Spectrophotometer (Varian, Houston, TX). Steady state fluorescence emission spectra were taken with a SPEX Fluorolog 2 model 212 spectrofluorimeter (SPEX Industries Inc., Metuchen, NJ) interfaced with an Epson Equity I through a Fluorometer Control System (Brown's Software, El Cerrito, CA). Monochromator slits were set to give a 4 nm bandwidth. Emission spectra were corrected for artifacts with an NBS standard lamp. All measurements were made with detection at 90 ° to excitation with a sample pathlength of 1 cm. Polarization measurements were made with Glan Thompson polarizers. The emission wavelength for the excitation polarization scan of AP monomer was 680 nm; the excitation wavelength for the emission polarization scan was 570 nm. For AP, AP-$L_C^{8.9}$ and APB-$L_C^{8.9}$ the emission wavelength for the excitation polarization scans was 720 nm and the excitation wavelength for the emission polarization scans was 580 nm. The dark noise of the photomultiplier (RCA31034A) was subtracted from all the measurements. This was important because the signal at wavelengths of low absorption was often just above the background.

Picosecond pulses for the time resolved measurements were provided by a tunable, cavity-dumped dye laser synchronously pumped by an argon ion laser (Spectra Physics, Mountain View, CA) which used DCM as the lasing dye. The excitation beam was vertically polarized and the fluorescence emission from the samples was monitored at the magic angle in order to eliminate anisotropic artifacts in the decay components. The emission wavelength was selected by a double monochromator (Instruments SA, Metuchen, NJ) with a 4 nm bandwidth. Data were collected from an ITT microchannel plate detector (model F4129) using the single photon timing technique and were analyzed as sums of exponential decays using an iterative deconvolution program. The time resolution of the instrument is limited by the response of the detector and is around 20 ps.

Table I

Absorbance maxima and shoulders, and fluorescence maxima of AP monomer and trimer, AP-$L_C^{8.9}$, and APB-$L_C^{8.9}$. Approximate position of shoulders is determined from the first derivative of the absorption spectra. Fluorescence maxima are independent of excitation wavelength.

Complex	Absorbance max.(nm)	Absorbance shoul.(nm)	Fluorescence max.(nm)
Monomer	615	565 (v.wk.)	645
AP	650	617, 590	660
AP-$L_C^{8.9}$	652	592	663
APB-$L_C^{8.9}$	654	595, 675	680

Results

Table I lists the absorption and emission maxima, and the approximate position of shoulders, of the four complexes. The absorption maximum of the monomer is at 615 nm with a weak shoulder at about 560 nm. The emission bands of all the proteins except APB-$L_C^{8.9}$ have only one maximum between 600 and 700 nm with no other features. AP trimer and AP-$L_C^{8.9}$ have absorption maxima at 650 and 652 nm respectively with pronounced shorter wavelength shoulders at around 600 nm. APB-$L_C^{8.9}$ has an absorption maximum at 654 nm and an emission maximum at 680 nm. Absorption and emission spectra display shoulders at 670 and 660 nm respectively. The 680 nm emission maximum of APB-$L_C^{8.9}$ is considerably red-shifted relative to those of AP and AP-$L_C^{8.9}$ due to the α^{APB} chromophore.

Figure 1 shows the absorption and polarization spectra of AP monomer and trimer. For 680 nm emission the polarization across the wavelength range 500 to 660 nm is 0.42±0.02 The emission polarization from 600 to 740 nm (not shown) using 570 nm excitation is also invariant within the noise of the experiment with a value of 0.41±0.02. The polarization spectra of AP trimer and AP-$L_C^{8.9}$ (excitation and emission) are almost identical, demonstrating a value of 0.08±0.05 (emission data not shown). The wide range included by the error is indicative of the difficulty of maintaining the integrity of the sample at dilute concentrations

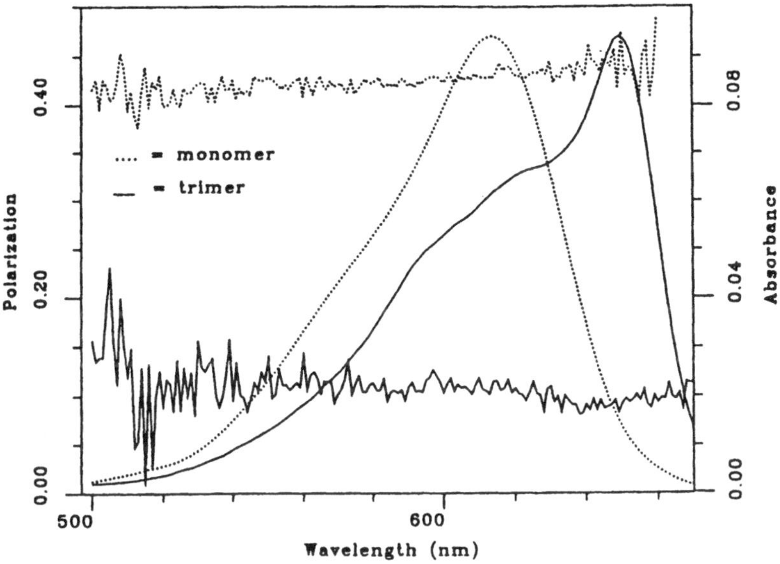

Fig. 1 Absorption and excitation polarization spectra of AP monomer and trimer. Emission is at 720 nm for the trimer polarization and 680 nm for the monomer polarization.

during the time required for these measurements.The polarization of AP trimer and AP-$L_C^{8.9}$ over both the excitation and emission wavelength ranges changes markedly with the time after dilution. After about 30 min the polarization at the long wavelength edge of the excitation scan and the short wavelength edge of the emission scan rises to a value greater than 0.2. We used the invariance of the polarization of the complexes during the time required for the lifetime measurements as a check of the integrity of the samples. In particular, we wanted to be sure that the time resolved results of APB-$L_C^{8.9}$ were not influenced by the presence of $(\alpha^{AP}\beta^{AP})$ and $(\alpha^{APB}\beta^{AP})$ monomers.

The time resolved fluorescence decay measurements on the AP monomer excited at 620 nm or 640 nm can be fit to sums of three exponentials: 1.8 ns and 0.8-1.0 ns components, which each constitute about 40% of the amplitude, and a 100-400 ps component which accounts for 10-20% of the amplitude. These components vary little over an emission wavelength range of 630 to 660 nm. Lifetime measurements of AP and AP-$L_C^{8.9}$ result in two-component fits to the data when the samples are excited at 640 nm and emission is viewed between 650 and 680 nm. The long-lived component of 1.6-1.7 ns accounts for 90% of the amplitude and the remaining 10% corresponds to a 0.25-0.4 ns component. The amplitudes for these components do not show a dependence on the emission wavelength. We do not see a fast lifetime as either a rise or a decay under these conditions. Table II provides a summary of lifetimes and amplitudes of the four complexes studied.

Table II

Summary of relative amplitudes (α) and lifetimes (τ) from time resolved measurements for selected excitation wavelengths on AP monomer and trimer, AP-$L_C^{8.9}$, and APB-$L_C^{8.9}$. Measurements were taken at ten nm intervals in the emission wavelength ranges listed.

	Monomer	AP	AP-$L_C^{8.9}$	APB-$L_C^{8.9}$	
λ_{ex}	620 nm	640 nm	640 nm	640 nm	670 nm
λ_{em}	630-650 nm	650-680 nm	650-680 nm	650-690 nm	680-690 nm
α_1	0.45	0.85-0.90	0.85-0.90	0.35-1.2	1.0
τ_1	1.8 ns	1.6 ns	1.7 ns	1.7ns	1.7ns
α_2	0.35-0.45	0.15-0.10	0.15-0.10	0.05	-
τ_2	0.77-1.0 ns	0.32-0.40 ns	0.32-0.35 ns	0.45-0.75 ns	-
α_3	0.25-0.11	-	-	0.60-[-0.25][a]	-
τ_3	0.14-0.35 ns	-	-	0.03-0.08 ns	-

[a] The brackets around this number are intended to point out that it is a rise time.

The APB-$L_C^{8.9}$ exhibits variation in the polarization over both excitation and emission wavelengths. The absorption spectrum and excitation polarization spectrum of APB-$L_C^{8.9}$ are shown in Figure 2. From 500 to 650 nm the polarization of APB-$L_C^{8.9}$ is almost constant at 0.05±0.03. At 650 nm it begins to rise and seems to approach a limiting value of about 0.3 by 690 nm. The emission polarization spectrum (not shown) behaves symmetrically; starting at a value of about 0.3 at 630 nm, the polarization drops to 0.05±0.03 by 650 nm where it remains to 780 nm.

In contrast to results for AP and AP-$L_C^{8.9}$, time resolved data from APB-$L_C^{8.9}$ fit to 1 to 3 components, depending on the excitation wavelength, and show a strong emission

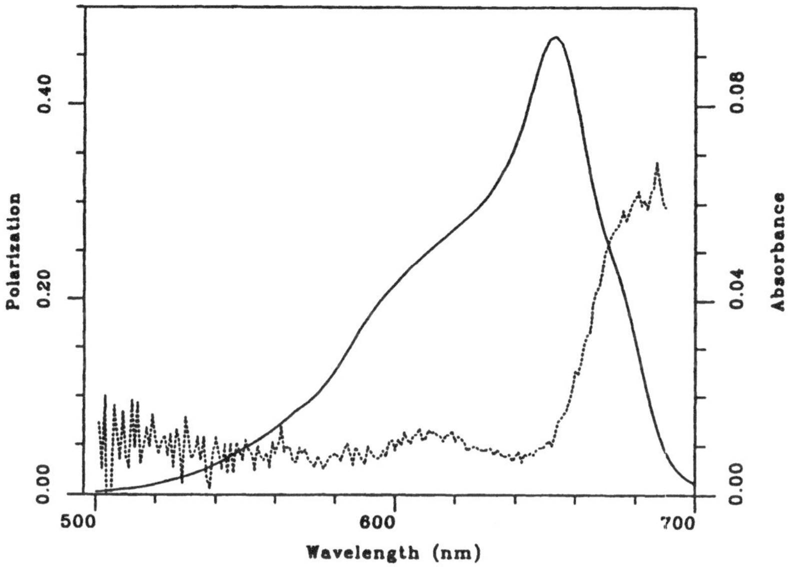

Fig. 2 Absorption and excitation polarization spectra of APB-$L_C^{8,9}$. For the polarization measurement emission is at 720 nm.

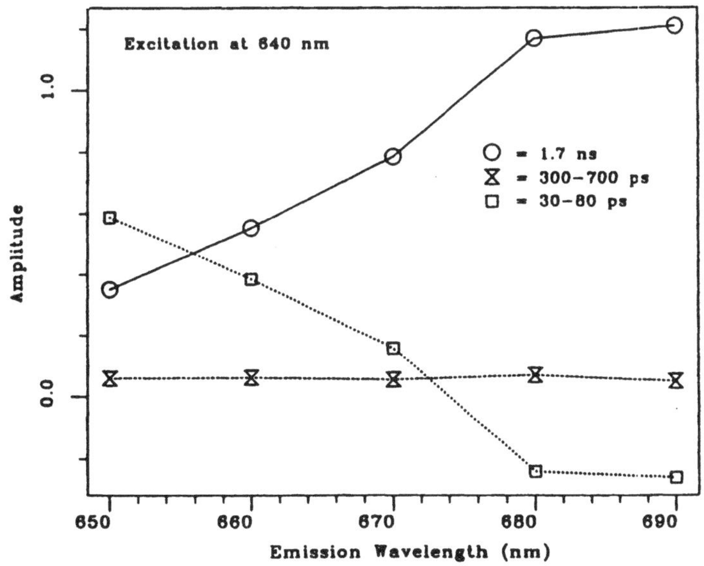

Fig. 3 Relative amplitudes of lifetime components vs emission wavelength for APB-$L_C^{8,9}$.

wavelength dependence in the amplitudes. Using 660 nm or 670 nm excitation wavelengths the emission decays at 680 nm and 690 nm fit to a single lifetime of 1.71±0.02 ns. Figure 3 shows the variation of amplitude with emission wavelength of data generated by 640 nm excitation. The same pattern is seen with an excitation of 650 nm. The long (1.7 ns) component is present with an amplitude that rises with increasing emission wavelength, as is a 300-700 ps component with constant and low (about 5%) amplitude. In addition we fit a fast (30 to 80 ps) component which is a decay at emission wavelengths shorter than 675 nm and a rise at longer wavelengths. The amplitude of this component is largest at 650 nm where it is positive and 690 nm where it is negative.

Discussion

The purpose of the spectroscopic studies described here is to try to gain an understanding of the interactions between the chromophores in the phycobilisome core complexes with the recognition that the isolation of the trimers and especially of the monomer may cause changes in their conformations and disruption of interactions with neighboring protein and chromophore groups that seriously affect their energetic properties. However, such complications are unavoidable.

The high polarization of the AP monomer corresponds to that expected for a protein with one isolated chromophore where the absorption and emission dipoles are at a small angle to each other. To reconcile this result with the two chromophores of AP monomer, the data can be interpreted as the result of: a) two non-interacting chromophores absorbing and emitting light independently of one another; or b) two interacting chromophores whose interaction is invisible to polarization measurements because the absorption and emission dipoles of the chromophores are nearly parallel. We see no fast fluorescence decay in the monomer which might be taken as evidence for energy transfer between the chromophores via an "s" to "f" mechanism but it is impossible to rule out such a mechanism for two reasons: 1) the transfer may take place at times faster than 20 ps, and 2) if the absorption bands completely overlap, any fast decay would be effectively cancelled by a complementary fast rise. The presence of two distinct long lifetimes (1.8 and 0.8-1.0 ns) with approximately the same amplitude

(40% each) does not support the notion that the chromophores are excitonically coupled. Also, circular dichroism studies [13] support the conclusion that the chromophores do not exhibit exciton interaction.

It is clear even from the red shift of the absorption and emission spectra of AP that new chromophore interactions are created in the formation of the trimer from the monomer. Two mechanisms of energy transfer have been proposed to account for the observed AP trimer polarization spectra: "s" to "f" transfer involving fast transfer times between distinct chromophores; and exciton coupling in which the excitation transfer is considered to be instantaneous because of delocalization of the electronic states. Our polarization data do not show a variation in the value of the polarization over the absorption wavelength range as is expected if "s" to "f" energy transfer is in effect. In most cases of "s", "f" chromophore pairs, the orientation of the absorption dipole of the "s" chromophore differs from the emission dipole of the "f" chromophore which results in a change of polarization over the absorption wavelength range. An example of this behavior occurs in the phycocyanin beta subunit which has two chromophores [14]. In the case of a protein with two chromophores interacting via exciton coupling, a change in the polarization spectrum is also expected since the two states created by the splitting of the exciton will be orthogonal to one another. Based on analogy to the crystal structure of the C-phycocyanin from *M. laminosus* and from *A. quadruplicatum* [15], trimer formation of AP creates 3 pairs of chromophores about 20 Å apart which are symmetric about the z axis of the trimer. If the chromophore pairs are involved in exciton coupling, their geometry could give rise to a net depolarization of the emission. Our time resolved data do not exhibit the fast decays or rises which are expected when a system is governed by an "s" to "f" mechanism of energy transfer but, again, we cannot rule out transfer times which are faster than 20 ps. The lifetime data are consistent with the model of three exciton pairs within the trimer which have the same energy level. We would not expect to see evidence of "s" to "f" transfer between the exciton pairs if the absorption bands overlap completely.

The results on the AP trimer and monomer complexes presented here agree with those of Yeh, et al. [6]. The authors of that work used the streak camera technique for the detection of time resolved fluorescence emission which has a time resolution of 8 ps. Their data fit to a single 1.7 ns component which indicates that if a fast decay due to "s" to "f" transfer occurs in either complex, it must be shorter than 8 ps.

Addition of the linker polypeptide to AP does not cause as profound an effect on the spectroscopic properties of the trimer as does conversion of the monomer to the trimer. There is a 2-3 nm red shift of the absorption and emission maxima between AP and AP-$L_C^{8.9}$, and the shoulder of AP-$L_C^{8.9}$ is smaller and at a shorter wavelength than in AP. The polarization and time resolved measurements for the two complexes are identical within the error of the experiments.

Our polarization results for AP and AP-$L_C^{8.9}$ differ from data published on the corresponding trimers isolated from *Mastigocladus laminosus* by similar techniques [7]. The excitation and emission polarization spectra for AP and AP-$L_C^{8.9}$ were reported to show rising values at the longer and shorter wavelength edges, respectively. This behavior was interpreted as evidence for "s" to "f" energy transfer. We see the same results for preparations from *Synechococcus* if the aggregation state of the complex is not carefully monitored but the difference in the data may also be due, in part, to the differences in amino acid sequence of the proteins between the two organisms. An important distinction between *Synechococcus* and *Mastigocladus* phycobilisomes is the presence of an additional core cylinder in *Mastigocladus*.

In contrast to AP and AP-$L_C^{8.9}$, both steady state and time resolved measurements of APB-$L_C^{8.9}$ support the "s" to "f" model of energy transfer. The rise in polarization throughout the shoulder of the absorption band indicates the transfer of excitation to a unique chromophore. The time resolved data are more explicit. The analysis of a fast decay which shows up as a rise at longer emission wavelengths matches the prediction of the "s" to "f" transfer model. It must be pointed out that the fit of the time resolved data to one short lifetime which follows a predicted pattern in the fall and rise of its amplitude is a simplification of analysis. In the APB-$L_C^{8.9}$ trimer there are two α^{AP} and three β^{AP} chromophores which are distinguishable due to their positions relative to the unique α^{APB} chromophore. What our instrument detects as one fast lifetime must, in fact, be a superposition of five, or more, lifetimes corresponding to the various, distinct excitation transfer events.

Acknowledgements. Work in the Laboratory of Chemical Biodynamics was supported by the Director, Office of Energy Research, Office of Basic Energy Sciences, Division of Biological Energy Conversion of the US Department of Energy under Contract DE-AC03-76SFO0098. Work in the Department of Microbiology and Immunology was supported by NSF grant DMB85-18066.

References

1. Glazer, A.N. 1984. *Biochim. Biophys. Acta* 768:29.
2. Yamanaka, G., Glazer, A.N., Williams, R.C. 1978. *J. Biol. Chem.* 253:8303.
3. Yamanaka, G., Glazer, A.N., Williams, R.C. 1980. *J. Biol. Chem.* 255:11004.
4. Glazer, A.N. 1985. *Ann. Rev. Biophys. Biophys. Chem.* 14:47.
5. Glazer, A.N., Yeh, S.W., Webb, S.,P., Clark, J.H. 1985. *Science* 227:419.
6. Yeh, S.W., Glazer, A.N., Clark, J.H. 1986. *J. Phys. Chem* 90:4578.
7. Füglistaller, P., Mimuro, M., Suter, F., Zuber, H. 1987. *Biol. Chem. Hoppe-Seyler* 368:353.
8. Dale, R.E., Teale, F.W.J. 1970. *Photochem. Photobiol.* 40:423.
9. Teale, F.W.J., Dale, R.E. 1970. *Biochem. J.* 116:161.
10. Yamanaka, G., Lundell, D.J., Glazer, A.N. 1982. *J. Biol. Chem.* 257:4077.
11. Glazer, A.N., Fang, S. 1973. *J. Biol. Chem.* 248:659.
12. Lundell, D.J., Glazer, A.N. 1983. *J. Biol. Chem.* 258:902.
13. Csatorday, K., MacColl, R., Csizmadia, V., Grabowski, J., Bagyinka, C. 1984. *Biochem.* 23:6466.
14. Switalski, S.C., Sauer, K. 1984. *Photochem. Photobiol.* 40:423.
15. Schirmer, T., Bode, W., Huber, R. 1987. *J. Mol. Biol.* 196:677.

PICOSECOND ENERGY TRANSFER KINETICS IN ALLOPHYCOCYANIN AGGREGATES FROM MASTIGOCLADUS LAMINOSUS

E. Bittersmann[a), W. Reuter[b), W. Wehrmeyer[b), and A.R. Holzwarth[a)

[a) Max-Planck-Institut für Strahlenchemie, Stiftstrasse 34 - 36, D-4330 Mülheim a.d. Ruhr, West-Germany

[b) Fachbereich Biologie-Botanik, Philipps-Universität Marburg, Lahnberge D-355 Marburg/Lahn, West-Germany

INTRODUCTION

Allophycocyanin (APC) is one of the phycobiliproteins which form the core of the so-called phycobilisomes (PBS) which function as accessory light-harvesting complexes in cyanobacteria and red algae. We have recently presented a detailed kinetic study of the energy transfer in C-phycocyanin (C-PC) as a function of the aggregation state of this chromoprotein (1,2; for recent reviews see 3,4). It has been shown there how the symmetry properties of C-PC, as determined from the X-ray structure (5) greatly simplify the observable energy transfer kinetics. C-PC and APC bear many similarities both with regard to their chromophores as well as their apoproteins (6-8). Thus the comparison of kinetic data of the more simple APC with those of C-PC should also allow for a further check of some of the conclusions drawn earlier (1). So far our understanding of the energy transfer kinetics and pathways between the APC-aggregates within the PBS core is much less detailed than the one within PBS rods and from the rods to the core (9). Therefore, the present study of isolated APC is intended as a first step towards a better understanding of the energy transfer processes in the PBS core.

RESULTS AND DISCUSSION

The APC complexes have been isolated from the PBS of the cyanobacterium Mastigocladus. The detailed description of the isolation procedures as well as the

Photosynthetic Light-Harvesting Systems

biochemical characterization of these complexes will be published elsewhere (Reuter and Wehrmeyer, unpublished results). According to this characterization the APC complexes studied here are a) a trimeric complex free of linker peptides with a composition $(\alpha_3^{AP}\beta_3^{AP})$ which will be called APC 660 according to its fluorescence emission maximum at about 660 nm; b) a trimeric complex containing in addition a 10 kD linker peptide of composition $(\alpha_3^{AP}\beta_3^{AP})\ L_c^{10}$ called APC 665; and c) an allophycocyanin B heterotrimer of composition $(\alpha^{APB}\alpha_2^{AP}\beta_3^{AP})\ L_c^{10}$ called APC 681. Here, the superscript AP and APB refer to the chromophores of allophycocyanin and allophycocyanin B, respectively. All complexes were dissolved in 0.1 M K phosphate buffer of pH= 7.0 and all measurements were carried out at ambient temperature. Picosecond fluorescence decays and decay-associated fluorescence spectra (DAS) were measured as described previously (1) on a single-photon timing apparatus with 5 ps resolution. The various sets of decay data were analyzed by a global data analysis procedure (1). The underlying kinetic model for the fluorescence kinetics F(t) as a function of the excitation (λ_{exc}) and emission (λ_{em}) wavelength was

$$F(t,\lambda_{exc},\lambda_{em}) = \sum_{i=1}^{n} A_i(\lambda_{exc},\lambda_{em}) \cdot \exp(-t/\tau_i).$$

A plot of the amplitude $A_i(\lambda_{exc},\lambda_{em})$ as a function of the emission wavelength represents a decay-associated emission spectrum for the kinetic component i. A positive amplitude represents a decay term and a negative amplitude a rise term. The latter indicates an indirect population of chromophores excited due to an energy transfer process between different types of chromophores. The shape of the decay-associated spectrum will in such a case be given by the weighted difference spectra of the donor and acceptor fluorescence (see 1).

The **decay-associated emission spectra** (DAS) with picosecond time resolution were measured for all three APC complexes. Fluorescence decays were recorded at 10 nm intervals as a function of the emission wavelength. The complete set of decay curves from each trimer was subjected to a global deconvolution procedure. The results of the kinetic analysis for the three APC complexes calculated by the global analysis procedure are given in Table 1. For **APC 660** the shortest-lived component of 27 ps shows a positive amplitude (decay term) at short emission wavelengths. At long wavelengths the DAS has a negative amplitude (rise term) and a broad spectral shape. The appearance of a negative sign in the DAS of the 27 ps component identifies it clearly as an energy transfer component. The two long-lived components of 720 ps and 1700 ps have positive amplitude across the whole emission spectrum. They have the same wavelength maximum of 660 nm and very similar shape but largely different amplitudes. The maximum of these two components coincides furthermore with the maximum of the steady state emission spectrum. For **APC 665** only two kinetic components are necessary for a good fit. The short-lived one with $\tau = 14$ ps has positive amplitude at short wavelength and negative amplitude at long wavelength. As compared to the corresponding DAS of APC 660 the bandwidth of the negative part is much narrower. The long-lived component has only positive amplitude and shows a spectral maximum very close to that of the steady state emission spectrum (~662 nm). The DAS of **APC 681** contain two short-lived components ($\tau_1 = 23$ ps, $\tau_2 = 66$ ps), both of which show negative amplitude contributions. Their zero-crossing wavelengths differ by about 20 nm. The third, long-lived component ($\tau_3 = 1700$ ps) shows only positive amplitudes. The spectrum peaks at ~680 nm, again very close to the maximum of the steady state emission and it also has a shoulder around 660 nm.

Table 1: Fluorescence lifetimes (τ, ps) of the kinetic components of the APC complexes from Mastigocladus laminosus

Sample	$\tau_1^{\neq}$	$\tau_2^{\neq}$	$\tau_3^{\neq}$
APC 660	27	720	1700
APC 665	14	—	1800
APC 681	23	66	1700

$^{\neq}$ Maximum errors in the short lifetimes are 10% or ± 3 ps, whichever is larger. For the long lifetimes the maximum error is 5%.

The protein α and β subunits of the different APC complexes carry one phycocyanobilin chromophore each at positions α-84 and β-84, respectively (6,7,10). Thus, the bilin chromophores are linked to the corresponding positions as in C-phycocyanin, except for the β-155 chromophore which is absent in APC. The amino acid sequences of APC α- and β-subunits are highly homologous to the sequences of the corresponding C-phycocyanin subunits (6,7). Therefore, similarities are expected in the folding of the respective polypeptide chains in APC and C-PC. We therefore are probably justified in assuming at least the same symmetry properties and also similar relative chromophore orientations for APC as has been determined from the X-ray data for C-PC (5). In fact, it will be entirely sufficient to assume a three-fold symmetry axis (C_3-symmetry) for APC 660 and APC 665. With this assumption for APC 660 and APC 665 the energy transfer kinetics can be treated analogous to that of C-PC which we have derived previously (1). These APCs contain two groups of identical chromophores Thus, we expect the fluorescence kinetics to be described theoretically by two exponential components. The kinetic component 1 will describe the radiative and non-radiative decay of the chromophores and it will be associated with a spectrum that should be very similar to the steady state emission spectrum. Component 2 describes the energy transfer process and it will be associated with a spectrum which is a weighted difference spectrum between the fluorescence spectra of the S and F chromophores (1). For this reason, it contains positive and negative amplitudes.

The theoretical prediction of two fluorescence lifetime components is met by the experimental fluorescence kinetics of APC 665. We find a short-lived energy transfer component of $\tau_1 = 14$ ps and the expected long-lived component ($\tau_2 = 1.8$ ns) for the decay of the equilibrium state of the excited S and F chromophores. According to the kinetic equations given above, the τ_1-component represents the time required for the equilibration of the excitation energy between S and F chromophores. The linker-free trimer APC 660 also shows one short-lived energy transfer component ($\tau_1 = 25$ ps). The basic interpretation for this component is the same as for the analogous one in APC 665. The well-defined chromophore conformation in APC 665 reflects the stabilizing interaction of the linker peptide. The absence of the linker in APC 660 apparently introduces some heterogeneity into this protein complex, in particular into the β-84 chromophore. The heterogeneity is reflected by the presence of two long-lived lifetime components. This situation is reminiscent of a similar observation which we have made earlier for C-phycocyanin (1). Also there only the aggregate which carried a linker peptide was found to be homogeneous.

In APC 681 the α-chain in one monomer is replaced by an α^{APB}-chain whose chromophore has different spectroscopic properties than that of α^{AP}. Thus, the C_3-symmetry is broken. This lowering in symmetry makes the energy transfer kinetics much more complex. We expect either three or four lifetime components, based on the number of groups of symmetrically equivalent chromophores. At least two of them should represent energy transfer processes between different chromophore types. The experimental time-resolved data clearly show two lifetimes and decay-associated spectra which must be assigned to energy transfer. The long lifetime τ_3 reflects the decay of the equilibrated system. Thus, the decay-associated spectra and lifetimes of APC 681 can also be understood well on the basis of its special structure and composition.

REFERENCES

(1) Holzwarth,A.R., Wendler,J., and Suter,G.W., 1987. Biophys. J. 51: 1-12.

(2) Wendler,J., John,W., Scheer,H., and Holzwarth,A.R., 1986. Photochem. Photobiol. 44: 79-85.

(3) Holzwarth,A.R., 1986. Photochem. Photobiol. 43: 707-725.

(4) Holzwarth,A.R., 1987.in Topics in Photosynthesis 8: The Light Reactions. Ed: J. Barber, Elsevier, Amsterdam.

(5) Schirmer,T., Bode,W., Huber,R., Sidler,W., and Zuber,H., 1985. J. Mol. Biol. 184: 257-277.

(6) Sidler,W., Gysi,J., Isker,E., and Zuber,H., 1981. Hoppe-Seyler's Z. Physiol. Chem. 362: 611-628.

(7) Rümbeli,R., Wirth,M., Suter,F., and Zuber,H., 1987. Biol. Chem. Hoppe-Seyler 368: 1-9.

(8) Füglistaller,P., Mimuro,M., Suter,F., and Zuber,H., 1987. Biol. Chem. Hoppe-Seyler 368: 353-367.

(9) Suter,G.W. and Holzwarth,A.R., 1987. Biophys. J. 52: in print.

(10) Glazer,A.N., 1983. Ann. Rev. Biochem. 52: 125-157.

PICOSECOND TIME-RESOLVED ENERGY TRANSFER KINETICS WITHIN C-PHYCOCYANIN AND ALLOPHYCOCYANIN AGGREGATES

T. Gillbro, Å. Sandström, V. Sundström

Department of Physical Chemistry
University of Umeå, 901 87 Umeå, Sweden.

R. Fischer, H. Scheer

Botanisches Institut der Universität
Menzinger Str. 67, D-8000 München 19, FRG

Introduction

The light-harvesting complexes of cyanobacteria and red algae, are supramolecular aggregates, so-called phycobilisomes (PBS), situated at the outer surface of the thylakoid membranes (1,2,3). They are composed of a central core of 2-3 cylinders to which usually six rods are connected. The core is mainly composed of allophycocyanin (APC) trimers, while the building blocks of the rods are hexameric units of phycocyanin (PC), phycoerythrocyanin (PEC) or phycoerythrin (PE) (1).

Due to the complex structure of phycobilisomes and the presence of several hundred chromophores that interact with each other in a complicated way it is difficult to determine experimentally the rate of each individual transfer step.One way to obtain a more detailed understanding of the excitation energy transfer between neighbouring chromophores would be to study the energy transfer in smaller biliprotein aggregates. Of special interest are C-phycocyanin (C-PC) aggregates, since the structures of C-PC trimers of Mastigocladus (M.) laminosus (5,6) and C-PC hexamers of Agmenellum quadruplicatum (4) have recently been determined at high (2.1-2.5 Å)resoultion by X-ray crystallography. From the crystallographic and spectroscopic (7,8) data it should in principle be possible to calculate the energy transfer rates in these systems assuming that the Förster merchanism (9) for energy transfer is in

Photosynthetic Light-Harvesting Systems

operation. Efforts in this direction have already been made by Sauer et al (10). The aim of this work was to study the energy transfer kinetics of C-PC and APC monomers and trimers of M. laminosus on the picosecond and femtosecond scale and to study the relaxation of the light-induced anisotropy. One interesting aspect would be to compare the energy transfer in C-PC and APC monomers and trimers. For structural and spectral reasons one might expect that the transfer rate between the α -84 and β -84 chromophores should be similar in C-PC and APC monomers.

Experimental

The C-PC and APC trimers of M. laminosus were prepared according to the method given in ref 11 and 12, respectively. Monomers were obtained by adding NaSCN to 1.2 M. No further check of the aggregation state for the monomers was made. The absorption maximum was at 615 and 611 nm for the C-PC trimer and the monomer, respectively. For APC trimers it was 652 nm and for monomers 615 nm. The picosecond measurements were made in a rotating cell of 1mm optical pathlength and the absorbance for both the trimers and the monomers were in the range 0.8-02 (in 1mm cells).

In order to follow the kinetics of energy transfer we employed the picosecond absorption recovery method with continuously tunable excitation and probing ligth. The laser system used to generate the picosecond pulses as well as the measuring technique have previously been described in detail (13). In short, the picosecond pulses were generated in a mode-locked and cavity-dumped dye laser, which was syncronously pumped by a mode-locked argon ion laser. The cavity dumper was operated in the 80-800 kHz range and typically gave ca. 10 ps long pulses (FWHM) of 1-2 nJ energy in the 580-670 nm wavelength range. The polarization of the pump and probe beams were controlled by a Soleil-Babinet compensator and prism polarizers, so that the absorption recovery kinetics could be measured with any relative orientation of the pump and probe polarizations. Measurements with parallel (I_{11}) and perpendicular ($I_{\perp}$) polarization were used to monotor the decay of induced anisotropy, $r(t)= (I_{11} - I_{\perp}) / (I_{11} + 2I_{\perp})$ and measurements at the magic angle (54.7°) were used to obtain the isotropic decay, free of depolarization effects. In some pump-probe experiments we used ca. 400 fs pulses from a fiber-grating pulse compressor.

Results and discussion

C-Phycocyanin monomers and trimers

In Fig. 1 we show the absorption recovery kinetics of C-PC monomers at 580 nm with different polarization of the excitation and the probe pulses. The data were analyzed by fitting them to a sum of two or three

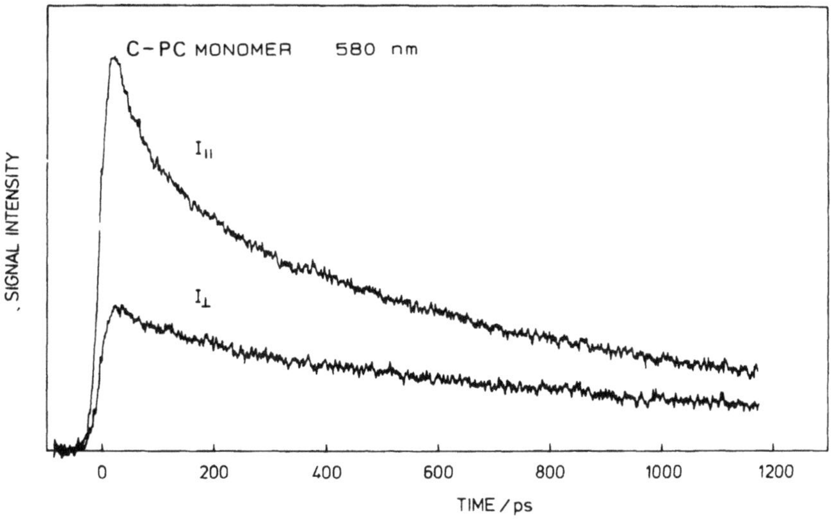

Fig.1

exponentials. The mean values of lifetimes and amplitudes obtained at different wavelength intervals are shown in table 1. In table 2 the corresponding anisotropy relaxation times and amplitudes are shown. From these data it is clear that besides a long lifetime in the ns range.

Table 1

Lifetimes (τ_1) and relative amplitudes (R_1) obtained from isotropic signals of C-PC monomers at different wavelength, intervals.

nm	τ_1(ps)	R_1 (%)	τ_2 (ps)	R_2 (%)	τ_3 (ps)	R_3(%)
580-600	57±4	27±2	665±101	62±1	4000	14±9
635-640	178±64	25±5	893±151	75±5		

Table 2

Anisotropy relaxation times (τ_{r_1}) and amplitudes (r_1) of C-PC monomers at different wavelength intervals.

nm	τr_1 (ps)	r_1	τr_2 (ns)	r_2	r(0)
580-590	52±19	0.13±0.03	2.6±1.5	0.28±0.03	0.41±0.01
635-640	36±8	0.08±0.04	4.4±0.1	0.29±0.04	0.37±0.02

we observe a lifetime of ca. 57 ps at shorter wavelength (∿ 590 nm), where the β -155 chromophore absorbes strongly (7,10). At longer wavelength, i.e. 635-640 nm, the fastest lifetime increases to 178 ± 64 ps. The most direct interpretation of these data is that the 57 ps lifetime is due to energy transfer between β -155 and β -84 within a C-PC monomer unit. As can bee seen in table 2 this transfer step is accompanied by an equally fast relaxation of the anisotropy from 0.41 + 0.01 to 0.28 + 0.03. This last anistropy is similar to that found in steady state fluorescence measurements (7).The centrum distance (R) between β -155 and β -84 in C-PC of M. laminosus 34.3 Å (5) and the orientation factor $\kappa = 0.84$. With the radiative lifetime $\tau_o = 2$ ns we obtain a Förster radius (R) of 61 Å, using the equation;

$$\frac{1}{\tau} = \frac{3}{2\tau_o}\kappa^2\left(\frac{R_o}{R}\right)^6$$

where τ is the measured lifetime of energy transfer and under the assumption that the rate of back transfer is comparatively small. Because the observed transfer rate (k) between two chromophores is the sum of the ratios in the forward (k_1) and back directions (k_{-1}) and inclusion of 25 % back transfer would just reduce the calculated value of R_o with 3.5 %. The calculated Förster radius is in fair agreement with litterature data and (10) thus one might conclude that the Förster mechanism for energy transfer is in operation in the C-PC monomers. A similar analysis of the long wavelength lifetime of 178 ps indicates that this is due mainly to the transfer step α-84 → β-84. Assuming similar rates for the forward (k_2) and back (k_{-2}) energy transfer, we calculate (with R = 50.2 Å and κ = 1.73) that R_o = 52 Å.This value is reasonable and as expected, due to the smaller overlap between donor emission and acceptor absorption, lower than for the β-155 → β-84 transfer.

Table 3

Lifetimes (τ_i) and relative amplitude (R_i) obtained from the isotropic signals of C-PC trimers at different wavelengths.

nm	τ_1(ps)	R_1 (%)	τ_2 (ps)	R_2 (%)	τ_3 (ps)	R_3 (%)
580-590	27±3	40±7	106±27	14±5	1162±67	46±5
616-625	27±6	40±7	173±85	27±8	1228±242	33±8
635-645	48±12	35±9	429±143	24±9	1190±287	41±17

Turning now to the C-PC trimers, we observe (see table 3) that the fastest process has a lifetime of 27 ps at 625 nm and increases to 48 ps at about 640 nm. There is also a long (fluorescence) lifetime of about 1.2 ns at all wavelengths. In addition, there is an intermediate lifetime that varies from about 200 to 400 ps when going to longer excitation wavelengths. Similar lifetimes were observed in the anisotropy measurements (table 4).

Table 4

Anisotropy relaxation lifetimes (τr_1) and amplitudes (r_1) of C-PC trimers at different wavelengths.

nm	τr_1(ps)	r_1	τ_2(ps)	r_2	r(0)	r(∞)
580-600	24±5	0.18±0.02	108±36	0.10±0.02	0.42±0.02	0.14±0.03
616-635	21±8	0.17±0.03	222±77	0.06±0.03	0.38±0.02	0.15±0.03

The final anisotropy at longer times of 0.05 is just about half the anisotropy found for the monomers. This of course is a reflexion of the final distribution of the excited state is over more chromophores in the trimers.

The interpretation of the observed lifetimes is of course more complex in the trimers than in the monomers, since the number of possible interactions is larger. From the crystallographic data, however, the by far closest pair of chromophores is α-84 and β-84 of adjacent monomers. With R = 20.8 Å and κ= -1.34 and R_o= 52 Å (see above) one would expect lifetime of about 1.5 ps (assuming that the backward and forward rates are similar). No such fast process was found, however, in our picosecond study. We therefore also performed some experiments on C-PC trimers at 618 nm with 400 fs pulses (Fig. 2)., but we were unfortunately

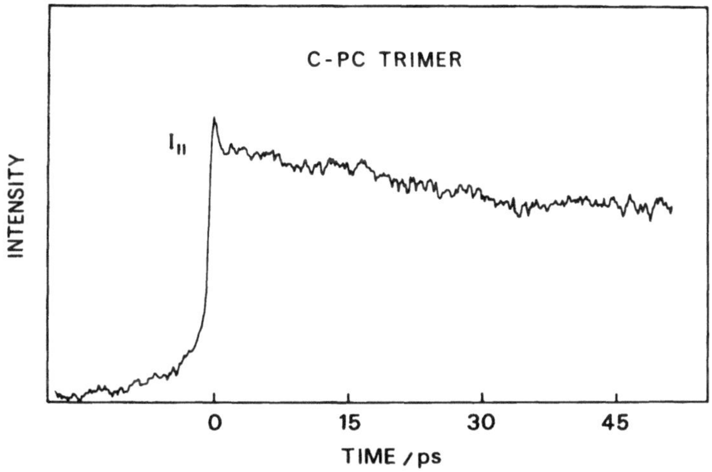

Fig.2

not able to resolve any lifetime (isotropic or anisotropic) in the interval 0.5-25 ps. The 27 ps component thus seems to be to slow for an α-84 → β-84 transfer. It has been attributed to transfer from β-155 to α-84 and/or β-84 from time-resolved fluorescence stdues (16) and our data at 580-590 nm would support this interpretation, however, going toward longer wavelengths, i.e. 616-625 nm, the relative amplitude of this component should decrease and it should only be about 10% at 640 nm. Where the absorption of β-155 is small (7,10). Since this is contrary to our data (see table 3) we must conclude that transfer among α-84 and β-84 chromophores or other processes also contribute to this component.

Allophycocyanin monomers and trimers

Again we will start with the monomer units. In Fig. 3 and absorption recovery measured at 610 nm is displayed cleary shows a biphasic decay. The fast small amplitude component has a lifetime of 144 ps, while the dominating decay has a lifetime of 1.3 ns. Since APC monomers only have two chromophores (α-80 and β-81) the short lifetime should be due to energy transfer between these chromophores. The fact that we can observe this signal means that the absorption spectra are not identical, however, the small amplitude of the signal indicates that the spectra are strongly overlapping. This might be expected when the spectral similarity to C-PC monomers is considered. We also note that within the experimental error the 144 ps component is the same as the corresponding energy transfer component found in C-PC monomers. Thus one may can conclude that the chromophores are situated on similar positions in the C-PC and APC monomers. This also what one would expect from the homology between the two proteins (12).

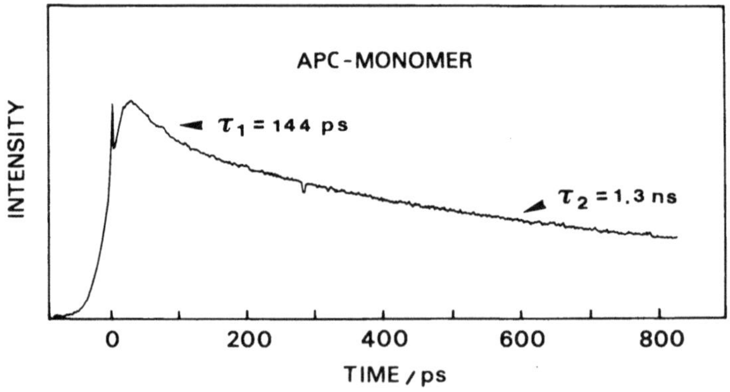

Fig. 3

APC trimers are the smallest aggregates of APC occuring in the core of the phycobilisomers. So far there has been no report on subnanosecond energy transfer processes in APC trimers. In this work we have observed a fast process with a lifetime of 45 $\pm$ 10 ps (see Fig. 4). Since the amplitude of this component is substantial (ca 60%) at 642 nm, the transfer has to take place between chromophores with different absorption spectra. The relative amplitude also increases in going from 670 to 630 nm as expected if this is a normal Förster (donor-acceptor) type of transfer. It is interesting to compare the APC with the C-PC trimer data about 640 nm, where a 48 $\pm$ 7 ps process was observed in CPC. This suggests that a transfer between α -80 and β -81 chromophores is responsible for this component. Since our preparation contained a small amount (< 10%) of chromophore with a red-shifted absorption spectrum (max $\sim$ 675 nm) it is however not possible to exclude that the process is partly due to a transfer of excitation energy to this chromophore in combination with a quenching process. We used 400 fs pulses at 648 nm in some experiments (Fig. 4) to investigate it there is any fast process in the range 0.5-10 ps. However, we could not observe such a process in the isotropic or anisotropic decay. One interesting finding was that the anisotropy at t=0 was only about 0.2 (Fig. 4) in stead of 0.4 as expected. This indicates that there is a fast (< 0.5 ps) unresolved anisotropy relaxation process. This might be a transfer of excitation between closely spaced states with differently directed transition dipole moments. Such states could for instance be formed in a strongly coupled dimer, (excitonic states), which has been suggested to give rize to the 652 nm absorption band in APC-trimers (15). The anisotropy at longer times is similar to the steady state value (r $\sim$ 0.05) (14).

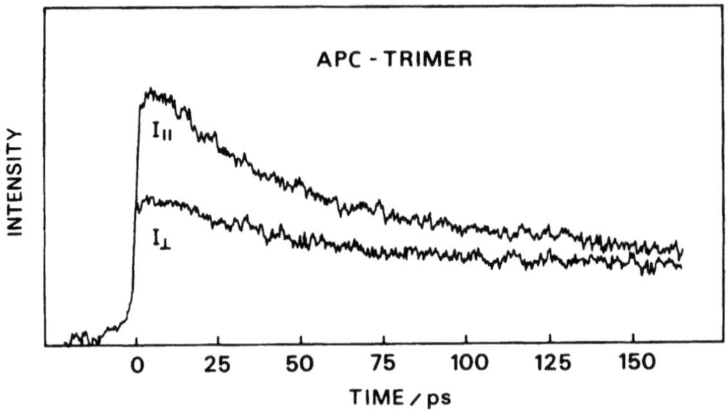

Fig. 4

It is clear from this study of APC trimers that the energy transfer and related processes are more complex than what one would expect considering the relative simplicity of the system. To understand the physical meaning of these processes further work is urgently needed.

Acknowledgements

We would like to thank the Swedish Natural Science Research Counsil and the Deutsche Forschungsgeinschaft for financial support. Substanstial support from the Erna and Victor Hasselblad foundation, the Kempe foundation and the Knut and Alice Wallenberg foundation is also gratefully acknowledged.

References

1. Gantt., E. 1986.: Encyclopedia of Plant Physiology, vol 19, 260-268.

2. Scheer, H. 1986, idibid, 327.

3. Holzwarth, A.R. 1986. Photochem. Photobiol. 43, 707.

4. Schirmer, T., W. Bode, R. Huber, W. Sidler, H- Zuber. 1985. J. Mol. Biol. 1984, 257.

5. Schirmer, T., W. Bode, R. Huber, 1987. J. Mol. Biol. 196, 677.

6. Schirmer, T., R. Huber, M. Schneider, W. Bode, M. Miller, M.L. Hackert 1986. J. Mol. 188, 651.

7. Mimuro, M., P. Füglistaller, R. Rümbeli, H. Zuber. Biochem. Biophys. Biophys. Acta 848, 155.

8. Scheer, H. 1986. progress in Photosynthesis Research (Ed. Biggins, J.) Martinus Nijoff Publ., Dodrech, Vol. 1, pp. 143-149.

9. Förster, Th. 1949, Ann. Physik.2, 55.

10. Sauer, K., H. Scheer, P. Sauer. Photochem Photobiol. 46, 427.

11. Hefferle, P., W. John, H. Scheer, S. Schneider. 1984. Photochem Photobiol 39, 221.

12. Frank, G., W. Sidler, H. Widmer, H. Zuber. 1978. Hoppe-Seyler's Z. Physiol. Chem. 362, 611.

13. Åkesson, E., V. Sundström, T. Gillbro. 1985. Chem. Phys. Letters 121. 153.

14. Yeh, S.W., A.N. Glazer, H.H. Clark. 1986. J. Phys. Chem. 90, 4578.

15. Mac Coll, R., K. Csatorday, D.S. Berns, E. Traeger. 1980. Biochem. 19, 2817.

16. Holzwarth, A.R., J. Wendler, G. Suter. 1987. Biophys. J. 52, 1.

ENERGY TRANSFER IN "NATIVE" AND CHEMICALLY MODIFIED C-PHYCOCYANIN TRIMERS AND THE CONSTITUENT SUBUNITS

S. Schneider, P. Geiselhart, F. Baumann
Institut für Physikalische und Theoretische Chemie der Technischen Universität München,
Lichtenbergstraße 4, D-8046 Garching

S.Siebzehnrübl, R. Fischer, H. Scheer
Botanisches Institut der Ludwig-Maximilians-Universität
Menzingerstraße 67, D-8000 München 19

Introduction

The light-harvesting complexes of blue-green and red algae, called phycobilisomes, are highly organized assemblies of the biliproteins allophycocyanin (APC) and phycocyanin (PC), often containing also phycoerythrin (PE) or phycoerythrocyanin (PEC) [1,2]. These biliproteins carry one or several tetrapyrrol chromophores bound covalently to polypeptides. In case of PC, which is discussed here, the monomeric unit consists of the α-polypeptide chain with one chromophore (A84) and the β-chain with two phycoyanobilin chromophores attached to it (B84 and B155) [3,4].

From X-ray work [5,6] on PC from two different organisms, the crystal structure of the chromoprotein, the conformations, distances and the approximate relative orientations of the chromophores have become known recently. Very similar results have been obtained for a trimer $(\alpha\beta)_3$ from Mastigocladus laminosus and a hexamer $(\alpha\beta)_6$ from Agmenellum quadruplicatum, which is composed of two trimeric units aggregated face-to-face. For the understanding of the energy transfer process within these units and in the whole phyco-

Photosynthetic Light-Harvesting Systems

bilisome, three features deduced from this analysis are important.

(i) The chromophore arrangement exhibits a 3-fold symmetry around the central axis orthogonal to the plane of the trimeric unit.

(ii) The shortest distance is found between the 1A84 chromophore of one monomeric unit and the 2B84 chromophore of the neighbouring monomer; it is only about 21 A center to center.

(iii) All three chromophores adopt an extended geometry; they exhibit hydrogen bonds to a conserved aspartate residue (A87, B87, B39, resp.) and are subject to strong coulombic interaction with the net charges localized on the surrounding amino acids.

If the threefold symmetry established in the crystal structure is also present in the native trimeric unit (i.e., there are only three types of non-equivalent chromophores), then theory predicts that any measurement, which monitors the time course of the excited state population, should be fittable by a sum of three exponentials provided that the evolution of the system can be described by a set of linear differential equations (radiative and intramolecular radiationless decay, energy transfer by dipole-dipole interaction (Förster transfer)).
Because of the short distance between pairs of chromophores, like 1A84 and 2B84, it was concluded that the energy transfer rate between these chromophores should be very high and/or the coupling so strong that Förster theory may not be applicable anymore [7]. Alternatively the model of localized excitonic states should then be used for the description of the energy transfer kinetics (see fig. 1). Besides of the very fast

(and eventually unresolvable) internal conversion process, only two decay times should then be observed in a picosecond time-resolved experiment: one related to the transfer between the s-chromophore and the excitonically coupled state of chromophores A84/B84 (devoted C^+ or C^- in fig.1) and intramolecular decay of the corresponding state (C^+). Since the important factors in excitonic coupling are the energy difference of the unperturbed states and their interaction energy, a modification of one of the chromophores by a distinct perturbation could result in an intermediate situation between the two above described limiting cases and therefore help to elucidate the situation in PC-trimers.

In a previous publication [8] it was shown that titration of solutions of biliproteins in various states of aggregation with the mercury salt p-chloromercurybenzenesulfonate (PCMS)

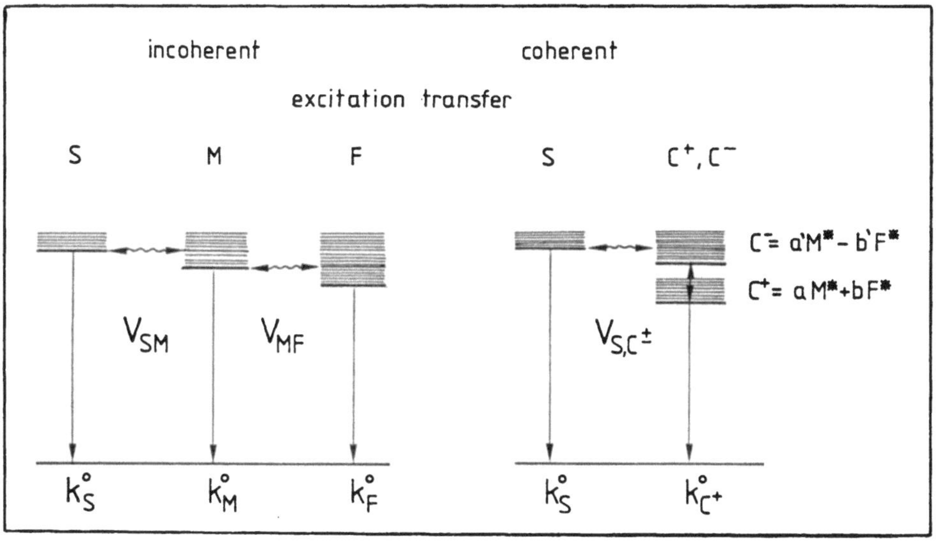

<u>Fig</u>. 1: Left : Förster type energy transfer
Right: Excitonic interaction between M and F chromophores

leads to a binding of this compound to the free cystein B111, and concomitantly to a selective change in the absorption and circular dicroism spectra of the close-by chromophore B84. Resonance-enhanced CARS spectra (Coherent Anti-Stokes Raman Scattering) of such modified PC-trimers confirm the evidence for a change in geometry of B84, which was deduced from the X-ray data [9]. In this contribution the effects of PCMS-binding on CARS spectra (structural information) and excited state kinetics are compared. It is concluded that in smaller aggregates (up to trimers) a microheterogeneity of the chromophore-protein arrangement exists, which produces a rather complex fluorescence decay pattern.

Materials and Methods

PC-trimers without linker peptides were prepared according to published procedures [10]. β-subunits were isolated by a modified flat-bed gel electrofocusing technique [11]. In particular, care was taken to avoid oxidative bleaching of the chromophores by gassing with Argon and work under subdued light. The pigment is present as a dimer (β_2).
Both types of samples were titrated with PCMS in approximately 1.1 fold excess with respect to the number of free cysteins to assure complete modification.

The fluorescence decay curves were recorded by means of a Hamamatsu synchroscan streak camera using interference filters for spectral narrowing of the monitored emission. Excitation was performed under magic angle conditions at low average intensity (≈10 mW or 10^{13} photons pulse^{-1}cm^{-2}) by the output of a synchronously pumped mode-locked dye laser (pulse repetition rate 84 MHz). The decay curves were analysed by applying a least-squares fit routine and assuming a 3-exponential decay law (More details are given in references 12-14).

Resonance-enhanced Coherent Anti-Stokes Raman Scattering (CARS) was measured with an apparatus described in detail in references 15 and 16. Two eximer laser-pumped dye lasers generate the so-called pump- and Stokes-beams, with wavelengths λ_p (fixed) and λ_s (variable), which are focused in the sample under a small angle. Recording the intensity of the generated anti-Stokes radiation as a function of the wavenumber difference $\tilde{\nu}_p - \tilde{\nu}_s$ yields a vibrational spectrum. Since λ_p coincides with an electronic transition (absorption band) of the chromophore, only the chromophore's vibrations are resonance-enhanced and, therefore, dominate the spectrum.

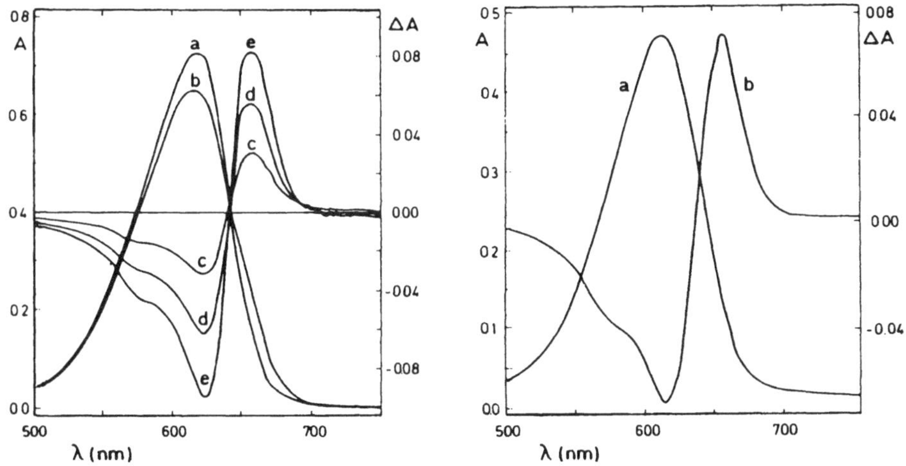

Fig. 2: Left: Absorption spectra of PC trimers before (a) and after titration with PCMS (b). Difference spectra recorded 30 (c), 60 (d) and 100 min. (e, final absorption) after PCMS addition.
Right: Absorption spectrum of β-subunit (a) and final difference spectrum after PCMS titration (b).

Results and Discussion

In figure 2, the effect of PCMS titration (binding) on the absorption spectra of β-subunits (present as dimers β_2) and trimeric PC are shown . In both cases the absorbance at the long-wavelength side of the absorption maximum (λ_{max} = 604 nm and 616 nm, resp.) is reduced by about 10 % and there is a concomitant increase in the long wavelength slope of this band (wavelength of maximum absorbance increase is ≈ 655 nm).

Figure 3 displays the CARS spectra recorded for the four samples in the fingerprint region. At first glance it is already apparent that the band centered at 1247 cm^{-1} in native PC-trimers disappears upon addition of PCMS, whereas the other bands remain essentially unchanged. The spectrum of β_2, on the other hand, exhibits no change which is beyond experimental uncertainty. Furthermore it is interesting to

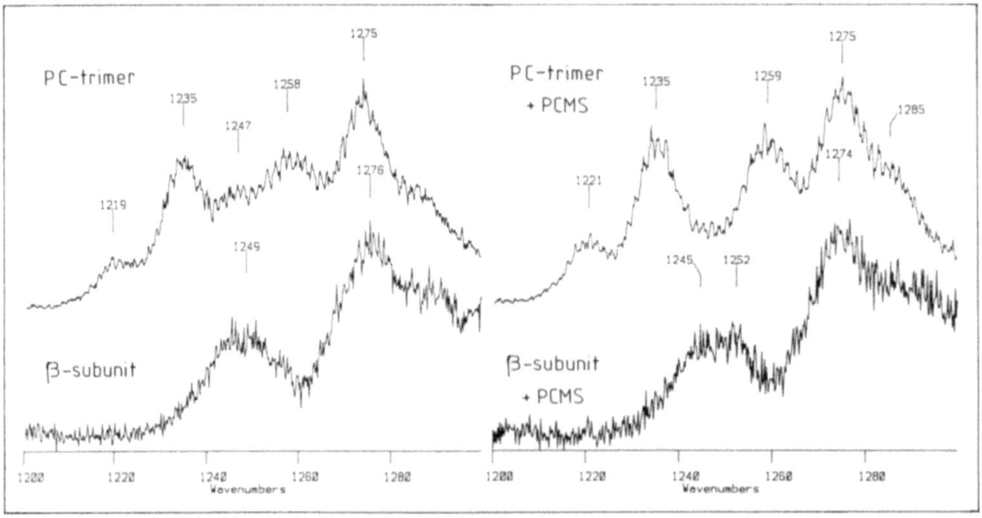

<u>Fig</u>. 3: Resonance-enhanced CARS spectra of PC trimers (left, top), PC β-subunits (left, bottom), PC trimers titrated with PCMS (right, top), PC β-subunits titrated with PCMS (right, bottom). Pumpwavelength 640 nm.

note that PC trimers, which have only one additional type of non-equivalent chromophore, show at least 3 more bands, namely at 1221, 1235 and 1259 cm^{-1}. The band which appears in both spectra (1245 - 1252 cm^{-1}) disappears upon PCMS binding in the trimer spectrum. This fact can be taken as additional evidence that it is the chromophore B84, whose geometry is changed predominantly upon PCMS binding. In the double bond stretching region (1550 - 1700 cm^{-1}) no changes upon PCMS titration have been observed for β_2 and only small, but significant ones are observed for PC trimers [16].

In order to demonstrate the large change induced in the fluorescence decay of PC trimers upon PCMS binding, the decay curves recorded with 600 nm excitation, but different observation wavelengths are shown in fig.4. Even without analysis it is obvious that the wavelength, above which a delayed onset of the fluorescence is observed, is increased from about 620 nm to 670 nm. Furthermore, it is apparent that the contribution of the long-lived component is much larger in the native system and this effect is increasingly pronounced at shorter wavelengths. In contrast to the pronounced effect found in PC-trimers, however, only a minor effect is observed for β_2 (small reduction in the lifetime of the terminal emmitter). In figures 5 to 8, the results (lifetimes deduced from a 3-exp. fit) are summarized for easier comparison. The following conclusions can be drawn:

In case of dimers of the β-subunit the lifetime of the long-lived component shows a slight reduction upon PCMS binding. Its apparent value varies with observation wavelength, but appears to be fairly independent of the excitation wavelength. A similar statement holds for the intermediate component. For the short-lived component, the situation is quite different. With increasing observation wavelength, the apparent decay time of the fast fluorescence increases by more than a factor of 2 up to 60 ps. In the red part of the emission, the derived rise times are generally speaking below 20 ps. In the measurements close to the "turning" wavelength, an excitation wavelength dependence is apparent.

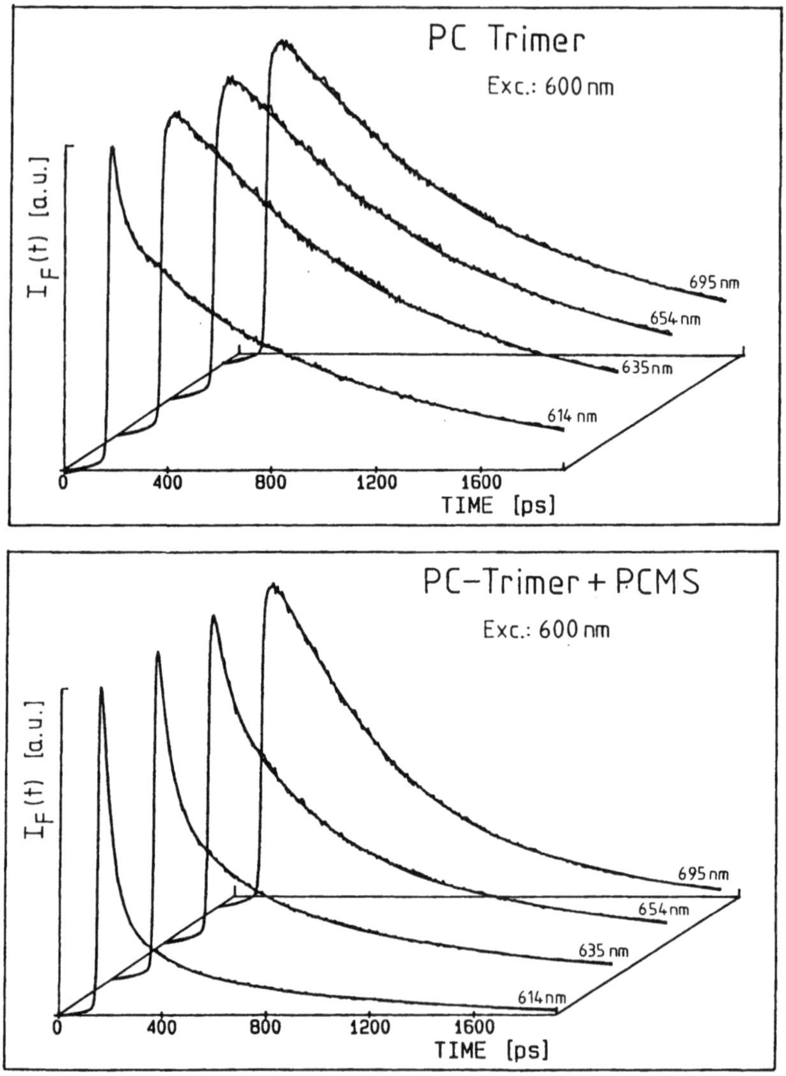

Fig. 4: Fluorescence decay of PC trimers and PC trimers titrated with PCMS.

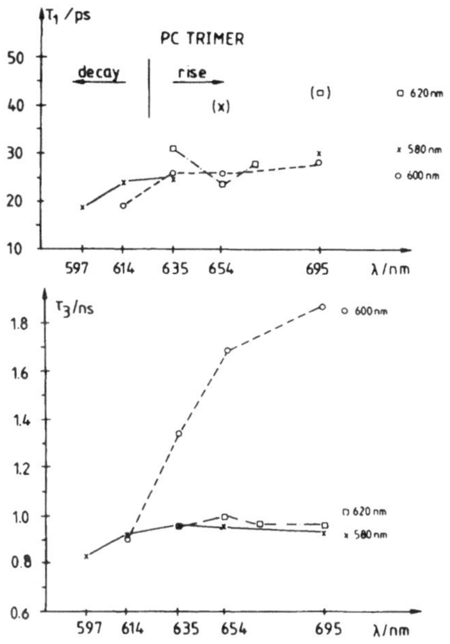

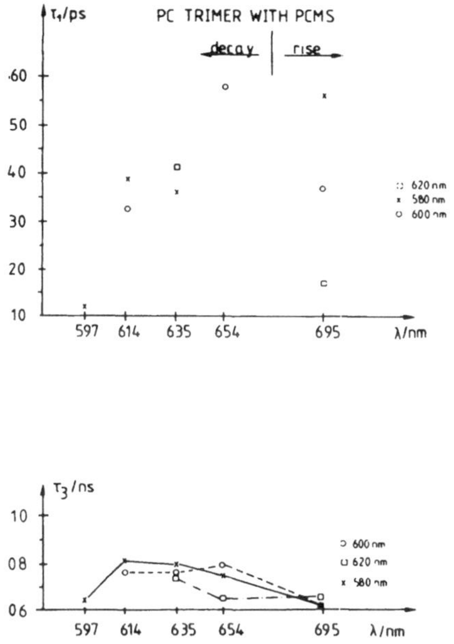

Fig. 5 ↑

Fig. 7 ↓

Fig. 6 ↑

Fig. 8 ↓

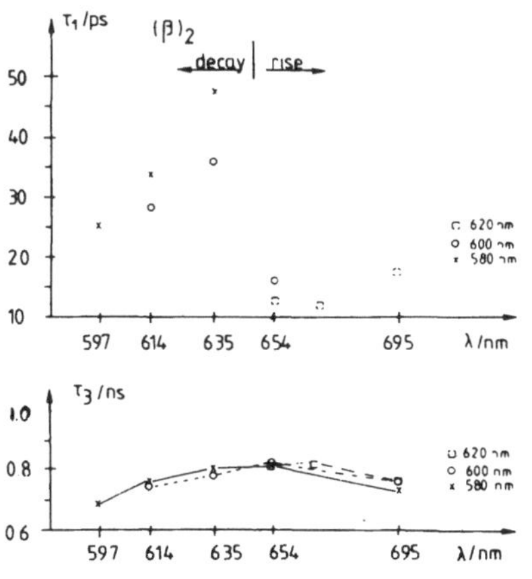

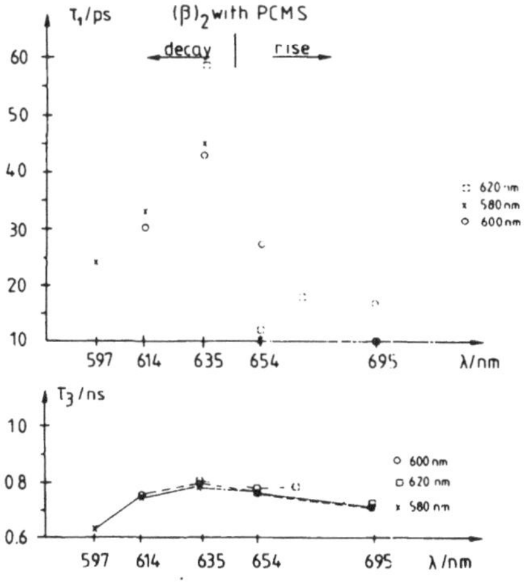

Fig. 5-8: Observed fluorescence lifetimes τ_1 and τ_3 of PC trimers in dependence of excitation (x=580nm, o=600nm, =620nm) and detection wavelength (absciss).

Fig. 5: PC trimers. Fig. 6: PC trimers titrated with PCMS.

Fig. 7: Dimers of β-subunits from PC.

Fig. 8: Dimers of β-subunits from PC titrated with PCMS.

As one might already expect from the behaviour of the CARS spectra, the effects of PCMS binding are substantially different in case of trimers. In the "native" system, excitation around 600 nm seems to preferentially select a chromophore-protein moiety, which is characterised by a very long fluorescence decay time (1.8 ns). By PCMS-binding the formation of this species must be prevented, because no such long time constant is observed for any of the excitation wavelengths (600, 580 or 620 nm).
The apparent lifetime of the predominant terminal emitter is reduced upon PCMS binding from about 0.9 to about 0.7 ns. Similarly, the decay time of the intermediate component is reduced from approx. 200 ps to approx. 100 ps.
The lifetime of the shortest-lived component experiences an interesting PCMS-effect. In the "native" system the decay times derived from the emission on the short-wavelength side of the fluorescence spectrum (20-30 ps) agree fairly well with the rise times found in the emission on the long-wavelength side. However, a situation similar to that described for β_2 is found in the PCMS treated trimers. There is an increase of the decay time with observation wavelength and a dependence of the rise time on excitation wavelength.

Because of the change in sign of its amplitude with increasing observation wavelength and the appearance of similar values in the decay of the β-subunit (in monomeric β-subunits prepared by application of detergent, a very similar decay pattern as in β_2 is observed) the 20 - 30 ps lifetime is assigned to the decay of the excited state population of B155 (in agreement with other authors [18, 19]).

The observed decrease in the lifetime of the terminal emitters upon PCMS binding could also be due to changes in radiationless decay rates, caused by the reduced energy gap ΔE ($S_1 \rightarrow S_0$). It can not be taken as unambigous indication for strong coupling between 1A84 and 2B84 chromophores.

The intermediate lifetime has been related in the past [20, 21, 22] to the energy transfer between 1A84 and 2B84 chromophores. Our present results do not support this interpretation for several reasons:

(i) A component exhibiting a lifetime of around 200 ps was found not only in the trimers, but also in monomers and both α-and β-subunits.

(ii) This component was always detected as a decaying component. Representing a transfer time, its amplitude should change sign for long wavelength detection similar to the one of the fast component or at least decrease in amplitude at long wavelength detection.

(iii) The amplitude was always above 10%. If it were related to a transfer between two chromophores with similar absorption and emission characteristics, the amplitude should be very small (it is expected to be proportional to the difference of the spectral properties, i.e. emission or absorption in case of pump-probe experiments).

We therefore believe that the intermediate component is connected with a modified chromophore-protein arrangement. E.g., we have found a lifetime in the 200 - 300 ps time domain in chromopeptides, where it is suspected that the chromophore adopts a ZZE configuration.

Before entering the discussion about the coupling of the 1A84 - 2B84 pair, it should be mentioned that, when pump-probe experiments were performed on PC trimers [14], all absorption recovery curves could be fit very well by assuming a biexponential decay law. The derived long lifetime varied around 950 ps, the shorter one between 20 and 100 ps, depending on wavelength of pump and probe beam.

Combining the results of transient absorption and fluorescence experiments we conclude that the energy transfer time between the 1A84 - 2B84 pair is on the order of or faster than the measured fast component. Then, the fast component is actually a superposition describing two processes (i.e. a s-to-f transfer from B155 and a f-to-f transfer A84/B84). Since the coupling strength between the A84 - B84 pair is sensitive to variations in the electronic properties of the chromophores, the pronounced wavelength dependence could be understood. That is to say we relate the observed dependence of the short lifetime on excitation and detection wavelength to the existence of a microheterogeneity in chromophore-protein arrangement which effects mainly the coupling between A84 and B84 of neighbouring monomers. Model calculations [23] have shown that small changes in the protein environment (e.g. formation of tautomeric forms of amino acid residues) are sufficient to shift the excitation energy of one chromophore by 10 nm.

Preliminary experiments using the CARS technique suggest that the microheterogeneity postulated here for the smaller aggregates is reduced when linker peptides are present as in phycobilisomes [24]. Deviations from exact threefold symmetry seem nevertheless to occur even in the larger systems. Because of the overall arrangement of the chromophores in the light-harvesting complexes a few "defects" should not reduce significantly the efficiency of energy transfer to the reaction center.

Acknowledgement

Financial support by Deutsche Forschungsgemeinschaft (SFB 143 and Forschergruppe 'Pflanzliche Tetrapyrrole') und Fonds der Chemie is gratefully acknowledged. Thanks are also due to Hamamatsu Photonics Europe for loan of the streak camera.

References

[1] Gant, E., 1981. Ann. Rev. Plant Physiol. 32, 327 - 347

[2] Mörschel E. and E. Riehl, 1987. In: Electron Microscopy of Proteins (vol 6): Membraneous Structures. (J. R. Harris, W. Home, eds.). Academic Press, London. 209-254.

[3] Scheer, H., 1982. In: Light Reaction Path of Photosynthesis. (F.K. Fong, ed.). Springer-Verlag, Berlin. 7 - 45.

[4] Frank G., W. Sidler, H. Widmer and H. Zuber, 1978. Hoppe-Seyler's Z. Physiol. Chem. 359, 1499 - 1507.

[5] Schirmer, T., W. Bode, R. Huber, W. Sidler and H. Zuber, 1985. J. Mol. Biol. 184, 257 - 277

[6] Schirmer, T., W. Bode, R. Huber, 1987. J. Mol. Biol., 196, 677-695.

[7] Sauer, K. and H. Scheer, 1987. Contribution in this volume.

[8] Siebzehnrübl, S., R. Fischer and H. Scheer, 1987. Z. Naturforsch. 42c, 258 - 262

[9] Duerring M. and R. Huber. Private communication.

[10] Füglistaller, P., H. Widmer, G. Frank and H. Zuber, 1981. Arch. Microbiol. 129, 268 - 274.

[11] Köst-Reyes E., S. Schneider, W. John, R. Fischer, H. Scheer, H.-P. Köst, 1987. Electrophoresis 8, 335-336.

[12] Hefferle, P, P. Geiselhart, T. Mindl, S. Schneider, W. John, and H. Scheer, 1984. Z. Naturforsch. 39c, 606 - 616.

[13] Geiselhart, P, 1987. PhD thesis, Techn. University Munich.

[14] Schneider, S., P. Geiselhart, S. Siebzehnrübl, R. Fischer and H. Scheer, 1987. Z. Naturforsch. c, in press.

[15] Klüter, U., 1986. PhD thesis, Techn. University Munich.

[16] Schneider, S, F. Baumann, U. Klüter, 1987. Z. Naturforsch. 42c(11/12).

[17] Schneider, S., P. Geiselhart, F. Baumann, H. Falk and W. Medinger, 1987. J. Photochem. Photobiol., (submitted).

[18] Holzwarth A.R.,J. Wendler and G.W. Suter, 1987. Biophys. J. 51, 1 - 12.

[19] Gillbro T., A. Sandström, V. Sundström, R. Fischer and H. Scheer, 1987. Contribution in this volume.

[20] Holzwarth, A.R., 1986. In: Topics in Photosynthesis (J. Barber, ed.). Elsevier, Amsterdam.

[21] Hefferle, P., W. John, H. Scheer and S. Schneider, 1984. Photochem. Photobiol. 39, 221 - 232.

[22] Wendler, J., W. John, H. Scheer and A.R. Holzwarth, 1986. Photochem. Photobiol. 44, 79 - 86.

[23] Schneider S., Ch. Scharnagl, M. Duerring, T. Schirmer and W. Bode, 1987. Contribution in this volume.

[24] Schneider S., F. Baumann, W. Steiner, S. Siebzehnrübl, R. Fischer and H. Scheer, 1987. Contribution in this volume.

EFFECT OF PROTEIN ENVIRONMENT AND EXCITONIC COUPLING ON THE EXCITED-STATE PROPERTIES OF THE BILINCHROMOPHORES IN C-PHYCOCYANIN

S. Schneider, Ch.Scharnagl
Institut für Physikalische und Theoretische Chemie der Technischen Universität München, D-8046 Garching

M. Duerring, T. Schirmer, W. Bode
Max-Planck-Institut für Biochemie, D-8033 Martinsried

Introduction

C-Phycocyanin (C-PC), one of the light-harvesting pigments of cyanobacteria, is composed of two subunits, α and β. The former contains one, the latter two open-chain tetrayrrole chromophores (phycocyanobilin), covalently bound to the apoprotein via thioether linkages (see e.g. (1)).
The geometries and individual environments of the three chromophores in the phycocyanin of Mastigocladus laminosus have recently been determined by peptide sequencing (2) and high resolution X-ray crystallography (3a,b,c). The crystal structure displays a strong similiarity of geometries, so that different spectroscopic, biochemical and functional properties must be mainly attributed to the different protein surroundings. Lately it has been possible to assign the absorption bands (maxima) to the different phycocyanin chromophores unambiguously in the following manner (4,5):

A84:	λ_{max}(abs)=	616 - 618 nm (4) ;	618 nm (5)
B84:		622 - 624 nm (4) ;	624 - 632 nm (5)
B155:		598 - 600 nm (4) ;	594 - 598 nm (5)
			582 - 598 nm (5,CD)

Photosynthetic Light-Harvesting Systems

CD-spectroscopy is an important source of information to elucidate the conformation of bilin chromophores and a sensitive probe to noncovalent interactions. Therefore we studied the effect of protein-chromophore and chromophore-chromophore interactions on the chiroptic properties by means of quantum mechanical model calculations, in order to improve our knowledge about the contribution of these interactions to the spectral and functional properties of the phycobilinproteins and to get an estimate of the significance of these interactions.

Methods

Our calculations are based on the high-resolution X-ray data for chromophore and protein (3c). Approximative wavefunctions result from a π-electron model calculation (PPP) with standard parametrisation and configuration interaction (CI) of singly excited states (CI-basis: 6 occupied, 6 virtual molecular orbitals; a larger CI-basis leads to a 10 to 15 nm redshift and small variations in the other calculated properties). The rotatory strengths and f-values are calculated without further approximations (6a,6b,7).
The chromophores and their surroundings are treated as one "supermolecule". Therefore, all interactions between permanent and transition charge distributions are included. Polar and aromatic amino acid residues lying in a sphere with 10 Å radius around C10 were taken into account.
Hydrogen bonds are simulated by varying the effective ionization potential (IP') of the heteroatom since a withdrawal of the hydrogen from a pyrrole nitrogen causes the π-electrons of this center to delocalize more easily. The parameter IP' was changed accordingly from 19.6 eV (no bonding) to 16.5 eV (strong bonding, IP' midway between the values for pyrrole and pyridine nitrogen). The values of all other parameters are given elsewhere (6,7).

Results and Discussion

Single Chromophores

Figures 1 and 2 show the variations of the absorption wavelength and rotatory strength R of the red band of B84 and B155, respectively, due to chromophore-protein interactions. We restrict our discussion to this band because it is well separated from other transitions and therefore easier to analyse - theoretically as well as experimentally.

Case I: isolated chromophore, neglecting all side chains
Case II: I + propionic acid side chains attached to C8 + C12
Case III: II + aspartate residue near pyrrole rings B and C (B87 and B39, respectively)
Case IV: III + residues of polar amino acids near the propionic side chains
Case V: IV + all polar and aromatic residues in a sphere of 10 Å around C10
Case VI: V + amide groups of the protein backbone within the same sphere (in brackets)

The calculated transition energies for the isolated chromophores (I) are around or below 550 nm. The origin for this hypsochromic shift, when compared to prior calculations (6a,6b,7), is the significant deviation from a planar geometry. Coulombic interaction (static coupling in terms of first order perturbation theory) effects the transition energies (bathochromic shift) mainly because of the different stabilization of the various molecular orbitals. The rotatory strength is altered because of changed coefficients in the CI-wavefunctions and the concomitant modified mixing of electric and magnetic transition moments.
Inclusion of the carboxylate groups of the propionic acid side chains (II) exhibits only small effects (the proton tautomerism is fixed by the interaction with adjacent amino acids (3c)).
Both chromophores (as well as the third, not displayed, at A84) show a common principle of interaction with the protein.

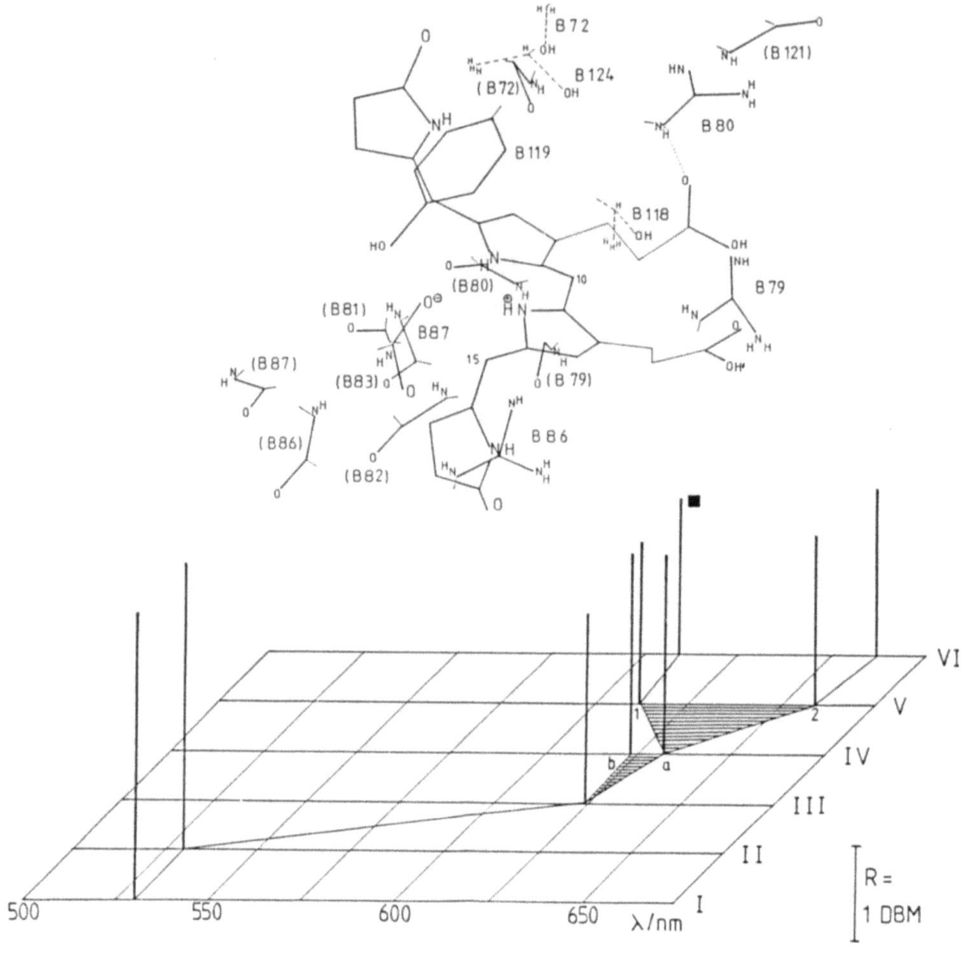

<u>Figure 1:</u> Calculated rotatory strength R of the longwave-length absorption band of the <u>B84</u> chromophore (1 DBM=0.93 10^{-38} cgs; case I to VI: see text). The shaded regions show the spectral variations with the tautomerism of the arginine groups:

a) B80 as shown b) tautomeric form

1) B79 and B86 as shown 2) tautomeric form

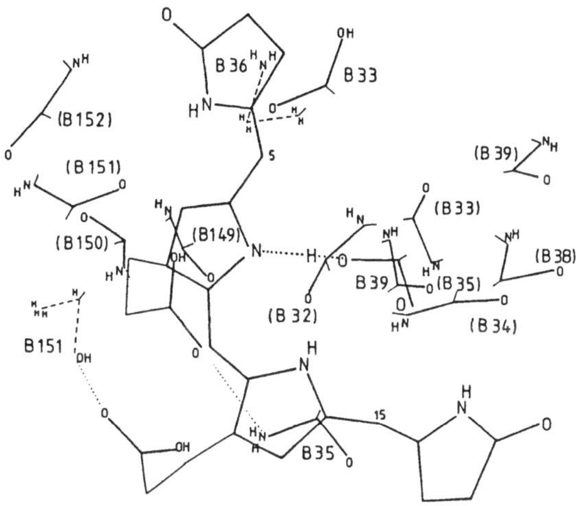

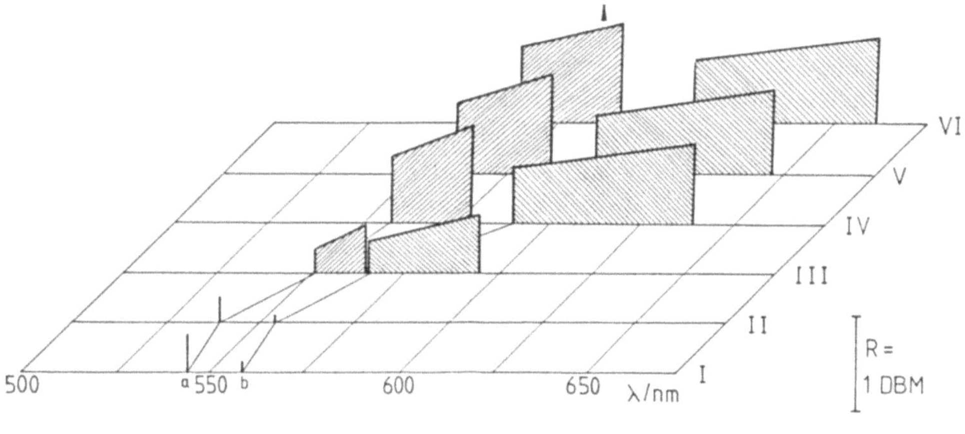

Figure 2: Calculated rotatory strength R of the longwave-length absorption band of the B155 chromophore. The shaded area reflects the bathochromic shift calculated for an increasing of the hydrogen bond strength to the B39 aspartate residue.
a) N23-H-tautomer b) N22-H-tautomer

They arch around an aspartate residue (B87 and B39, resp.), the nitrogens of pyrrole rings B and C (N22 and N23) being within hydrogen bonding distance of one of the carboxylate oxygens. This interaction dominates the wavelength adjustment (III).

For B84 (and A84) the assumption of a proton transfer to the chromophore, leaving a negative aspartate counterion, is necessary to shift the calculated electronic excitation energy into the experimentally established region.

B155 deviates from this behaviour; a proton transfer would result in a transition near 700 nm. Therefore we postulate that B155 and aspartate B39 are coupled via a hydrogen bond. Only a positive partial charge near N22 and N23 (OD2 treated as -O-H) yields a redshift. Depending on the methen tautomerism (N22-H or N23-H) and the hydrogen bond strength, the visual absorption band varies in a wavelength region of about 50 nm width (shaded areas in figure 2).

Inclusion of the charged and aromatic residues around B84 (case IV: Arg B80; case V: Arg B79, Arg B86, Ser B72, Tyr B119, Thr B124, Thr B118) leads to further shifts. There is a strong indication, that the arginine residues play a vital role in tuning the transition energy. Small variations in geometry (e.g. a rotation of the $-C(NH_2)(NH)$-groups by 180°, equivalent to an exchange of positive and negative partial charges) result in a drastic variation of the transition wavelength. Therefore, these residues may be responsible for the heterogeneity observed in the fluorescence kinetics of smaller aggregates (8).

Inclusion of the charges in the surrounding of B155 (case IV: Thr B151, Asn B35; case V: Lys B36, Glu B33) induces also additional redshifts of the first electronic transition. Only for the N23-H-tautomer is the theoretical result, however, in agreement with observation.

The interaction with the amide groups of the protein backbone (case VI) produces only small corrections to the excited state properties and can therefore be ignored in first approximation studies.

β-subunit

The results described so far predict that the CD-spectrum of two, non-interacting chromophores in the β-subunit should exhibit a large rotatory strength around 620 nm. This is, however, in contradiction to the experimental finding (4) (inactive longwavelength absorption band, attributed to B84). Therefore, we extended our model calculations to take into account both chromophores and their environments together. The selected status of chromophore-protein interaction according to approximation VI is marked in figures 1 and 2 for both chromophores. It yields transition energies in the correct wavelength region and also the required energy spacing for B84 and B155. Figure 3 shows the thereby obtained theoretical CD-spectrum of the β-subunit.

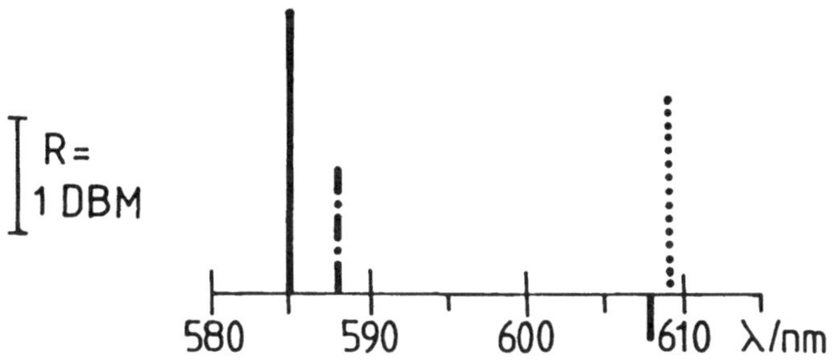

Figure 3: Calculated visible CD-spectrum of individual chromophores (B84 ········, B155 –·–·–·–·–) and the β-subunit with excitonic coupling included (————).

Due to excitonic coupling, the shorter wavelength transition ($\psi^{+}=0.9\ \psi^{*}$(B155) + 0.1 ψ^{*}(B84)) gains rotatory strength, which is lost in the corresponding longwavelength transition. This final result fits the experimentally found CD-bandshape very well and is a clear indication, that the interaction with charged residues controls also the chromophore-chromophore coupling within the phycobiliproteins. It is furthermore a warning that the properties of isolated chromophores may not be extracted from measurements on biological systems with interacting chromophores.

Acknowledgement

We want to thank Prof. H. Scheer for many fruitful discussions. Free computer time and the access to NOS/VE by the Leibniz Rechenzentrum der Bayerischen Akademie der Wissenschaften is greatly appreciated.

References

1. Scheer, H. 1983. In: Light Reaction Path of Photosynthesis (F. K. Fong, ed.). Springer Verlag, Berlin. pp. 7-45.

2. Frank, G., W. Sidler, H. Widmer and H. Zuber. 1978. Hoppe-Seyler's Z. Physiol. Chem. 359, 1491-1507.

3a. Schirmer, T., W. Bode, R. Huber, W. Sidler and H. Zuber. 1985. J. Mol. Biol. 184, 257-277

b. Schirmer, T., R. Huber, M. Schneider, W. Bode, M. Miller and M. L. Hackert. 1986. J. Mol. Biol. 188, 651-676

c. Schirmer, T., W. Bode and R. Huber. 1987. J. Mol. Biol. 196, 677-695

4. Siebzehnrübl, S., R. Fischer and H. Scheer. 1987. Z. Naturforsch. 42c, 258-262.

5. Mimuro, M., R. Rümbeli, P. Füglistaller and H. Zuber. 1987. Biochimica et Biophysica Acta 851, 447-456.

6a. Blauer, G. and G. Wagnière. 1975. J. Am. Chem. Soc. 97, 1949-1954.

b. Blauer, G. and G. Wagnière. 1976. J. Am. Chem. Soc. 98, 7806-7810.

7. Scharnagl, Ch., E. Köst-Reyes, S. Schneider, H.-P. Köst and H. Scheer. 1983. Z. Naturforsch. 38c, 951-959.

8. Schneider, S., P. Geiselhart, F. Baumann, S. Siebzehnrübl, R. Fischer and H. Scheer. Contribution in this volume.

EXCITATION ENERGY MIGRATION IN C-PHYCOCYANIN AGGREGATES ISOLATED FROM *PHORMIDIUM LURIDUM* - PREDICTIONS FROM THE FÖRSTER'S INDUCTIVE RESONANCE THEORY.

J. Grabowski

Institute of Environmental Engineering, Poznan Technical University, Piotrowo 5 St., 60-965 Poznan, Poland.

G. S. Björn

Department of Plant Physiology, University of Lund, Box 7007, S-220 07 Lund, Sweden.

Abstract

Absorption and circular dichroism spectra, as well as excitation and emission spectra of polarized fluorescence of monomeric and hexameric C-phycocyanin from *Phormidium luridum* have been measured. From contributions of the three spectrally different chromophores of phycocyanin to overall absorption and circular dichroism have been calculated from Gaussian approximations of these spectra. By combining the results with data on interchromophore distances and orientations available for *Agmenellum quadruplicatum*, the rates of excitation energy transfer between phycocyanin chromophores predicted by Forster's inductive resonance theory were determined.

Introduction

Since the discovery [1] and isolation in intact form [2,3,4] of phycobilisomes a remarkable progress has taken

Photosynthetic Light-Harvesting Systems

place in understanding their structure and function in the photosynthesis of red and blue-green algae [5 - 14].

As phycocyanin is a necessary phycobiliprotein component of phycobilisomes, knowledge of the paths and rates of energy transfer in phycocyanin is essential for the understanding of phycobilisome function. Until now they were based on speculation [5,8,15,16], since interchromophore distances and orientations were unknown. Such data are now available [17]. It was predicted by Zickendraht-Wendelstadt et al. [19] and shown by Schirmer et al. [17] that all three phycocyanin chromophores (designated B84, A84 and B155) must be considered spectrally different due to different microenvironments. Thus it is necessary when using Förster's theory to determine these spectral characteristics (absorption and emission spectra) for the three chromophores separately.

The purpose of the present investigation was to determine the spectra of the B84, A84 and B155 chromophores in monomeric and hexameric phycocyanin, to evaluate (using Forster's theory) the rates of excitation energy transfer between the chromophores, and to find the main paths of energy funneling through the phycocyanin component of phycobilisomes.

Materials and methods

The organism, *Phormidium luridum var. olivacea* Boresch and the cultivation conditions were as described by Björn and Björn [20]. The cells were disrupted in 0.01 M potassium phosphate buffer, pH 7.0, by using a French press (Aminco) at 41 MPa and centrifuged for 1 h at 120,000×g. The separation of the pigments was described earlier [20].

The molecular weight of the hexamer (designated PC-1 by

Brown et al. [21] and in our eariel paper [20]) was determined by gel filtration on a Sephacryl S-200 (Pharmacia, Sweden) column (2.5×35 cm) equilibrated with 0.01 M potassium phosphate buffer, pH 7.0, containing 0.1 M sodium chloride and 0.02 % (w/v) sodium azide. The phycocyanin monomer was obtained from the hexamer by treatment with 1 M sodium perchlorate in phosphate buffer.

Absorption spectra were determined on an Aminco DW-2a spectrophotometer, unpolarized and polarized fluorescence excitation and emission spectra on a Fluorolog 5 spectrofluorimeter (Spex, NJ USA) equipped with polarizing filters (Oriel) when required, and circular dichroism spectra on a Jasco J 41A spectropolarimeter.

Since there is no direct way of determining absorption and emission spectra of B84, A84 and B155 chromophores separately, the Gaussian deconvolution method [22] in combination with Weber's matrix method [23] was applied to the phycocyanin spectra.

Results

The main problem with the Gaussian deconvolution method is that one cannot expect an unambiguous solution if the number of Gaussian components exceeds two. Thus use is usually restricted to testing results obtained by other methods. As stated, this is not possible in our case, but fortunately the emission spectra from both monomeric and hexameric phycocyanin could be closely approximated by two Gaussian components (Table 1, Figs 1 and 2). Similar results have been obtained eariel for allophycocyanin [16] and phycoerythrin [24], so it may be a general feature of phycobiliprotein emission spectra.

Table I.

Weber's matrix [17] for C-phycocyanin hexamer fluorescence (data for monomer in parenthesis).

λ_{exc} \ λ_{obs}	630-640	640-650	650-660	660-670	670-680	680-690
560-570	– (.037)	– (.010)	– (-.000)	– (-.013)	– (-.007)	– (–)
570-580	– (.006)	.014 (-.007)	.006 (-.003)	.003 (-.001)	-.004 (.001)	-.012 (–)
580-590	– (-.005)	.004 (.003)	.004 (.006)	-.006 (-.002)	.000 (-.007)	-.014 (–)
590-600	– (.025)	-.002 (-.007)	-.000 (-.009)	.005 (.005)	-.007 (.008)	.011 (–)
600-610	– (.007)	.008 (-.000)	.009 (-.004)	.023 (.003)	-.007 (-.007)	-.013 (–)
610-620	– (–)	.049 (–)	.069 (–)	.019 (–)	.048 (–)	.050 (–)

To make the deconvolution as reliable as possible, analysis was performed "globaly", i.e. the deconvolutions of all spectra (absorption spectra, circular dichroism spectra and polarized fluorescence excitation spectra of monomer as well as hexamer) were treated simultaneously as one problem. Thus, only those computer-generated bands were accepted that were reasonably consistent with all of the experimental spectra. To this end, the following set of equations was analysed numerically:

$$\left\{ \begin{aligned} A(\bar{\nu})/\bar{\nu} &= \sum_i A_{oi} \exp\left[-4\cdot(\ln 2)\cdot\left(\frac{\bar{\nu} - \bar{\nu}_{oi}}{\sigma_{oi}} \right)^2 \right] \\ \Theta(\bar{\nu}) &= \sum_i \Theta_{oi} \exp\left[-4\cdot(\ln 2)\cdot\left(\frac{\bar{\nu} - \bar{\nu}_{oi}}{\sigma_{oi}} \right)^2 \right] \\ r(\bar{\nu}) &= \sum_i \frac{r_{oi}\ A_i(\bar{\nu})}{A(\bar{\nu})} \end{aligned} \right. \qquad (1)$$

where $A(\bar{\nu})$: absorbance at wavenumber $\bar{\nu}$,

A_{oi} : amplitude of i-th Gaussian absorption component band,

$\bar{\nu}_{oi}$: location of i-th band,

σ_{oi} : half-band width of i-th band,

$\Theta(\bar{\nu})$: circular dichroism,

Θ_{oi} : amplitude of i-th Gausian component circular dichroism band,

$r(\bar{\nu})$: anisotropy of emission ($r=2\cdot P/(3-P)$; where P is the degree of fluorescence polarization),

r_{oi} : limitting emission anisotropy characteristic for a given kind of phycocyanin chromophore.

Table II.

Results of Gaussian deconvolution of absorption (expressed in terms of $A(\bar{\nu})/\bar{\nu}$ [20]), circular dichroism and emission (expressed in terms of $F(\bar{\nu})/\bar{\nu}^4$ [20]) of C-phycocyanin hexamer (data for monomer in parenthesis).

band no.	band location [$\cdot 10^4 cm^{-1}$]	half-band width [$\cdot 10^4 cm^{-1}$]	intensity [rel.]	CD intensity [rel.]
1	1.555 (1.587)	0.078 (0.091)	62 (62)	102 (36)
2	1.678 (1.702)	0.156 (0.140)	26.5 (32.7)	28 (28)
3	1.589 (1.672)	0.078 (0.090)	48 (56)	12 (58)
4	1.710 (1.742)	0.156 (0.140)	20.5 (29.5)	6 (44)
5	1.643 (1.678)	0.078 (0.090)	43 (23.5)	8 (40)
6	1.764 (1.797)	0.156 (0.140)	18.3 (12.4)	4 (29)

For the two fluorescence bands the locations were 1.526 (1.542) and 1.405 (1.425), the half-band widths 0.078 (0.091) and 0.156 (0.140), and the intensities 110 (110) and 47 (58).

The bands obtained in this way (Table II) have been assigned to phycocyanin chromophores according to a suggestion

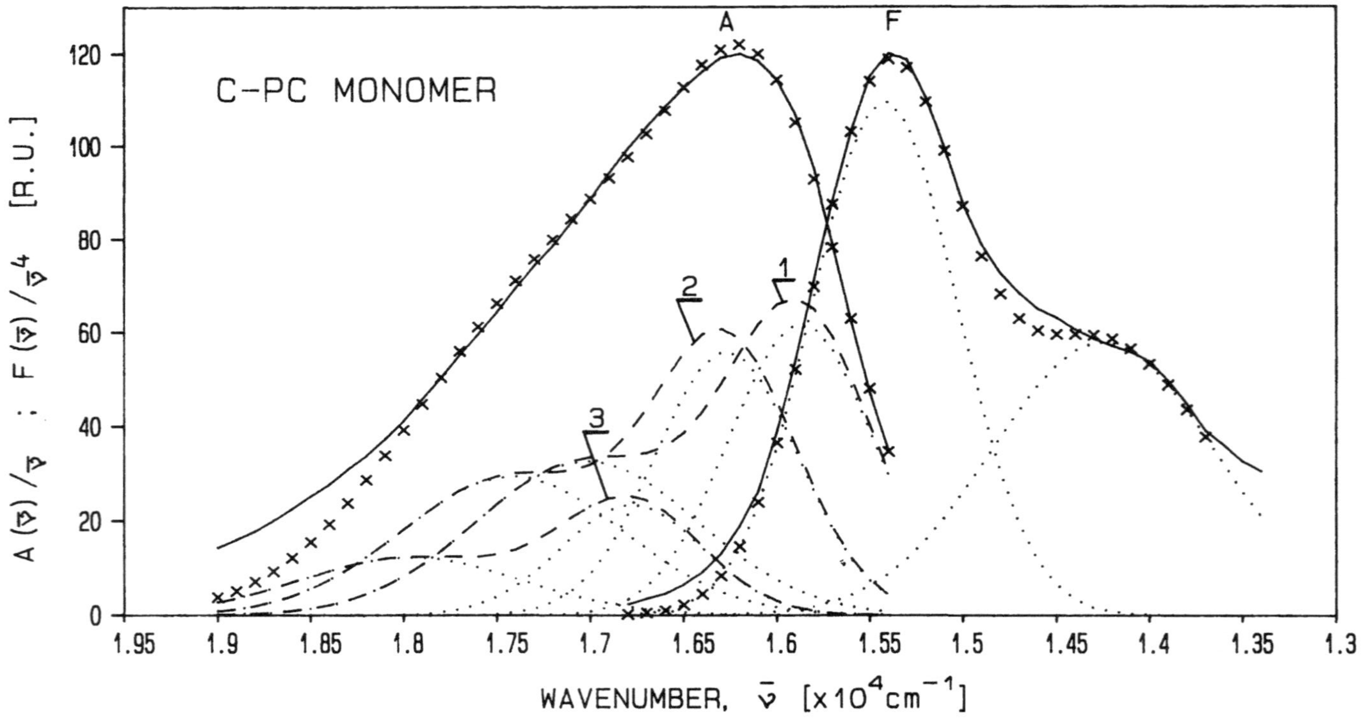

Fig. 1 Deconvolution into Gaussian bands of absorption (A, expressed in terms of $A(\bar{\nu})/\nu$ [16]) and emission (F, expressed in terms of $F(\bar{\nu})/\bar{\nu}^4$ [16]) spectra of C-phycocyanin monomer from *Phormidium luridum* (...). The crosses (x) represent the sums of Gaussian bands, and the goodnes of fit can be judged by comparison to the measured spectra (——). Curves 1, 2 and 3 (---) are the assumed contribution from B84, A84 and B155 chromophores, respectively.

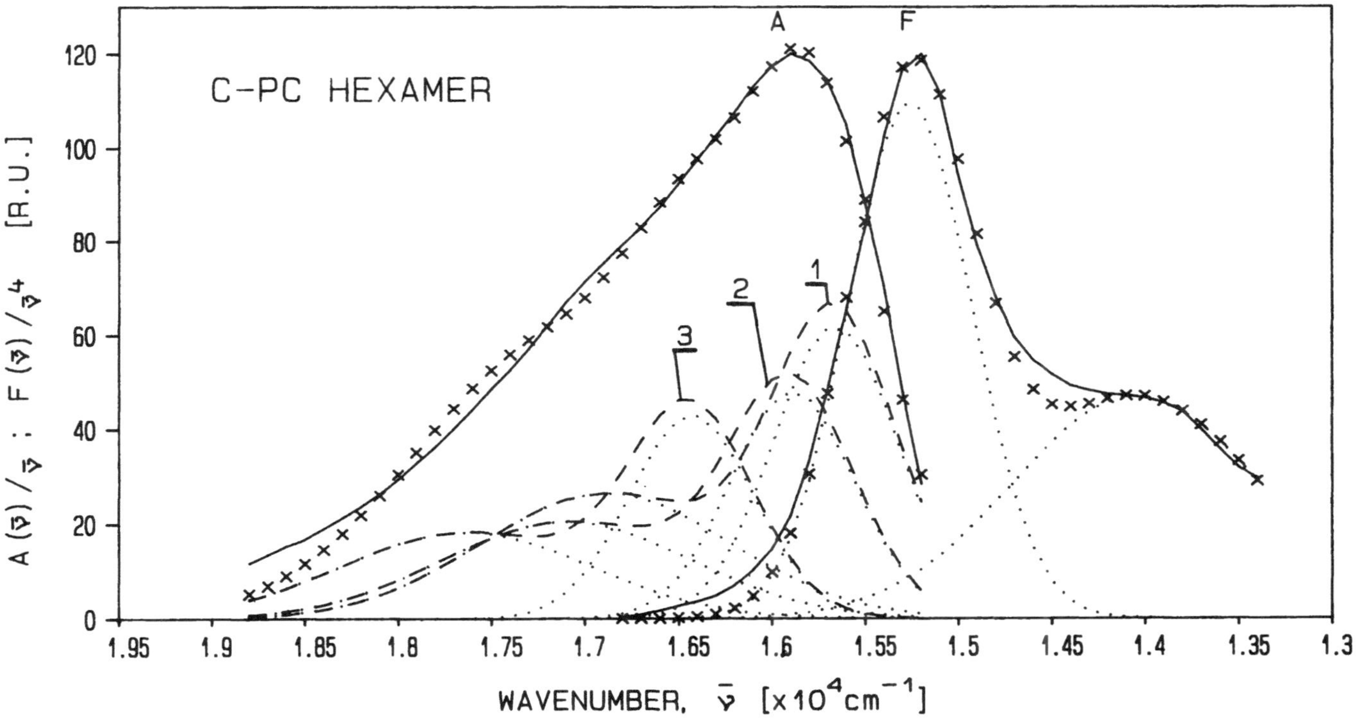

Fig. 2 Deconvolution into Gaussian bands of absorption (A, expressed in terms of $A(\nu)/\nu$ [16]) and emission (F, expressed in terms of $F(\nu)/\nu^4$ [16]) spectra of C-phycocyanin hexamer from *Phormidium luridum* (...). The crosses (x) represent the sums of Gaussian bands, and the goodnes of fit can be judged by comparison to the measured spectra (——). Curves 1, 2 and 3 (---) are the assumed contribution from B84, A84 and B155 chromophores, respectively.

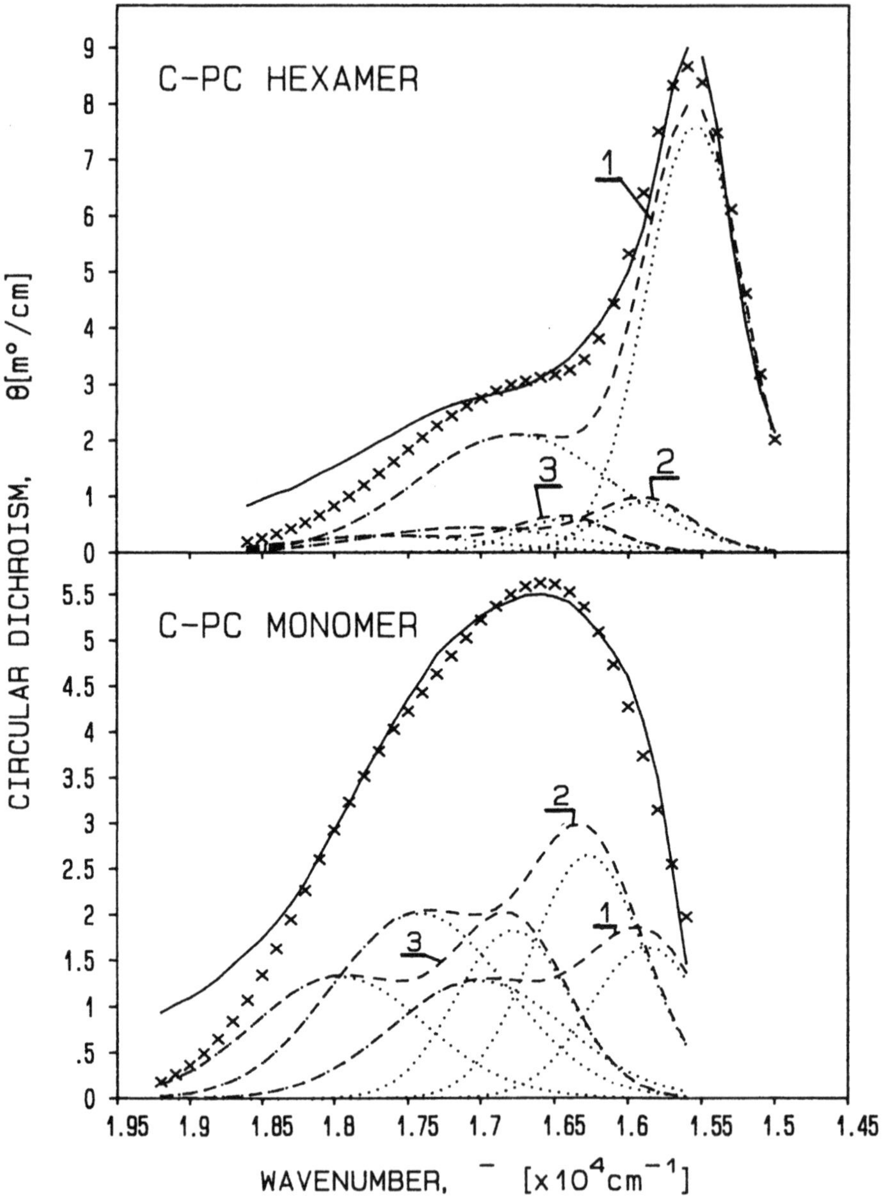

Fig. 3 Deconvolution of circular dichroism spectra of C-phycocyanin hexamer (upper panel) and monomer from *Phormidium luridum*. Otherwise as Fig. 1.

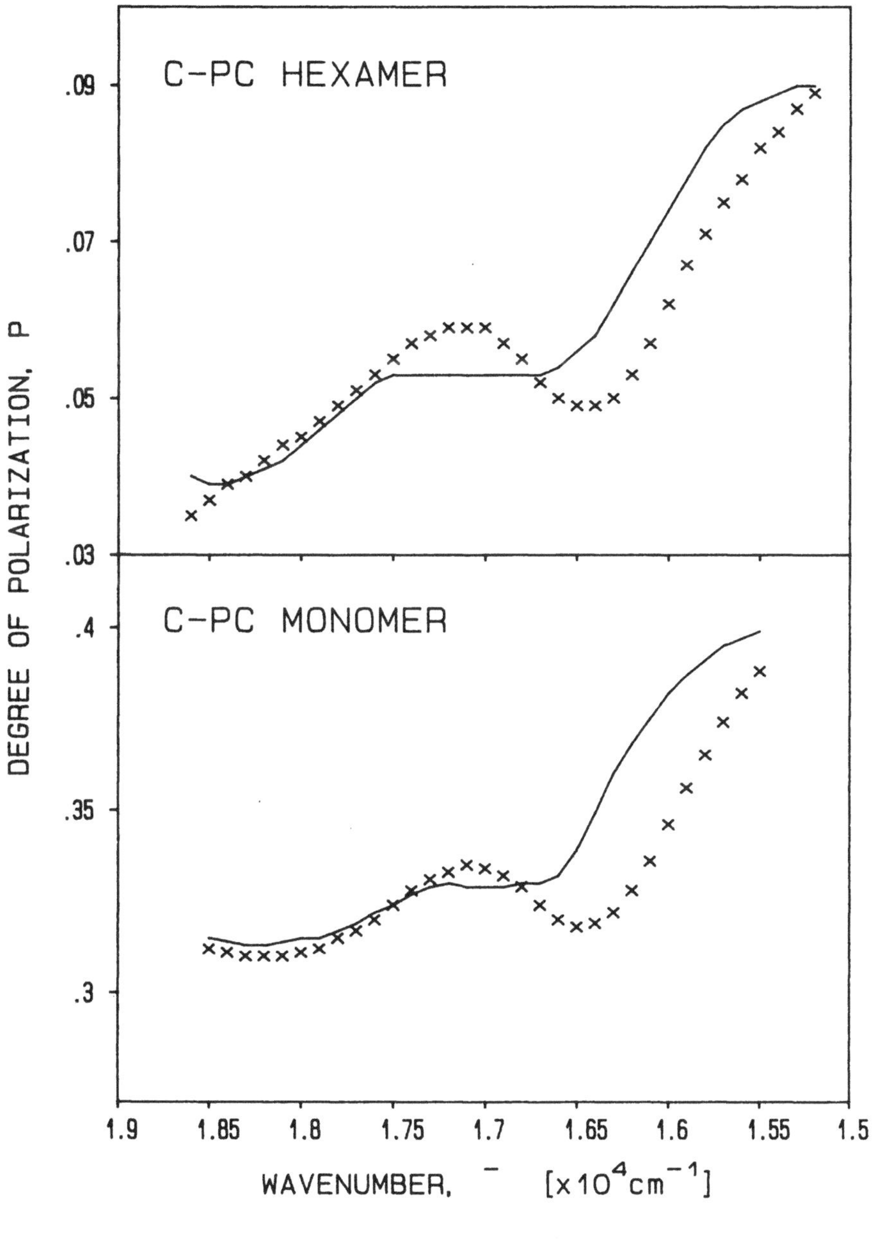

Fig. 4 Comparison of measured (——) an calculated (x) excitation spectra of polarized fluorescence of C-phycocyanin hexamer (upper panel) and monomer from *Phormidium luridum*.

by Siebzenrubl et al. [25]: The long wavelength absorbing species is B84, the short wavelength one B155, and A84 is intermediate. The spectra obtained for the chromophores are shown in Fig's 1 - 3.

The rates constants for energy transfer in phycocyanin aggregates were then calculated, using interchromophore distances and orientations from Schirmer et al. [17] the quantum yield and lifetime 0.53 and 2.2 ns, respectively [5] (same for all chromophores) and Förster's [18] formula:

$$n_{D-A} = \frac{9000\cdot(\ln 10)\cdot\varkappa^2\cdot\phi}{128\cdot\pi^5\cdot n^4\cdot N\cdot R^6\cdot\tau}\cdot\int_0^\infty \frac{f(\bar{\nu})\varepsilon(\bar{\nu})}{\bar{\nu}^4}\,d\bar{\nu} \qquad (2)$$

where
- ϕ : fluorescence quantum yield of donor in absence of acceptor,
- n : refraction index,
- N : Avogadro's number,
- τ : lifetime of donor excited state in absence of acceptor,
- $f(\bar{\nu})$: normalized fluorescence ($\int_0^\infty f(\bar{\nu})d\bar{\nu} = 1$) intensity of donor
- $\varepsilon(\bar{\nu})$: molar extinction coefficient of acceptor,
- $\varkappa$: orientation factor,
- R : donor-acceptor distance.

For peak absorption coefficients we used values obtained for related organisms: $3.33\cdot10^5\ M^{-1}cm^{-1}$ for the monomer [25] and $2.63\cdot10^5\ M^{-1}cm^{-1}$ [26] for the hexamer.

Rate constants were calculated for energy transfer from each of the three chromophores in one phycocyanin molecule to the six chromophores in the two other phycocyanin molecules in the same trimer (12 values, Table III, upper part), and also from the three chromophores in one trimer to the nine chromophores in the other trimer (27 values, Table III, lower part).

Table III.

Rate constants of excitation energy transfer between chromophores within one C-phycocyanin trimer (upper part) and between chromophores in one trimer of a hexamer (lower part), as calculated from Förster theory. The unit is s^{-1}, and powers of ten designated by E.

TRIMER (an upper one):

	2A84	2B84	2B155	3A84	3B84	3B155
1A84	1.17 E10	2.00 E12	6.95 E10	1.17 E10	6.16 E9	2.20 E8
1B84	4.46 E9	4.23 E12	2.55 E9	1.44 E12	4.23 E12	1.02 E10
1B155	1.64 E9	3.66 E11	1.13 E9	5.17 E11	9.20 E10	1.13 E9

HEXAMER:

	4A84	4B84	4B155	5A84	5B84	5B155	6A84	6B84	6B155
1A84	3.43 E11	4.80 E8	1.31 E9	1.17 E10	2.16 E10	2.62 E8	1.51 E9	1.38 E9	2.81 E10
1B84	3.47 E8	1.86 E11	5.08 E8	1.64 E10	2.31 E13	1.18 E9	1.28 E11	2.01 E11	1.03 E9
1B155	9.74 E9	1.83 E10	1.99 E8	1.77 E9	4.27 E10	5.34 E10	2.09 E11	3.23 E10	5.43 E9

IV. DISCUSION.

The absorption maxima of the phycocyanin hexamer and monomer were found at 633 and 615 nm, respectively. The maxima in the CD spectra were displaced to 643 (red shift) and 605 nm (blue shift), respectively. Gaussian deconvolution of these spectra (Fig. 3) shows that the long wavelength chromophore (presumably B84) dominates the CD spectrum of hexamer, while this does not hold for the monomer. On the other hand, the contribution of the short wavelength chromophore (presumably B155) to the CD spectrum of the hexamer is very weak, while it is considerable in the monomer. This probably explains the discrepancy in location of maxima in absorption and CD spectra for both monomer and hexamer. The strong elipticity observed at the high wavelength side of the CD spectrum of the hexamer (observed also by Glazer [25]) indicates a particular state of B84 in the hexamer. Whether this strong ellipticity is due to a modified chromophore configuration induced by protein or simply reflects the helical distribution of these chromophores in the hexamer cannot be decided from our data. The values of limitting polarizations that characterize the chromophores in the monomer and the hexamer can be read from Fig. 4 and are, for B84, A84 and B155, respectively, as follows: 0.10, 0.04 and 0.025; monomer: 0.415, 0.238, 0.342. They have to be regarded as "effective values" reflecting not only the physical structures of the various chromophores, but also their orientation in relation to other chromophores within the aggregate. The values obtained for interchromophore energy transfer rate constants are summarized in Fig. 5 (cf. Table III). It is seen that the highest values are of the order of 10^{10} to 10^{11} s^{-1}, while some values are in the range 10^{8} to 10^{10} s^{-1} - see Table III - and thus similar to the rates measured for the α-subunit of phycoerythrin [26]. The data indicate that energy is not exchanged indiscriminately between the chromophores (especially in the phycocyanin hexamer, which is a natural phycobilisome component), but there are "preferred paths": from the outside to the inside chromophores and then from "upper" or distal (in distal trimer) to "lower" or proximal (in proximal trimer) chromophores. Since the phycocyanin hexamer is composed of two trimers, there are (at least) three "expressways" for funneling the excitation energy through phycocyanin to the next phycobiliprotein, allophycocyanin.

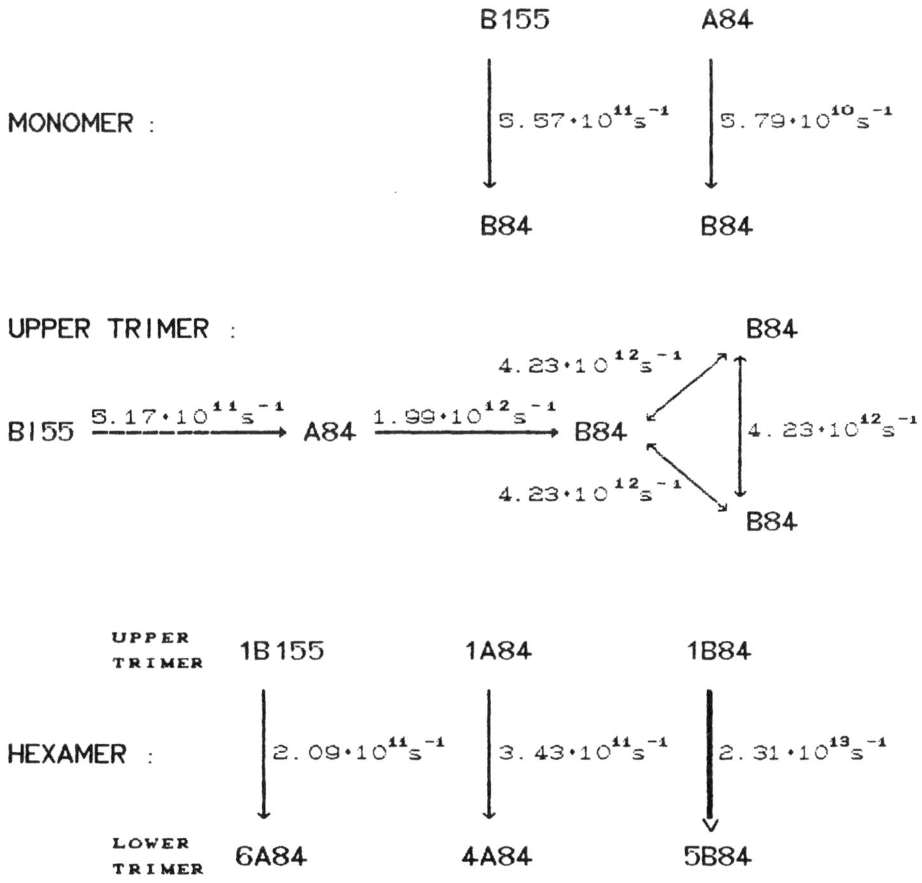

Fig. 5 The excitation energy migration paths (numbers represent calculated rates of transfer) in C-PC aggregates forms.

Acknowledgements

We thank Ms Eva Olsson for technical assistance and Dr Janet Bornman for linguistic revision. CD spectra were measured by Mr Jan Glans.

References

1. Gant E., S.F. Conti: "The ultrastructure of *Porphyridium cruentum*"; J. Cell Biol. 26, 365-381 (1965).
2. Gantt E. and C.A. Lipschultz: "Phycobilisomes of *Porphyridium cruentum* (Isolation)"; J. Cell Biol. 54, 313-324 (1972).
3. Gray B.H. and E. Gantt: "Spectral properties of phycobilisomes and phycobiliproteins from the blue-green alga - *Nostoc sp.*"; Photochem. Photobiol. 21, 121-128 (1975).
4. Gantt E., C.A. Lipschultz, J. Grabowski and B.K. Zimmerman: "Phycobilisomes from blue-green and red algae - Isolation criteria and dissociation characteristics"; Plant Physiol. 63, 615-620 (1979).
5. Grabowski J. and E. Gantt: "Photophysical properties of phycobiliproteins from phycobilisomes: fluorescence lifetimes, quantum yields, and polarization spectra"; Photochem. Photobiol. 28, 39-45 (1978).
6. Ohki K. and T. Katoh: "Excitation transfer from dark-synthesized phycocyanin directly to pigment system I chlorophyll in *Anabaena variabilis*"; Photosynth. Organells - special issue of Plant and Cell Physiol., 15-21 (1977).
7. Mörchel E., Koller K.P., Wehrmeyer W. and Schneider H.: "Biliprotein assembly in the disc-shaped phycobilisomes of *Rhodella violacea*. I. Electron microscopy of phycobilisomes in situ and analysis of their architekture after isolation and negative staining"; Cytobiologie 16, 118-129 (1977).

8. Grabowski J. and E. Gantt: "Excitation energy migration in phycobilisomes: comparison of experimental results and theoretical predictions"; Photochem. Photobio. 28, 47-54 (1978).

9. Kume N., T. Isono and T. Katoh: "Stability of cyanobacterial phycibilisomes in reference to their concentration"; Photobiochem. Photobiophys. 4, 25-37 (1982).

10. Grabowski J., E. Gantt, C.A. Lipschultz and L.G. Erokhina: "Spectral and dissotiation characteristics of heterophycobiliprotein complexes separated from phycobilisomes of *Nostoc sp.* and *Anabaena variabilis*"; in Photosynthetic solar energy conversion and storages (Ed. J. Poskuta), Wyd. Uniw. Warszawskiego, Warsaw 1983.

11. Searle G.F.W., J. Barber, G. Porter and C.J. Tredwell: "Picosecond time-resolved energy transfer in *Phorphyridium cruentum* (part II); in the isolated light harvesting complex (phycobilisomes); Biochim. Biophys. Acta 50, 246-256 (1978).

12. Wanner G. and H.P. Köst: "Investigation of arrangement and fine structure of *Porphyridium cruentum* phycobilisomes", Protoplasma 102, 97-109 (1980).

13. Björn L.O. and G.S. Björn: "Studies on energy dissipation in phycobilisomes using the Kennard - Stiepanov relation between absorption and fluorescence emission spectra"; Photochem. Photobiol. 44, 535-542 (1986).

14. Chereskin B.M. and E. Gantt: "Enrichement of 50-kilodalton polipeptide in a photosystem II -phycobilisome particle from *Porphiridium cruentum*"; Arch. Biochem. Biophys. 250, 286-293 (1986).

15. Dale R.E. and F.W.J. Teale: "Number and distribution of chromophore types in native phycobiliproteins"; Photochem. Photobiol. 12, 99-117 (1970).

16. MacCall R. and D.S. Berns: "Energy transfer studies on cryptomonad biliproteins"; Photochem. Photobiol. 27, 343-349 (1978).

17. Schirmer T., K. Huber, M. Schneider, W. Bode, M. Miller and M.L. Hackert: "Crystal structure analysis and refinement at 2.5 Å of hexameric C-phycocyanin from the cyanobacterium *Agmenellum quadruplicatum* (The molecular model and its implication for light-harvesting)"; J. Mol. Biol. __188__, 651-676 (1986).

18. Förster Th.: "Fluoreszenz organischer Verbindungen", Vandenhoeck Ruprecht, Götingen 1951.

19. Zichendraht - Weendelstadt B., J. Friedrich and W. Rüdiger: "Spectral characterization of monomeric C-phycoerythrin from *Pseudoanabaena* W 1173 and its α and β submits and C-phycoerythrin"; Photochem. Photobiol. __31__, 367-376 (1980).

20. Brown A.S., J.A. Foster, P.V. Voynow, C. Franzblau and R.F. Troxler: "Allophycocyanin from the filamenous cyanophyte, *Phormidium luridum*."; Biochemistry __14__, 3581-3588 (1975).

21. Kazaczienko L.P.: "Moliekuliarnaia spiektroskopia zidkostiei"; Ed. Izd. B.G.U. im. . I. Lienina, Minsk 1978.

22. Weber G.: "Enumeration of components in complex systems by fluorescence spectrometry"; Nature __190__, 27-29 (1961).

23. Grabowski J. and E. Gantt; unpublished.

24. Siebzenrubl S., R. Fisher and H. Scheer: "Chromophore assigment in C-phycocyanin from *Mastigocladus laminosus*"; submited to FEBS - Lett. (1986).

25. Glazer A. N.: "Phycocyanins: Structure and function"; in "Photochemical and Photobiological Reviews" Ed. K.C. Smith), Vol. 1, pp. 71-115, Plenum Publishing Corp. (1976).

26. Dagen A. J., R. R. Alfano, B. A. Zilinskas and C. E. Swenberg: "Analysis of fluorescence kinetics and energy transfer in isolated α subunits of phycoerythrin from *Nostoc sp.*"; Photochem. Photobiol. __43__, 71-76 (1986).

ENERGY TRANSFER CALCULATIONS FOR TWO C-PHYCOCYANINS BASED ON REFINED X-RAY CRYSTAL STRUCTURE COORDINATES OF CHROMOPHORES

K. Sauer

Laboratory of Chemical Biodynamics, University of California, Berkeley, CA94720, USA

H. Scheer

Botanisches Institut der Universität, Menzinger Str. 67, D-8000 München 19, FRG

Photosynthetic antennas are capable of efficient light absorption and excitation energy transfer to the reaction centers over distances of several tens of nanometers. Light capture is realized by a variety of strongly absorbing chromophores embedded in a protein matrix. These chromophores have absorption maxima that are typically closely spaced. Energy transfer proceeds in a stepwise fashion among these chromophores. For these steps, two extreme mechanisms are generally considered (1). Weak exciton coupling is operative over short distances (typically ≤1nm), and Förster transfer remains efficient over much longer distances (typically ≤6nm). In both cases, orientation and distances among the chromophores are critical parameters, on which little information is available in most antenna systems.

C-Phycocyanin (PC) is one of the few cases where the necessary structural information is known. Based on the preliminary crystal structure of PC from the cyanobacterium Agmenellum quadruplicatum (2), rate constants for pairwise energy transfer and the derived sets of differential equations have been determined (3) by using exclusively the Förster model (4). The fair agreement with experimentally determined absorption and

Transfer Step	Coordinates New		Old	
β-Subunit				
1B155 --> 1B84	34.4	(7.6)	28.5	(6.3)
αβ-Monomer				
1A84 --> 1B84	13.9	(10.6)	0.037	(0.028)
1B155 --> 1A84	2.15	(0.52)	3.44	(0.84)
$(\alpha\beta)_3$-Trimer				
3A84 --> 1B84	1527.	(1160.)	59.0	(44.8)
2B84 --> 1B84	8.8	(8.8)	46.2	(46.2)
2A84 --> 1B84	2.14	(1.62)	0.18	(0.138)
2B155 --> 1A84	2.4	(0.58)	16.9	(4.1)
2A84 --> 1A84	1.25	(1.25)	0.42	(0.42)
3B155 --> 1B84	0.82	(0.18)	6.2	(1.37)
2B155 --> 1B84	0.59	(0.131)	1.55	(0.35)
2B155 --> 1B155	0.29	(0.29)	0.37	(0.37)
3B155 --> 1A84	0.092	(0.023)	0.054	(0.0131)
$(\alpha\beta)_6$-Hexamer				
4A84 --> 1A84	391.	(391.)	185.4	(185.4)
6B155 --> 1B155	186.3	(186.3)	174.8	(174.8)
6B155 --> 1A84	68.8	(16.7)	20.7	(5.0)
6A84 --> 1B84	40.7	(30.9)	1.64	(1.24)
6B84 --> 1B84	10.6	(10.6)	10.9	(10.9)
6B155 --> 1B84	9.2	(2.04)	2.01	(0.45)
5B84 --> 1B84	9.0	(9.0)	53.6	(53.6)
4B84 --> 1B84	4.5	(4.5)	10.2	(10.2)
5B155 --> 1B84	3.8	(0.83)	1.11	(0.244)
4A84 --> 1B84	1.31	(0.99)	0.059	(0.045)
6A84 --> 1A84	0.94	(0.94)	0.58	(0.58)
4B155 --> 1B84	0.79	(0.176)	1.31	(0.29)
5A84 --> 1B84	0.72	(0.54)	0.044	(0.033)
4B155 --> 1A84	0.55	(0.134)	3.5	(0.85)
5B155 --> 1A84	0.52	(0.127)	0.0033	(0.0009)
5A 84 --> 1A84	0.37	(0.37)	0.134	(0.134)
5B155 --> 1B155	0.0112	(0.0112)	0.035	(0.035)
4B155 --> 1B155	0.0094	(0.0094)	0.045	(0.045)

fluorescence kinetic data suggested that this model might be sufficient to describe energy transfer in PC.

However, more recent experimental results (5-7) indicated a conflict. Whereas theory predicted transfer between the two chromophores of lowest energy, e.g. α84 and β84, as the fastest process, wavelength-resolved decay kinetics suggested that transfer from the high-energy β155 chromophore to the low-energy ones contributes strongly to the fastest resolved component of appx. 20-40 ps.

The recent refinement of the crystal structures of trimeric PC from Mastigocladus laminosus and hexameric PC from A. quadruplicatum (8), prompted us to reinvestigate this problem. The refined structures gave a much better definition of the chromophores, in particular of the previously only poorly resolved β-84 chromophore. This resulted in minor changes in the distances, but considerable changes in the orientation parameters. We have now recalculated the energy transfer and shown that the Förster mechanism may not be sufficient to describe the system, and that strong coupling has to be invoked at least for three symmetrically equivalent pairs of chromophores.

The rate constants for pairwise energy transfer according to Förster theory are listed in Table 1 ("new") for revised coordinates of PC from A. quadruplicatum. The rate constants for

Table 1: Rate Constants for Förster Excitation Transfer in PC

Rate constants calculated using the Förster inductive resonance transfer expression and spectroscopic overlap integrals, as described previously (3), but using a refractive index value n=1.34 (10). Results which are compared are based on the "old" (2) and "new" (8) coordinates. Rate constants for backward transfer are indicated in parenthesis. For increasing extent of aggregation, values from the top of the table are to be included in a cumulative manner. All other transfer constants are related by symmetry (2,8) to the ones shown.

M. laminosus, are very similar (not shown), in agreement with the very similar structures. The largest differences are calculated for rates involving chromophore β155, whose conformation seems to be somewhat affected by crystal packing (9). As compared with the rate constants based on the previous structural data Table 1, "old"), changes up to a factor of 40 are observed. Although this increase was largest for some of the minor rate constants, some of the fastest and hence most important ones are affected as well. In aggregates, a rate of greater than $1 ps^{-1}$ is calculated for transfer between adjacent α-84 and β-84 chromophores situated on different monomeric units. This rate is now by far the fastest, which is even more pronounced

Table 2: Excited State Exponential Decay Time Constants Calculated for Förster Excitation Transfer in C-Phycocyanin

	β-subunit	Monomer	Trimer	Hexamer
τ_1	24	21	0.37	0.36
τ_2	--	45	22	7.5
τ_3	1500	1500	1500	1500

Calculated values listed are time constants in ps for the fast (τ_1) and intermediate (τ_2) components of a three exponential analysis, with k_F input as $0.67 ns^{-1}$ for the rate constant of radiative and non-radiative decay corresponding to τ_3.

Table 3: Exciton Energies of Pairwise Interaction Calculated for Chromophores of C-Phycocyanin (only values ≥$5 cm^{-1}$ listed)

	Chromophores [$V_\pm$, cm^{-1}]	
β-Subunit	1B155/1B84	7.3
(αβ) -Monomer	1A84/1B84	5.3
$(\alpha\beta)_3$-Trimer	1A84/2B84	56.0
$(\alpha\beta)_6$-Hexamer	1A84/4A84	32.0
	1B155/6B155	23.8
	1A84/6B155	10.8
	1A84/6B84	9.1

Exciton energies calculated from the relation $V_\pm = Dk/\varepsilon R^3$ (11), with orientation factors, k, and interchromophore distances, R, taken from Schirmer, et al (8). Dipole strength (D, in $Debye^2$) determined from the spectroscopic deconvolution (3), with dielectric constant n=1.34 (10).

because the originally second fastest rate within the trimer (β-84->β-84) is strongly decreased.

The occurence of these fast rates raised the possibility that other mechanisms of energy transfer must be considered. This is supported by calculations of exciton interactions using point-dipole approximation (table 3). Pairwise interactions are in particular substantial for α84/β84 on adjacent monomers, and for α84/α84 on different trimer disks in the hexamer. Similar results were reported using point-monopole approximation (12).

Acknowledgment

This work was supported by the Deutsche Forschungsgemeinschaft (SFB 143, Forschergruppe "Pflanzliche Tetrapyrrole", and by the CIP botany computer facilities, H.S.), by the Alexander-von-Humboldt Stiftung, Bonn (award to K.S.) and by the Director, Office of Energy Research, Office of Basic Energy Sciences, Division of Biological Energy Conversion of the U.S. Department of Energy under Contract DE-AC03-76SF00098 (K.S.).

References

1. Sauer, K. 1975. In: Bioenergetics of Photosynthesis (Govindjee, ed.). Academic Press New York, p. 115.
2. Schirmer, T., R. Huber, M. Schneider, W. Bode, M. Miller, M.L. Hackert. 1986. J.Mol.Biol. 188, 651.
3. Sauer, K., H. Scheer, P. Sauer. 1987. Photochem.Photobiol. 46, 427.
4. Förster, T. 1967. In: Comprehensive Biochemistry, Vol 22, (M. Florkin and E.H. Stotz, eds.). Elsevier Amsterdam, p.61.
5. Holzwarth,A.R., J.Wendler, G.W.Suter. 1987. Biophys.J. 51,1.
6. Sandström, A., T. Gillbro, V. Sundström, R. Fischer, H. Scheer: Biochim.Biophys.Acta. (in press)
7. Schneider, S., P. Geiselhart, S. Siebzehnrübl, R. Fischer, and H. Scheer, contribution to this volume
8. Schirmer, T., W. Bode, R. Huber. 1987. J.Mol.Biol. 196, 677.
9. Dürring, M., R.Huber (1987) Private communication
10. Moog, R.S., A. Kuki, M.D. Fayer, S.G. Boxer. 1984. Biochemistry 23, 1564.
11. Tinoco, I,Jr. 1963. Radiation Research 20, 133.
12. Schneider,S., C.Scharnagl, M.Dürring, T.Schirmer, W.Bode, contribution to this book.

ENERGY TRANSFER IN LIGHT-HARVESTING ANTENNA OF PURPLE BACTERIA, STUDIED BY PICOSECOND SPECTROSCOPY

V. Sundström, H. Bergström, T. Gillbro
Department of Physical Chemistry, University of Umeå, S-901 87 Umeå, Sweden

R. van Grondelle
Department of Biophysics, Physics Laboratory of the Free University, De Boelelaan 1081, 1081 HV Amsterdam, The Netherlands

W. Westerhuis, R.A. Niederman
Department of Biochemistry, Rutgers University, Piscataway, NJ 08854, USA

R. Cogdell
Department of Botany, University of Glasgow, Glasgow G12 8QQ, UK

Introduction

The dynamics of energy transfer within the bacteriochlorophyll antenna and its constituents of purple bacteria has been studied at room temperature and 77 K, by means of low intensity picosecond absorption spectroscopy. Whole membrane preparations of Rb. sphaeroides and R. rubrum were studied with open and closed reaction centers at room temperature and with closed reaction centers at 77 K. The isolated pigment-protein complexes B800-850 and B875 were also investigated at room temperature and 77 K. All studies taken together yield a coherent and detailed picture of the energy transfer dynamics. The composition of the light-

Photosynthetic Light-Harvesting Systems

harvesting antenna seems to be more complex than previously assumed. Thus, clear evidence is obtained for a long-wavelength spectral species in both R. rubrum and Rb. sphaeroides. This species absorbs in the red wing of the main antenna absorption band, and is assigned to a minor redshifted antenna component, B896.

Results

Measurements of isotropic picosecond absorption recovery as well as decay of induced anisotropy, using a sync-pumped and cacity dumped dye laser system, is a powerful technique to monitor energy transfer dynamics in these systems. With the aid of infrared laser dyes picosecond pulses continuously tunable in the region 700-900 nm could be obtained. Due to the powerful signal averaging inherent in the high repetition rate of the laser pulses and the lock-in detection used, kinetics could be measured at very low light intensity; minimum detectable absorbance change was about 10^{-6}.

Energy transfer in chromatophores

Fig. 1 summarizes in a condensed way the energy transfer kinetics measured for chromatophores of Rb. spaeroides at 77 K with closed reaction centers (1). Normalized absorbance difference spectra for the various antenna pigments are deduced from these kinetic results. The wavelength variation of the measured kinetics is consistent with an energy transfer sequence involving the following steps and time constants. Bchl 800-->Bchl850 ($\tau = 2 \pm 1$ ps)-->Bchl 875 ($\tau = 40 \pm 5$ ps)-->Bchl 896 ($\tau = 20 \pm 5$ ps)-->RC quenching ($\tau = 200 \pm 10$ ps), for Rb. sphaeroides. For R. rubrum the energy transfer kinetics is very similar to the last two steps of the sequence for Rb. sphaeroides (not shown). When the same experiments are repeated at room temperature (2) a very similar picture is obtained, with the only difference that each transfer step now contains information about the equilibration of excitation energy between the different pigment pools.

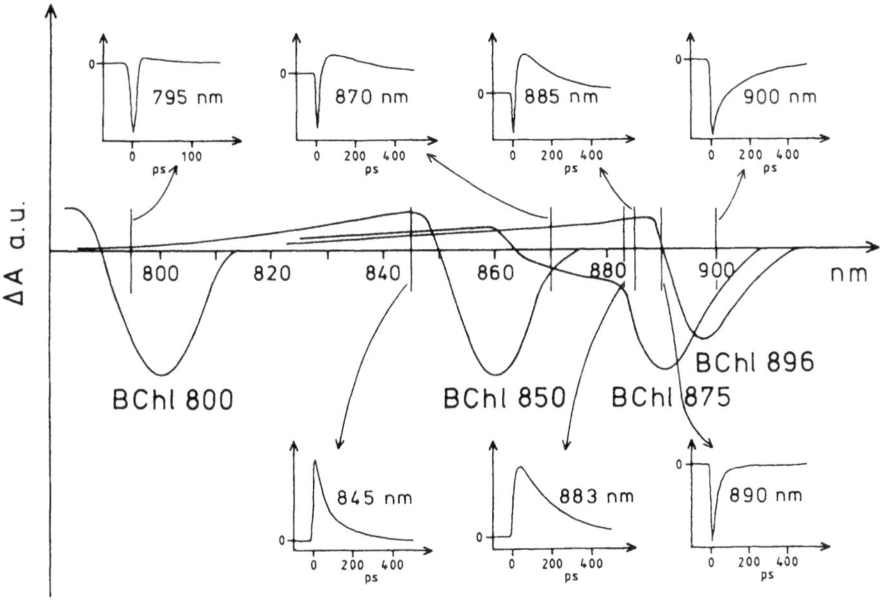

Experiments at room temperature with open reaction centers showed that the overall quenching time of the energy by open reaction centers is about 60 ps (2). This short time warrants a high efficiency of the energy transfer. Measurements of time resolved fluorescence under similar conditions have also been performed (3). In cases where results obtained with both techniques exist there is seen to be a quite good overall agreement.

Carotenoid excitation

The transfer of excitation energy from carotenoids to Bchl was studied in the detergent isolated pigment-protein complex

B800-820 of R. acidophila, by exciting the carotenoid at 540 nm and observing the decay of the induced bleaching as the energy is transferred to bacteriochlorophyll. From the observed decays and known rate of radiationless decay of carotenoids a time constant of about 3 ps was obtained for the energy transfer from carotenoids to Bchl (4).

Energy transfer in isolated pigment-protein complexes

Absorption recovery curves measured at 77 K are shown for the isolated pigment-protein complexes B800-850 and B875 of Rb. sphaeroides, in Fig. 2 and 3 respectively. For the first complex we can clearly resolve the initial very fast Bchl800-->Bchl850 transfer step (τ= 2±1 ps). From measurements of decay of induced anisotropy it was further concluded that fast Bchl800--Bchl800 transfer occurs on the same timescale as the Bchl800-->Bchl850 transfer (5,6). The decay of the Bchl850 excited state is seen to have a time constant of about 1 ns, which is typical for a terminal Bchl pigment of an isolated Bchl pigment-protein complex.

B 800 - 850 complex, 77 K

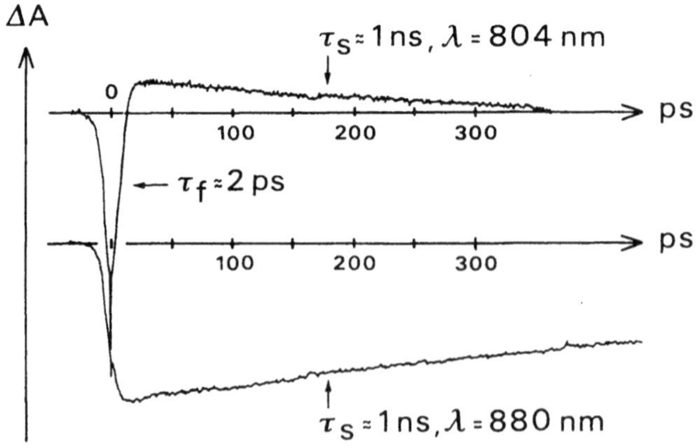

The decay obtained for the isolated B875 complex of Rb. sphaeroides is also biphasic; a short 15 ps time constant along with a much longer 650 ps component is observed in most of the absorption band. At the isosbestic point of B896 only the fast component is observed, and can thus be determined accurately. In the very red wing (λ >905 nm) of the absorption band, only the slow 650 ps component is observed. We assign these lifetimes to the B875-->B896 transfer (15 ps) and quenching of the Bchl896 excited state (650 ps), respectively. These results suggest that B896 is a minor redshifted antenna component which is strongly coupled to B875. Various possibilities for the origin of B896 have been discussed (1,7).

The energy transfer dynamics as it is observed in the different isolated complexes is generally seen to be in good agreement with the corresponding part of the overall process as it is observed in chromatophores. This suggest that the results obtained from studies of isolated complexes are very useful when analyzing the total dynamics of membrane preparations.

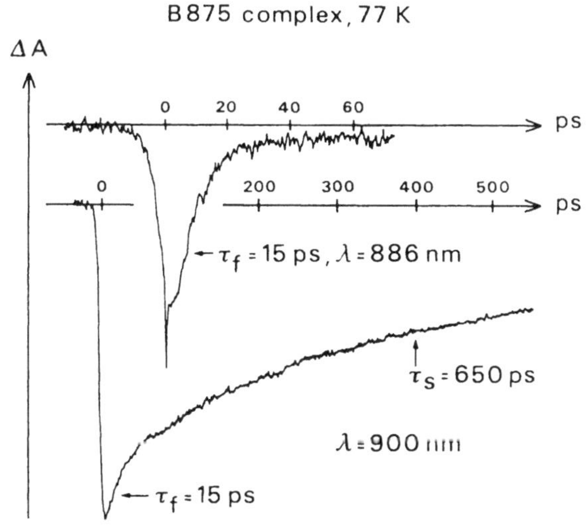

References

1. van Grondelle, R., H. Bergström, V. Sundström, T. Gillbro 1987. Biochim. Biophys. Acta. (in press)

2. Sundström, V., R. van Grondelle, H. Bergström, E. Åkesson, T. Gillbro. 1986. Biochim. Biophys. Acta 851, 431-446.

3. Borisov, A.Yu., A.M. Freiberg, V.I. Godik, K.K. Rebane, K.E. Timpmann. 1985. Biochim. Biophys. Acta 807, 221-229.

4. Gillbro, T., R. Cogdell, V. Sundström. To be published.

5. Bergström, H., V. Sundström, R. van Grondelle, T. Gillbro. 1986. Biochim. Biophys. Acta 852, 279-287.

6. Bergström, H., V. Sundström, R. van Grondelle, R. Cogdell, T. Gillbro. To be published.

7. Bergström, H., W. Westerhuis, V. Sundström, R. van Grondelle, R. Niederman, T. Gillbro. To be published.

EXCITATION ENERGY TRANSFER IN THE LIGHT-HARVESTING ANTENNA OF PHOTOSYNTHETIC PURPLE BACTERIA: THE ROLE OF THE LONG-WAVE-LENGTH ABSORBING PIGMENT B896

R. van Grondelle[a], H. Bergström[b], V. Sundström[b], R.J. van Dorssen[c], M. Vos[c] and C.N. Hunter[d]

a Department of Biophysics, Physics Laboratory of the Free University, Amsterdam, The Netherlands
b Department of Physical Chemistry, University of Umeå, Umeå, Sweden
c Department of Biophysics, State University, Leiden, The Netherlands
d Department of Pure and Applied Biology, Imperial College, London, United Kingdom

Introduction

Since the early work of Duysens [1] it is generally accepted that excitations, which are absorbed by the various light-harvesting pigments can be transferred to the reaction center (RC) for useful photosynthesis [1-6]. The efficiency of this process is often extremely high. In photosynthetic purple bacteria the pigment molecules involved are bacteriochlorophyll (Bchl) and carotenoid (Car). Usually different spectral forms of Bchl and Car occur, dependent upon the species, and these spectral forms are assumed to be associated to different pigment-protein complexes. For instance, in Rhodobacter (Rb.) sphaeroides, a central species in our work, the conventional absorbance spectrum of the wild-type species directly reveals the existence of three main Bchl a forms. The first two are associated to the B800-850 complex, characterized by absorption peaks at 800 nm ($Bchl_{800}$) and 850 nm ($Bchl_{850}$). The third form is associated to the B875 complex, characterized by an absorbance peak at 875 nm ($Bchl_{875}$).

The common feature of the pigment-binding proteins is that they

Photosynthetic Light-Harvesting Systems

are larger aggregates of a pair of polypeptides α and β. For all purple bacteria these α and β subunits show a rather homologous sequence, with a central stretch of about 20 aminoacids that is proposed to be transmembrane, α-helical and with the Bchl binding site at about 1 nm from the membrane surface [7-10]. It is believed that the minimum packing unit of these α and β subunits is a 'so called' αβ-trimer (i.e. 3α+3β) binding 6 $Bchl_{875}$ in the case of B875 and 6 $Bchl_{850}$ plus 3 $Bchl_{800}$ in the case of B800-850.
The structural organization of the bacterial light-harvesting system is as follows. RC's are thought to be surrounded by B875. The RC-B875 complex forms a network that spreads over at least three other RC's plus associated B875. In total such a 'core' contains four RC's and 100-150 $Bchl_{875}$'s. Those cores can be further connected either via direct contact between different cores or indirectly via B800-850 [11-14]. The B800-850 is most likely found at the periphary of the core [14,15], it occurs in units of 30-50 Bchl molecules that may further aggregate into larger units.
In a minimal $(\alpha\beta)_3$-unit energy transfer is extremely fast. Within a few picoseconds the excitation is localized on one of the pigments having the lowest excited state energy (i.e. $Bchl_{850}$ in B800-850 or $Bchl_{875}$ in B875). Energy transfer between different $(\alpha\beta)_3$-units is slower, 10-20ps, while overall energy transfer from B800-850 to B875 takes close to 40-50ps. Transfer from B875 to the RC and trapping takes 60-70ps at room temperature. At room temperature the equilibration of the excitation density among the different spectral forms dominates the kinetics. In contrast, at low temperatures ($\lesssim$ 77 K) energy transfer occurs only energetically down-hill, but the timeconstants for the forward reactions are consistent with those extracted at 293 K [16-21].
A few years ago we have proposed the existence of another spectral component at the far red side of the absorbance spectrum of B875. In low temperature fluorescence measurements this component could be distinguished and was called B896,

according to its proposed 4 K absorbance maximum. It was thought that about 10-20% of the long-wavelength Bchl a contributed to B896. [22]

In this paper we shall review the recent history of B896, discuss some possible structures and propose future plans for the study of its properties and function.

Results and Discussion

Low temperature polarized absorption and fluorescence

In fact the first discovery of a spectral inhomogeneity in B875 was published in 1981 by Bolt et al [23], who described a strong increase of the measured fluorescence polarization upon scanning the excitation wavelength over the B875 absorption band. At that time the "spectral inhomogeneity" was thought to be due to exciton effects. Later we repeated these experiments at room and low temperature with membranes from a variety of purple bacteria and isolated B875 complexes [23,24]. At 4 K in all investigated preparations the fluorescence polarization increased from a low value, typically of the order of 0.05, upon excitation below 880 nm to a value close to 0.4 upon excitation above 900 nm. An example of such an experiment recently done with a mutant of Rb. sphaeroides containing only B875 and RC's is shown in fig. 1. We have concluded from these experiments that at the red side of the B875 absorption band a component absorbs with $\lambda_{max} \simeq 896$ nm (at 4 K), which upon direct excitation gives rise to a strongly polarized emission. In the case of direct excitation of $Bchl_{875}$ the excited state becomes rapidly depolarized, mainly due to energy transfer among identical $Bchl_{875}$ molecules within the $(\alpha\beta)_3$-unit but in addition caused by transfer between different $(\alpha\beta)_3$-units and transfer to B896. In contrast, the B896 emission is highly polarized because B896 represents either several parallel interacting Bchl a monomers within a pigment-protein complex or isolated red-shifted Bchl a monomers.

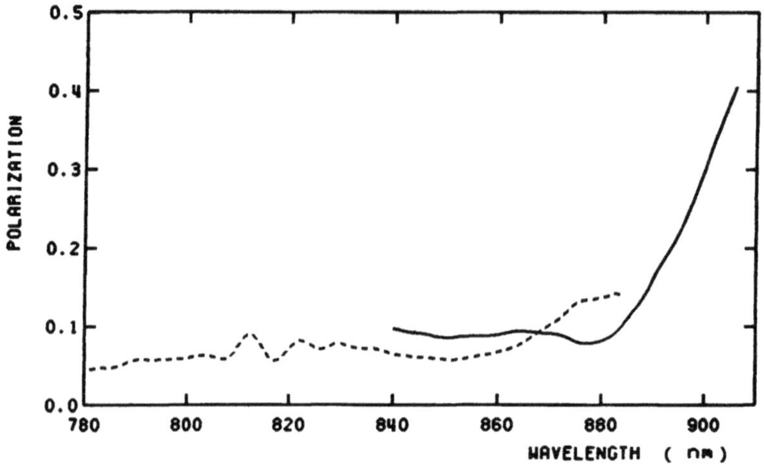

Fig. 1 Fluorescence polarization spectra of M21 (——), a mutant of Rb. sphaeroides containing B875 plus RC's and NF57 (----), a mutant of Rb. sphaeroides containing only B800-850. The fluorescence was detected at 940 nm for (M21) and at 920 for NF57. The experiment was performed at 4 K. Note that the increase in fluorescence polarization occurs in M21 and in NF57 and in both cases is due to a minor spectral form absorbing in the red wing of the main absorption band (taken from ref. 24).

If these experiments are performed at room temperature the increase in fluorescence polarization upon excitation in the red wing of B875 is undetectable in membranes of purple bacteria [22]. This is most likely due to back transfer from B896 to B875 followed by transfer to and emission from another B896. In isolated B875 complexes this depends strongly on the aggregation state of the complex.

Recently, we found additional spectroscopic evidence for the existence of B896 [24]. Linear dichroism experiments showed a well-ordered component at the red side of the B875 absorption band (see fig. 2).

The dominant orientation of all the Bchl a Q_y transition moments is parallel to the membrane plane. However, as may be concluded from fig. 2, B896 is the component with the highest

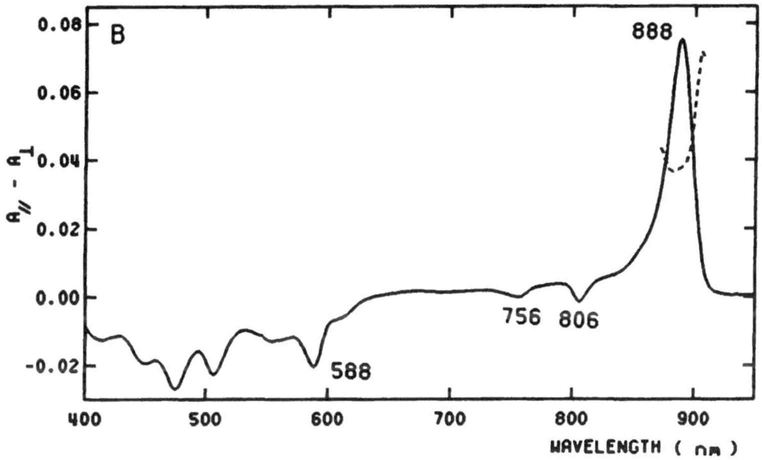

Fig. 2 Linear dichroism ($A_{||}-A_{\perp}$) spectra of M21, measured at 4 K. The broken line shows the reduced dichroism, $(A_{||}-A_{\perp})/A$, on an arbitrary scale. For the preparation used the absorbance at 875 nm was 0.4 at room temperature (taken from ref. 24).

dichroic ratio in the 800-900 nm region of the spectrum. Finally, an analysis of the absorptionspectrum and its second and fourth derivative also indicated the presence of a long-wavelength component. In fact, Van Dorssen et al [24] could present a consistent fit of all this data using the components obtained via Gaussian deconvolutian of the absorbance spectrum.

Picosecond absorption and absorption anisotropy measurements

The illumination of intact photosynthetic systems with weak picosecond infrared excitation and probing flashes allows the study of the excited state dynamics due to energy transfer and trapping in the light-harvesting antenna [18,19,21]. We have performed such experiments on membranes and isolated light-harvesting complexes of photosynthetic purple bacteria. At room temperature the excited state decay is dominated by the equilibration of the excitation density among the various Bchl species, before any appreciable trapping or loss takes place. These experiments have shown that the excitation density equilibrates between B875 and B896 with a time constant of about 50 ps [18].

At low temperatures (77 K) excitation energy transfer is unidirectional from the pigments absorbing at higher energy to those absorbing at lower energy. This is shown is fig. 3 where

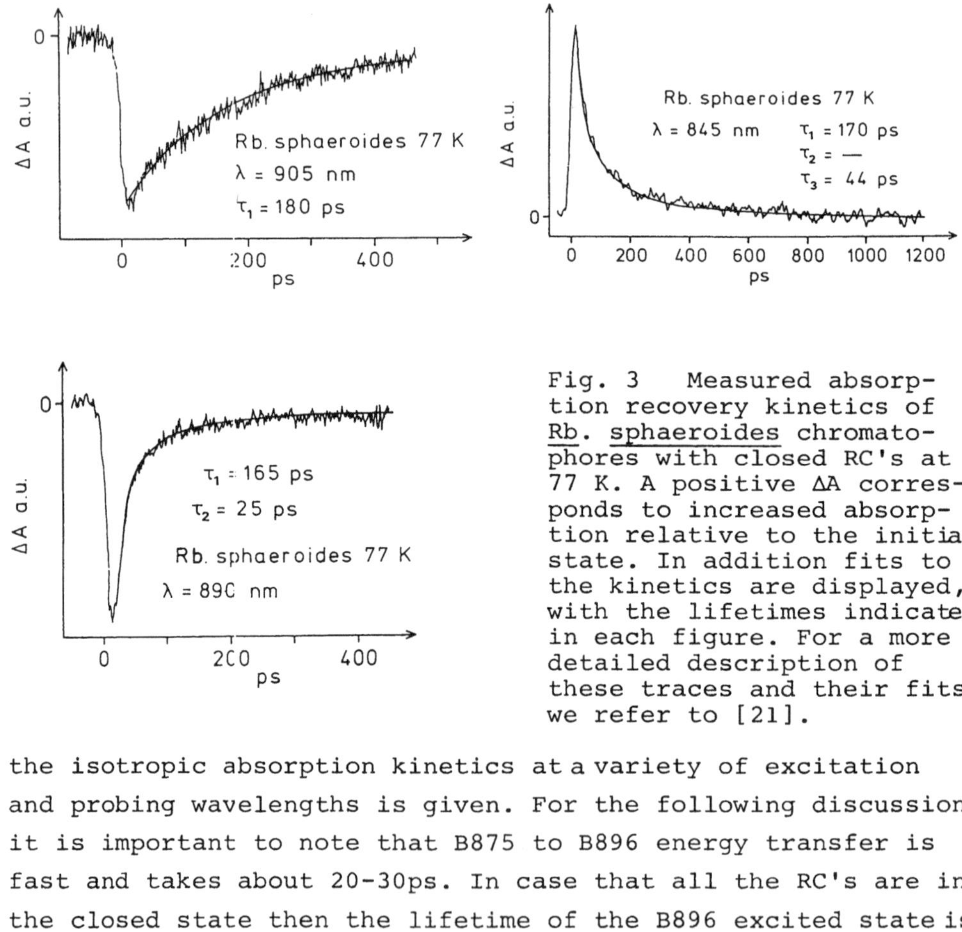

Fig. 3 Measured absorption recovery kinetics of Rb. sphaeroides chromatophores with closed RC's at 77 K. A positive ΔA corresponds to increased absorption relative to the initial state. In addition fits to the kinetics are displayed, with the lifetimes indicated in each figure. For a more detailed description of these traces and their fits we refer to [21].

the isotropic absorption kinetics at a variety of excitation and probing wavelengths is given. For the following discussion it is important to note that B875 to B896 energy transfer is fast and takes about 20-30ps. In case that all the RC's are in the closed state then the lifetime of the B896 excited state is about 150-200ps [18,20,21]. Sofar, all attempts to perform similar low temperature kinetic experiments with all the RC's active have failed, mainly for technical reasons. However, it is well known that at these temperatures the fluorescence yield with active RC's is about four fold lower as compared to the situation with inactive (closed) RC's. The simplest explanation of all these experimental results is that: (1), B896 is an intermediate between the main B875 antenna and the RC and (2), the excitation transfer from B896 to the RC is surprisingly slow and takes of the order of 40-50ps. This finding is unexpected and must be related to the structural organization of B896 in the neighbourhood of the RC.

The absorption anisotropy upon excitation of B896 is high in contrast to what is observed upon excitation and probing in the main B875 band.

Excitation annihilation

The study of the annihilation of excitations generated by an intense picosecond flash in a photosynthetic antenna system can give valuable information about the organization of the photosynthetic antenna and the dynamics of energy transfer and trapping [5]. We have performed such experiments in <u>Rhodospirillum</u> (<u>R</u>.) <u>rubrum</u> and <u>Rb</u>. <u>sphaeroides</u> and in a variety of mutants of <u>Rb</u>. <u>sphaeroides</u> at room temperature and at 4 K [12-14,25]. The results indicate that the large photosynthetic domains that are observed at room temperature are separated into much smaller 'cores' at low temperature. In addition it is clear that at 4 K excitation annihilation occurs predominantly in B896 and not in B875. In figure 4 it is shown that at high laser flash intensities the emission spectrum is markedly blue-shifted as compared to the spectrum at low intensity illumination.

An attempt has been made to simulate the excitation transfer and annihilation processes on a computer using a random walk model [26]. Sofar, the only arrangement of $Bchl_{875}$ molecules,

$Bchl_{896}$ molecules and RC's that yields a reasonable fit to the fluorescence quenching curve, the observed fluorescence lifetimes and the intensity dependence of the emission spectra is a structure in which each RC is surrounded by about four $Bchl_{896}$ molecules. These groups of $Bchl_{896}$ molecules are only connected with each other via the 'lake' of B875.

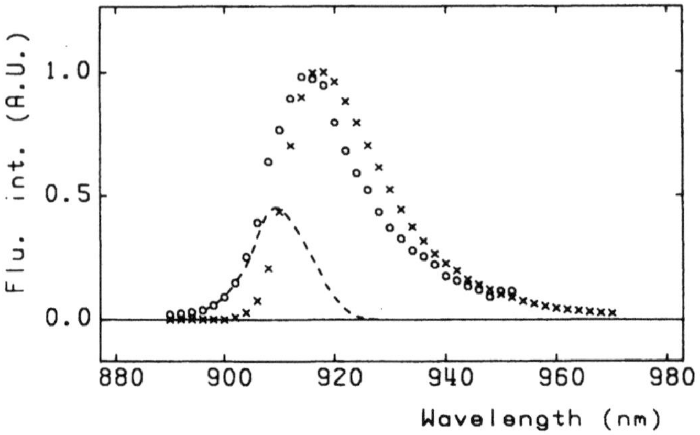

Fig. 4 Emission spectra of R. rubrum at 4 K. The RC's were closed by continuous background illumination. High intensity emission spectrum (o-o); excitation at 532 nm by a picosecond laser flash of 4-5 mJ cm^{-2}; low intensity emission spectrum (x-x), excitation by a xenon flash at 519 nm with an energy of 0.1 mJ cm^{-2}. The spectra were normalized at the emission peak. The dashed spectrum (---) was obtained by normalizing the low-intensity and high-intensity emission spectra at 930 nm and subtracting (taken from ref. 14).

Conclusions and future prospects

The evidence in favour of the existence of a special Bchl species, absorbing at the long-wave side of the B875 absorption band is becoming very strong. Various techniques have all led to the same conclusion: B896 exists. It is important to remark that this view is strongly supported by fluorescence

lifetime measurements in purple photosynthetic bacteria performed by Freiberg et al (Freiberg, A.M., personal communication). In addition we conclude that if B896 exists, it must be positioned between the RC and the main B875 antenna.
What may be the function of B896? We propose here that B896 exists because a 'focus' is needed in the antenna for efficient transfer to the RC. In all calculations, either by ourselves or by others, it was generally assumed that the transfer time to and from the RC was of the same order of magnitude as the effective transfer time among different antenna molecules. For this latter process the experimental results indicated that it took on the average about one picosecond to transfer an excitation between two neighbouring $Bchl_{875}$ molecules [12,14]. However, if in such a system transfer to the RC is about 50ps, trapping by the RC would take 0.5 ns, assuming 40 $Bchl_{875}$ per RC and four $Bchl_{875}$ molecules as possible excitation donors. If the RC is accessible from less than four sides this would reduce the trapping time even further. The presence of B896 'concentrates' the excitation density on pigment molecules relatively close to the RC. With B896 in the proposed stoichiometry (10-20%), at room temperature about half the excitation density would reside on B896 and the other half on B875 (i.e. in the equilibrium!). With a true B896 to RC transfer time of 50ps, the effective overall transfertime would be about 100ps, which is not unreasonable. Thus, the fact that transfer to the RC is so slow absolutely requires a 'focus' in the antenna to achieve a reasonable efficiency.
To explain the high polarization of the B896 excited state and its red shifted absorption, we propose that B896 represents (at least) a dimer of strongly interacting Bchl a molecules with parallel (head to tail) transition dipoles. The distance between B896 and the P870 dimer must be of the order of 3 nm to account for the 50ps transfertime. Although the Q_y transitions that contribute to B896 and those that make up P870 are parallel to the membrane, they are not necessarily mutually parallel.

Our proposed function for B896 neatly explains an old experiment. It was always a puzzle why the direct excitation of the RC-pigments did not lead to fluorescence from the antenna [27]. In contrast , fluorescence yield and recombination luminescence measurements all suggested that back transfer to the antenna did occur. However, with a 50ps transfer time from B896 to the RC, the back transfer time will be about 20ps at 293 K. Therefore, if a RC becomes excited charge separation will be the dominant process and consequently fluorescence due to direct RC illumination will be weak. Of course the emission due to recombination luminescence will be present but weak.

All arguments raised sofar about RC's from which escape could occur, were in fact referring to B896 and not to the RC. B896 localizes the excitations to some extent, and escape from B896 can easily occur in view of the low transfer rate to the RC. In contrast, the probability that an excited RC will back transfer to B896 is low.

The following approaches shall be used in the near future to further elucidate the role of B896 and the organization of the light-harvesting antenna in the neighbourhood of the RC.

(1) Biochemical isolation of B896-RC complexes. It is of crucial importance to establish whether B896 exists as a separate pigment-protein complex or if it is an integral part of the B875 antenna. Experiments with RC-less mutants of Rb. sphaeroides suggest that B896 exists even in these mutants, but with slightly different spectroscopic properties. Apparently, the RC has a modulating influence on the detailed properties of the surrounding antenna. An alternative possibility, namely that B896 arises due to aggregation of B875 complexes, also needs careful investigation.

(2) Spectroscopic measurements can be used to obtain information about the association state of the B896-RC complex. Excitation annihilation and excited state lifetime experiments upon selective excitation in the B896 absorption band, especially under conditions that energy back transfer is excluded

can be used to probe the association state of the B896-RC complex. In addition photoselection experiments can provide the orientation of B896 with respect to the P_{870} dimer in the RC. Hopefully these and other experiments will lead to a clear picture of the structural association of the antenna pigments with the RC into a functional unit.

References

1. Duysens, L.N.M.. 1952. Doctoral Thesis, State University of Utrecht. The Netherlands.

2. Duysens, L.N.M.. 1964. Progr. Biophys. 14, 1-104.

3. Knox, R.S.. 1975. In: Bioenergetics of Photosynthesis (Govindjee, ed.) pp. 183-221. Academic Press, New York.

4. Pearlstein, R.M.. 1982. In: Photosynthesis, Energy Conversion by Plants and Bacteria (Govindjee, ed.), Vol. 1. Academic Press, New York. pp. 293-329.

5. Grondelle, R. van. 1985. Biochim. Biophys. Acta 811, 147-195

6. Grondelle, R. van and J. Amesz. (1986). In: Light Emission by Plants and Bacteria (Govindjee, J. Amesz and D.C. Fork, eds.), chapter 8. Academic Press, Inc., Orlando, Florida 32887. pp. 191-223.

7. Zuber, H. 1985. Photochem. Photobiol. 42, 821-844.

8. Theiler, R., F. Suter, H. Zuber and R.J. Cogdell. 1984. FEBS. Lett. 175, 231-237.

9. Theiler, R., F. Suter, J.D. Pennoyer, H. Zuber and R.A. Niederman. 1985. FEBS. Lett. 184, 231-236.

10. Theiler, R. and H. Zuber. 1984. Hoppe-Seyler's Z. Physiol. Chem. 365, 721-729.

11. Vredenberg, W.J. and L.N.M. Duysens. 1963. Nature 197, 355-357.

12. Bakker, J.G.C., R. van Grondelle and W.Th.F. den Hollander. 1983. Biochim. Biophys. Acta 725, 508-518.

13. Hunter, C.N., R. van Grondelle and H.J.H. Kramer. 1985. Biochim. Biophys. Acta 807, 44-51.

14. Vos, M., R. van Grondelle, F.W. van der Kooy, P. van der Poll, J. Amesz and L.N.M. Duysens. 1986. Biochim. Biophys. Acta 850, 501-512.

15. Monger, T.G. and W.W. Parson. 1977. Biochim. Biophys. Acta 460, 393-407.

16. Grondelle, R. van, H.J.M. Kramer and C.P. Rijgersberg. 1982. Biochim. Biophys. Acta 682, 208-215.

17. Kramer, H.J.M., R. van Grondelle, C.N. Hunter, W.H.J. Westerhuis and J. Amesz. 1984. Biochim. Biophys. Acta 765, 156-165.

18. Sundström, V., R. van Grondelle, H. Bergström, E. Åkesson and T. Gillbro. 1986. Biochim. Biophys. Acta 851, 431-446.

19. Bergström, H., V. Sundström, R. van Grondelle, E. Åkesson and T. Gillbro. 1987. Biochim. Biophys. Acta 852, 279-287.

20. Borisov, A.Yu., A.M. Freiberg, V.I. Goaik, K.K. Robane and K.E. Timpman. 1985. Biochim. Biophys. Acta 807, 221-229.

21. Grondelle, R. van, H. Bergström, V. Sundström and T. Gillbro. Biochim. Biophys. Acta, in press.

22. Kramer, H.J.M., J.D. Pennoyer, R. van Grondelle, W.M.J. Westerhuis, R.A. Niederman and J. Amesz. 1984. Biochim. Biophys. Acta 767, 335-344.

23. Bolt, J.D., C.N. Hunter, R.A. Niederman and K. Sauer. 1981. Photochem. Photobiol. 34, 653-656.

24. Dorssen, R.J. van, C.N. Hunter, R. van Grondelle, A.H. Korenhof and J. Amesz. Biochim. Biophys. Acta, submitted for publication.

25. Vos, M., R.J. van Dorssen, J. Amesz, R. van Grondelle and C.N. Hunter. Biochim. Biophys. Acta, submitted for publication.

26. Vos. M.. 1987. Doctoral Thesis, State University Leiden. The Netherlands.

27. Wang, R.T. and R.K. Clayton. 1971. Photochem. Photobiol. 13, 215-224.

Acknowledgement

This work was financially supported by the Dutch Foundation of Biophysics, the Swedish Natural Science Research Council and the European Molecular Biology Organization.

THE FUNCTION OF CHLOROSOMES IN ENERGY TRANSFER IN GREEN PHOTOSYNTHETIC BACTERIA

R.J. van Dorssen, M. Vos and J. Amesz

Department of Biophysics, Huygens Laboratory of the State University
P.O. Box 9504, 2300 RA Leiden, The Netherlands

Introduction

The antenna system of green photosynthetic bacteria differs in important respects from that of purple bacteria. Green bacteria lack the invaginations of the cytoplasmic membrane observed in purple bacteria. Instead they contain large numbers of chlorosomes, which constitute the main antenna of these organisms (1). These chlorosomes are oblong bodies, about 100 nm long and several tens of nm wide, which are attached to the cytoplasmic membrane. Chlorosomes of the green sulfur bacterium *Chlorobium limicola* and of the green non-sulfur (green filamentous) bacterium *Chloroflexus aurantiacus* contain approximately 10,000 bacteriochlorophyll (BChl) *c* molecules and serve as a common antenna for several reaction centers (1). Although the organization of the antenna pigments in the membrane and the properties of the reaction center of the green sulfur and non-sulfur bacteria are clearly different, the chlorosomes appear to be quite similar. Electron microscopic studies (2,3) have shown that in both groups of organisms the interior of the chlorosome consists of rod-shaped structures, which are thought to be composed of aggregates of the BChl *c*-binding proteins (4,5). Comparison of the spectroscopic properties of the chlorosomes with those of aggregates of BChl *c* indicates that the pigment is present in oligomeric form (6,7).

In addition to BChl *c*, chlorosomes of *Cfl. aurantiacus* and of *C. limicola* contain small amounts of BChl *a* (8,9). Since the fluorescence emission band of this pigment shows a more favorable overlap with the absorption spectrum of the BChl *a* associated with the membrane, it has been postu-

Photosynthetic Light-Harvesting Systems

lated that the chlorosomal BChl *a* is an intermediate for energy transfer from the chlorosome to the membrane (10-12).

In this communication we shall present some data concerning energy transfer within the chlorosomes and between the chlorosomes and the cytoplasmic membrane in green sulfur and non-sulfur bacteria.

Material and Methods

Prosthecochloris aestuarii, strain 2K, was grown anaerobically in a mixed culture known as 'Chloropseudomonas ethylica' (13) as described by Holt et al. (14). After harvesting by centrifugation the cells were resuspended in 10 mM Tris buffer, pH = 8.0. Cells were disrupted by sonication and large debris and unbroken cells were removed by centrifugation at 30,000 x g for 10 min. The resulting supernatant was layered on a 35 - 40 % discontinuous sucrose gradient. After centrifugation for 16 h at 144,000 x g three diffuse bands were observed. The middle band was enriched in chlorosomes. This fraction was incubated for 30 min with Miranol (0.2 % v/v) and layered on a 15 - 45 % continuous sucrose gradient. After centrifugation for 16 h at 144,000 x g chlorosomes were collected from a dark green band in the gradient. Chlorosomes ('medium density') of *Chlorobium limicola* and of *Chloroflexus aurantiacus* were prepared as described in refs. 9 and 11, respectively. Low temperature absorption and fluorescence spectroscopy was performed using an apparatus described by Kramer and Amesz (15). To reduce scattering and to obtain clear samples at low temperature 66 % glycerol (v/v) was added to the samples.

Results

Energy transfer between BChl *c* molecules

Information on the rate of energy transfer between BChl *c* molecules was obtained by measurements of excitation annihilation. Fig. 1 shows the

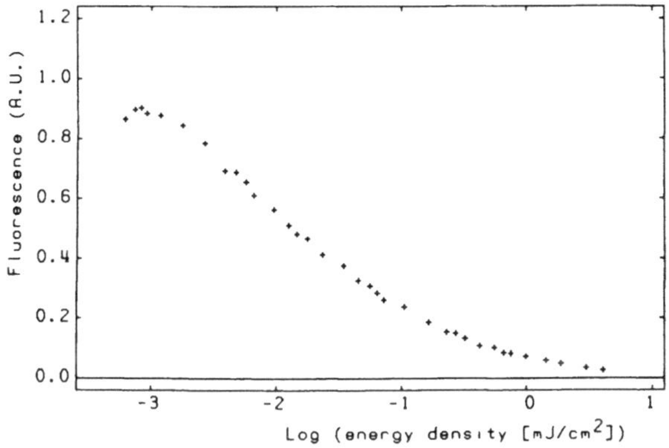

Fig. 1. Time-integratred fluorescence yield of BChl c in chlorosomes of C. limicola induced by 35 ps, 532 nm laser flashes, plotted as a function of the energy density of the flash, and measured in the presence of 10 mM dithionite. The fluorescence was measured at 780 nm. A.U., arbitrary units.

time-integrated fluorescence yield of BChl c in chlorosomes of C. limicola, excited by 35 ps flashes at 532 nm, plotted as a function of the energy density of the flash. It can be seen that the fluorescence yield decreased strongly with increasing flash intensity, due to singlet-singlet annihilation. Such annihilation can take place when more than one excitation is present in the same domain, i.e. in the ensemble of pigment molecules among which energy transfer can occur. Analysis of the data of Fig. 1 by means of the theory described by Paillotin et al. (16) showed that in C. limicola chlorosomes such a domain contains at least 7500 BChl c molecules (17). Since the chlorosome contains about 10,000 BChl c molecules, this indicates that energy transfer can take place throughout the whole chlorosome, and that no internal barriers for energy transfer exist within this organelle.

The domain size and the shape of the annihilation curve, together with the lifetime for mono-excitation decay allow a calculation of the rate constant for energy transfer between BChl c molecules. Lifetimes of 21 ps (18) and of 85 ps (19) have been reported as main components in the decay of excited BChl c in chlorosomes of C. limicola. An average lifetime of 40 ps was estimated from the fluorescence yield (17). If we take the last value, then, by use of the random walk model of den Hollander et al. (20), a rate constant $k_h \approx 2 \cdot 10^{13}\ s^{-1}$ is obtained. A 50 times smaller rate constant was obtained at 4 K.

Our data thus indicate that the average rate of energy transfer between BChl c molecules is very high, one to two orders of magnitude higher than between BChl a molecules in the antenna of purple bacteria (21), as would be expected from the high BChl c pigment density (8). However, as mentioned in the introduction, there is evidence that BChl c occurs in clusters among which strong interaction and consequently very rapid energy transfer does occur. This consideration does not affect the calculated domain size, but it does affect the calculated transfer rate. If it is assumed that the BChl c is present in clusters of, e.g., 10 molecules, with very rapid energy transfer within these clusters, then the average rate of energy transfer between these clusters is $k_h^c \approx 2 \cdot 10^{12}\ s^{-1}$ at room temperature.

Energy transfer between BChl c and BChl a within the chlorosomes

The low temperature absorption spectrum of chlorosomes of the green sulfur bacterium *P. aestuarii* is shown in Fig. 2. Bands of BChl c are observed at

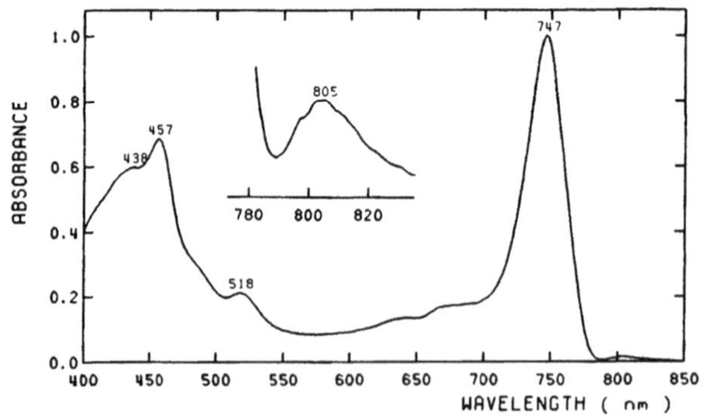

Fig. 2. Absorption spectrum of isolated chlorosomes of *P. aestuarii* at 4 K. The inset shows the region around 800 nm on a 20-fold expanded scale.

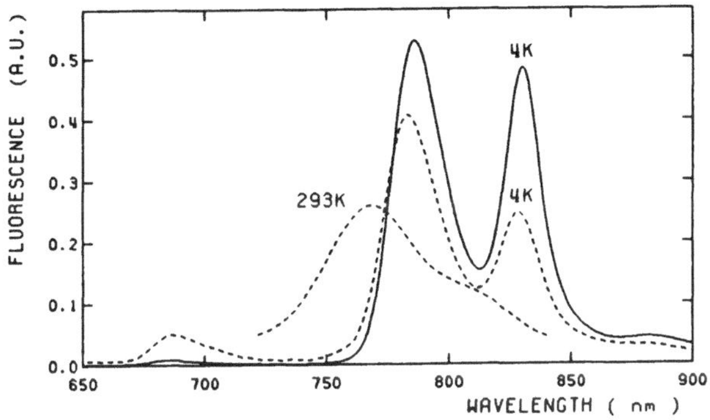

Fig. 3. Emission spectra at two temperatures of chlorosomes of P. aestuarii in the presence (———) and the absence (-----) of 10 mM dithionite and excited at 460 nm. The spectra measured in the absence of dithionite are shown on a 10-fold expanded scale.

747 (Q_y transition), 457 and 438 nm (Soret transitions). The presence of carotenoid is reflected by the band at 518 nm and a shoulder at 489 nm. A weak band is observed at 805 nm. The inset of Fig. 2 shows the minor component at 805 nm on a 20-fold expanded scale. In analogy to results obtained with C. limicola (12), this component is attributed to BChl a. Since the BChl c and BChl a bands are better resolved than in C. limicola (12) the BChl c : BChl a ratio can be determined with a reasonable accuracy from the absorption spectrum. Using molar in vivo extinction coefficients of 102 for BChl c (11) and 137 for BChl a (22) a ratio of 85 is obtained. This value is very similar to that obtained for C. limicola (9).

Emission spectra of the isolated chlorosomes are shown in Fig. 3. Upon excitation in the Soret band of BChl c (460 nm) the room temperature spectrum showed a main band at 769 nm (BChl c) and a shoulder at 816 nm (BChl a). Upon cooling to 4 K these bands shifted to 783 and 828 nm, respectively. The solid line shows the emission after incubation with 10 mM dithionite. An increase by a factor of about 14 was observed for the total integrated fluorescence. A similar effect has been observed for membrane preparations of P. aestuarii (23) and chlorosomes of C. limicola (12) and has been attributed to the removal of non-fluorescing quenchers by the dithionite treatment. The shift of the BChl a emission band upon cooling occurred gradually with decreasing temperature in a fashion very similar to that observed for C. limicola chlorosomes (12). This behavior clearly differs from that of the fluorescence of the 'soluble' BChl a protein (24). We therefore conclude that, as in chlorosomes of C. limicola (12) and Cfl. aurantiacus (10,11), BChl a is an integral component of the chlorosome of P. aestuarii.

The emission spectra demonstrate that transfer of excitation energy occurs from BChl c to BChl a 805. This was confirmed by the excitation spectra of the BChl a fluorescence given in Fig. 4. In the absence of dithionite a very low contribution of BChl c relative to that of BChl a 805 was observed, corresponding to a transfer efficiency to BChl a 805 of only 17 %. Upon addition of dithionite the efficiency of transfer from BChl c to BChl a 805 increased by a factor of 2.4 to 40 %. It may be noted here that the efficiency obtained under the same conditions with chlorosomes of C. limicola was higher (12), but for these chlorosomes the emissions from BChl c and BChl a are less well separated, and as noted earlier (12), this may have resulted in a too high apparent efficiency. The efficiency of energy transfer from carotenoid to BChl a 805 was 27 %. From the excitation spectrum of the BChl c emission (not shown) the transfer efficiency from carotenoid to BChl c was calculated to be 75 %, both in the presence and absence of dithionite. This number, together with the values calculated for

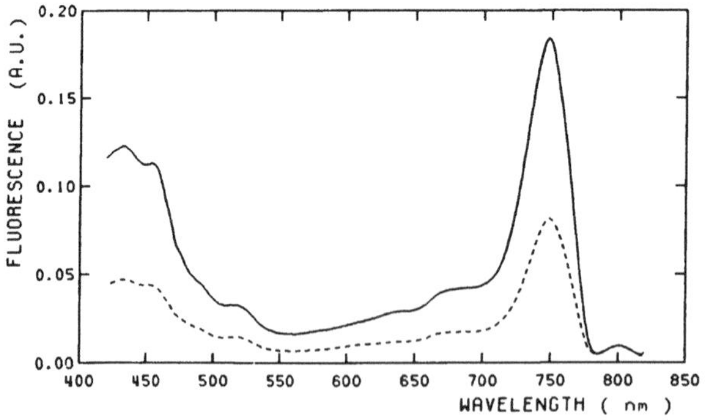

Fig. 4. Fluorescence excitation spectra, measured at 4 K, of chlorosomes of P. aestuarii in the presence (———) and in the absence (-----) of dithionite, detected at 830 nm and normalized at 803 nm.

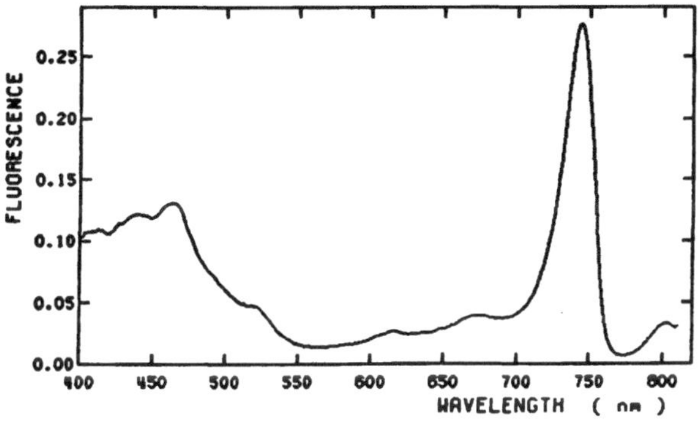

Fig. 5. Fluorescence excitation spectrum (arbitrary units) of chlorosomes of Cfl. aurantiacus at 4 K upon detection at 835 nm. The absorbance at 742 nm was 0.3.

transfer from BChl c to BChl a 805 agrees well with the observed overall carotenoid to BChl a 805 transfer efficiency and strongly suggests that excitation energy is transferred from carotenoid to BChl a 805 via BChl c.

Chlorosomes of Cfl. aurantiacus contain about 4 % BChl a (25). This pigment shows an absorption band at approximately 795 nm (798 nm at 4 K) and a fluorescence band at 805 nm, which intensifies and shifts to 819 nm at 4 K (11). Comparison of the excitation spectra for BChl a (Fig. 5) and BChl c fluorescence with the absorption spectrum at 4 K indicated an efficiency of approximately 65 % for energy transfer from carotenoid to BChl c and of 55 % from BChl c to BChl a at 4 K in isolated chlorosomes. Approximately the same efficiencies were observed at room temperature (11).

Energy transfer in whole cells

The above experiments, together with those earlier reported (11,12), indicate that in chlorosomes of both green sulfur and non-sulfur bacteria, BChl a, albeit present in small amounts, is an integral component of the chlorosomes. In order to obtain information on its possible role in energy transfer from the chlorosome to the membrane, fluorescence excitation

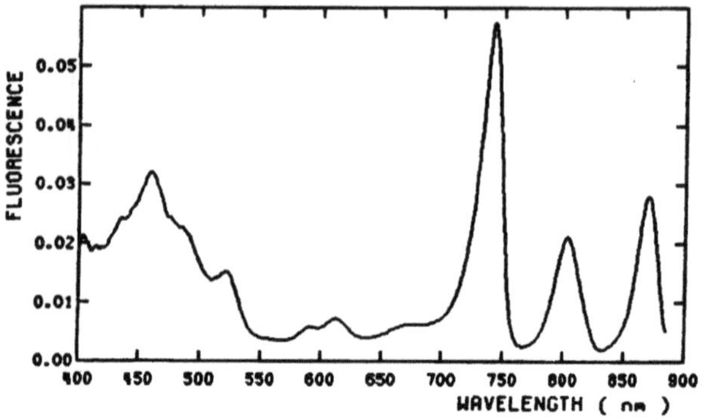

Fig. 6. Excitation spectrum of the fluorescence of BChl a 866, measured in whole cells of Cfl. aurantiacus at 4 K. Detection wavelength 900 nm.

spectra were measured of whole cells of Cfl. aurantiacus. At 4 K three emission bands were observed, with maxima at 755, 817 and 891 nm, which can be ascribed to BChl c, BChl a 798 (the chlorosomal BChl a) and BChl a 866, respectively. The latter pigment is a component of the B808-866 light-harvesting complex of the membrane (22). The excitation spectrum of the fluorescence from BChl a 866 is shown in Fig. 6. Comparison with the absorption spectrum gave an overall efficiency of energy transfer upon excitation at 750 nm of 15 %. An average efficiency of 57 % was obtained at 798 nm. Comparison of the absorption spectrum with that of isolated membranes (22) showed that 39 % of the absorption at 798 nm is due to BChl a 808. Since it is known that the latter pigment, which is part of the B808-866 complex, transfers its energy with an efficiency of close to 100 % to BChl a 866 (21), it can be calculated that the efficiency of energy transfer from BChl a 798 to BChl a 866 is 30 % at 4 K. The excitation spectrum for BChl a 798 fluorescence gave an efficiency of transfer from BChl c to BChl a 798 of 50 %, in good agreement with the results obtained with isolated chlorosomes (see Fig. 5). These results strongly indicate that energy transfer from BChl c to the membrane indeed occurs via BChl a 798: The overall efficiency of 15 % for transfer to the membrane equals the product of the yields of the individual steps (BChl c to BChl a 798, 50 %; BChl a 798 to BChl a 866, 30 %).

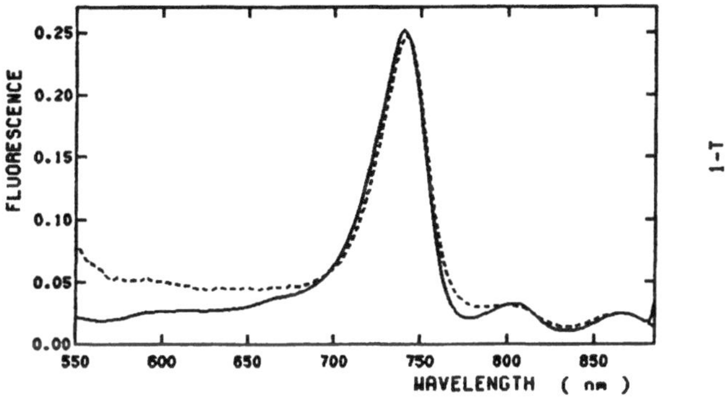

Fig. 7. Absorption (-----) and excitation (———) spectra of BChl a 866 fluorescence of whole cells of Cfl. aurantiacus at 50 °C.

In view of the low efficiencies for energy transfer observed at low temperature it was of interest to extend the experiments to more physiological conditions. Fig. 7 shows the absorption and excitation spectra of Cfl. aurantiacus at 50 °C, close to its growth temperature. At this temperature an overall efficiency of energy transfer from BChl c to the membrane of close to 100 % was observed. However, below 40 °C the efficiency started to drop already, and at 10 °C it reached a value of 60 % (Fig. 8). This decrease appeared to be due to a lowering mainly of the efficiency of energy transfer from BChl c to BChl a 798, since the efficiency upon excita-

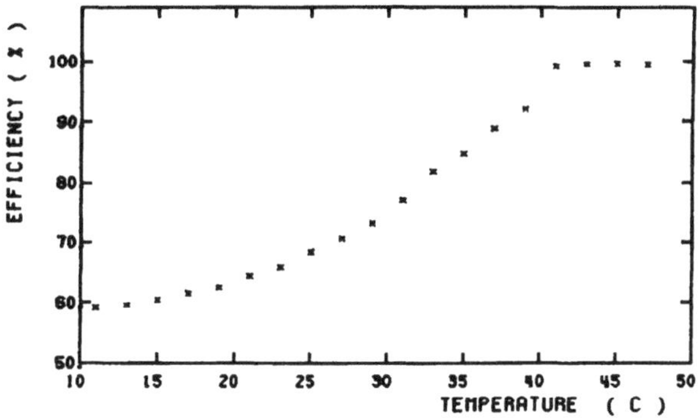

Fig. 8. Efficiency of energy transfer from BChl c to BChl a 866 in whole cells of Cfl. aurantiacus as a function of temperature.

tion at 800 nm did not change. The drop may be related to a phase transition in the lipid monolayer which surrounds the chlorosome (26). The efficiency for transfer betweeen BChl a 798 and the membrane appears to be affected only upon further cooling.

Acknowledgement

This investigation was supported by the Netherlands Foundation for the Advancement of Pure Research (ZWO) via the Netherlands Foundations for Chemical Research (SON) and for Biophysics.

References

1. Olson, J.M. 1980. Biochim. Biophys. Acta 594, 33-51.
2. Staehelin, L.A., J.R. Golecki, R.C. Fuller and G. Drews. 1978. Arch. Microbiol. 119, 269-277.
3. Staehelin, L.A., J.R. Golecki and G. Drews. 1980. Biochim. Biophys. Acta 589, 30-45.
4. Zuber, H., R. Brunisholz and W. Sidler. 1987. In: Photosynthesis (J. Amesz, ed.). Elsevier, Amsterdam. pp. 233-271.
5. Gerola, P., J. Højrup, J. Knudsen and J.M. Olson. 1988. In: Proc. EMBO Workshop on Green Photosynthetic Bacteria (J.M. Olson, ed.). Plenum Press, New York (in press).
6. Blankenship, R.E., D.C. Brune, J.M. Freeman, G.H. King, J.H. McManus, N. Tsusenoiri, J.T. Trost and B.P. Wittmershaus. 1988. In: Proc. EMBO Workshop on Green Photosynthetic Bacteria (J.M. Olson, ed.). Plenum Press, New York (in press).
7. Lutz, M. and G. van Brakel. 1988. In: Proc. EMBO Workshop on Green Photosynthetic Bacteria (J.M. Olson, ed.). Plenum Press, New York (in press).
8. Schmidt, K. 1980. Arch. Microbiol. 124, 21-31.
9. Gerola, P.D. and J.M. Olson. 1986. Biochim. Biophys. Acta 848, 69-76.
10. Betti, J.A., R.E. Blankenship, L.V. Natarajan, L.C. Dickinson and R.C. Fuller. 1982. Biochim. Biophys. Acta 680, 194-201.

11. Van Dorssen, R.J., H. Vasmel and J. Amesz. 1986. Photosynth. Res. 9, 33-45.

12. Van Dorssen, R.J., P.D. Gerola, J.M. Olson and J. Amesz. 1986. Biochim. Biophys. Acta 848, 77-82.

13. Olson, J.M. 1978. Int. J. Syst. Bacteriol. 28, 128-129.

14. Holt, S.C., S.F. Conti and R.C. Fuller. 1966. J. Bacteriol. 91, 311-323.

15. Kramer, H.J.M. and J. Amesz. 1982. Biochim. Biophys. Acta 682, 201-207.

16. Paillotin, G., C.E. Swenberg, J. Breton and N.E. Geacintov. 1979. Biophys. J. 25, 513-533.

17. Vos, M., A.M. Nuijs, R. van Grondelle, R.J. van Dorssen, P.D. Gerola and J. Amesz. 1987. Biochim. Biophys. Acta 891, 275-285.

18. Gillbro, T., Å. Sandström, V. Sundström and J.M. Olson. 1988. In: Proc. EMBO Workshop on Green Photosynthetic Bacteria (J.M. Olson, ed.). Plenum Press, New York (in press).

19. Freiberg, A.M., K.E. Timpmann and Z.G. Fetisova. 1988. In: Proc. EMBO Workshop on Green Photosynthetic Bacteria (J.M. Olson, ed.). Plenum Press, New York (in press).

20. Den Hollander, W.T.F., J.G.C. Bakker and R. van Grondelle. 1983. Biochim. Biophys. Acta 725, 492-507.

21. Vos, M., R. van Grondelle, F.W. van der Kooij, D. van de Poll, J. Amesz and L.N.M. Duysens. 1986. Biochim. Biophys. Acta 850, 501-512.

22. Vasmel, H., R.J. van Dorssen, G.J. de Vos and J. Amesz. 1986. Photosynth. Res. 7, 281-294.

23. Karapetyan, N.V., T. Swarthoff, C.P. Rijgersberg and J. Amesz. 1980. Biochim. Biophys.Acta 593, 254-260.

24. Swarthoff, T., J. Amesz, H.J.M. Kramer and C.P. Rijgersberg. 1981. Israel J. Chem. 21, 332-337.

25. Feick, R.G., M. Fitzpatrick and R.C. Fuller. 1982. J. Bacteriol. 150, 905-915.

26. Cox, R.P., M.T. Jensen, M. Miller and J.P. Pedersen. 1988. In: Proc. EMBO Workshop on Green Photosynthetic Bacteria (J.M. Olson, ed.). Plenum Press, New York (in press).

ENERGY TRANSFER IN *CHLOROFLEXUS AURANTIACUS*: EFFECTS OF TEMPERATURE AND ANAEROBIC CONDITIONS

Bruce P. Wittmershaus, Daniel C. Brune, and Robert E. Blankenship

Department of Chemistry, Arizona State University, Tempe, AZ 85287-1604 USA

Introduction

The green bacterium *Chloroflexus aurantiacus* is a thermophilic, facultative photosynthetic organism which grows optimally from 50 to 65 °C under anaerobic conditions (1). The light-harvesting antenna pigments are separated into two structures, the chlorosomes and the cytoplasmic membrane (2,3). The chlorosomes are ellipsoidal structures of ~100x30x10 nm attached to the cytoplasmic side of the cell membrane (4). Within them are arrays of rod-like structures containing bacteriochlorophyll *c* (BChl *c*) (4,5). It has been proposed that either pigment-protein (6) or pigment-pigment (7-11) interactions result in BChl *c* oligomers. Also within the chlorosomes is a small amount of protein-bound BChl *a* (ratio to BChl *c* 1:25-30 (3,12)) that is proposed to be part of an attachment site or 'baseplate' through which the chlorosome interacts with the membrane surface (2,3,5,12,13). Carotenoids are also found in the chlorosomes. Dichroism measurements on isolated chlorosomes show that BChl *c* is highly oriented with an ~40° angle to the chlorosome's long axis (12,13). The BChl *a*'s are nearly perpendicular to the long axis. The cytoplasmic membrane contains the reaction centers (RC) and an antenna protein complex with two spectral forms of BChl *a*, BChl *a* (808,866). The relative proportions of antenna pigments are dependent on growth conditions (1,14).

Photosynthetic Light-Harvesting Systems

The proposed excitation transfer scheme (2,3,12,13), with each element's absorption maximum indicated in nm, is:

Chlorosome	**Membrane**

BChl *c* (740)--->BChl *a* (795)------>BChl *a* (808,866)--->RC (870)

Energy transfer within a BChl *a* (808,866) complex is essentially 100% efficient (15). The efficiency of transfer from this complex to the RC's is unknown. Excitations in the BChl *c*'s of the chlorosome are transferred among BChl *c*'s very rapidly (16) and transferred to BChl *a* (795) in the baseplate in ~14 ps (17,18). Reports of the energy transfer efficiency from BChl *c* to BChl *a* (795) have ranged from 55% in isolated chlorosomes (12) to 100% in cells (17) at 23 °C. The reported overall of energy transfer efficiency from BChl *c* to BChl *a* (808,866) range from 65% (19) to 100% (17) at 23 °C to 100% at 55 °C (19).

Using absorption and fluorescence excitation spectra, we have measured efficiencies of excitation transfer within whole cells of *Chloroflexus aurantiacus* from 10 to 55 °C. We observe that fluorescence from BChl *a* (808,866) increases with decreasing temperature. We propose this is caused by a phase transition in the cytoplasmic membrane that has been observed to occur in this temperature region (20). By varying the light intensity of the excitation beam and chemically closing the RC's, we here give evidence that all temperature dependent changes in the spectra simply reflect changes in the state of the RC's caused by this phase transition. Excitation transfer among the various components of the light-harvesting network is not affected by temperature in this region. This is a somewhat different conclusion than that of van Dorssen and Amesz (19).

Upon isolation of chlorosomes from the membrane, the excitation transfer scheme given above predicts a large increase in fluorescence from BChl *a* (795). This increase has not been observed in previous work (12,13,17). Our investigation of excitation transfer in isolated chlorosomes has led to the development of an anaerobic preparation that has significantly increased BChl *a* (795) fluorescence indicating that indeed a chemical quenching mechanism is responsible for the low fluorescence yield of isolated chlorosomes (17).

Materials and Methods

Absorption, fluorescence emission and fluorescence excitation spectra were taken using equipment designed in our laboratory (17). An integrating sphere was used in absorption measurements to correct for scattered light. The spectral bandwidths used were: absorption, 1.8 nm; fluorescence emission, 18 nm excitation and 11 nm emission; fluorescence excitation, 9 nm excitation and 14 nm emission. The intensity of the excitation beam during fluorescence excitation measurements was varied using neutral density filters. A jacketed quartz cuvette of 10 mm pathlength was used in the temperature controlled studies. The temperature of the sample was held constant to within 0.2 °C and monitored using a thermocouple placed in the sample.

Chloroflexus aurantiacus cells, strain J-10-fl, were grown photoheterotropically at 50-55 °C as described in (13). Ten separately grown batches of cells were examined for this study. The ratio of absorption of these cells at 740 versus 866 nm ranged from 3.2 to 4.5 with ~4 being typical. The cells were removed from the growth bottles and placed directly in the temperature controlled cuvette to minimize any thermal shock. Cells were diluted to < 0.15 absorbance at 740 nm for fluorescence measurements. Cells were diluted in 150 mM NaCl, 10 mM Tris (pH 8 at 25 °C) or 150 mM NaCl, 10 mM phyrophosphate (pH 8 at 25 °C) at the temperature of the cells, usually 55 °C.

Chlorosomes were isolated by the procedure in (3) except for the following modifications. All solutions were degassed and back-filled with argon gas. To ensure the absence of oxygen, 2mM ascorbate was added to all solutions. Sodium dithionite (1mM) was also added to the dialysis buffer and to samples placed on sucrose step gradients. At each step, exposure of the preparation to air was limited by filling containers with argon. The chlorosomes were diluted in degassed 10mM Tris, pH 8 buffer under argon for spectral measurements.

Results and Discussion

Figures 1a and b illustrate the temperature dependence of fluorescence from whole cells of *Chloroflexus aurantiacus* using the full excitation intensity of the light source at 460 nm. Fluorescence from the membrane antenna complex BChl *a* (808,866), with a maximum at 883 nm, is observed to increase with decreasing temperature, starting at ~40 °C and leveling off at ~20 °C at 3 to 4 times its value at 50-55 °C. Fluorescence from the two chlorosome pigments, BChl *c* (740) at 750 nm and BChl *a* (795) at 805 nm, varies less than ±5%. The change in 883 nm fluorescence upon cooling whole cells to 10 °C is reversible and returns to its original value upon returning the cells to 55 °C.

A differential scanning calorimetry study of *Chloroflexus aurantiacus* (20) found that a lipid phase transition occurs in the 40 to 45 °C region for cells grown at 50 to 55 °C. Figs. 1a and b indicate excitation transfer between BChl *c* and BChl *a* (795) in the chlorosome and quenching of excitations in the chlorosome by pigments in the membrane do not vary with temperature and are unaffected by such a phase transition. The trapping of BChl *a* (808,866)'s excitations is reduced at lower temperature. Following the proposed energy transfer scheme, the trapping rate of the RC's must be decreasing.

Figure 2 illustrates the temperature dependence of fluorescence from BChl *a* (808,866) at different intensities of excitation light. Our results show that the excitation beam and temperature work together to close the RC's. At a low excitation intensity, the recovery rate of the RC's is fast enough to utilize the excitations received even as the temperature is lowered. The RC's appear mostly open to BChl *a*

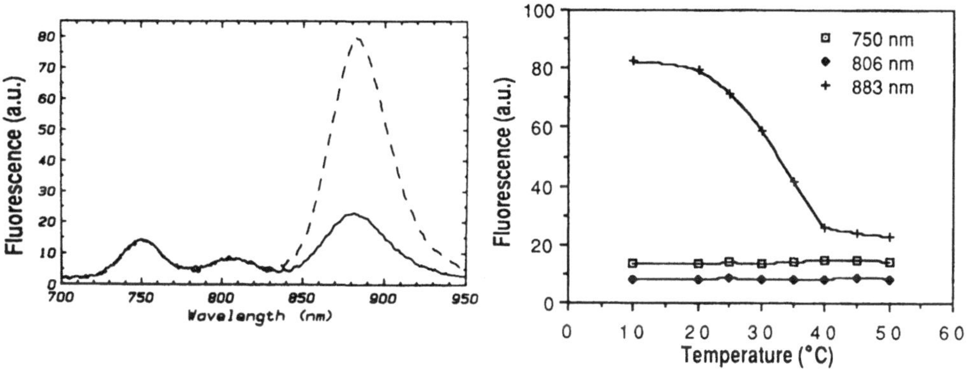

Figure 1 Fluorescence from *Chloroflexus aurantiacus* whole cells at a) 50°C (——) and 10°C (– – –) and b) at the major emission peaks versus temperature using 460 nm excitation.

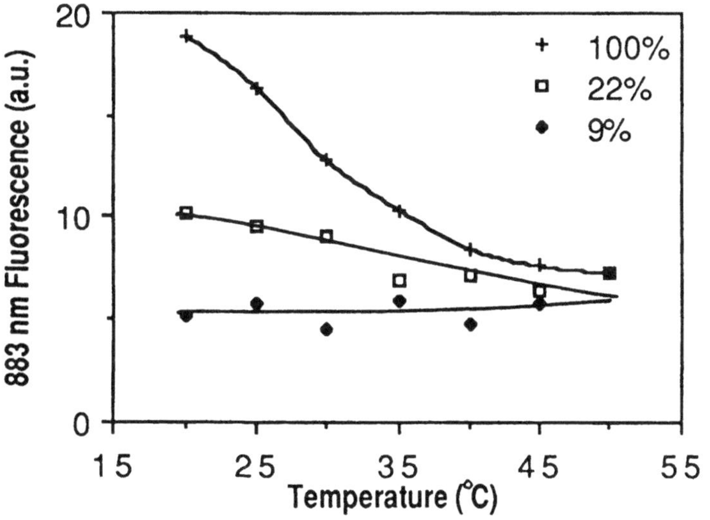

Figure 2 Fluorescence at 883 nm from BChl *a* (808,866) in whole cells as a function of temperature and intensity of the excitation beam (100% = 0.26 mw cm^{-2} at 868 nm).

(808,866) even at 20 °C (Fig. 2). At full excitation intensity, the RC turnover rate still has the same temperature dependence, but now the RC's have to accept ten times the number of excitations per unit time. The fraction of closed RC's at one temperature is higher for a higher excitation intensity, in turn causing more 883 nm fluorescence from BChl *a* (808,866).

To check the hypothesis that closed RC's are the only cause of the increase in fluorescence at 883 nm, the RC's were closed using chemical treatment. Dithiothreitol (10mM) reduces the quinones in the electron transfer chain, thereby blocking electron transfer and closing the primary acceptor. Orthophenanthroline (10mM) inhibits electron transfer between the quinones Q_A and Q_B, similarly closing the RC. When either chemical was added to whole cells at 55 °C, the fluorescence spectrum changed to one similar to that at 10 °C (Fig.1a) with a three fold increase in emission at 883 nm and no change at 750 or 805 nm (not shown). Adding either chemical to the cells at 20 °C, causes virtually no change in fluorescence with emission at 883 nm increasing <10%. Cells with 10 mM orthophenanthroline added at 55 °C and then lowered to 20 °C experienced an increase in 883 nm fluorescence of < 15%. The chemical closure of RC's showed nearly the same magnitude of effect on fluorescence intensity as lowering the temperature. This observation indicates that RC closure is the likely cause of the increase in fluorescence from BChl *a* (808,866). The tripling of fluorescence is similar to that observed from antenna proteins of purple photosynthetic bacteria when the RC's are closed (21).

We found the efficiency of excitation transfer from BChl *c* in the chlorosomes to the BChl *a* (808,866) complex in the membrane at 55 °C to be somewhat sample dependent. Figure 3 shows typical absorption and fluorescence excitation spectra. The average efficiency over 10 batches of cells was 69%±13 with values ranging from 51 to 89%. These values were obtained by comparing the excitation spectrum of 900 nm fluorescence to the absorption maxima at 740 nm (BChl *c*) and at 868 nm (BChl *a*

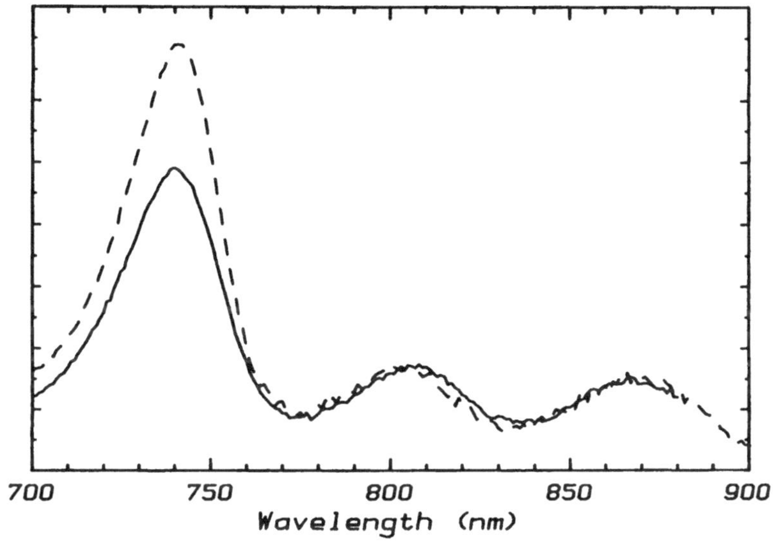

Figure 3 Absorption (– – –) compared to fluorescence excitation of 900 nm fluorescence (——) from cells at 55°C.

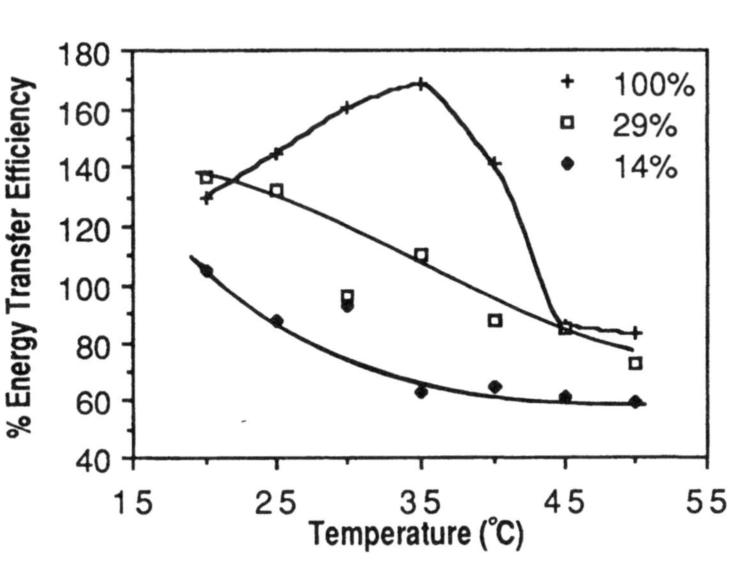

Figure 4 Apparent efficiency of energy transfer from BChl *c* to BChl *a* (808,866) in whole cells versus temperature. Fluorescence at 900 nm was measured by exciting into BChl *c* at 740 nm (100% = 0.41 mw cm^{-2}).

(808,866)). At 55 °C the full intensity of the excitation beam was not sufffficient to cause significant closure of the RC's. This was checked by repeating the excitation spectrum at 29% of the full intensity and comparing the spectra.

The RC's receive approximately four times the number of excitations with 740 nm light than with 868 nm light because of the difference in absorption (Fig. 3). This is equivalent to changing the excitation intensity at one wavelength by a factor of four, the effect of which is shown in Fig. 2. At a given temperature, exciting at 740 nm causes more RC's to be closed, increasing the relative yield of 900 nm fluorescence from BChl *a* (808,866) when compared to exciting BChl *a* (808,866) directly. For measurement of the 900 nm fluorescence excitation spectrum at various temperatures (Fig. 4), the excitation beam intensity was decreased to study how the RC turnover rate and the difference in absorption at 740 and 868 nm affects the apparent energy transfer efficiency. The 14% excitation intensity was low enough so that the RC turnover rate was sufficient to keep nearly all of the RC's open at both excitation wavelengths until <30 °C. Using low intensity excitation, the apparent energy transfer efficiency from BChl *c* to BChl *a* (808,866), remains constant from 55 to 30 °C. The increase in apparent efficiency at <30 °C using 14% excitation intensity and the temperature dependence of the efficiency using 29 or 100% intensities are caused by different amounts of RC's closed for 740 and 868 nm exciting light. The 900 nm fluorescence quantum yield varies with the amount of RC's closed resulting in artifactually large values of the apparent energy transfer efficiency. This produces impossible values of >100% when using 29 or 100% excitation intensity (Fig. 4). We infer from the 14% excitation intensity curve that the true energy transfer efficiency from BChl *c* to BChl *a* (808,866) does not change from 55 to 10 °C.

Energy transfer within the chlorosome as a function of temperature was measured by monitoring 823 nm fluorescence from BChl *a* (795) in the baseplate, exciting from 700 to 810 nm. The magnitudes at 740 and 795 nm in the excitation spectra remained

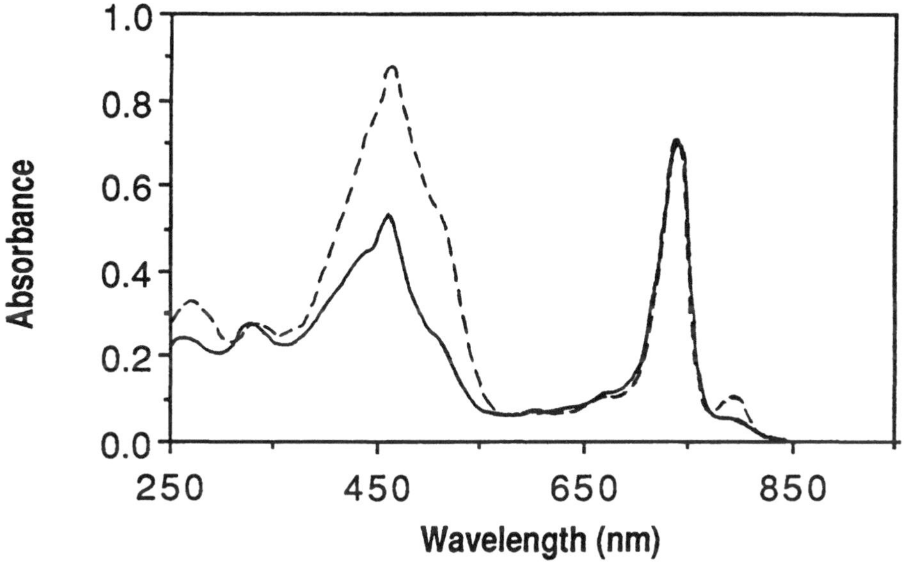

Figure 5 Comparison of absorption spectra from an anaerobic preparation of isolated chlorosomes (– – –) and those prepared using the normal technique (——) (3).

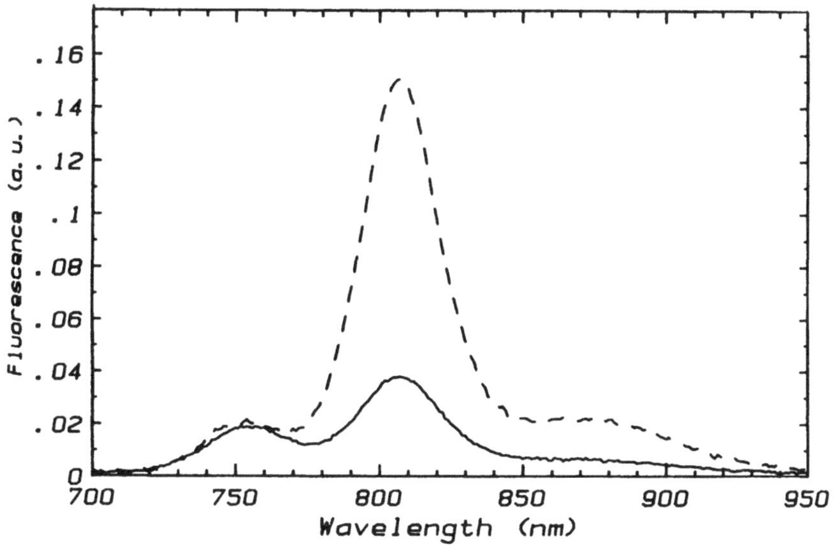

Figure 6 Fluorescence from chlorosomes isolated anaerobically (– – –) and by the normal technique (——) (3) using 460 nm excitation.

constant from 55 to 10 °C and their ratio independent of excitation intensity. We conclude from this observation and the results in Fig. 1b that energy transfer from BChl *c* to BChl *a* (795) within the chlorosome does not depend on temperature. Thus all the excitation transfer rates linking the light-harvesting antenna pigments remain nearly constant from 55 to 10 °C and are unaffected by the temperature dependent phase transition of the membrane. We were unable to calculate the exact efficiency of transfer from BChl *c* to BChl *a* (795) because the contribution of BChl *a* (795) to the absorption around 800 nm could not be reliably determined.

Figures 5 and 6 compare the absorption and fluorescence spectra, respectively, of chlorosomes prepared by the normal technique of Feick and Fuller (3) to those prepared by the same method but under anaerobic conditions. The anaerobic chlorosomes show a relatively higher absorption by carotenoids. The BChl *a* (795) peak is higher and better defined than in the normal preparation. The ratio of 750 nm fluorescence from BChl *c* to 805 nm fluorescence from BChl *a* (795) is 1:(~7) compared with 1:(<2) from our study (Fig. 6) and in published spectra for the normal preparation (12,13).

Chlorosomes isolated from their apparent excitation traps, the BChl *a* (808,866) of the membrane, using the normal aerobic procedure (3) do not show a large increase in BChl *a* (795) fluorescence yield (12,13) or lifetime (17) compared to attached chlorosomes. It has been proposed this might be caused by a quenching mechanism introduced during the isolation procedure (17). The increase in 805 nm fluorescence of the anaerobic preparation suggests a quencher of BChl *a* (795) fluorescence is indeed present, with the amount dependent on sample preparation. The anaerobic preparation also indicates the normal preparation may cause some degradation of the baseplate antenna pigments, BChl *a* (795), and carotenoids within the chlorosome.

Our work shows that the photosynthetic apparatus of *Chloroflexus aurantiacus* is very sensitive to oxygen, temperature, and probably growth conditions. We propose that the phase transition occurring at ~40 °C in the cytoplasmic membrane interferes with the mobility of the quinones of the electron transport chain, slowing down the turnover of the system and its ability to carry out photosynthesis.

The efficiency of the chlorosomes in transfering excitations to the membrane antenna complex is variable and is typically less than 100%. Why and how the cells exhibit such variablity in this rate is still an open question and may possibly be related to growth conditions. We have not ruled out the possibility that the RC's themselves trap excitations directly from the chlorosomes, bypassing the BChl *a* (808,866) antenna.

Acknowledgements

Supported by a grant from the United States Department of Energy Biological Energy Storage Program (DE-FG02-85ER13388). The authors would like to thank George H. King for his work in designing, constructing, and testing the spectroscopic equipment and operating software. We also thank Drs. R.J. van Dorssen and J. Amesz for suppying us with a prepublication copy of their work on energy transfer in *Chloroflexus aurantiacus*.

References

1. Pierson, B.K. and R.W. Castenholz. 1974. Arch. Microbiol. 100, 283-305.

2. Blankenship, R.E., and R.C. Fuller. 1986. In: Encyclopedia of Plant Physiology, New Series Vol. 19 (L.A.Staehelin and C.J.Arntzen,eds.) Springer-Verlag Berlin-NewYork.pp.390-399.

3. Feick, R.G., and R.C. Fuller. 1984. Biochem. 23, 3693-3700.

4. Staehelin, L.A., J.R. Golecki, R.C. Fuller, and G. Drews. 1978. Arch. Microbiol. 119, 269-277.

5. Sprague, S.G., L.A. Staehelin, M.J. DiBartolomeis, and R.C. Fuller. 1981. J. Bacteriol. 147, 1021-1031.

6. Wechsler, T., F. Suter, R.C. Fuller, and H. Zuber. 1985. FEBS Lett.181, 173-177.

7. Brune, D.C. and and R.E. Blankenship. 1987. In: Progress in Photosynthesis Research, Vol. 1 (J. Biggins, ed.). Martinus Nijhoff, Dordrecht. pp. 419-422.

8. Bystrova,M.I., I.N. Mal'gosheva, and A.A. Krasnovskii. 1979. Mol. Biol. 13, 582-594.

9. Smith, K.M., L.A. Kehres, and J. Fajer.1983. J. Amer. Chem. Soc.105, 1387-1389.

10. Lutz, M. and G. van Brakel. 1987. In: Proc.of EMBO Workshop on Green Photosynthetic Bacteria (J.Olson, ed.) Nyborg, Denmark. (in press).

11. Brune, D.C., T. Nozawa, and R.E. Blankenship. 1987. Biochem. (in press).

12. van Dorssen, R.J., H. Vasmel, and J. Amesz. 1986. Photosynth. Res. 9, 33-45.

13. Betti, J.A., R.E. Blankenship, L.V. Natarajan, L.C. Dickinson, and R.C. Fuller. 1982. Biochim. Biophys. Acta 680, 194-201.

14. Oelze, J. and R.C. Fuller. 1987. Arch Microbiol.148,132-136.

15. Vasmel, H., R.J. van Dorssen, G.J. deVos, and J. Amesz. 1986. Photosynth. Res. 7, 281-294.

16. Vos, M., A.M. Nuijs, R. van Grondelle, R.J. van Dorssen, P.D. Gerola, and J. Amesz. 1987. Biochim. Biophys. Acta 891, 275-285.

17. Brune, D.C., G.H. King, A. Infosino, T. Steiner, M.L.W. Thewalt, and R.E. Blankenship. 1987. Biochem. (in press).

18. Blankenship, R.E., D.C. Brune, J.M. Freeman, G.H. King, J.H. McManus, T. Nozawa, J.T. Trost, and B.P. Wittmershaus. 1987. In: Proc. of EMBO Workshop on Green Photosynthetic Bacteria (J. Olson, ed.) Nyborg, Denmark. (in press)

19. van Dorssen,R.J. and J.Amesz.1987. Photosyth.Res.(in press).

20. Oelze, J. and R.C. Fuller. 1983. J. Bacteriol. 155, 90-96.

21. Zankel, K.L., D.W. Reed, and R.K. Clayton. 1968. Proc. Nat. Acad. Sci. USA 61, 1243-1249.

INTERPRETATION OF OPTICAL SPECTRA OF BACTERIOCHLOROPHYLL ANTENNA COMPLEXES

Robert M. Pearlstein

Physics Department, Indiana-Purdue University, 1125 East 38th Street, P.O. Box 647
Indianapolis, IN 46223 U.S.A.

Introduction

Understanding the optical spectra of photosynthetic antenna complexes is important for at least two reasons. First, these spectra can provide information relating to the structure of the complexes and to the nature of pigment-pigment and pigment-protein interactions. Second, they should reveal the parameters of the exciton states, the delocalized excited electronic states of the pigment molecules from which energy transfer begins. The interpretation of optical spectra has been considerably hampered by the long-standing failure of theory (1) to explain the spectra of the water-soluble bacteriochlorophyll (BChl) *a*-protein complex from the green sulfur bacterium *P. aestuarii*, the so-called Fenna-Matthews (FM) complex (2,3). Because the structure of this complex is known to atomic resolution from x-ray diffraction analysis (4,5), and because its low temperature solution spectra have the narrowest linewidths of any photosynthetic pigment-protein complex (6), the FM complex presents an extraordinary opportunity in this field for the confrontation of theory and experiment. This has made the situation all the more frustrating. While the problem remains unsolved, theoretical analysis of the spectra of any BChl-protein complex (including the reaction center) remains open to question. Although a theory that gave reasonably good agreement with the experimental spectra was published a decade ago (7), one of the tenets of that theory -- that the Q_Y transition dipoles (responsible for the longest wavelength bands) of the individual BChl molecules in the complex are rotated by 90° relative to the direction assigned by porphyrin molecular orbital theory -- has never been justified. Thus, it is satisfying to present here a theory with standard Q_Y dipole directions of the solution optical spectra of the FM structure.

This new theory has been made possible by the recent presentation of both the complete linear sequence of the FM complex's ~39-kD polypeptide (8), and the atomic coordinates

Photosynthetic Light-Harvesting Systems

of the entire complex (5), i.e., polypeptide backbone and side chains as well as BChls. From this wealth of information, two pertinent findings emerge. First, some of the BChls in each subunit are extremely close to the outer, beta-sheet wall of the polypeptide. This raises the possibility that if two subunits each belonging to a different complex (each of which is a trimer of identical subunits) are in contact through their beta-sheets, additional strong BChl-BChl exciton interactions can occur between subunits. This is known not to happen for subunits of different trimers in either hexagonal or trigonal crystals of the FM complex (9). However, the fact that solution linear dichroism (LD) spectra are observed (10,11) proves that the complexes are aggregated; individual trimers cannot align along their only symmetry axis because that axis corresponds to the shortest dimension of the complex. The 140-μs relaxation of LD measured in pulsed electric fields shows that the aggregates consist of at least 36 trimers (10). Although much larger aggregates that occur at higher protein concentration were described long ago (12), and a model for their structure based on the known crystal structures has been proposed (9), the validity of the model remains uncertain. In any case, the structure of the lower-concentration aggregates, whose high-resolution spectra have been reported (6,11,13), is unknown. Inter-beta-sheet contact between FM trimers in solution must be considered a possibility, and is proposed here. Second, each BChl molecule is in essential van der Waals contact with at least one aromatic amino acid (R. E. Fenna, Person Communication). This appears to be the only universal pigment-protein interaction in the complex (5), and thus is the only candidate to provide a mechanism to explain the ubiquitous, protein-induced portion of the BChl Q_Y-band red shift. The specific mechanism proposed here, and elsewhere for the reaction center (14), is modification by the amino acid's pi-electron system of the intra-BChl configuration interaction (CI), leading simultaneously to Q_Y red shift and borrowing of intensity into it from the Soret.

Some details of the modified-CI mechanism are provided in the Theoretical Methods section, along with a basic description of the exciton theoretic techniques used. The Results include BChl-beta-sheet distances and a specific proposal for inter-trimer interaction; numerical output of the modified-CI model applied to each FM subunit; exciton interactions and stick spectra for a pair of strongly interacting subunits; and theoretically derived low-temperature absorption and circular dichroism (CD) spectra. The first part of the Discussion considers what must be done to improve and verify this new theory of the FM spectra. The final part suggests some conclusions for purple bacterial antennas that can already be drawn from this theory.

Theoretical Methods

Modified-CI method. The modified-CI hypothesis (14) is that protein-induced spectral red shifts of the BChl Q_y band result from an increase of the (negative) CI energy between the two basis states (15,16) contributing to the BChl Q_Y and B_Y transitions. This produces intensity borrowing from the B_Y to the Q_Y. The increase of CI energy is assumed to be caused by pi electron-pi electron interactions of aromatic amino acid rings in van der Waals contact with each BChl macrocycle. The method consists first, of assigning the protein-induced red shift of a given BChl empirically on the basis of number of aromatic amino acids in van der Waals contact; and then calculating the dipole strength of that BChl's Q_Y transition from (14)

$$\mu_{Q_Y}^2 = (\sqrt{x}\ \mu_1 + \sqrt{x+1}\ \mu_2)^2/(2x+1), \tag{1}$$

where

$$x = (\nu_o - \nu_{Q_Y})/\Delta. \tag{2}$$

In Eqs. (1) and (2), μ_1 and μ_2 are the dipole strengths of the upper- and lower-energy (respectively) Y-polarized basis states contributing to the Q_Y and B_Y transitions of BChl in a solvent, ν_{Q_Y} is the Q_Y energy of BChl in the protein (without exciton interaction); ν_o is the energy of basis state 2 (the lower-energy state); and Δ is the state 1-state 2 energy difference. Usually, $x \ll 1$, so that

$$\mu_{Q_Y}^2/\mu_2^2 \cong 1 + 2\sqrt{x}\ (\mu_1/\mu_2), \tag{3}$$

and in the special case, $\mu_1 = \mu_2$, which holds approximately for BChl a,

$$\mu_{Q_Y}^2/\mu_2^2 \cong 1 + 2\sqrt{x}\,. \tag{4}$$

Equation (4) states that the fractional borrowed intensity is just twice the square-root of the red shift energy expressed as a fraction of what is essentially the Soret-red energy difference. For the values of parameters used here, i.e., $\mu_1^2 = \mu_2^2 = 40D^2$, and $\gamma_o =$ 13 kcm^{-1} (~770 nm), the value of $2\sqrt{x} \cong 0.4$. This agrees fairly well with the early observation (2) that, per molecule of BChl, the integrated Q_Y-intensity of the FM structure is ≈ 30% greater than that of BChl in a solvent.

Exciton calculations. These are done by standard numerical techniques (1,7) except as noted. Input parameters include the coordinates and orientation vectors for the 14 strongly-interacting BChl molecules, the diagonal transition energies and strengths of the individual BChls as determined from the modified-CI procedure, and the distribution of transition monopoles over the atoms of each BChl's macrocycle. The absolute value of each transition monopole is then determined by the computer on the basis of the input diagonal transition strength for each BChl. For these calculations only one electronic transition, the Q_Y, is used, because it has previously been shown (1,7) that for the FM structure higher electronic states have a negligible perturbing effect on the Q_Y exciton transitions.

Generation of theoretical spectra. Spectra are produced first as tabular values generated from sums of symmetric Gaussian curves. Each spectrum includes 14 Gaussian components, one for each exciton transition. The width of each Gaussian component is expressed (7) as a "monomer" width divided by $\sqrt{N}$, where N is the approximate number of BChls contributing to the corresponding exciton transition (estimated from the eigenvector; see Table 3, column 6). For absorption, the monomer width is taken to be 167 cm^{-1}, a value previously determined (7) to be appropriate for the FM structure at 5 K. For CD, the monomer width is set at 200 cm^{-1} (the appropriate value at 77 K) for the 6 longest wavelength exciton transitions, but is arbitrarily increased to 300 cm^{-1} for the remaining 8 transitions. For the ith component, the peak molar extinction per BChl is given by (7)

$$\varepsilon_i = 73.05 \times 10^3 \mu_i^2/(W_i E_i). \qquad (5)$$

In Eq. (5), the exciton transition dipole strength μ_i^2 is in D^2, the exciton transition energy E_i is in kcm^{-1}, and W_i, the full-width at half maximum determined as just described, is in nm. The peak molar ellipticity per BChl is given by (7)

$$\Theta_i = 8942 \times 10^3 R_i/(W_i E_i), \qquad (6)$$

for the ith component, where the rotational strength R_i is in DBM. The theoretical spectra in Figs. 2 and 3 are drawn by computer from the generated tabular values.

Results

The second column of Table 1 gives the distances from the center of each BChl molecule to the plane of the flat portion of beta-sheet in the same FM subunit, calculated from the1.9

Table 1. Properties of BChl Molecules in the Fenna-Matthews Structure.

BChl No.	R_{normal} (a) (Å)	$2R_{normal}$ + 3.4 Å	AAA (b) No.	E (c) (kcm^{-1})	λ (c) (nm)	μ^2 (c) (D^2)
1	8.7	20.8	4	12.41	806	56.0
2	8.0	19.4	3	12.41	806	56.0
3	7.7	18.8	4	12.41	806	56.0
4	6.2	15.8	4	12.41	806	56.0
5	5.0	13.4	2	12.56	796	54.0
6	11.6	26.6	6	12.26	816	57.6
7	13.0	29.4	1	12.56	796	54.0

(a) Distance to flat beta-sheet
(b) Number of aromatic amino acids in approximate van der Waals contact
(c) Theoretically assigned diagonal energy, wavelength, or strength

(a)

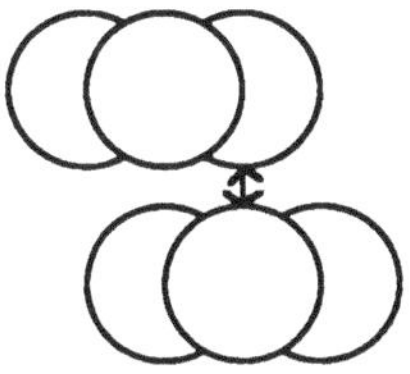

(b)

Fig. 1. Juxtuposition of two Fenna-matthews trimers in (a) a polar configuration as in the trigonal (shown) or hexagonal crystal structures, on (b) an equatorial configuration as proposed here.

Å-refined coordinates (5). The third column shows the resulting minimum, center-center separations of corresponding BChls in two subunits (not part of the same trimer) whose flat beta-sheets are placed in van der Waals contact. Actual minimum separation of a given pair only occurs if the line joining the centers of the two BChls is normal to the beta-sheet plane. The minimum separation of the two BChls 5, 13.4 Å, is very little larger than the mean nearest neighbor distance (4) between BChls within a subunit (~12Å). The minimum separation of the two BChls 4, 15.8 Å, is also not too much larger than the intra-subunit nearest neighbor distance. Other separations are significantly bigger, especially in view of the inverse cube dependence of exciton (dipole-dipole) interactions on distance.

The possibility of strong, inter-subunit exciton couplings, in addition to the strong intra-subunit couplings, leads to the proposal that FM trimers, prepared for spectroscopy in solution, form aggregates through mutual contacts of the flat portions of the beta-sheets of juxtaposed subunits belonging to different trimers. Specifically, it is assumed that two such juxtaposed subunits are related by a 2-fold axis drawn through the midpoint of the line (normal to both beta-sheet planes) joining the centers of the two BChls 5, and parallel to the projection on either beta-sheet plane of the line joining the centers of BChls 4 and 5. This particular transformation places both BChl 4 and BChl 5 pairs at essentially minimum separations. As shown in Fig. 1, this sort of trimer-trimer interaction is equatorial with respect to the trimer 3-fold axes, rather than polar as is the case for inter-trimer contacts in both crystal structures that have been described (9).

The fourth column of Table 1 lists the number of aromatic amino acids (phenylalanines, tryrosines, or tryptophans) in or near van der Waals contact with each of the 7 BChls in a FM subunit. Most of these contacts involve substantial overlap of pi electrons in the aromatic ring with those in the BChl macrocycle. Based on the number of interacting aromatics, the BChls have been divided into 3 spectral classes: BChls 1-4, moderately red-shifted; BChls 5 and 7, less red-shifted; and BChl 6, more red-shifted. The actual magnitudes chosen for these red shifts are indicated by the diagonal energies and wavelengths in Table 1 (columns 5 and 6), which should be compared with the corresponding values for the Q_Y transition of BChl a in an organic solvent, 12.99 kcm^{-1} and 770 nm, respectively (13). While these red-shift magnitudes were chosen to optimize the spectral fits, the dipole strength given in the last column of Table 1 is derived from Eq. (4), i.e., is correlated with the red shift on the basis of the modified-CI hypothesis.

Table 2 indicates the exciton interaction matrix elements for all pairs of BChls, both within the same subunit and in two different subunits that are juxtaposed through their flat beta sheets as described above. Because the matrix is symmetric, only upper-half off-diagonal elements are included. The upper left and lower right quadrants of Table 2 are identical; each gives the exciton interactions of the 7 BChls within one subunit. The upper right quadrant contains the inter-subunit interaction energies. Most notable is the interaction, 178 cm^{-1}, between BChls 5 and 5', which is almost as large as the 204-cm^{-1} interaction between BChls 1 and 2 in the same subunit. The inter-subunit interactions of BChl 3 with 3' and 4', of BChl 4 with 4' and 5', and of BChl 5 with 6' are also significant. The wavelengths, dipole strengths, and rotational strengths of the BChl Q_Y exciton band for two strongly interacting FM subunits are displayed in Table 3. The last column of Table 3 lists only the major contributing BChls of one subunit; the corresponding "primed" BChls contribute to the same

Table 2. Q_Y-Exciton Interaction Matrix for Two Fenna-Matthews Subunits with Inter-Beta-Sheet van der Waals Contacts (see text) (a)

BChl	2	3	4	5	6	7	1'	2'	3'	4'	5'	6'	7'	BChl
	-204	11	-10	12	-26	-10	2	-0.5	-4	1	-6	3	3	1
		56	14	3	21	6	-0.5	2	2	8	15	-8	0.2	2
			-110	-2	-18	15	-4	2	26	-31	-1	-3	-12	3
				-141	-30	-104	1	8	-31	53	38	-9	10	4
					112	0.8	-6	15	-1	38	178	-30	2	5
						77	3	-8	-3	-9	-30	4	2	6
							3	0.2	-12	10	2	2	8	7
								-204	11	-10	12	-26	-10	1'
									56	14	3	21	6	2'
										-110	-2	-18	15	3'
											-141	-30	-104	4'
												112	0.8	5'
													77	6'

(a) Elements in cm^{-1}, derived using renormalized Weiss transition monopoles (16) extent. The theoretical absorption spectrum is shown in Fig. 2. It is simply the stick spectrum of Table 3 dressed in 14 symmetric Gaussian envelopes, whose individual widths are set as described in Theoretical Methods. An experimental 5 K absorption spectrum is shown for comparison. Note that the vertical scale is given, not in arbitrary, but in definite units. Thus, the agreement of theory and experiment, which is somewhat better here with standard Q_Y-dipole assignments than for the previous model with rotated dipoles (7), is all the more significant. This is because the monomer dipole strengths, determined here from the modified-CI equation and not as independent parameters, simultaneously determine both height and width of the band. Agreement of theoretical and experimental CD spectra, shown in Fig. 3, is not as good as for absorption, but is perhaps no worse than that obtained with the previous rotated-dipole, theory (7). Again, the theoretical spectrum is the stick spectrum of Table 3 dressed in 14 symmetric Gaussian envelopes. As for absorption, the ordinate is given in definite units.

Table 3. Calculated Q_Y Stick Spectra for Two Interacting Fenna-Matthews Subunits

Exciton Transition	E (kcm^{-1})	λ (nm)	Absorption (D^2)	CD (DBM)	Contributing BChls(a)
1	12.806	780.9	3.5	0.6	4,5
2	12.686	788.3	6.1	-2.6	4,5,7
3	12.633	791.6	0.5	1.3	1,2,5,7
4	12.620	792.4	33.0	-6.1	1,2
5	12.618	792.5	2.0	5.8	1,2,7
6	12.526	798.3	66.3	-6.8	3,4,7
7	12.492	800.5	125.0	-26.0	5,7
8	12.413	805.6	156.0	37.9	3
9	12.295	813.3	205.9	6.4	1,3,4
10	12.249	816.4	31.7	17.1	3,4,6
11	12.226	817.9	4.0	-10.2	6
12	12.189	820.4	6.2	18.0	1,2,3
13	12.188	820.5	56.5	-20.4	1,2
14	12.098	826.6	81.8	-15.0	4,5,6

(a) Numbers of the BChl molecules in each subunit that contribute significant (>5%) squared amplitude to the transition

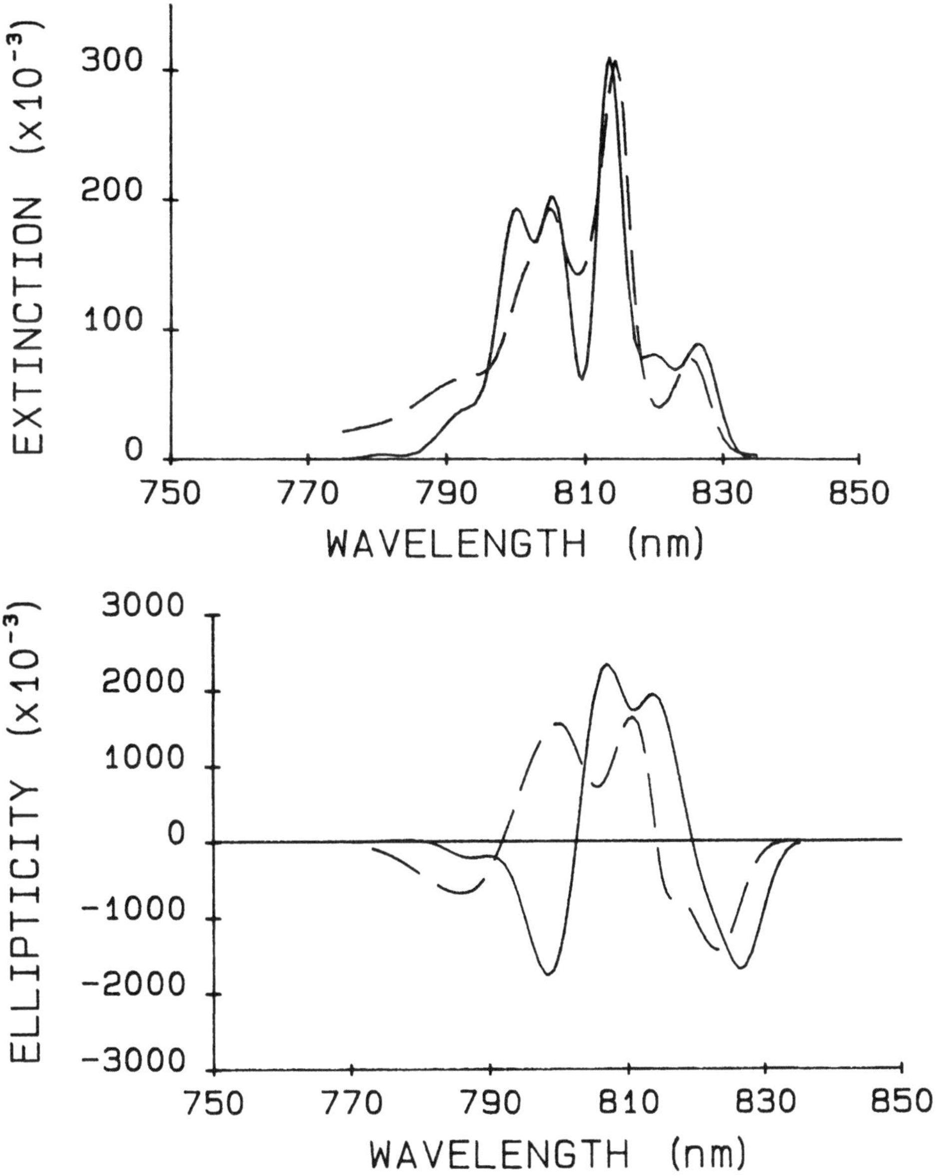

Fig. 2 (top). Molar extinction per BChl molecule in the Q_Y band of the FM complex. Solid curve: Theory for two FM subunits with their flat beta-sheets in van der Waals contact (see text). Dashed curve: 5 K experimental spectrum (6).

Fig. 3 (bottom). Molar ellipticity per BChl molecule in the Q_Y band of the FM complex. Solid curve: Theory, as in Fig. 2. Dashed curve: 77 K experimental spectrum (13).

Discussion

Fenna-Matthews structure

The results presented here show that the absorption and CD spectra of the FM structure may be explainable in terms of standard exciton theoretic (1) and porphyrin molecular orbital concepts (16) after all. This is an important vindication for these theories, which hopefully will lend greater credibility to their use in the analysis of the optical spectra of other BChl-protein complexes. However, the results must be viewed as still preliminary for the optical spectra of the FM structure itself, since the proposed equatorial trimer-trimer interaction remains to be substantiated. Possible inter-subunit bonding can be explored on the computer. Also, the inter-trimer geometry can be varied theoretically in an attempt to improve spectral fits. For example, the crossing angle of the two sets of beta-sheet strands on the two juxtaposed subunits is 78° for the geometry used here. This is close to but perhaps a bit larger than the 60°-70° range that is considered "usual" (H. Zuber, personal communication).

Following optimization of the proposed trimer-trimer interaction geometry, it would be worthwhile to attempt optimization of the BChl Q_Y diagonal energies and dipole strengths. In this regard, the modified-CI theory is important but not crucial. While it may be significant that the square-root-of-red-shift-energy dependence of diagonal dipole strength predicted by the theory appears to agree well with experiment, it must be noted that optimal values of the 14 diagonal parameters can be determined by numerical search procedures, independent of any such theory. It is puzzling that the diagonal red-shift energies themselves seem to correlate so well with the number of BChl-aromatic amino acid contacts, and not with other presumably red-shifting interactions, e.g., hydrogen-bonding of BChl acetyl or keto carbonyl oxygens. Thus, for example, BChls 1 and 3 with 4 aromatics each in contact are assigned the same diagonal red shift energy; but BChl 3 has both carbonyl oxygens hydrogen-bonded, which BChl 1 has neither (5).

Once geometrical and diagonal parameters are better determined, more precise spectral predictions can be made. These should include LD, as well as absorption and CD. A detailed aggregation model of the trimers in solution is necessary in order to make a definite prediction of the solution LD spectrum. This will be a stringent test of any such model. Of equal importance is the prediction of the LD spectrum of thin (trigonal) crystals, not yet determined experimentally at low temperature. Since equatorial interactions do not occur in the crystal structure (9), it can already be predicted that the low-temperature crystal LD spectrum should be quite different in detail from that of the solution aggregates.

Finally, a remark about the positive sign of the large BChl 5-BChl 5' exciton interaction in the equatorial model is in order. In the old, single-subunit model, rotating all the Q_Y transition dipoles by 90° induces a net increase of ~100 cm^{-1} for the exciton interactions of BChl 5 compared to the case of standard Q_Y dipoles (7). The large positive interaction brought about by juxtaposition of the second subunit in the equatorial model mimics this effect. This may explain, in part, why a 90°-dipole-rotation gives the appearance of accounting for the spectral anomalies of the FM structure.

Purple bacterial antennas

This theoretical model of the FM structure permits certain inferences to be drawn regarding spectral features of BChl a-containing purple bacterial antennas. Nonexcitonic red shifts at least to 820 nm, presumably due to interactions with aromatic amino acids, appear possible. Thus, neither B800 nor B820 forms requires a very strongly coupled BChl-exciton-dimer. On the other hand, B850 and B880 forms almost certainly do involve such dimers, which are absent in the FM structure. In these longer wavelength forms, the degree of protein-induced versus exciton-induced red shift probably varies from species to species. In those species having CD bands of large rotational strength the protein-induced portion of the red shift is probably greater than it is in those species with small rotational strength CD, because large CD requires a BChl-BChl dimer geometry somewhat less favorable to exciton-induced red shift. The partial loss of exciton contribution to the red shift must then be compensated by increased protein interaction. Indeed, Brunisholz *et al.* (17) have evidence for more aromatic amino acid interaction with the BChls of large-CD species than with those of small-CD species. It has recently been proposed that the CD of B850 and B880 arises primarily from the distant (~25 Å) interaction of pairs of exciton dimers (18,19). Although a BChl tetramer, corresponding to a polypeptide tetramer, almost certainly occurs in both B850 and B880 (17,19), the distant-interaction-CD model has certain difficulties (1). It is simpler to assume that the CD arises primarily from individual exciton dimers.

Acknowledgments

The author thanks R. E. Fenna for help with locating BChl contacts with aromatic amino acids in the FM structure, and B. W. Matthews for providing the high resolution atomic coordinates of the strucutre in a convenient form.

References

1. Pearlstein, R. M. 1987. In: Photosynthesis (J. Amesz, ed.). Elsevier, p. 299.

2. Olson, J. M. 1966. In: The Chlorophylls (L. P. Vernon and G. R. Seely, eds.). Academic Press, p. 413.

3. Fenna, R. E., B. W. Matthews, J. M. Olson, E. K. Shaw. 1974. J. Mol. Biol. *84*, 231.

4. Fenna, R. E., B. W. Matthews. 1975. Nature (London) *258*, 573.

5. Tronrud, D. E., M. F. Schmid, B. W. Matthews. 1986. J. Mol. Biol. *188*, 443.

6. Whitten, W. B., J. A. Nairn, R. M. Pearlstein. 1978. Biochim. Biophys. Acta *503*, 251.

7. Pearlstein, R. M., R. P. Hemenger. 1978. Proc. Natl. Acad. Sci. USA *75*, 4920.

8. Daurat-Larroque, S. T., K. Brew, R. E. Fenna. 1986. J. Biol. Chem. *261*, 3607.

9. Matthews, B. W., R. E. Fenna, S. J. Remington. 1977. J. Ultrastruct. Res. *58*, 316.

10. Whitten, W. B., R. M. Pearlstein, E. F. Phares, N. E. Geacintov. 1978. Biochim. Biophys. Acta *503*, 491.

11. Swarthoff, T., B. G. de Grooth, R. F. Meiburg, C. P. Rijgersberg, J. Amesz. 1980. Biochim. Biophys. Acta *583*, 51.

12. Olson, R. A., W. H. Jennings, C. H. Hanna. 1969. Arch. Biochem. Biophys. *130*, 140.

13. Philipson, K. D., K. Sauer. 1972. Biochemistry *11*, 1880.

14. Pearlstein, R. M. 1988. In: Structure of the Photosynthetic Bacterial Reaction Center (J. Breton, ed.). Plenum Press (in press).

15. Gouterman, M., G. H. Wagniere, L. C. Snyder. 1963. J. Mol. Spectrosc. *11*, 108.

16. Weiss, C. Jr. 1972. J. Mol. Spectrosc. *44*, 37.

17. Brunisholz, R., H. Zuber. 1988. In: Organization and Function of Photosynthetic Antennas (H. Scheer and S. Schneider, eds.). Walter de Gruyter Publishers (This Volume).

18. Scherz, A., W. W. Parson. 1986. Photosynth. Res. *9*, 21.

19. Rozenbach-Belkin, V., P. Braun, P. Kovatch, A. Scherz. 1988. In: Organization and Function of Photosynthetic Antennas (H. Scheer and S. Schneider, eds.) Walter de Gruyter Publishers (This Volume).

TIME RESOLUTION AND KINETICS OF "F680" AT LOW TEMPERATURES IN SPINACH CHLOROPLASTS

Robert Knox and Su Lin

Department of Physics and Astronomy and Laboratory for Laser Energetics, University of Rochester, Rochester, NY 14627 USA

Introduction

The excitation energy transport system of chloroplasts has been modelled in increasing detail as a result of picosecond fluorescence studies. At room temperature the principal emission band at 685 nm (F685) is broad and spectrally inhomogeneous and has numerous components with distinguishable lifetimes. At low temperatures, fluorescence in the 680-695 nm region has clear structure but there is much less information available on its kinetics. Using streak camera techniques, we have resolved the shortest-wavelength component in this region and it is identifiable with a lifetime of 9.5 ± 1.5 ps over the range of temperature 20-110 K. This is consistent with the assignment of F680 to light-harvesting chlorophyll proteins (LHC) in steady-state measurements on subchloroplast particles (1) and chloroplasts (2).

Low-intensity pulse-excited time-resolved chloroplast fluorescence data are traditionally reduced in terms of several, typically three, components exponential in time. Research in the field has been concerned with understanding the relative amplitudes and the lifetime parameters of these components under various conditions. The most extensive work along these lines has been done by Sauer (3-5), Moya (6,7), and Holzwarth (8,9) and their colleagues. F685 dominates the fluorescence

Photosynthetic Light-Harvesting Systems

and its three components have the following general characteristics. The τ_i are lifetime parameters and the A_i are amplitudes in the sense that fluorescence intensity is $I_f(t) = \Sigma_i A_i \exp(-t/\tau_i)$, with $\Sigma_i A_i = 100\%$.

Fast component. $\tau_i \sim$ 50–150 ps. Depending on the wavelength and bandpass of emission, one may be seeing either or both of two distinct sources of the fast component. One is very fast (~80 ps) PS-I fluorescence at longer wavelengths, ca. 700 nm (10,11). This short lifetime is the reason that PS-I is usually described as nonfluorescent. The other source is PS-II emission (~180 ps) as identified by preferential excitation at 680 nm and maximum emission at 685 nm. This emission probably originates in the "proximal antenna," i.e., the chlorophylls closest to reaction center II (P680).

Medium component. $\tau_2 \sim$ 400–800 ps. This is the primary contributor to fluorescence from PS-II antenna pigments, with a lifetime determined by trapping at reaction center II, as generally described in the literature; however, adopting the proximal antenna model mentioned above, it may be more accurate to say that this component's lifetime is determined by pathways into the proximal antenna. The medium component is seen by many workers (12–15) as two components, one originating in PS-IIα (200–300 ps) and the other in PS-IIβ (500–650 ps). However, this interpretation is controversial (16). These lifetimes are increased significantly and A_2 decreases when the reaction centers become closed (reduced). These medium-component excitations are affected by exciton annihilation under high-excitation conditions (5).

Slow component. $\tau_3 \sim$ 1500–2500 ps. A component also not free from controversy, attributed to recombination luminescence (15,17) or to a change in the initial trapping rate when traps are closed (11,18). both of these interpretations are consistent with the fact that the yield of slow component increases with trap closure. However, the recombination mechansism implies a primary trapping process invariant with trap closure and is therefore inconsistent with the decrease of A_2 on trap closure. Consequently the current (but not firmly established) view favors the variable trapping rate model with $A_2 + A_3 \approx$ constant.

In the range 685-700 nm, the chloroplast's fluorescence is extremely complex. For example, in spinach cloroplasts at 77K Reisberg and Sauer considered up to five kinetic components necessary.

At low temperatures the chloroplast emission spectrum changes significantly; a shoulder at 680 nm is identified with LHC (2). Considering the complexity mentioned above, the band at 695 nm which separates out and nearly dominates the emission at 77K is deceptively distinct. It has been the object of numerous hypotheses over the past 25 years [see the discussion by Breton (19)].

For a complete kinetic description it is desirable to have some knowledge about the sources as well as decay channels of various kinetic components. As we have found in the case of F735(20), well-resolved rise-time measurements can supply very useful information of this kind. In the work reported here, the rising part of F685 is considered explicitly in the process of sorting out a new component on its short-wavelength side.

LHC fluorescence

Indications of LHC's contribution to normal chloroplast fluorescence appeared first in the analyses of low-temperature emission spectra (2). Rijgersberg found that as T dropped to 34K a small shoulder (F680) developed, at 680 nm, on the short-wavelength side of the F685 band. The question remained, however, whether F680 was a strong band with a short lifetime or a weak band with a long lifetime. The latter could exist, for example, if some free chlorophyll were induced by low temperature damage. The reproducibility of the band and its absence in mutants without LHC pointed toward a non-artifactual origin. At 77K, time-resolved spectra of spinach chloroplasts show a tendency for a shift from 680 to 685 nm over a period of some 80 ps (21), and it is encouraging to note that the LHC-II emission may be distinguishable at very short times in PS-II enriched fractions at room temperature (22).

We recently reported our first results on the kinetics of F680 (23). The characteristics of the emission did support the LHC hypothesis. Here we report a fuller analysis, showing that F680 maintains its kinetic character (obvious at 20K) as the temperature is raised, becoming increasingly hidden by the broadening and increasing F685 band.

Materials and methods

Samples were prepared following the procedure of Breton et al. (24). Spinach chloroplasts were suspended in a buffer solution of 0.4M sucrose, 20 mM kcl, and 20 mM Tris buffer-HCl. Samples were kept in the dark and on ice until use. For low temperature measurements, the chloroplast suspension was mixed with 70% glycerol. A closed-cycle helium refrigerator was used to maintain the temperature.

Fluorescence was excited at 532 nm by 30-ps pulses from a Nd:YAG laser and detected with the low-jitter streak-camera system described earlier (20,25). The system configuration for these experiments is shown in Fig. 1.

Time-dependent traces were analyzed by integrating kinetic equations directly, using the actual excitation pulse shape, as in (5), except that we did not need nonlinear terms. Low jitter and signal averaging enable this method to distinguish short lifetime components to an accuracy of the order of magnitude of the jitter (~ 2 ps) (26). Because of this technique and the nature of the noise, best fits were found by trial and error rather than by least-squares curve-fitting.

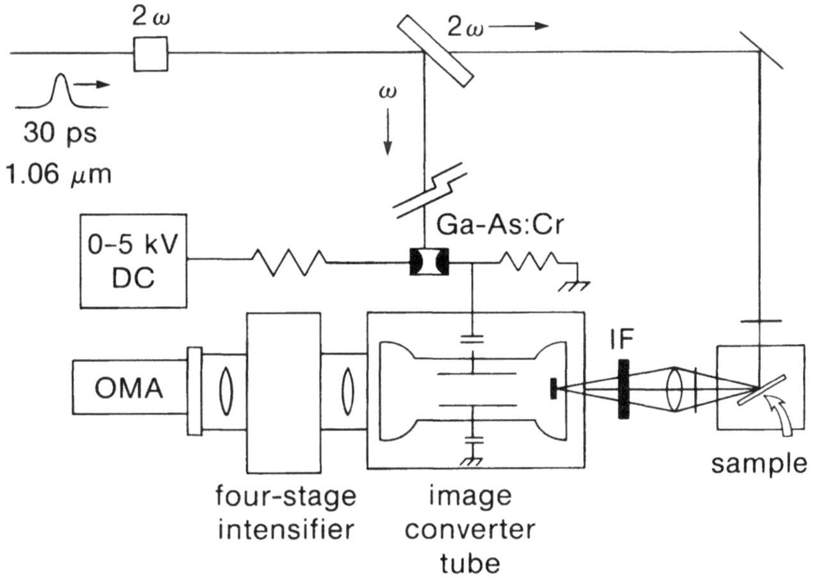

1. Setup for excitation of the sample and detection of fluorescence by a low-jitter streak camera. The system has been described in more detail previously (20,25). Features essential to this experiment are the semiconducting switch (Ga-As:Cr) that allows extensive signal averaging and the filter IF used to wavelength-resolve the output.

Results and analysis

Figure 2 shows a sequence of spinach-chloroplast time-resolved emission selected by 10-nm interference filters at 680 nm and 295K, 670 nm and 77K, and 660 nm and 20K. A full set of data

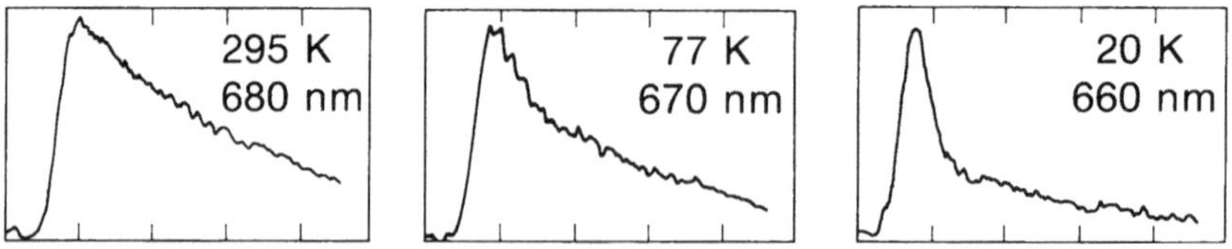

2. Emergence of a fast component of chloroplast fluorescence as temperature (T) and wavelength (λ) are changed. The trends are similar as T and λ are changed individually. For scales, see caption of Fig. 3. (After Lin and Knox (23).)

at all three wavelengths and temperatures has been published (23). A short component is clearly building in at low temperatures and is best resolved at the shortest wavelengths.

If one uses a simple decomposition of the response in terms of two pools of emitters with exponential lifetimes, the longer of the two lifetimes is about 200ps and the shorter one must be taken as 9 ± 2 ps at 20K and 14 ± 2 ps at 77K. This is one possible basis for understanding the data, but it brings with it the need to explain the 14 ps lifetime. Such an increase is unusual. At higher T there should be, if anything, a decrease in the lifetime (increase in the rate of transfer) because of improved overlap of emission and absorption bands.

As an alternative and more physical model, we have chosen coupled kinetic equations for the "F680" and F685" species with an activated back-transfer. if we let k_1 be the rate constant for transfer from F680 to F685, then the reverse rate constant k_{-1} is to be taken as $k_1 \exp(-\Delta E/kT)$. k_1 dominates the decay of the short-wavelength component and we need make no distinction between this rate and the total rate of decay of F680. In our fits (Fig. 3), $k_1 = (9.5 \pm 1.5\ \text{ps})^{-1}$, and, as in the two-exponential analysis, the longer component (which probably corresponds mainly to a PS-II short component of the standard literature) requires a T-independent basic decay rate of $(200\ \text{ps})^{-1}$. k_1 needs no T dependence in contrast with the pure exponential case. We worked as follows: at 20K, virtually no back transfer was needed. Then at 77K we found that $k_{-1}/k_1 = 0.109$ gave the best fit, implying a ΔE of 14.7 meV (5.5 nm at 680 nm). Thus $k_{-1}/k_1 =$

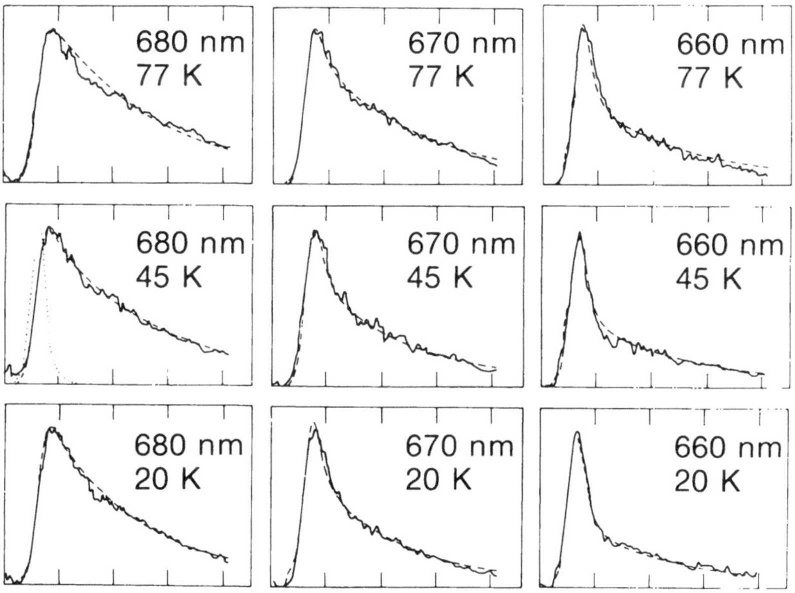

3. Emission (solid line) and theoretical fit (dashed line) at the wavelengths and temperatures shown. The experimental curves are averages of many data sets (100 in the case of 680 nm, 200 in the others). Horizontal scale divisions 96 ps. Vertical scale arbitrary units. Excitation is the same in each frame but is shown only at 45K, 680 nm. Because of the low jitter, application of the kinetic equations results in accuracies of the order of ±1.5 ps in the short lifetimes involved, despite the width of the excitation pulse (26).

0.0002 at 20K and 0.023 at 45K. The fit at 45K is predicted from the others with no free kinetic parameters, and is seen to be quite reasonable. The only parameter was the relative contribution of the two kinetic populations, as in each of the fits. The worst fit is in the 680 region, as might be expected because of the increased contribution from the complex bands at longer wavelengths.

Low-resolution time-resolved spectra can be generated from our results. At 20K the short- and long-time spectra were reasonable and consistent with each other (23). At 77K the bands are affected by low signal-to-noise and the above-mentioned probable intrusion of long-wavelength bands into F685. They still indicate, as did the 20K data, that the short-wavelength band we have called "F680" is more nearly at 676 nm and "F685" is at 683 nm (not to be taken as seriously as the 676 nm result, even at 20K).

Conclusions and remarks

A very short-lived (~9 ps) chloroplast fluorescence band at 676 nm has been identified and it is proposed that it corresponds to the steady-state band "F680" originally identified at low temperatures by Satoh and Butler and by Rijgersberg. The band maintains its lifetime as the temperature is raised, acquiring reexcitation by activated back-transfer from the species ("F685") to which it donates.

Short lifetime and intense emission of "F680" are of course necessary but not sufficient to demonstrate its origin as LHC. We have several further options. If the emission originates in LHC, its yield (amplitude divided by excitation fluence) should be reduced at very high excitation intensities because of exciton annihilation. We plan to extend the studies to mutants, with the following rationale. In a barley mutant without chlorophyll b, the band should be absent [as it is in steady state (2)]. In mutants that have no photosystem I (27), there should be no change or at most a lengthening of the lifetime by the order of a factor of two to account for loss of any LHC-to-system-I transfer in the chloroplast.

Acknowledgment

This research was supported in part by the United States Department of Agriculture under grant 87-CRCR-1-2424 and in part by the sponsors of the Laser Fusion Feasibility Project, of the Laboratory for Laser Energetics. The authors gratefully acknowledge helpful discussions with B. Wittmershaus, L. France, and N. Geacintov.

References

1. K.Satoh and W.L.Butler, Plant Physiol **61**, 373-389 (1978).

2. C. P. Rijgersberg, J. Amesz, A. P. G. M. Thielen, and J. A. Swager, Biochim. Biophys. Acta **545**, 473-382 (1979); C. P. Rijgersberg, Fluorescence and Photochemistry at Low Temperatures in Photosynthetic Systems, Ph.D. thesis, Leiden, 1980.

3. P. Reisberg and K. Sauer, Photochem. Photobiol, **36**, 657-661 (1982).

4. J. A. Nairn, W. Haehnel, P. Reisberg, and K. Sauer, Biochim. Biophys. Acta **682**, 420-429 (1982).

5. W. Haehnel, J. A. Nairn, P. Reisberg, and K. Sauer, Biochim. Biophys. Acta **680**, 161-173 (1982).

6. I. Moya, J. E. Mullet, J. -M. Briantais, and R. Garcia, in Proc. 5th Intl. Congr. Photosyn. (G. Akoyunoglou, ed., Balaban Intl. Scientific Services, Philadelphia, 1981), pp. 163-172.

7. I. Moya and R. Garcia, Biochim. Biophys. Acta **722**, 480–491 (1983).

8. A. R. Holzwarth, Photochem. Photobiol. **43**, 707–725 (1986).

9. A. R. Holzwarth, in Topics in Photosynthesis (ed. by J. Barber; Elsevier, 1986), in press.

10. R. J. Gulotty, L. Mets, R. S. Alberte, and G. R. Fleming Photochem. Photobiol. **42**, 51–57 (1985).

11. A. R. Holzwarth, J. Wendler, and W. Haehnel, Biochim. Biophys. Acta **807**, 155–167 (1985).

12. W. L. Butler, D. Magde, and S. J. Berens, Proc. Natl. Acad. Sci. (USA) **80**, 7510–75 (1983).

13. S. J. Berens, J. Scheele, W. L. Butler, and D. Magde, Photochem. Photobiol. **42**, 51–57 (1985).

14. K. Karukstis and K. Sauer, Biochim. Biophys. Acta **725**, 384–393 (1983).

15. W. Haehnel, J. A. Nairn, P. Reisberg, and K. Sauer, Biochim. Biophys. Acta **680**, 161–173 (1982).

16. M. Hodges and I. Moya, Biochim. Biophys. Acta **849**, 193–202 (1986).

17. V. V. Klimov, in Advances in Photosynthesis Research, vol. 1 (C. Sybesma, ed.; W. Junk Publishers, The Hague, 1984), pp. 131–138.

18. W. Haehnel, A. R. Holzwarth, and J. Wendler, Photochem. Photobiol **37**, 435–443 (1983).

19. J. Breton, FEBS Lett. **147**, 16–20 (1982); see also FEBS Lett. **159**, 1–5 (1983).

20. B. P. Wittmershaus, T. M. Nordlund, W. H. Knox, R. S. Knox, N. E. Geacintov, and J. Breton, Biochim. Biophys. Acta **806**, 93–106 (1985).

21. M. Mimuro, N. Tamai, T. Yamazaki, and I. Yamazaki, FEBS Lett. **213**, 119–122 (1987).

22. M. Mimuro, I. Yamazaki, N. Tamai, T. Yamazaki, and Y. Fujita, VII Intl. Congr. on Photosynthesis, Brown U. (1986), paper 105-089.

23. S. Lin and R. S. Knox, J. Luminescence, in press (Proc. 1987 Intl. Conf. on Luminescence, Beijing, 1987).

24. J. Breton, E. Roux, and J. Whitmarsh, Biochem. Biophys. Res. Comm. **64**, 1274-1277 (1975).

25. W. H. Knox and G. A. Mourou, Opt. Comm. **37**, 203-206 (1981).

26. W. Knox and L. Forsley, in Y. Talmi, ed., Multichannel Image Detectors, vol. 2 (American Chemcial Society ACS Symposium Series no. 236, 1983), pp. 221-231, esp. Fig. 1.

27. J. Olive, F. A. Wollman, P. Bennoun, and M. Recouvreur, Biol. Cell, **48**, 81-84 (1983).

PICOSECOND STUDIES OF FLUORESCENCE AND ABSORBANCE CHANGES IN PHOTOSYSTEM II PARTICLES FROM SYNECHOCOCCUS SP.

Alfred R. Holzwarth, Günther H. Schatz and Helmut Brock

Max-Planck-Institut für Strahlenchemie, Stiftstraße 34 - 36
D-4330 Mülheim a.d. Ruhr, Federal Republic of Germany

INTRODUCTION

The measurement of excitation trapping, charge separation and primary charge stabilization of photosystem II (PS II) is complicated by the fact that available preparations enriched in PS II and still active in charge stabilization ($I^-Q_A \rightarrow IQ_A^-$) contain a relatively large amount of antenna pigments bound to the RC. This imposes several experimental difficulties in particular to absorbance change measurements with picosecond time resolution: i) the relative signal amplitude is small due to strong background absorption and ii) the threshold for singlet - singlet annihilation within the antenna is shifted to much lower values of photon fluence per pulse as compared to isolated RCs. The latter problem was evident in a recent study (1), which for the first time indicated the transient reduction of pheophytin a as intermediate electron acceptor I in PS II as proposed originally by Klimov et al. (2) on the basis of steady state difference spectra. The kinetics of excitation trapping and $P_{680}^+I^-$ formation could not be resolved in that study. It has been proposed that in the RCs of PS II the reduction of Q_A should induce a delayed fluorescence after charge recombination of $P_{680}^+ I^-$ to which the long-lived (2-4 ns) fluorescence component observed with closed PS II was attributed (2). A direct experimental test of this hypothesis by absorbance spectroscopy has to differentiate between the contributions from the radical pair ($P_{680}^+I^-$) and the excited states of the primary donor (P_{680}^*) and/or antenna chlorophyll (Chl^*). However, all these contributions show much closer spectral overlap in PS II (1) as compared to the RC of purple bacteria. We have therefore chosen a new approach in order to gain more insight into the primary processes in PS II: i) We correlated measurements of picosecond kinetics of absorbance changes with those of fluorescence decays. The two types of

Photosynthetic Light-Harvesting Systems

measurement are carried out under almost identical conditions using intact, oxygen-evolving, PS II particles with about 80 Chl a/PS II. ii) In order to avoid artificial contributions by singlet - singlet annihilation we have applied a novel type of picosecond spectrometer which allows us to measure picosecond absorbance transients at extremely low excitation fluence (< 2 $\mu J/cm^2 \cdot$pulse).

RESULTS and DISCUSSION

The **PS II particles** were prepared from a thermophilic cyanobacterium Synechococcus sp. as described in (4,5), and were characterized by about 80 Chl/PS II-RC (5,6) and full oxygen-evolving capacity (4) . **Fluorescence decay kinetics** were measured at 685 nm with a single-photon-timing system as described in (3) after excitation with laser pulses (675 nm, 10 ps FWHM) of 400 kHz repetition rate and a fluence E <0.1 $\mu J/cm^2 \cdot$pulse. **Absorbance transients** were determined by a novel type of pump-probe apparatus. A scheme of the transient absorption spectrometer is shown in Fig. 1. The pump pulse (675 nm, 10 ps FWHM, 2 $\mu J/cm^2$ pulse, if not stated otherwise) [Aand the probe pulse (620 - 700 nm, <15 ps FWHM, 0,02 - $\mu J/cm^2$ pulse) were provided by two synchronized cavity-dumped (400 kHz) dye lasers (Spectra-Physics), which were both pumped by a mode-locked CW Nd:YAG laser (Spectra-Physics). In all measurements F_o-conditions (open PS II centers) were established by addition of 2 mM $K_3Fe(CN)_6$ in darkness, F_{max}-conditions (closed PS II centers) by addition of 2 mM dithionite under nitrogen at ambient light. The temperature was stabilized at 15°C. **Data analysis** of the absorbance changes was performed by a deconvolution procedure based on a global fitting algorithm (7,8) using the cross correlation of the two laser pulses as system response. The absorbance signals measured at different wavelengths (λ_k) are simultaneously fitted to a sum of exponential functions plus a constant

$$A(\lambda_k,t) = A_o(\lambda_k) + \sum_{i=1}^{n} A_i(\lambda_k)\cdot \exp(-t/\tau_i) \qquad (1)$$

by free running optimizations for all amplitudes and lifetimes (8). Plots of the resulting amplitudes $A_i(\lambda_k)$ versus the detection wavelength yield the absorbance difference spectra (cf. Figs. 2 and 3).

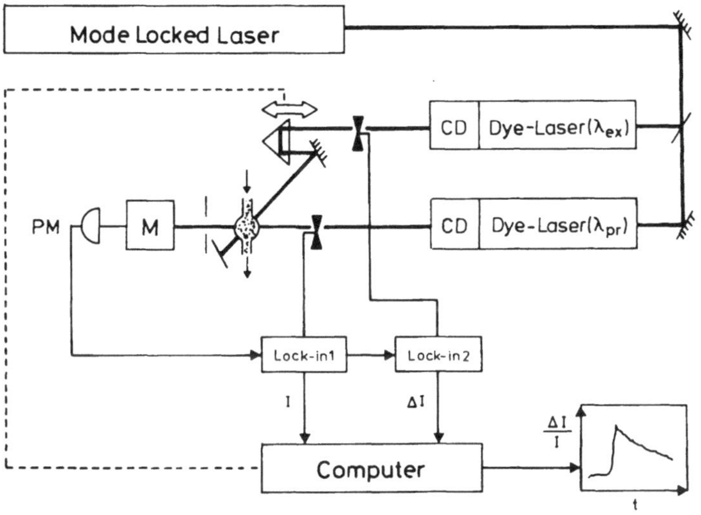

Fig. 1: Schematic diagram of the picosecond absorption spectrometer used in this work. Full details of this apparatus will be given elsewhere.

TABLE 1: Lifetimes τ_i and relative amplitudes α_i (in percent) from a three-exponential analysis of the fluorescence decay components in PS II particles at F_o and F_{max}. (λ_{exc} = 675 nm, λ_{em} = 685 nm)

	F_o		F_{max}	
τ_1/ps	80 ± 20	(72 ± 5)	220 ± 30	(44 ± 4)
τ_2/ps	520 ± 120	(20 ± 3)	1300 ± 150	(47 ± 4)
τ_3/ps	1800 ± 100	(8 ± 2)	4000 ± 500	(9 ± 2)

Fluorescence decay kinetics. Characteristic results from three exponential fits of independent measurements on six sample preparations are given in Table 1. They show that the main contribution to the F_o-kinetics derives from a fast decaying (60 -100 ps) component. The 1.8 ns component of minor amplitude (< 10%) is attributed to sample impurities, possibly allophycocyanin (APC) which was always present to a small extent in our PS II preparations (4). Ignoring such long-lived component(s) with less than 10% of the total amplitude, the fluorescence decay kinetics of the chlorophylls coupled to the RC appears to be biexponential. Closing the RCs by reduction of Q_A with dithionite (or 20μM DCMU plus light) results in an increase of the lifetimes by approximately a factor of 2 - 3 and in a decrease of the amplitude ratio α_1/α_2 by a factor of 3 - 4 (see Table 1). Concomitantly the total chlorophyll fluorescence yield increases by a factor of 3 to 4.

Absorbance changes in open PS II centers were probed in the wavelength range from 620 to 700 nm. All kinetic data were decovoluted by the global analysis procedure. A fit to a sum of three exponentials was sufficient to describe the kinetics. The resulting lifetimes (see Table 2) show good agreement with the corresponding fluorescence lifetimes (cf. Table 1). The resulting amplitudes (cf. eq.(1)) are plotted as the decay-associated absorbance difference spectra for each kinetic component in Fig. 2. At a first glance the difference spectra of (A) the long- and (B) the middle-lived component show the typical features of the (P^+_{680} - P_{680}) and ($Pheo^-$ - Pheo) difference spectra, respectively, which are well known ((6,9) and references therein). The sum of all component spectra is shown in Fig. 4D. It is defined (cf. eq.(1)) as the absorbance difference spectrum at zero time delay and is therefore attributed to the (S_1 - S_0) difference spectrum of Chl<u>a</u>, the initially excited pigment. Similar spectra were reported in (1) and (10). Thus, in a first approach, one can assign the different components to the processes listed in Table 2. The real situation is more complex, however.

TABLE 2: Lifetimes τ_i associated with the absorbance changes in PS II particles with open (F_o) and closed (F_{max}) RCs. (λ_{exc} = 675 nm)

	F_o	F_{max}
τ_1/ps	80 - 120	170 - 260
τ_2/ps	460 - 560	1600 - 1750
τ_3/ns	>> 10 (a)	

(a) beyond our accessible time range.

Due to the increase of the mean exciton lifetime in F_{max} by a factor of 3 - 4 with respect to F_o, the probability of singlet-singlet annihilation is also expected to increase by a similar factor. Thus, for measurements of the **absorbance changes with closed PS II centers** (F_{max}) the excitation fluence had to be reduced by a factor of 3 as compared to that used for open PS II centers in order to avoid annihilation effects. The absorbance signals with closed centers were well fitted (global analysis) by two exponentials with the lifetimes τ_1 = 170 - 260 ps and τ_2 = 1.6 - 1.75 ns. These are again in good accordance with the fluorescence lifetimes (cf. Table 1). The associated absorbance difference spectra are shown in Fig. 3. The fast (170 - 260 ps) phase is associated with a difference spectrum (Fig. 3B) similar to that of the 80 - 120 ps phase in open centers (Fig. 2C). This may indicate the process of trapping and charge separation in closed PS II. The spectra representing the initial absorbance difference at zero time delay which are due to (Chl^* - Chl) are almost identical for open (Fig. 2D) and closed (Fig. 3C) PS II.

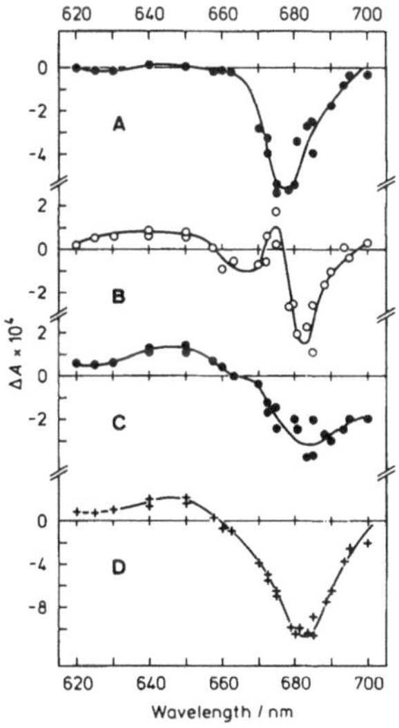

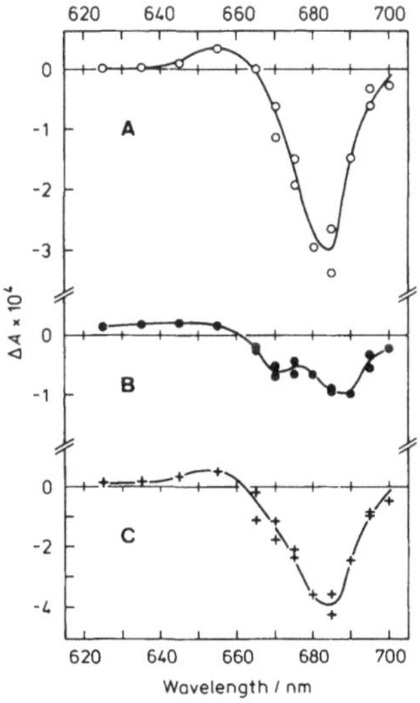

Fig. 2: (left) Absorbance difference spectra of the three exponential components in PS II particles with 2 mM K_3 Fe $(CN)_6$ (open centers) associated with the lifetimes: (A) $\tau_3 >> 10$ ns, (B) $\tau_2 = 510 \pm 50$ ps, (C) $\tau_1 = 100 \pm 20$ ps. (D) shows the initial amplitude identical to the sum of A, B and C.

Fig. 3: (right) Absorbance difference spectra of the two exponential components in PS II particles with 2 mM dithionite (closed centers) associated with the lifetimes: (A) $\tau_2 =$ 1.6-1.75 ns, (B) $\tau_1 =$ 170-260 ps. (C) shows the initial amplitude identical to the sum of A and B.

The kinetic phases. With open RCs the slowest phase in absorbance has a lifetime of many nanoseconds and can be attributed to the reduction of P_{680}^+ by the intact donor side due to its typical difference spectrum (Fig. 4A). A similar spectrum was reported earlier with the same material as used in this work (6). From (6) it is also known that the P_{680}^+-reduction proceeds within 20 ns $\leq t_{1/2} \leq$ 300 ns. Such lifetimes are far beyond our time range which is limited by the length of the optical delay line and are fitted with any time constant larger than 10 ns. The second kinetic phase which was observed in open PS II is attributed to the electron transfer step $P_{680}^{+} I^{-} Q_A \rightarrow P_{680}^{+} I Q_A^{-}$ due to its absorbance difference spectrum (Fig. 2B). Its decay with a time constant of $\tau_2 = 510 \pm 50$ ps (corresponding to $t_{1/2}$ = 350 ps) is slower by about 30% than reported by Nuijs et al. for another PS II particle (1). It remains to be clarified by further work whether this may be related to the thermophilic properties of our PS II particles (11). Our observation that the lifetime of the I^{-}-reoxidation appears also in the fluorescence decay kinetics (Table 1) is essential since it implies a partial reversibility of the energy partition between excited states (Chl^{*}) and the radical pair ($P^{+}I^{-}$). The fastest component with a lifetime of 80 - 100 ps is observed both in absorbance (Table 2) and in fluorescence (Table 1). It is attributed to the trapping of an exciton from the antenna and the concomitant charge separation. After closing the RC by reduction of Q_A, that process was slower and had a smaller relative amplitude (cf. Table 1). Furthermore, the absorbance changes in closed RCs are only biexponential. The apparent absence of a long-lived component under these conditions indicates that except for fluorescence and non-radiative decay there is no other significant pathway of energy dissipation in closed RCs of PS II. This corresponds well with the low (<15%) triplet yield which was reported for PS II (12).

A kinetic model which accounts for the described observations is proposed in Fig. 4. Our data support the concept of a trap-limited charge separation kinetics (13). This means that the process of charge separation is rate limiting while the reversible exciton transfer between the antenna- and the reaction center Chl is much faster. Consequently, the excitation created in the antenna-Chl is always in equilibrium with P^{*}_{680}. This concept has been proposed in (14) and receives further support by our observation of a feedback from the RC towards the antenna which is evident from the strong decrease of both the rate constant and the relative amplitude of the fast Chl^{*}-trapping kinetic upon closing the RC. This is indicative of a shallow, reversible trap P_{680}. Our kinetic scheme can be used as a general model for the primary processes in both open and closed RCs of PS II by solely varying the k_j-values. In principal analogy to the model discussed in (14) the model of Fig. 4 predicts biexponential kinetics of Chl^{*} and of pheophytin (I) associated with the lifetimes τ_1 and τ_2. The P_{680}^{+} kinetics are predicted with three exponentials (τ_1, τ_2 and τ_3). A full quantitative analysis of the results will be given later.

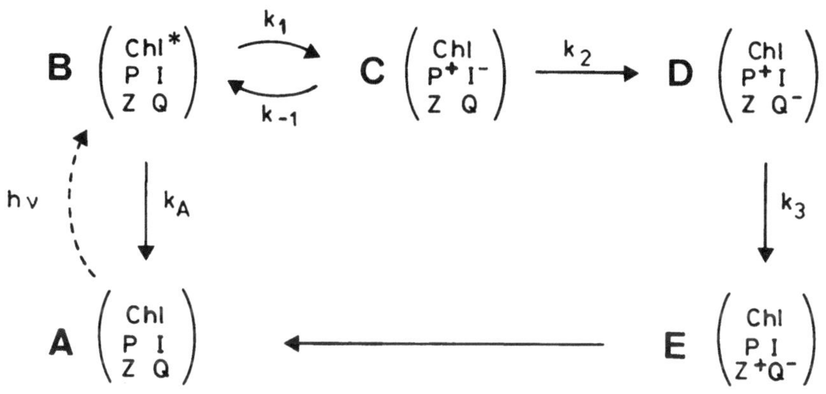

Fig. 4: Kinetic model of the primary reactions in open PS II. In this scheme the rate constants k_j describe: $k_A = k_d + k_{rad}$, the radiationless plus the radiative decay of Chl^*; k_1, the apparent primary charge separation; k_{-1}, charge recombination to the excited state; k_2 and k_3, the processes of charge stabilization. For closed PS II the rate constant k_2 is replaced by k_2' leading to as yet unspecified product(s) instead of D.

TABLE 3. Correlation of typical lifetimes of absorbance change kinetics, the associated difference spectra, their extrapolated maximal difference absorption coefficients and the main process which they reflect as obtained with open PS II (cf. Fig. 2).

τ	spectrum (Fig. 2)	$\Delta\varepsilon$ / $mM^{-1}cm^{-1}$	major process
100 ± 20 ps	C	--	$Chl^*\ P\ I \rightarrow Chl\ P^+\ I^-$
510 ± 50 ps	B	- (45 -55)	$I^-\ Q \rightarrow I\ Q^-$
>> 10 ns	A	- (60 -75)	$P^+\ Z \rightarrow P\ Z^+$
at t = 0	D = A+B+C	-(120 -140)	$Chl^* \rightarrow Chl$

From a qualitative analysis it is clear that the lifetime τ_3 in absorbance is almost exclusively associated with the process of P^+_{680}-reduction, since the contributions by Z^+ formation to the absorbance changes in the observed spectral region are negligible (5) . Correspondingly, the absorbance difference spectrum shown in Fig. 2A is a pure ($P_{680}{}^{+}$-P_{680}) spectrum in good accordance with a previously published spectrum (6). The lifetimes τ_1 and τ_2 are associated with changes in the state of all three constituents: Chl, P and I. Hence, the time-resolved absorbance difference spectra from Figs. 2B,C and 3A,B are expected to be composite spectra with contributions which depend on the relative values of the rate constants k_j. However, such additional contributions are unlikely to explain the difference between the spectrum in Fig. 2B and the (I^--I)-spectra reported from steady state measurements (2,15), particularly the zero transition near 670 nm. Such differences could be attributed to electrochromic shifts induced by the presence of $P_{680}{}^{+}$ and/or the concomitant change in the electric charge on Q_A which are inherent to the transient (I^--I) spectrum of Fig. 2B. The spectrum shown in Fig. 2C (associated with τ_3) is apparently the most complex one. It contains contributions by the decay of Chl^* and by the rise of both P and I. At approximately 697 nm, where the latter two contributions vanish (cf. Figs. 2B,C), the amplitude is a measure of Chl^* which becomes trapped. The energy-normalized amplitudes at this wavelength show (Fig. 3B) that in closed PS II centers about 50-70% less excitons become trapped as compared to open centers (Fig. 2C). This means that the yield of $P_{680}{}^{+}I^-$-formation is reduced by about the same factor as a consequence of Q_A being reduced. This result is of major importance for the understanding of the mechanisms of primary processes in PS II. It indicates that the yield of the radical pair ($P_{680}{}^{+}I^-$) formation is controlled by the redox state of Q_A. A similar conclusion was already derived from a model analysis of Chl-fluorescence kinetics of unfractionated thylakoids (14). The control of $P_{680}{}^{+}I^-$ formation by the redox state of Q_A may be a consequence of the electrical field created by the negative charge on Q_A^- and the smaller distance from Q_A^- to I than from Q_A^- to P_{680}.

In conclusion our data show that the Chl excited state kinetics with open and closed RCs of small PS II particles are biexponential. In both cases the rapid phase reflects the equilibration of energy between the antenna and the radical pair while the slow phase is associated with the overall decay of the resulting equilibrium by deactivation of the antenna and the forward reaction of the radical pair. Pheophytin clearly functions as the first electron acceptor in PS II. The characteristic changes upon reduction of Q_A are the drastically increased lifetime of antenna Chl^* and the decreased yield of transient $P_{680}{}^{+}I^-$-formation. Consequently, one can conclude that the emission from PS II with closed RCs is primarily prompt fluorescence of antenna-Chl^*. This result is in conflict with the hypothesis by Klimov (2,16) that a delayed luminescence after charge recombination from $P_{680}^{+}I^{-}$ should account for the long-lived fluorescence decay component in closed PS II. The latter hypothesis has already been questioned solely on the basis of the fluorescence decay properties in green algae (14). A full account of this work will be given elsewhere (17).

REFERENCES

(1) Nuijs,A.M., van Gorkom,H.J., Plijter,J.J., and Duysens,L.N.M., 1986. Biochim. Biophys. Acta 848: 167-175.

(2) Klimov,V.V., Klevanik,A.V., Shuvalov,V.A., and Krasnovsky,A.A., 1977. FEBS Lett. 82: 183-186.

(3) Holzwarth,A.R., Wendler,J., and Haehnel,W., 1985. Biochim. Biophys. Acta 807: 155-167.

(4) Schatz,G.H. and Witt,H.T., 1984. Photobiochem. Photobiophys. 7: 1-14.

(5) Schatz,G.H. and van Gorkom,H.J., 1985. Biochim. Biophys. Acta 810: 283-294.

(6) Schlodder,E., Brettel,K., Schatz,G.H., and Witt,H.T., 1984. Biochim. Biophys. Acta 765: 178-185.

(7) Knorr,F.J. and Harris,J.M., 1981. Anal. Chem. 53: 272-276.

(8) Holzwarth,A.R., Wendler,J., and Suter,G.W., 1987. Biophys. J. 51: 1-12.

(9) Parson,W.W. and Ke,B., 1982. in Energy Conversion by Plants and Bacteria: 331-385. Academic Press; New York.

(10) Klug,D.R., Gore,B.L., Giorgi,L.B., and Porter,G., 1987. in Progress in Photosynthesis Research. Eds: J. Biggins, Nijhoff; Dordrecht, 1: 95-98.

(11) Schatz,G.H. and Witt,H.T., 1984. Photobiochem. Photobiophys. 7: 77-89.

(12) Kramer,H. and Mathis,P., 1980. Biochim. Biophys. Acta 593: 319-329.

(13) Pearlstein,R.M., 1982. Photochem. Photobiol. 35: 835-844.

(14) Schatz,G.H. and Holzwarth,A.R., 1986. Photosynth. Res. 10: 309-318.

(15) Klimov,V.V., Shuvalov,V.A., and Heber,U., 1985. Biochim. Biophys. Acta 809: 345-350.

(16) Klimov,V.V. and Krasnovsky,A.A., 1982. Biophysics 27: 186-198.

(17) Schatz,G.H., Brock,H., and Holzwarth,A.R., 1987. Proc. Natl. Acad. Sci. USA in press.

ANALYSIS OF EXCITATION ENERGY TRANSFER IN THYLAKOID MEMBRANES BY THE TIME-RESOLVED FLUORESCENCE SPECTRA

Mamoru MIMURO
National Institute for Basic Biology, Myodaiji, Okazaki, Aichi 444, Japan

Introduction

The chain of reactions in photosynthesis is triggered by absorption of light. Light energy is then transferred to reaction center (RC) where photochemical reaction takes place. On the energy transfer in photosynthetic pigments, some mechanisms are proposed (1); Forster mechanism and exciton interaction are the representative examples. However, the energy flow in the photosynthetic pigments has not yet been strictly described by the formula based on a specific mechanism.

For the analysis of the energy transfer, each process should be described by two terms; components and their time behaviour. Two methods are usually adopted for this purpose. One is the combination of fluorescence decay curve at fixed wavelength and global analysis. The other is the combination of the time-resolved fluorescence spectra and deconvolution of those spectra. For a full analysis by either method, however, some assumptions are made; the number of component and decay kinetics in the former, and the number of component and band shape (Gaussian or Lorentzian) in the latter.

We have been working on the analysis of energy transfer mechanism by using time-resolved fluorescence spectroscopy and deconvolution of the spectra (2,3). We adopted this method based on following reasons. (1) Time-resolved spectroscopy directly shows the energy flow, (2) it does not require any assumption for the decay kinetics and (3) discrimination on

Photosynthetic Light-Harvesting Systems

the presence of components is more objective than by the global analysis. By using this method, we analyzed the energy transfer on phycobilin-chl a system and proposed the decay kinetics proportional to the square root of time (2,3). An alternative decay kinetics, exponential decay, is also proposed on the system (4,5,6). This kinetics has been applied to chl a-b system (7,8,9). The energy transfer in thylakoid membranes, however, is much more ambiguous because of the overlap of component bands in a narrow wavelength region.

In this report, we described the changes in the time-resolved spectra of spinach chloroplasts at -196°C, and the rise and decay curves of individual components. This is regarded as the first step for the analysis of the transfer mechanism in thylakoid membranes. When the components and their time behaviour will be described properly, a probable mechanism can be selected and a new mechanism might be proposed.

Materials and Method

Spinach chloroplasts were isolated basically by the method of Jensen and Bassham (10). The buffer solution contains 0.33 M sucrose, 50 mM Tricine-NaOH (pH 7.5), 10 mM NaCl and 2 mM $MgCl_2$. Separation of chloroplasts into subparticles were done by the method of Sane et al. (11); a combination of mechanical disruption and differential centrifugation (cf. ref. 12).

Measurements of time-resolved fluorescence spectra were carried out by using the apparatus reported before (2,3). Briefly, the light source was composed of a synchronously pumped, cavity-dumped dye laser and a mode-locked Ar^+-laser. The repetition rate of the dye laser was 800 kHz with a pulse width of 6 ps (FWHM). The excitation light intensity of laser at 630 nm was in a range of 10^8 to 10^9 photons/cm^2 per pulse which is low enough to avoid singlet-singlet annihilation. Fluorescence was detected by a time-correlated single-photon counting method. By using a 6-um microchannel plate photomultiplier (R2809U-01, S-20 type photocathode, Hamamatsu Photonics, Japan), time resolution of 9.92 ps was achieved in terms of the spectrum. Time zero was set to the time when the excitation pulse was the strongest. A 6-ps pulse (FWHM) was usually observed as a pulse of 30 to 40 ps (FWHM) due to a response function of the detector system, thus spectrum before time zero was obtainable.

Results and Discussion

Fluorescence components detected by a steady state spectrum.

Fig. 1 shows a steady state fluorescence spectrum of spinach chloroplasts at -196°C. In this spectrum, at least four components are known; the component whose maximum is located at 680 nm (called F680 hereafter), F683, F693 and F735. These are assigned as the emissions from LHC II, photosystem (PS) II, PS II and PS I chl a, respectively (13). In addition to these, another component in PS I is detected at 720 nm (F720) by fluorescence polarization spectrum (14). The second derivative spectrum confirmed the presence of the above mentioned compo-

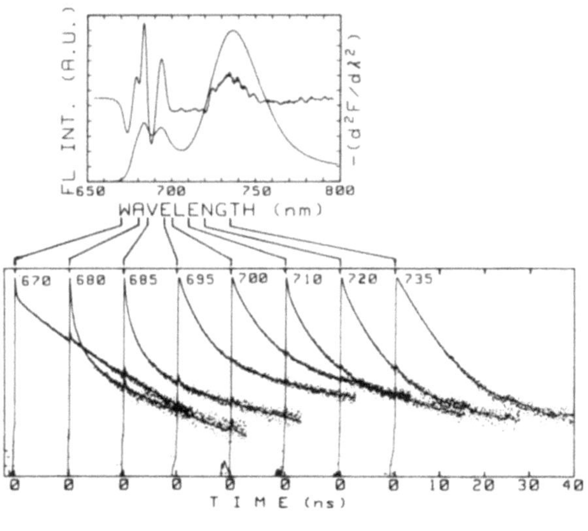

Fig. 1. Steady state fluorescence spectrum of spinach chloroplasts at -196°C and its second derivative spectrum (upper) and decay curves at various wavelength (lower). Samples were suspended in the buffer solution containing 15 % polyethylene glycol 4000 to obtain homogeneous ice at -196°C (cf. ref. 12).

nents including F720, and further suggests the presence of another bands around 740 and 750 nm.

Fig. 1 also shows the fluorescence decay curves at various wavelength between 670 and 735 nm. A fast decaying component was found between 670 and 685 nm, and also at 710 nm, probably corresponding to the energy transfer. A remarkable feature in these decay curves is the presence of a long lifetime components, especially clear around 695 and 700 nm. Its lifetime was estimated to be about 20 ns, thus it cannot be assigned as a prompt fluorescence but as a delayed fluorescence.

The time-resolved fluorescence spectra were reconstructed from the decay curves between 650 and 760 nm in 0.625 nm interval. The original data points corresponding to a particular time after the pulse were newly expressed as a function of wavelength.

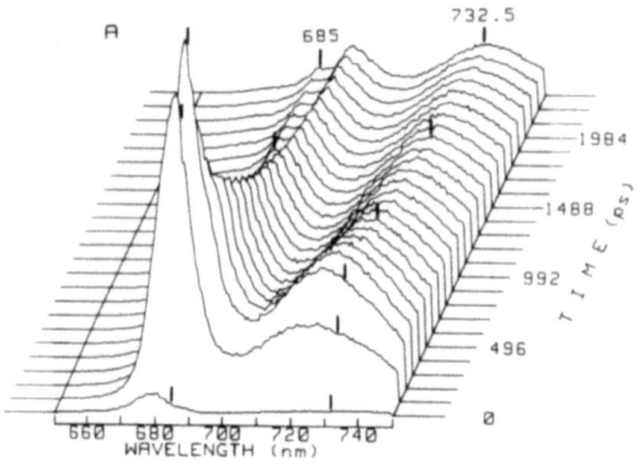

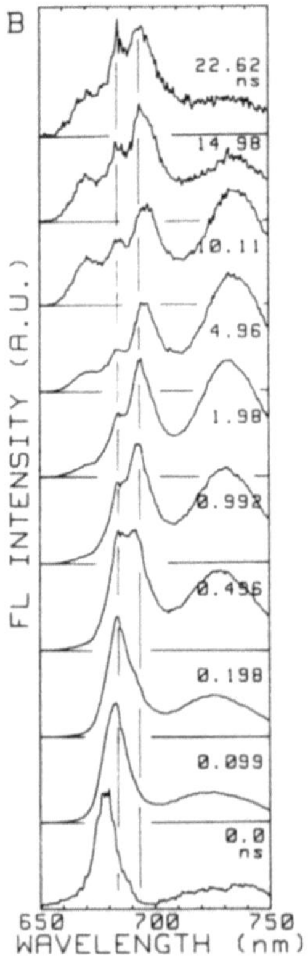

Fig. 2. (A) Time-resolved fluorescence spectra of spinach chloroplasts at -196°C. Tic marks represent particular wavelength. (B) Normalized time-resolved fluorescence spectra.

Time-resolved fluorescence spectra in a long time range.

The time-resolved fluorescence spectra of spinach chloroplasts at -196°C were shown in Fig. 2 as a bird's-eye view. The excitation was done at 630 nm where all the chl's in chloroplasts can absorb light. Selective excitation of the pigment in the highest energy level could not be attained by the α-band excitation.

On excitation, fluorescence was initially observed at 679 nm (F679) (Fig. 2). The maximum shifted to 683 nm within 200 ps. This change corresponds to the energy flow from LHC II to PS II chl a. The energy on the F683 is further transferred to the F693, the component assigned as PS II chl a. The delay time from F683 was about 200 ps. In the wavelength region longer than 700 nm, changes in the spectra were not simple. The maximum was first observed around 705 nm, and it shifted to 725 nm within 100 ps. Further shift was observed for a next few ns. The number of components in the wavelength region longer than 700 nm was not clear based on these spectra.

Overall changes in the time-resolved spectra can be seen much clearly in the normalized spectra (Fig. 2B). During first 2 ns, the energy flow from LHC II to PS II chl a was clearly monitored by the shift of the maximum. A red shift of the maximum of PS I chl a is also evident. After 2 ns, the shift of the F693 was remarkably observed. About 10 ns, the maximum was located at 696 nm and it shifted back to 693 nm in a later time range as seen in the spectrum at 22.6 ns. Contrary to this, the location of the F683 was constant throughout the measuring time. These results suggest the presence of a component band longer than but close to the F693. Its lifetime seems to be shorter than that of the F693.

In the time longer than 15 ns of the excitation, two bands were evident at 683 and 693 nm when all the other component almost disappeared. The locations of these components were identical to those of the F683 and F693 found in the initial time range. Their lifetimes were longer than that of free chl a (about 6 ns) found at 670 nm, as seen in Fig. 1.

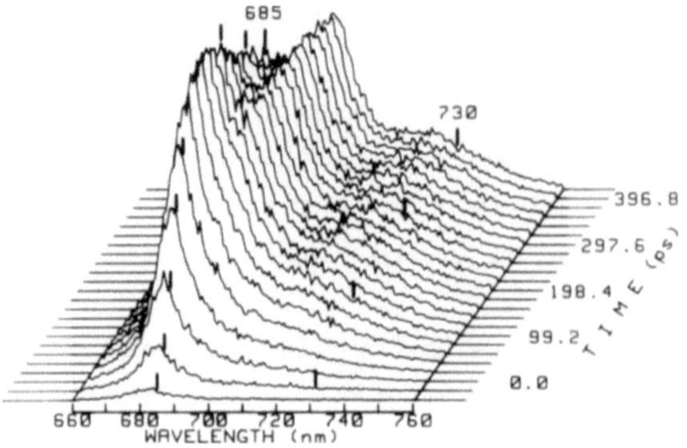

Fig. 3. Time-resolved fluorescence spectra of spinach chloroplasts at -196°C with a shorter time resolution.

Therefore, those can be regarded as a delayed fluorescence. These long lifetime components are also found in intact cells of the cyanobacterium Anabaena variabilis (M-3) (Mimuro et al., unpublished data). In this sense, this phenomenon is general and the most probable origin of those components is the charge recombination in RC II.

Time-resolved spectra of chloroplasts in a short time range.

When the time-resolved fluorescence spectra were measured in a short time range, the rise of individual components became clear (Fig. 3). On excitation, the emission from F679 was initially observed. Fluorescence from chl b could not be detected even with the time resolution of 9.92 ps. The energy transfer from chl b to chl a in LHC II was reported to be a much faster process (15). The F679 shifted to F683 within 100 ps, and further to F693, within 250 ps. At room temperature, the shift of the maximum from 680 to 683 nm occurred within 50 ps (16). This indicates the occurrence of the temperature effect on this transfer process.

Time-resolved fluorescence spectra of subparticles and identification of component bands by deconvolution.

For detection of intrinsic fluorescence components in each PS, spinach chloroplasts were separated to subparticles by mechanical disruption and differential centrifugation (11,

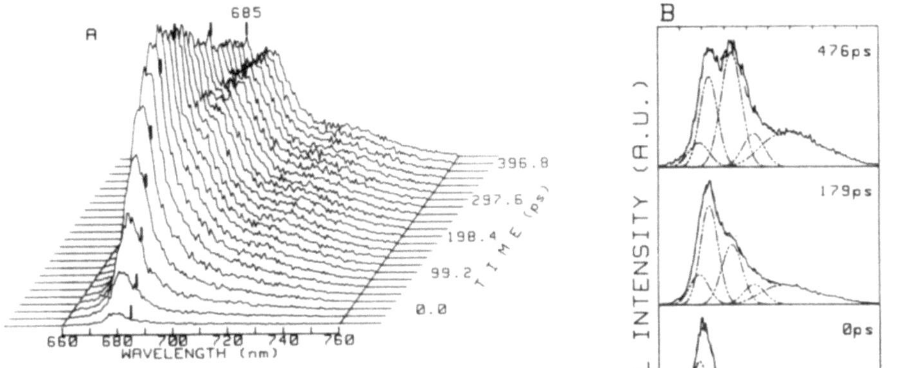

Fig. 4. (A) Time-resolved fluorescence spectra of PS II particles at -196°C and (B) their deconvolution patterns. In (B), (———), original spectrum, (—·-·—), simulated component spectra and (- - - -) sum of simulated curves.

12). The purities of our PS I and II particles were about 85 and 75 %, respectively (12). The changes in the spectra in PS II particles (Fig. 4A) were similar to those of whole chloroplasts. One remarkable difference was the rise time of the F693; it was slightly longer than that in chloroplasts (Fig. 3), indicating that the energy transfer to F693 is affected by the isolation procedure. This result suggests that the F693 originates from a polypeptide different from that carrying the F683, because if both originate from the same polypeptide, the transfer to F693 is expected not to be affected by this kind of mild treatment.

Fig. 4B shows a typical deconvolution pattern of time-resolved fluorescence spectra of PS II particles. The spectrum between 660 and 710 nm was simulated by five components with Gaussian band shape; those are F679, F683, F693, F703 and F715. One exception was the assumption of the F688 only in the spectrum at 0 ps. This is ragarded as the contribution of PS I chl a with a very short lifetime. Each spectrum was reasonably fitted with these five components, thus the changes in the spectra can be expressed by the changes in the intensities of

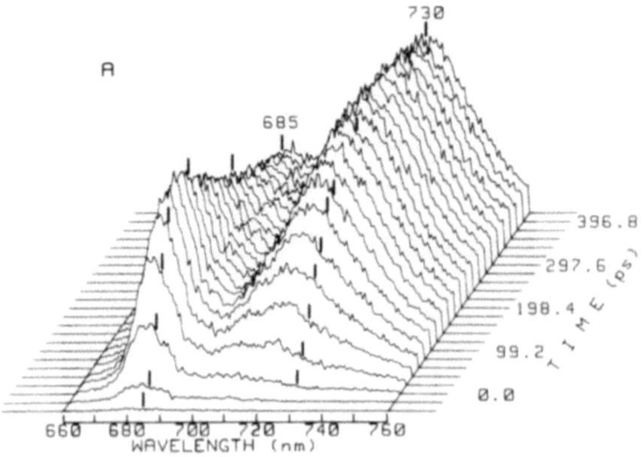

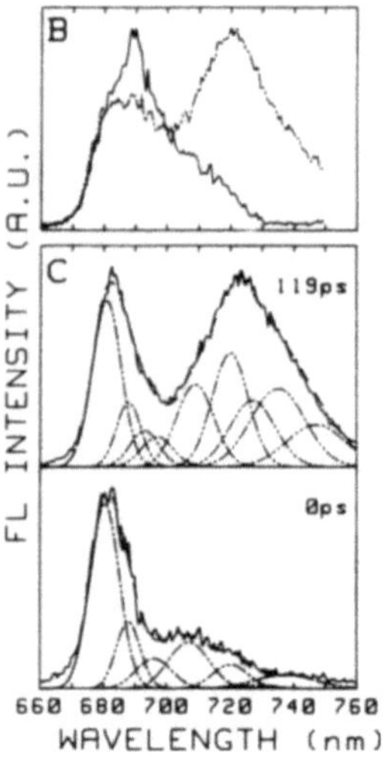

Fig. 5. Time-resolved fluorescence spectra at -196°C of PS I particles prepared by mechanical disruption and differential centrifugation (A), and normalized spectra of PS I particles prepared with Triton X-100 (B). In (B), (———) at 0 ps and (------) at 59.4 ps. (C) deconvolution patterns of time-resolved fluorescence spectra.

individual components. At 0 ps, the F679 was the dominant component. Accompanying the shift of the maximum, the F683 and F693 became dominant progressively. The F703 is a component of CP-47 (Mimuro et al., unpublished data). The components locating at longer than 710 nm, however, are not necessarily clear, because of the overlap of vibrational bands.

In the case of PS I particles, the changes in the spectra were drastically different from those of whole chloroplasts (Figs. 5A and 3). On excitation, the maximum was observed at 680 nm and a shoulder, around 690 nm. Two minor bands at 707 and 720 nm were also observed. The rise time of the F680 was about 100 ps and the maximum shifted to 685 nm due to a contamination of PS II particles. The increase in the intensity at longer than 700 nm was clearly observed. The rise time was about 200 ps and the red shift of the maximum was ascribed to the increase in the intensity of the F720. The shift of the maximum from 720 to 730 nm seemed gradual.

The F680 in PS I particles may correspond to the emission

from LHCIa (17), however it might be due to a contamination of LHC II. PS I particles prepared with Triton X-100 (18), which is equivalent to PS I-200 by Mullet et al. (19), also showed the presence of F680 and F690 (Fig. 5B) with a higher intensity of the F690. These results indicate that the F680 and F690 are intrinsic components of PS I and the coupling between these two is weakened by the detergent treatment. This suggests that the origins of the two components are different. F690 might be located close to RC I, because the rise and decay is much faster in the preparation obtained by detergent treatment.

The components locating at longer than 700 nm are assigned as PS I chl _a_. Identification of components in this region was attained much clearly by deconvolution (Fig. 5C). The spectrum at 0 ps was well simulated with the components peaking at 680, 688, 696, 707, 720, 727 and 735 nm. Of these, the F696 was not clear in the spectrum, however it most probably corresponds to F698 found in our previous study (12). At longer than 700 nm, the increase in the intensity can be ascribed to the increase in those of the F720 and F735, as seen in the spectrum at 119 ps. This spectrum was not well fitted only with the above mentioned components. Additional components at 693 and 747 nm were required. F693 is due to the contamination of PS II. Since the amplitude of F747 was high, it can be regarded as an intrinsic component, even if the vibrational structure will be taken into account.

Rise and decay curves of the component bands in PS II particles.

Based on the deconvolution pattern of the time-resolved fluorescence spectra, the rise and decay curves of individual components can be estimated. As the simplest case, PS II particles were choosen, and the rise and decay curves of F679, F683 and F693 were obtained (Fig. 6). Compared with the rise and decay curves at corresponding wavelength, time behaviour of the F679 was very different, prorbably due to a fast energy

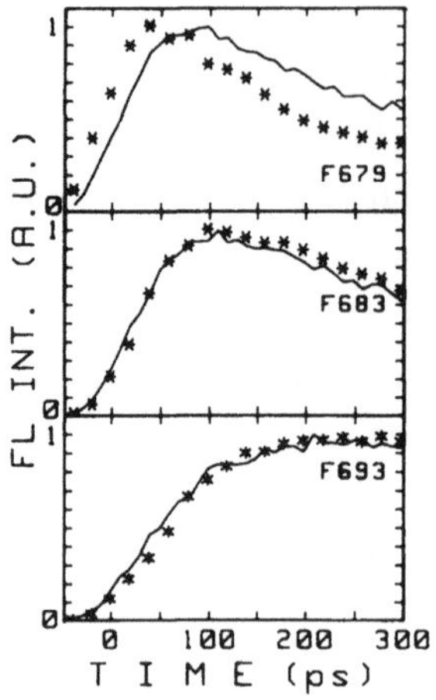

Fig. 6. Rise and decay curves of the fluorescence components in PS II particles. From upper to lower, the curves for the F679, F683 and F693. In each figure, asterisks indicate relative intensity of the components based on the deconvolution, and full lines, the observed fluorescence intensities at 679, 683 and 693 nm, respectively.

transfer. This suggests that the estimation of decay kinetics by the curves at fixed wavelength is not necessarily adequate. The rise of the F679 was fast, giving rise to the maximum at 50 ps, and the delay times from the maximum of the F679 to those of F683 and F693 were 60 and 200 ps, respectively. As clearly shown in the Fig. 6, the rise of the F683 did not correspond to the decay of the F679. These time behaviours are essentially the same as in the case of phycobilin-chl a system (2,3). Decay kinetics depend on the three dimensional structure of the system, thus it is not easy to apply a specific function to the present observation. However, it seems that simple exponential decay is not necessarily the case.

The same analysis is possible on PS I particles and whole chloroplasts. However, the number of component is too many. Feasibility on the presence of the component band is the primary point.

Estimation of components by decay curves at fixed wavelength.

On the fluorescence decay curves of spinach chloroplasts at various wavelength (Fig. 1 lower), we tried simulation calculation to detect components and their lifetimes. Exponential decay of fluorescence was assumed and each decay curve was expressed as multi-exponential function. Two parameters were used as the criteria of the best fit; chi-square and the Durbin-Watson (DW) parameter (20). The latter means the sequential correlation of data set. By using this parameter, the artificial mimimum of chi-square can be avoided.

All the decay curves can be fitted by at most five components with the DW parameter larger than 1.6. However, we could not obtain any indication consistent with the results by the time-resolved fluorescence spectra. As seen by the spectra, the number of components is more than five in chloroplasts. Increase in the number of component insures a better fit, however, the reason for the increase is not necessarily clear in our system. Contradiction should be checked by using more simple experimental system.

Acknowledgement

The author expreeses his cordial thanks to Drs. I. Yamazaki and N. Tamai, the Instrumental Center, Institute for Molecular Science, Okazaki, Japan for their operation of picosecond spectroscopy and stimulative discussion. He acknowledges Dr. S. Hoshina (Kanazawa Univ.) for his supply of Triton-PS I particles. This work was partly supported by a Grant-in-aid for the Scientific Research from the Ministry of Education, Science and Culture, Japan (62540520).

References

1. Borisov, A.Y. 1978. In: Photosynthetic Bacteria, (R.K. Clayton and W.R. Sistrom eds.). Plenum Press, New York, pp. 323-331.

2. Yamazaki, I., M. Mimuro, T. Murao, T. Yamazaki, K. Yoshihara and Y. Fujita. 1984. Photochem. Photobiol., 39, 233-240.

3. Mimuro, M., I. Yamazaki, T. Yamazaki and Y. Fujita. 1985. Photochem. Photobiol., 41, 597-603.

4. Holzwarth, A.R., J. Wendler and W. Wehrmeyer. 1982. Photochem. Photobiol., 36, 479-487.

5. Suter, G.W., P. Mazzola, J. Wendler and A.R. Holzwarth. 1984. Biochim. Biophys. Acta, 766, 269-276.

6. Wendler, J., A.R. Holzwarth and W. Wehrmeyer. 1984. Biochim. Biophys. Acta, 765, 58-67.

7. Haehnel, W., J.A. Nairn, P. Reisberg and K. Sauer. 1982. Biochim. Biophys. Acta, 680, 161-173.

8. Holzwarth, A.R., J. Wendler and J. Haehnel. 1985. Biochim. Biophys. Acta, 807, 155-167.

9. Hodges, M. and I. Moya. 1986. Biochim. Biophys. Acta, 849, 193-202.

10. Jensen, R.G. and J.A. Bassham. 1964. Proc. Natl. Acad. Sci. USA., 56, 1095-1101.

11. Sane. P.V., D.J. Goodchild and R.B. Park. 1970. Biochim. Biophys. Acta, 216, 162-178.

12. Mimuro, M., N. Tamai, T. Yamazaki and I. Yamazaki. 1987. FEBS Letters, 213, 119-122.

13. Murata, N. and Ki. Satoh. 1986. In: Light Emission by Plants and Bacteria. (Govindjee, J. Amesz and D.C. Fork, eds.). Academic Press, New York. pp. 137-159.

14. Garab, G.I. and J. Breton. 1976. Biochem. Biophys. Res. Commun., 71, 1095-1102.

15. Gillbro, T., V. Sundstrom, A. Sandstrom, M. Spangfort and B. Anderson. 1985. FEBS Letters, 93, 267-270.

16. Mimuro, M., I. Yamazaki, N. Tamai., T. Yamazaki and Y. Fujita. 1987. In: Primary Process in Photobiology. (T. Kobayashi ed.). Academic Press, Tokyo. In press.

17. Lam., E., W. Ortiz and Malkin, R. 1984. FEBS Letter, 168, 10-14.

18. Hoshina, S. and S. Itoh. 1987. Plant & Cell Physiol., 28, 599-609.

19. Mullet, J.E., J.J. Burke and C.J. Arntzen. 1980. Plant Physiol., 65, 814-822.

20. O'Connor, D.V. and D. Phillips. 1984. Time-correlated Single Photon Counting. Academic Press, London.

V. Concluding Remarks

FUTURE PROBLEMS ON ANTENNA SYSTEMS AND SUMMARY REMARKS

Elisabeth Gantt, Smithsonian Institution/Department of Botany, University of Maryland, College Park, MD 20742

The present workshop on the organization and function of photosynthetic antennas is the first one devoted exclusively to this topic. It is extremely timely, especially now that several bacterial reaction centers have been characterized which appear to have great similarities with the photosystem II reaction centers of green plants, we can begin to concentrate on the complexity of antenna systems.

It is in the antenna systems where the greatest diversity can be found. One cannot help but be impressed by the strategic manifestations, remembering that all of them evolved to enhance the light gathering capacity of the organisms. Structurally there are two major classes. In one class, to which most groups belong, the main antenna pigments are integral components of the photosynthetic membranes. In the other class the antenna complexes project into the stroma and are probably only attached to the reaction centers. The chlorosomes of the green bacteria, and phycobilisomes of the red algae and cyanobacteria are examples of the latter. Organisms with such structural separation of the antennas and reaction centers may be suitable models for identification of the specific binding polypeptides involved in the attachment.

Except for certain purple bacterial complexes, little comprehensive data exists on primary protein structure, chromophore attachment, and orientation of the complexes in the membranes. Considerable progress is also being made in phycobiliprotein complexes, as is evident from this workshop. In phycobilisomes the pigments and energy migration path from peripheral pigments to specific terminal pigments have been identified. In chlorosomes the binding of bacteriochlorophyll _c_ to the apoprotein and its in vivo arrangement remain to be resolved in both _Chlorobium_ and in _Chloroflexus_. In _Chloroflexus_ temporal uncoupling of the chlorosomes from the reaction center seems to be readily accomplished, thus this

Photosynthetic Light-Harvesting Systems

organism would appear to be one of the most promising organisms for studies on photoinhibition.

Advances in phycobiliproteins have been aided by the covalent attachment of the chromophores, the relative stability of phycobiliprotein complexes, and above all the data from the three-dimensional crystal structure of phycocyanin. Yet, there is no clear concensus in several areas. Clearly some phycocyanobilin chromophores are coupled, but which ones, and by what mechanism? It appears that energy transfer from s to f chromophores, according to the Foerster mechanism is too simplistic an explanation, and other mechanisms have to be considered. Comparisons of data from different laboratories on similar phycobiliproteins is complicated by several factors. Variables such as state of aggregation, the presence or absence of linker polypeptides, and species varieties require resolution. A systematic study using mild protein cross-linkers could avoid some of the more significant variations. Crystallization of allophycocyanin, which has the fewest chromophores, is still lacking.

The largest variety of antenna systems occurs in the algal groups, but these are generally unknown to active researchers in photosynthesis. A very large group of algae contain chlorophyll c, and sometimes b, in addition to a, fucoxanthin, peridinin and other accessory pigments. Among these are diatomes, brown algae, dinoflagellates, cryptophytes, prymnesiophytes, xanthophytes and others. Separation of the complexes, and stabilization of the chlorophylls and carotenoids has been a major drawback. In green plants conditions for isolating chlorophyll a/b complexes have been steadily improving, and additional complexes are being discovered. At least four chlorophyll a/b complexes exist in green plants, of which three are associated with photosystem II. Of these only the most common (24-28 kD apoprotein range) has thus far been studied in any detail. Here also, production of three dimensional crystals remains as one of the major challenges.

Sophisticated analysis methods have been applied to many of the pigment antenna complexes but most of the pigment-proteins, even very simple ones are still too complex to make meaningful functional inferences. Selected purple bacteria complexes are the exception. Additional major questions which require elucidation are:

-- What are the binding domains of chlorophyll *a*, and other chlorophylls?

-- How and where are carotenoids bound, and in what proportion to chlorophyll?

-- What is the significance of membrane stacking in green plants, and are there functional equivalents in chlorophyll-*c* containing algae?

-- Are the antenna domains around the reaction centers homogenous or heterogeneous?

-- Is the stoichiometry of reaction center and antenna clusters fixed, or variable with growth conditions? In some of the models of purple bacteria a grouping of four reaction centers surrounded by antenna complexes has been suggested. In some cyanobacteria there appear to be two photosystem II reaction centers per phycobilisome, while in some red algae there are ca. 4 photosystem II centers per phycobilisome. An integrated development sequence is expected, since genetic deletions of reaction center components often result in non-assembly of the reaction centers and also of the attendant antenna complexes.

The organizers had selected a catalytic mixture ranging from theoreticians, to biochemists, biophysicists, molecular and structural biologists. It is at meetings such as this, where those in various disciplines learn to understand the jargon and appreciate the experimental potentials of the others. I predict that the long term benefits derived from this workshop will indeed be significant, and I commend the organizers for their foresight.

INDEX OF AUTHORS

SUBJECT INDEX

* look for entry to this term